T0182186

Thermal and Catalytic Processing in Petroleum Refining Operations

This book presents the thermal and catalytic processes in refining. The differences between each type of process and the types of feedstock that can be used for the processes are presented. Relevant process data is provided, and process operations are fully described. This accessible guide is written for managers, professionals, and technicians as well as graduate students transitioning into the refining industry.

Key Features:

- Describes feedstock evaluation and the effects of elemental, chemical, and fractional composition.
- Details reactor types and bed types.
- Explores the process options and parameters involved.
- Assesses coke formation and additives.
- Considers next generation processes and developments.

Petroleum Refining Technology Series

Series Editor:
James G. Speight

This series of books is designed to address the current processes used by the refining industry and take the reader through the various steps that are necessary for crude oil evaluation and refining. Technological advancements and processing innovations are highlighted in each of the volumes.

Refinery Feedstocks
James G. Speight

Dewatering, Desalting, and Distillation in Petroleum Refining
James G. Speight

Thermal and Catalytic Processes
James G. Speight

Thermal and Catalytic Processing in Petroleum Refining Operations

James G. Speight

CRC Press

Taylor & Francis Group

Boca Raton London New York

CRC Press is an imprint of the
Taylor & Francis Group, an **informa** business

Designed cover image: © Shutterstock.com

First edition published 2023
by CRC Press
6000 Broken Sound Parkway NW, Suite 300, Boca Raton, FL 33487-2742

and by CRC Press
4 Park Square, Milton Park, Abingdon, Oxon, OX14 4RN

© 2023 Taylor and Francis Group, LLC

CRC Press is an imprint of Taylor & Francis Group, LLC

Library of Congress Cataloging-in-Publication Data
Names: Speight, James G., author.
Title: Thermal and catalytic processing in petroleum refining operations /
James G. Speight.
Description: First edition. | Boca Raton : CRC Press, 2023. | Series:
Petroleum refining technology series | Includes bibliographical
references and index.
Identifiers: LCCN 2022048254 | ISBN 9781032027203 (hbk) |
ISBN 9781032027319 (pbk) | ISBN 9781003184904 (ebk)
Subjects: LCSH: Petroleum–Refining. | Cracking process. |
Catalytic reforming.
Classification: LCC TP690 .S7455 2023 | DDC 665.5–dc23/eng/20230111
LC record available at https://lccn.loc.gov/2022048254

ISBN: 978-1-032-02720-3 (hbk)
ISBN: 978-1-032-02731-9 (pbk)
ISBN: 978-1-003-18490-4 (ebk)

DOI: 10.1201/9781003184904

Typeset in Times
by codeMantra

Contents

Preface

As the third book of the refinery series, this book presents a description of the various thermal conversion processes and the catalytic conversion processes that are used to convert the refinery feedstocks (after the dewatering, desalting, and distillation processes have been applied) to higher-value products, with emphasis on the viscous feedstocks (such as heavy crude oil, extra heavy crude oil, tar sand bitumen, and distillation residua).

The fractions of the refinery feedstock produced by distillation are rarely ready for sales as high-value products. The purpose of the thermal and catalytic processes is to convert the feedstocks to product with more desirable products that are focused on the production of intermediate products that can be sued by further refining to produce the valuable products that are ultimately for sale to the consumers.

Chapter 1 focuses on the general layout of a refinery and the relationship between the various processes. In order to fully comprehend the function of the various process units within refinery operations, it is necessary to understand the place at which the process units employed and the reason for this employment. This will assist the reader to place thermal cracking processes and the catalytic cracking process in the correct perspective of the refinery, not forgetting that prior to any conversion processes, it is necessary to desalt, dewater, and distill the raw (i.e., unrefined) refinery feedstock, as described in the second book of this series.

Chapter 2 introduces the concept of refinery design to ensure that the design is adequate to process the feedstocks of different composition. For example, a heavy oil refinery would differ somewhat from a conventional refinery and a refinery for tar sand bitumen would be significantly different from both. Therefore, it is the purpose of this chapter to provide a description of the processes that can be used to convert the viscous feedstocks (i.e., heavy crude oil, extra heavy crude oil, and tar sand bitumen) into more useful distillates from which the high-value products can be produced.

Chapter 3 presents an outline of the tests that may be applied to crude oil, heavy crude oil, extra heavy crude oil, and tar sand bitumen as well as biomass and their respective products as a pointer to the chemical properties and physical properties from which a feedstock or product can be evaluated. For this purpose, data relating to various chemical physical properties have been included as illustrative examples, but theoretical discussions of the physical properties of hydrocarbons were deemed irrelevant and are omitted.

Chapter 4 presents a description of the thermal cracking processes that are applied to a feedstock in a refinery which removes the volatile constituents with the concurrent production of a residuum that can be used as a cracking (coking) feedstock or as a precursor to asphalt. Thus, the purpose of this chapter is to provide information that will assist the practicing engineer/crude oil refiner to: (1) determine if a particular thermal cracking process would be suitable for a specific application and could fit into the overall operation, (2) develop a basic design for a thermal cracking process, and (3) operate an existing or proposed process.

Chapter 5 presents an overview of the application of catalytic cracking processes that are used to convert a variety of feedstocks (including viscous feedstocks) or more valuable products. The feedstocks range from naphtha fractions (included in normal heavier feedstocks for upgrading) to viscous feedstocks. Descriptions are presented of the process types and the process variables, such as temperature, pressure, catalyst-oil ratio (ratio of the weight of catalyst entering the reactor per hour to the weight of oil charged per hour), and space velocity (weight or volume of the oil charged per hour per weight or volume of catalyst in the reaction zone). Wide flexibility in product distribution and quality is possible through control of these variables along with the extent of internal recycling is necessary.

Chapter 6 presents a description of the concept of fouling in refineries which can occur as deposit formation, encrustation, deposition, scaling, scale formation, slagging, and sludge formation, all of which can have an adverse effect on operations. It is the accumulation of unwanted material within a processing unit or on the solid surfaces of the unit to the detriment of function. For example, when it does occur during refinery operations, the major effects include (1) loss of heat transfer as indicated by charge outlet temperature decrease and pressure drop increase, (2) blocked process pipes, (3) under-deposit corrosion and pollution, and (4) localized hot spots in reactors, all of which culminate in production losses and increased maintenance costs. Thus, the separation of solids occurs whenever the solvent characteristics of the liquid phase are no longer adequate to maintain polar and/or high molecular weight material in solution.

Chapter 7 presents the means by which the thermal and catalytic processes are integrated into the refinery through the various process innovations that have evolved over the decades of the 20th century. This includes recognition of the evolution of the processes through design and integration of the active, supporting, and promoting components that give the desired results.

Each chapter will be presented to the reader through the various steps that are necessary for the thermal and catalytic conversion of various feedstocks. The book brings the reader further up to date and adds more data as well as processing options that may be the processes of the 21st century and into the 21st century.

By understanding the evolutionary changes that have occurred to date, this book will satisfy the needs of engineers and scientists at all levels from academia to the refinery and help them understand the initial refining processes and prepare for the new changes and evolution of the industry.

The target audience includes engineers, scientists, and students who want an update on petroleum processing and the direction of the industry in the next 50 years. Non-technical readers, with help from the extensive glossary, will also benefit from reading the book.

Dr. James G. Speight,
Laramie, Wyoming, USA
September 2022

Author

Dr. James G. Speight has a BSc and PhD in Chemistry; he also holds a DSc in Geological Sciences and a PhD in Petroleum Engineering. He has more than 50 years of experience in areas associated with (1) the properties, recovery, and refining of conventional petroleum, heavy oil, and tar sand bitumen; (2) the properties and refining of natural gas; and (3) the properties and refining of biomass, biofuels, biogas, and the generation of bioenergy. His work has also focused on environmental effects, environmental remediation, and safety issues associated with the production and use of fuels and biofuels. He is the author (and coauthor) of more than 95 books related to petroleum science, petroleum engineering, biomass and biofuels, and environmental sciences.

Although he has always worked in private industry which focused on contract-based work, Dr. Speight has served as a Visiting Professor in the College of Science, University of Mosul (Iraq), Visiting Professor in Chemical Engineering at the Technical University of Denmark, and the University of Trinidad and Tobago as well as adjunct appointments at various universities. He has also served as a thesis examiner for more than 25 theses.

As a result of his work, Dr. Speight has been honored as the recipient of the following awards:

- Diploma of Honor, United States National Petroleum Engineering Society. *For Outstanding Contributions to the Petroleum Industry.* 1995.
- Gold Medal of the Russian Academy of Sciences. *For Outstanding Work in the Area of Petroleum Science.* 1996.
- Einstein Medal of the Russian Academy of Sciences. *In Recognition of Outstanding Contributions and Service in the field of Geologic Sciences.* 2001.
- Gold Medal – Scientists without Frontiers, Russian Academy of Sciences. *In Recognition of His Continuous Encouragement of Scientists to Work Together across International Borders.* 2005.
- Gold Medal – Giants of Science and Engineering, Russian Academy of Sciences. *In Recognition of Continued Excellence in Science and Engineering.* 2006.
- Methanex Distinguished Professor, University of Trinidad and Tobago. *In Recognition of Excellence in Research.* 2007.
- In 2018, he received the American Excellence Award for Excellence in Client Solutions from the United States Institute of Trade and Commerce, Washington, DC.

1 Overview of Refining Processes

1.1 INTRODUCTION

Crude oil refining (also commonly referred to as petroleum refining) is achieved through the application of a series of chemical processes, physical processes, chemical engineering processes, and other processes (all of which are crude oil dependent) that are employed in a refinery to convert (transform) crude oil into useful products such as liquefied petroleum gas, naphtha solvents, gasoline (often referred to as petrol in some countries), kerosene, jet fuel, diesel fuel, fuel oil, lubricating oil, wax, and asphalt as well as feedstocks for the petrochemical industry. Thus, crude oil refining is a series of unit processes that are used to convert crude oil into a variety of products (Parkash, 2003; Gary et al., 2007; Speight, 2011a, 2011b, 2014; Hsu and Robinson, 2017; Speight, 2017, 2019).

On a historical basis, the first systematic refinery for converting crude oil to saleable products crude oil refinery was built in Ploiești (Romania) in 1856 using the crude oil that was available in that country. Shortly thereafter, in North America, the first well to recover crude oil was drilled in 1858 by James Miller Williams in Oil Springs, Ontario, Canada, followed by the drilling of a well in Titusville, Pennsylvania, United States, in 1859 (Larraz, 2021).

The refining industry has been the subject of the four major forces that affect most industries and which have hastened the development of new crude oil refining processes: (1) the demand for products such as gasoline, diesel, jet fuel, and fuel oil; (2) feedstock supply, specifically the changing quality of crude oil and geopolitics between different countries and the emergence of alternate feed supplies such as heavy crude oil, extra heavy crude oil, and bitumen from tar sand; (3) environmental regulations that include more stringent regulations in relation to sulfur in gasoline and diesel fuel; and (4) technology development such as new catalysts and processes.

In the early days of the 20th century, refining processes were developed to extract kerosene for lamps. Any other products were considered to be unusable and were usually discarded, and thus, the first refining processes were developed to purify, stabilize, and improve the quality of kerosene. However, the invention of the internal combustion engine led (at approximately the time of World War I) to a demand for naphtha (for the production of gasoline) for use in increasing quantities as a starting liquid for use in motor fuel for cars and trucks. This demand on the lower-boiling products increased, particularly when the market for aviation fuel developed leading to the need for improved refining processes to meet the quality requirements and needs of automobile engines and aircraft engines.

Since then, the general trend throughout refining has been to produce more products from each barrel of crude oil and to process those products in different ways to meet the product specifications for use in modern engines. Overall, the demand for gasoline has rapidly expanded and demand has also developed for gas oils and fuels for domestic central heating and fuel oil for power generation, as well as for low boiling distillates derived from crude oil for the petrochemical industries.

As the need for the lower-boiling products developed, crude oil yielding the desired quantities of the lower-boiling products became less available, and refineries had to introduce conversion processes to produce greater quantities of lighter products from the higher-boiling fractions. The means by which a refinery operates in terms of producing the relevant products depend not only on the nature of the crude oil feedstock but also on the refinery configuration (i.e., the number of types of the processes that are employed to produce the desired product slate), and the refinery configuration

DOI: 10.1201/9781003184904-1

is, therefore, influenced by the specific demands of a market. In fact, refineries need to be constantly adapted and upgraded to remain viable and responsive to ever-changing patterns of crude supply and product market demands.

In general, refining crude oil consists of two major phases of production which are (1) dewatering and desalting followed by distillation or separating of the feedstock into various fractional components based on boiling range and involves heating, vaporization, fractionation, condensation, and cooling of feedstock, and (2) cracking processes in which the larger molecular constituents of the feedstock are converted (broken down) into products that are of a lower molecular size and have increased volatility.

As first performed using a high temperature and a high pressure, the process involved into the use of catalysts to reduce the production of less valuable products such as heavy fuel oil. As the 20th century matured, cracking technology evolved from the need to solve a series of technical issues such as (1) carbon deposition on the catalyst surface, (2) catalyst breakage, and (3) equipment failure which led to the development of the fluid catalytic cracking process in the early 1940s.

In the modern refinery, the refining activities commence with the receipt of a variety of feedstocks (which can be labeled as crude oil, heavy crude oil, extra heavy crude oil, and tar sand bitumen) for storage. To accomplish the production of saleable products, the flow of the feedstock through a processing flow scheme is determined by the composition of the feedstock and the desired slate of petroleum products (Figure 1.1), but the arrangement of these processes does vary among refineries, and a few, if any, employ all of these processes.

First and foremost, a refinery is a group of manufacturing plants that vary in number according to the variety of products produced (Parkash, 2003; Gary et al., 2007; Speight, 2011a, 2011b, 2014, 2017; Hsu and Robinson, 2017). The processes – in the current context of the thermal cracking processing units and the catalytic cracking processing units – are selected to convert crude oil (or fractions thereof) into products that are manufactured according to market demand. Moreover, to meet the current market demand for high-octane gasoline, jet fuel, and diesel fuel, any high-boiling fraction of the crude oil and heavy crude oil (such as the gas oil fraction, the residuum, as well as extra heavy crude oil and tar sand bitumen) must be converted to naphtha and other low-boiling fractions. It is in this respect that the older process options such as the thermal cracking process, the visbreaking process, and the various coking processes are used to convert the crude oil constituents into products that are more volatile and can be prepared for sale to the various consumers.

The feedstocks that are difficult to refine but, of necessity, are being accepted by refineries are the viscous feedstocks (often referred to as heavy feedstocks) such as heavy oil, extra heavy oil, tar sand bitumen, and residua (atmospheric residua and vacuum residua) which are characterized by low API gravity (high density) and high viscosity, high initial boiling point, high carbon residue, high nitrogen content, high sulfur content, and high metals content. In addition to these properties, the viscous feedstocks also have an increased molecular weight and reduced hydrogen content as well as a relatively low content of volatile saturated constituents and aromatic constituents. These feedstocks also contain a relatively high content of asphaltene constituents plus a relatively high content of resin constituents that is accompanied by a high heteroatom (nitrogen, oxygen, sulfur, and metals) content. Thus, the criteria for selection of upgrading option options for the viscous feedstocks depend on several factors which must be analyzed in detail before an upgrading sequence is set into motion. The choice, put simply, is to apply (1) a carbon rejection technology sequence or (2) a hydrogen addition technology sequence.

However, the deposition of coke can lead to fouling, and subsequent corrosion (Chapter 6) as well as coke deposition on any catalysts employed for the process can (and will) happen during thermally driven catalytic upgrading which deactivates the catalysts within short time interval. In addition, the presence of heavy metals (such as vanadium and nickel), sulfur, nitrogen, and other contaminants severely reduces the catalytic activity rapidly. On the other hand, hydrogen addition technologies typically produce a high yield of products with a higher commercial value than the products from the carbon rejection technologies but require a larger investment to produce the amounts of

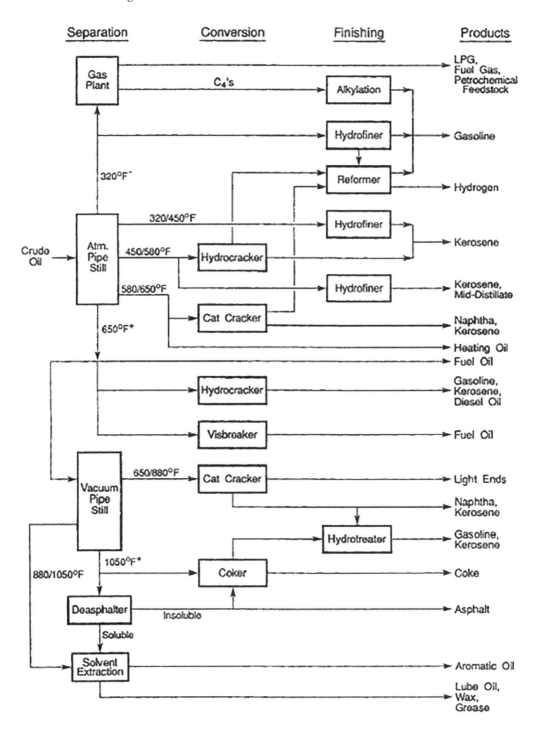

FIGURE 1.1 Schematic overview of a refinery. (Speight, J.G. 2014. *The Chemistry and Technology of Petroleum*, 5th Edition. CRC Press, Taylor & Francis Publishers, Boca Raton, Florida. Figure 15.1, page 392.)

hydrogen and steam required for these processes. In both cases, at some stage of the refining, the minimization of unwanted heteroatom species (such as sulfur, nitrogen, oxygen, and metals) and coke deposition is necessary to sustain not only the catalyst activity but also the product value.

In this respect, the presence of hydrogen not only elevates the extent of the removal of hetero-atom contaminants that are present in the original refinery feedstock but also increases the ability of the process to accelerate the breakdown of higher molecular weight constituents to lower-boiling products. Furthermore, thermal cracking in the absence of concurrent (or consequent) hydrogen transfer without a catalyst may lead to a product that is unstable and incompatible not only with the other products but also with any unchanged feedstock in the reactor. For such feedstocks, a choice of different technologies may help address such challenges.

By customizing the approach to refining the viscous feedstocks through an understanding of the properties and behavior of such feedstocks, a refinery can meet the challenges that occur during heavy feedstock upgrading (Casey, 2011). For example, by understanding the behavior of the feedstock under thermal conditions or by the development of an improved catalyst system with appropriate reactor internals, it is possible to provide optimal feedstock catalyst. In many cases, the successful selection of an emerging technology for viscous feedstock upgrading should consider at least (1) the type and properties of the feedstock to be processed, (2) the quality of upgraded product, (3) the suitability of the process, (4) the ease of application of the process by way of the flexibility of the process, (5) the operating parameters in the reactor, and (6) the properties of the products and any by-products.

Thus, a major decision at the time of acceptance of and the selection of processes for viscous feedstocks (or a blend of viscous feedstocks) is to acknowledge the practical aspects of upgrading which can include partial upgrading or full upgrading in a conversion refinery (Colyar, 2009). In addition, the regulatory need for the continuing reduction of the amount of sulfur in refinery products has created critical issues in relation to upgrading the viscous heavy feedstocks into acceptable, low-sulfur products. Thus, a refinery must also be flexible and adaptable to changing the operations on an as-needed basis, especially if the more viscous feedstocks (such as heavy crude oil, extra heavy crude oil, tar sand bitumen, and distillation residua) are the primary feedstocks accepted and used by the refinery for thermal cracking processes and for catalytic cracking processes.

The conversion of conventional crude oil to products is well established by the evolution of the thermal cracking processes and the catalytic cracking processes, but the conversion of viscous feedstocks is now in a significant transition period since the demand is increasing for saleable products such as, for example, transportation fuels and solvents.

Thus, in order to satisfy the changing pattern of product demand, there is an increased need for conversion processes that can be applied to the viscous feedstocks which will take the conversion of the viscous feedstocks beyond current limits and, at the same time, reduce the amount of coke and other non-essential products. Such conversion schemes may require the use of two or more technologies in series rather than an attempt to develop a whole new one-stop conversion technology. Processes for the conversion of viscous feedstocks, such as the visbreaking and coking options, augment the deasphalting units in many refineries operations. An example is the application of a mild visbreaking technology (Chapter 4) which takes the feedstock just beyond the point of coke formation (and deposition) such that the initial coke forms on any particles of mineral matter in the feedstock, thereby encouraging demineralization of the feedstock along with removal of the constituents that readily form coke and removal of these forming constituents a minor de-coking operation (with concomitant demineralization) with the result that the product is amenable for use as a feedstock for a fluid catalytic cracking unit (Speight and Moschopedis, 1979; Speight, 1987).

However, in spite of any potential drawbacks or the difficulty of upgrading viscous feedstocks, originally referred to as resid upgrading (bottom of the barrel processing, which was shunned by many refineries in favor of asphalt production) offers a means by which refiners can lower the feedstock cost by using a heavy feedstock as part of a crude oil blend. Although the resid fluid catalytic cracking process (often referred to by the acronym RFCC process) remains one of the most attractive ways to economically upgrade heavy feedstocks into transportation fuels and decrease fuel-oil production, there are alternative processing schemes that are applied according to the properties of the feedstock (Phillips and Liu, 2002; Kressmann et al., 2000; Ross et al., 2005; Hsu and Robinson, 2006; Gary et al., 2007; Stratiev and Petkov, 2009; Bridjanian and Khadem Samimi, 2011; Speight, 2011a).

Nevertheless, there remains the need to understand the thermal cracking processes and the catalytic cracking processes in order to (1) reduce the formation of hydrocarbon gases, (2) inhibit the formation of polynuclear aromatic products that are not originally present in the feedstock bur are products of the process, and (3) separate an intermediate that produces low amounts of coke before (or during) conversion. An additional challenge for the conversion processes is to take advantage of the nickel and vanadium in the heavy feedstocks to generate an in situ dispersed catalyst and to eliminate catalyst cost.

Once the refinery feedstock has been cleaned and subjected to the distillation processes (atmospheric distillation and vacuum distillation), the conversion activities start with the cracking processes, of which thermal cracking processes (including visbreaking) and catalytic cracking are the prime examples (Parkash, 2003; Hsu and Robinson, 2006; Gary et al., 2007; Speight, 2011a, 2011b, 2014; Hsu and Robinson, 2017; Speight, 2017, 2022). It is at this stage that the processing flow scheme (Figure 1.1) is largely determined by the composition of the crude oil feedstock and the chosen slate of crude oil products.

Thus, the options for feedstock processing must be selected and products manufactured to give a balanced operation in which the refinery feedstock is converted into a variety of products in amounts that are in accord with the demand for each. For example, the manufacture of products from the lower-boiling portion of crude oil automatically produces a certain amount of higher-boiling components (which is feedstock dependent) due to the complex chemistry that occurs during the conversion processes. If the latter cannot be sold as, for example, heavy fuel oil, these products will accumulate until refinery storage facilities are full. To prevent the occurrence of such a situation, the refinery must be flexible and be able to change operations as needed. This usually means more processes such as (1) thermal processes to change an excess of heavy fuel oil into more naphtha with coke as the residual product, or (2) vacuum distillation to separate the heavy oil into lubricating oil blend stocks and asphalt.

In the feedstock, the basic elements of crude oil, hydrogen, and carbon form the main input into a refinery, combining into thousands of individual constituents and the economic recovery of these constituents varies with the individual crude oil according to its particular individual qualities, and the processing facilities of a particular refinery. In general, crude oil, once refined, yields three basic groupings of products that are produced when it is broken down into fractions (often referred to as cuts) (Table 1.1). The gas and naphtha fractions form the lower-boiling products and are usually more valuable than the higher-boiling fractions and provide gas (liquefied petroleum gas), naphtha, aviation fuel, motor fuel, and feedstocks, for the petrochemical industry (Table 1.2).

Also, another valuable product, naphtha (which is a precursor to gasoline and solvents) is produced from the low boiling and the middle range of distillate cuts (sometimes referred to collectively kerosene and/or as light gas oil) and is also used as a feedstock for the petrochemical industry (Table 1.3).

The middle distillates refer to products from the middle boiling range of crude oil; include kerosene, diesel fuel, distillate fuel oil, and light gas oil; and emerge from the atmospheric distillation unit between the naphtha and the heavy atmospheric gas oil (Figure 1.1). Waxy distillate (produced from the atmospheric distillation unit) and lower-boiling constituents of lubricating oils are sometimes included in the middle distillates. The remainder of the crude oil includes light-vacuum gas oil and heavy-vacuum gas oil (from which the higher-boiling constituents of lubricating oil can be produced) and residuum (the nonvolatile fraction of the atmospheric residuum) (Figure 1.1). The vacuum residuum with further deep distillation may also produce useful products but is more commonly used for asphalt production. The complexity of crude oil is emphasized insofar as the actual proportions of light, medium, and heavy fractions vary significantly from one crude oil to another.

Refinery process for crude oil are generally divided into three categories: (1) separation processes, of which distillation is the prime example; (2) conversion processes, of which coking and

TABLE 1.1
Boiling Fractions of Conventional Crude Oil

Fraction	Boiling Range	
	°C	°F[a]
Low-boiling naphtha	30–150	30–300
High-boiling naphtha	150–180	300–400
Middle distillates[b]	180–290	400–500
Kerosene	180–260	355–500
Fuel oil	205–290	400–550
Atmospheric gas oil	260–315	500–800
Vacuum gas oil	425–600	800–1,100
Light-vacuum gas oil	315–425	600–800
Heavy-vacuum gas oil	425–600	425–600
Residuum	>510	>950

[a] For convenience, boiling ranges which can vary from refinery to refinery are approximate and, for convenience, are converted to the nearest 5°C.

[b] Obtained in the "middle" boiling range which is on the order of 180°C–260°C (355°F–500°F) during the crude oil distillation process. The middles distillate is so named because the fraction is removed at mid-height in the distillation tower during the multi-stage process of thermal separation.

TABLE 1.2
Production of Starting Materials for Petrochemical Processing

Feedstock	Process	Product
Crude oil	Distillation	Light ends
		Methane
		Ethane
		Propane
		Butane
	Catalytic cracking	Ethylene
		Propylene
		Butylene derivatives
		Higher-boiling olefins
	Coking	Ethylene
		Propylene
		Butylene derivatives
		Higher-boiling olefins

catalytic cracking are prime examples; and (3) finishing processes, of which hydrotreating to remove sulfur is a prime example.

The simplest refinery is designed to prepare feedstocks for petrochemical manufacture or for the production of industrial fuels and is often referred to as the topping refinery. The topping refinery consists of tankage, a distillation unit, recovery facilities for gases and light hydrocarbons, and the

TABLE 1.3
Sources of Naphtha

Process	Primary Product	Secondary Process	Secondary Product
Atmospheric distillation	Light naphtha	Cracking	Petrochemical feedstock
	Heavy naphtha	Catalytic cracking	Light naphtha
	Gas oil	Catalytic cracking	Light naphtha
	Gas oil	Hydrocracking	Light naphtha
Vacuum distillation	Gas oil	Catalytic cracking	Light naphtha
		Hydrocracking	Light naphtha
	Residuum	Coking	Light naphtha
		Hydrocracking	Light naphtha
Cracking processes	Light naphtha	Cracking	Petrochemical feedstock
	Heavy naphtha	Catalytic cracking	Light naphtha
	Gas oil	Catalytic cracking	Light naphtha
	Gas oil	Hydrocracking	Light naphtha

necessary utility systems (steam, power, and water-treatment plants). Topping refineries are highly dependent on local markets, but the addition of hydrotreating and reforming units to this basic configuration results in a more flexible hydroskimming refinery, which can also produce desulfurized distillate fuels and high-octane gasoline. These refineries may produce up to half of their output as residual fuel oil, and they face increasing market loss as the demand for low-sulfur (even no-sulfur) high-sulfur fuel oil increases.

The most versatile refinery configuration is the conversion refinery which incorporates all of the basic units found in both the topping and hydroskimming refineries, but it also features gas oil conversion plants such as catalytic cracking and hydrocracking units, olefin conversion plants such as alkylation or polymerization units, and, frequently, coking units for sharply reducing or eliminating the production of residual fuels (Parkash, 2003; Gary et al., 2007; Speight, 2014; Hsu and Robinson, 2017; Speight, 2017). Modern conversion refineries may produce two-thirds of their output as unleaded gasoline, with the balance distributed between liquefied petroleum gas, jet fuel, diesel fuel, and a small quantity of coke. Many such refineries also incorporate solvent extraction processes for manufacturing lubricants and petrochemical units with which to recover propylene, benzene, toluene, and xylenes for further processing into polymers.

Finally, the yields and quality of refined crude oil products produced by a refinery depend on the mixture of crude oil used as feedstock and the configuration of the refinery. Light/sweet (conventional low-sulfur) crude oil has inherently greater yields of higher value low-boiling products such naphtha and kerosene. Heavy-sour (high-viscosity, high-sulfur) crude oil produces higher yields of lower-value higher-boiling products that must be converted into the more preferable lower-boiling products (Speight, 2013, 2014).

Since a refinery is a group of integrated manufacturing plants (Figure 1.1), each of which is selected to give a balanced production of saleable products in amounts that are in accord with the demand for each product, it is necessary to prevent the accumulation of non-saleable products, the refinery must be flexible and be able to change operations as needed. The complexity of crude oil is emphasized insofar as the actual amounts of the products vary significantly from one crude oil to another (Speight, 2014, 2016). In addition, the configuration of refineries may vary from refinery to refinery. Some refineries may be more oriented toward the production of gasoline (large reforming and/or catalytic cracking), whereas the configuration of other refineries may be more oriented toward the production of middle distillates such as jet fuel and gas oil.

Moreover, it is only by comprehensively considering factors such as the (1) feedstock properties, (2) catalyst performance, (3) product specifications, (4) chemical kinetics, (5) operating conditions, and (6) running period that optimal results can be achieved. Therefore, further improvement of the various upgrading process for the viscous feedstocks and the development of catalysts, which can tolerate a high content of impurities and metals in the feedstock, are two major challenges for the future. Catalyst activity, selectivity, particle size and shape, pore size and distribution, and the type of the reactor have to be optimized according to the properties of the heavy oils and to the desired purification and conversion levels. In fact, a particularly crucial issue that has arisen within the latter half of the 20th century (and continues in the present) relates to the use of blends of any viscous feedstock with other viscous feedstocks as well as blends of any viscous feedstock with the more typical conventional refinery feedstock.

Blending is one of the typical operations that a refinery must pursue not only to prepare a product to meet sales specifications but also to blend the different crude oils and viscous feedstocks to produce a total feedstock (Parkash, 2003; Gary et al., 2007; Speight, 2014a; Hsu and Robinson, 2017; Speight, 2017). Although simple in principle, the blending operation must be performed with care and diligence based on an understanding of the properties and composition of the feedstock to be blended and whether or not the constituents in the blend will be compatible and not result in the formation of a semi-solid or solid separate phase (Speight, 2014, 2017). Lack of attention to the properties of the individual feedstocks prior to the blending operations can (in the case of the high-asphaltene viscous feedstocks) lead to phase separation (fouling) prior to and during the refining operations because of the phenomenon of incompatibility of the different components of the blend (Chapter 6) (Schermer et al., 2004; Speight, 2014).

Briefly, fouling as it pertains to the crude oil industry, is manifested in the form of formation, encrustation, deposition, scaling, scale formation, slagging, and sludge formation) which has an adverse effect on the refinery operations (Chapter 6). It is the accumulation of unwanted material within a processing unit or on the on solid surfaces of the unit to the detriment of function. For example, when it does occur during refinery operations, the major effects include (1) loss of heat transfer as indicated by charge outlet temperature decrease and pressure drop increase, (2) blocked process pipes, (3) under-deposit corrosion and pollution, and (4) localized hot spots in reactors, all of which culminate in production losses and increased maintenance costs. Thus, the separation of solids occurs whenever the solvent characteristics of the liquid phase are no longer adequate to maintain polar and/or high molecular weight material in solution.

If attention is not paid to the phenomenon of incompatibility of the feedstocks either at the time of blending or because of the effects of elevated temperatures involved in the process, the result will be the occurrence of fouling deposits in heat transfer equipment and reactors (Stark and Asomaning, 2003; Van den Berg et al., 2003). Therefore, it is advisable for the refiner to be able to predict the potential for incompatibility by determining not only the appropriate components for the blend but also the ration of individual crude oils and heavy feedstocks in the blend.

Such blends raise the issues of the compatibility of the constituents of the blends which is not always assured and there can (as a result of blend instability) is excessive laydown of coke (and metals) on the catalyst. Thus, the conversion processes must be chosen accordingly, and only after satisfactory test methods have been applied not only to the individual constituents of a blend but also to the blend itself. Once this has been accomplished, the viscous feedstock can proceed to the conversion step which typically involves the application of thermal cracking processes and/or catalytic cracking process to the feedstock.

In addition, incompatibility and fouling can lead to corrosion which is especially relevant when high-sulfur feedstocks are processed. In such cases, corrosion can occur where metal temperatures are on the order of 230°C–480°C (450°F–900°F). Above 480°C, the coke forms a protective layer on the metal. The furnace, soaking drums, lower part of the tower, and high-temperature exchangers are usually subject to corrosion.

Thus, the basic processes introduced to bring about thermal decomposition of the higher-boiling streams and the more viscous feedstocks are known as cracking processes. In these processes, the higher-boiling fractions are converted to lower-boiling products. Thermal cracking processes were the original processes of choice which led to the beginnings and evolution of catalytic cracking processes in which viscous feedstocks or cuts are broken down or changed by being heated and reacted with catalysts.

Thus, after presentation of the types of feedstocks that are currently acceptable to a refinery and the future feedstocks (such as the various types of biomass) that will also be acceptable to the future refinery, this chapter presents an introduction to the general layout of a crude oil refinery (with a brief description of the various processes) in order for the reader to place the conversion processes, notably (1) the visbreaking process; (2) the thermal cracking processes, which are often referred to as coking processes and include delayed coking, fluid coking, and flexicoking; and (3) the catalytic cracking process in the context of the overall refinery operations.

The chapter focuses on the thermal cracking processes and the catalytic cracking processes and the relationship of these processes to the other processes in the refinery, which are described more fully in the subsequent volumes of this series. The thermal cracking processes and the catalytic cracking processes are important aspects of the conversion of the viscous feedstocks (heavy crude oil, extra heavy crude oil, tars and bitumen, and distillation residua), and while they are the focus of this book, it would be remiss not to include the use of naphtha, kerosene, and gas oil as feedstocks to thermal processes.

In order to fully comprehend the function of the various process units within refinery operations, it is necessary to understand the place at which the process units employed for into the refinery and the reason for this employment. This will assist the reader to place thermal cracking processes and the catalytic cracking process in the correct perspective of the refinery, not forgetting that prior to any conversion processes it is necessary to desalt, dewater, and distill the raw (i.e., unrefined) refinery feedstock.

1.2 THERMAL CRACKING PROCESSES

The term "thermal cracking," as employed in the crude oil industry, is the thermal decomposition, under pressure, of high molecular weight crude oil constituents (typically, but not always, higher molecular weight constituents of naphtha) to form lower molecular weight (lower-boiling-range and higher-value) products, and there are several such processes used in a refinery (Table 1.4).

TABLE 1.4
Commonly Used Thermal Cracking Processes

Process	Description
Visbreaking	A form of mild thermal cracking
	Feedstock is typically a viscous feedstock such as residuum, heavy crude oil, extra heavy crude oil, and tar sand bitumen
	Used to reduce the viscosity of a viscous feedstock
	Product is a less viscous fluid stock that is easier to handle
	Typically used to produce a fuel-oil blend stock
Thermal cracking	Feedstock may be a gas oil or a viscous feedstock or blends thereof
	Products are gases, naphtha, kerosene, fuel oil, and resid (or coke)
	Includes vapor-phase thermal cracking to convert low-boiling hydrocarbon feedstocks (gases and low-boiling naphtha) to olefinic feedstocks for the production of petrochemicals
Coking	Thermal decomposition of viscous feedstocks
	Products are gases, naphtha, kerosene, fuel oil, and coke
Thermal reforming[a]	Used to convert of low-octane naphtha to higher-octane product
	Also used to produce ethylene from naphtha; may involve conversion of naphthene derivative to aromatic derivatives by dehydrogenation of the naphthene ring systems

[a] Also includes thermal polymerization, thermal alkylation.

More specifically, the thermal cracking process is designed to produce naphtha from higher-boiling charge stocks, and any unconverted or mildly cracked charge components (components which have partially decomposed but are still higher boiling than gasoline) are usually recycled to maximize gasoline production. As the thermal cracking process proceeds, any reactive unsaturated products (i.e., olefin derivatives) that are formed will ultimately higher molecular weight products, some of which may be suitable from inclusion in asphalt. These higher molecular weight product molecules are hydrogen deficient and tend to form coke readily and cannot be recycled without excessive coke formation and are typically removed from the reactor system (often as fuel oil).

The precise origins of cracking distillation are unknown. Whether or not any credence should be given to such a rumor, it is rumored that, in 1861 while distilling crude oil, the attending stillman had to leave his charge for a longer time than he intended (the actual reason for this departure is not known) during which time the still overheated. When he returned, he noticed that the distillate in the collector was much more volatile than anticipated at that particular stage of the distillation. Further investigation leads to the development of cracking distillation (i.e., thermal degradation with the simultaneous production and removal of distillate from the reactor). However, cracking was used commercially in the production of oils from coal and shales before the crude oil industry began, and the discovery that the higher molecular weight materials products could be decomposed to lower molecular weight (and distillable) product was used to increase the production of kerosene.

The process known as thermal cracking distillation has been used continuously as a process for the production of valuable lower-boiling products (such as naphtha and kerosene) from higher-boiling nonvolatile feedstocks. In fact, in the early days of the process (i.e., in the three decades from 1870 to 1900), the technique was very simple – a batch of crude oil was heated until most of the kerosene had been distilled from it and the overhead material had become dark in color. At this point, distillation was discontinued and the heavy oils were held in the hot zone, during which time some of the high molecular weight components were decomposed to produce lower molecular weight products. After a suitable time, distillation was continued to yield light oil (kerosene) instead of the heavy oil that would otherwise have been produced. As the need for gasoline arose in the early 1900s, the necessity of prolonging the cracking process became apparent, and a process known as pressure cracking evolved.

Pressure cracking was a batch operation in which, as an example, a batch of gas oil (on the order of 200 bbl) was heated to approximately 425°C (800°F) in stills (reactors) that had been reinforced to operate at pressures as high as 95 psi. The gas oil was held under maximum pressure for 24 hours, while fires maintained the temperature and distillation was commenced. During the next 48 hours, a distillate (on the order of 100 bbl) was produced which contained the naphtha (gasoline) constituents which was then treated with sulfuric acid to remove any unstable potential gum-forming components and then redistilled to produce a cracked naphtha in the gasoline boiling range.

Following this, in 1912 the large-scale production of cracked gasoline was first developed by Burton. The process employed batch distillation in horizontal shell stills and operated at approximately 400°C (ca. 750°F) and 75–95 psi and was the first successful method of converting higher-boiling oils into gasoline. Nevertheless, heating a bulk volume of oil was soon considered cumbersome, and during the years 1914–1922, a number of successful continuous cracking processes were developed. In these processes, a higher-boiling feedstock (i.e., gas oil) was continuously pumped through a unit that heated the feedstock to the required temperature, held it for a time under pressure, and then discharged the cracked product into distillation equipment where it was separated into gases, naphtha, gas oil, and tar.

Another process, the tube-and-tank cracking process was not only typical of the early (post-1900) cracking units but also is one of the first units on record in which the concept of reactors (soakers) being on-stream/off-stream is realized. Such a concept departs from the true batch concept and allowed a greater degree of continuity. In fact, the tube-and-tank cracking unit may be looked upon as a forerunner of the delayed coking operation.

In the tube-and-tank process, a feedstock (at that time a gas oil) was preheated by exchange with the hot products from the unit pumped into the cracking coil, which consisted of several hundred feet of very strong pipe that lined the inner walls of a furnace where oil or gas burners raised the temperature of the gas oil to 425°C (800°F). The hot gas oil passed from the cracking coil into a large reaction chamber (soaker) where the gas oil was held under the temperature and pressure conditions long enough for the cracking reactions to be completed. The cracking reactions formed coke which, in the course of several days, filled the soaker. The gas oil stream was then switched to a second soaker, and the first soaker was cleaned out by drilling operations similar to those used in drilling an oil well. The cracked product (other than coke) left the on-stream soaker to enter an evaporator (tar separator) maintained under a much lower pressure than the soaker where, because of the lower pressure, all of the cracked material, except the tar, became vaporized. The vapor left the top of the separator where it was distilled into separate fractions – gases, gasoline, and gas oil. The tar that was deposited in the separator was pumped out for use as asphalt or as a heavy fuel oil.

Early in the development of tube-and-tank thermal cracking, it was found that adequate yields of naphtha (gasoline) could not be obtained by one passage of the stock through the heating coil; attempts to increase the conversion in one pass brought about undesirable high yields of gas and coke. It was better to crack to a limited extent, remove the products, and recycle the rest of the oil (or a distilled fraction free of tar) for repeated partial conversion. The high-boiling constituents once exposed to cracking were so changed in composition as to be more refractory than the original feedstock.

With the onset of the development of the automobile, the most important part of any refinery became the gasoline-manufacturing facilities. Among the processes that have evolved for gasoline production are thermal cracking, catalytic cracking, thermal reforming, catalytic reforming, polymerization, alkylation, coking, and distillation of fractions directly from crude oil.

Briefly, the thermal reforming of naphtha is a severe, vapor-phase thermal conversion process conducted under pressure (Parkash, 2003; Gary et al., 2007; Speight, 2014; Hsu and Robinson, 2017; Speight, 2017). The purpose of thermal reforming is to increase the octane number of gasoline charge stock. A recycle stream is not normally used. Significant quantities of light hydrocarbon gases, pentanes, and lighter compounds are produced. These gases contain sizable quantities of light olefins which are used as alkylation, polymerization, or petrochemical feedstocks. In the thermal reforming process, the conditions are sufficiently severe to cause some polymerization of the olefin derivatives and condensation to polynuclear compounds. These compounds form a stream with a higher boiling range than the naphtha feedstocks and, typically, is removed and can be used as a fuel-oil component.

When kerosene was the major product, the naphtha fraction (used as a precursor to gasoline) was the portion of crude oil too volatile to be included in kerosene. The refiners of the 1890s and early 1900s had no use for it and often dumped an accumulation of gasoline into the creek or river that was usually nearby. As the demand for gasoline increased with the onset of World War I and the ensuing 1920s, more crude oil had to be distilled not only to meet the demand for gasoline but also to reduce the overproduction of the heavier (higher-boiling, higher-density) crude oil fractions, including kerosene.

The problem of how to produce more gasoline from less crude oil was solved in 1913 by the incorporation of cracking units into refinery operations in which fractions heavier than gasoline were converted into gasoline by thermal decomposition. The early (pre-1940) processes employed for gasoline manufacture were processes in which the major variables involved were feedstock type, time, temperature, and pressure and which need to be considered to achieve the cracking of the feedstock to lighter products with minimal coke formation.

As refining technology evolved throughout the 20th century, the heavy distillate from a distillation unit (heavy gas oil, vacuum gas oil) and residuum from distillation units were used as feedstocks for cracking processes. In addition, the residual oils produced as the end-products of the distillation processes and even some of the heavier virgin oils often contain substantial amounts of asphaltic materials, which preclude use of the residuum as fuel oils or lubricating stocks. However, subjecting these residua directly to thermal processes has become economically advantageous, since, on the one hand, the end result is the production of lower-boiling salable materials; on the other hand, the asphaltic materials in the residua are regarded as the unwanted coke-forming constituents.

As the thermal processes evolved and catalysts were employed with more frequency, poisoning of the catalyst with a concurrent reduction in the lifetime of the catalyst became a major issue for refiners. To avoid catalyst poisoning, it became essential that as much of the nitrogen and metals (such as vanadium and nickel) as possible should be removed from the feedstock. The majority of the heteroatoms (nitrogen, oxygen, and sulfur) and the metals are contained in, or associated with, the asphaltic fraction (residuum). It became necessary that this fraction be removed from cracking feedstocks.

With this as the goal, a number of thermal processes, such as tar separation (flash distillation), vacuum flashing, visbreaking, and coking, came into wide usage by refiners and were directed at upgrading feedstocks by removal of the asphaltic fraction. The deasphalting of viscous feedstocks using liquid hydrocarbon gases such as propane and/or butane was widely adopted in many refineries in the 1950s and was very effective for the preparation of residua for cracking feedstocks (the deasphalted oil fraction) from residua. In this process, the desirable oil in the feedstock (the deasphalted oil product) is dissolved in the liquid hydrocarbon and asphaltic materials remain insoluble.

Operating conditions in the deasphalting tower depend on the boiling range of the feedstock and the required properties of the product. Generally, extraction temperatures can range from 55°C to 120°C (130°F to 250°F) with a pressure of 400–600 psi. Hydrocarbon/oil ratios on the order of 6:1 to 10:1 by volume are typically used.

1.2.1 VISBREAKING

Visbreaking is a mild form of thermal cracking that is employed cause a significant lowering of the viscosity of a viscous feedstock (that was originally developed to treat distillation residua) that can then be used for blending with low-viscosity fuel oils. The process is also used to reduce the pour point of a waxy feedstock residues and to produce middle distillates, depending on product demand

Visbreaking (viscosity breaking, viscosity reduction) is a process of the post-1940 era and was initially introduced as a mild thermal cracking operation that employs a mild cracking operation (insofar as the thermal reactions are not allowed to proceed to completion) used to reduce the viscosity of viscous feedstocks (Parkash, 2003; Hsu and Robinson, 2006; Gary et al., 2007; Speight, 2011a, 2011b, 2014; Hsu and Robinson, 2017; Speight, 2017).

The process, evolved from the older and now obsolete thermal cracking processes, is classed as mild because the thermal reactions are not allowed to proceed to completion, and the process can be used to reduce the viscosity of residua to allow the products to meet fuel-oil specifications. Also, the product from the visbreaking of residua could be blended with lighter product oils to produce fuel oils of acceptable viscosity. By reducing the viscosity of the residuum, visbreaking reduces the amount of light heating oil that is required for blending to meet the fuel-oil specifications. In addition to the major product, (fuel oil), products in the gas oil and gasoline boiling range is produced. The gas oil may be used as additional feed for catalytic cracking units, or as heating oil.

In the process, the viscous feedstock is heated to a temperature on the order of 425°C–510°C (800°F–950°F) at atmospheric pressure after which the mildly cracked product is then quenched with cool gas oil to control any additional cracking (sometimes referred to as over-cracking) and flashed in a distillation tower. The thermally cracked residue tar, which accumulates in the bottom of the fractionation tower, is vacuum flashed in a stripper and the distillate recycled.

The processes use two major configurations which are (1) the short-contact option and (2) the soaker option. In the short-contact visbreaking option, the feedstock is heated to a temperature on the order of 480°C 900°F and sent to a reactor (referred to as the soaking zone) under a pressure in the range of 140–300 psi. The elevated pressure allows cracking to occur while restricting coke formation, and to avoid over-cracking (leading to the formation of coke), the residence time in the soaking zone is short – on the order of several minutes compared to the residence time in a delayed coking unit (typically, several hours) after which the hot oil is quenched (usually with cold gas oil) to inhibit further cracking and sent to a vacuum tower for product separation.

The soaker visbreaking process maintains the feedstock at an elevated temperature for a longer time to increase the yield of middle distillates whereas the low-viscosity visbreaker gas oil can be used as heavy fuel oil or sent to a fluid catalytic cracking (FCC) unit or to a hydrocracking unit for further processing.

Using fuel oil as an example, residua are sometimes blended with lighter heating oils to produce fuel oils of acceptable viscosity. Thus, by reducing the viscosity of the nonvolatile fraction, visbreaking reduces the amount of the more valuable distillate light heating oil that is required for blending to meet the fuel-oil specifications. The process is also used to reduce the pour point of a waxy residue. Visbreaking conditions range from 455°C to 510°C (851°F to 950°F) and 50 to 300 psi at the heating coil outlet. Liquid-phase cracking takes place under these low severity conditions.

In a typical visbreaking operation, a viscous feedstock is passed through a furnace where it is heated to a temperature of 480°C (895°F) under an outlet pressure of approximately 100 psi. The heating coils within the furnace are arranged to provide a soaking section where the feedstock remains until the visbreaking reactions are completed and the cracked products are then passed into a flash-distillation chamber. The overhead material from this chamber is then fractionated to produce a low-quality gasoline as an overhead product and light gas oil as bottom. The liquid products from the flash chamber are cooled with a gas oil flux and then sent to a vacuum fractionator. This yields a heavy gas oil distillate and a residual tar of reduced viscosity.

1.2.2 Thermal Cracking

One of the earliest conversion processes used in the crude oil industry is the thermal decomposition of higher-boiling materials into lower-boiling products. The process was developed in the early 1900s to produce naphtha (often referred to as gasoline but not the gasoline of the modern refinery) from the unwanted higher-boiling products of the distillation process. However, it was soon learned that the thermal cracking process also produced a wide slate of products varying from highly volatile gases to nonvolatile coke.

The concept behind thermal cracking is the thermal decomposition of higher molecular weight constituents of crude oil to produce lower molecular weight, normally more valuable, products. The first commercial process was in 1913 and is known as the Burton process. Even though catalyst cracking generally replaced thermal cracking in the 1940s, noncatalytic cracking processes using high temperature to achieve the decomposition are still in operation (as illustrated by the continued use of coking processes). Catalytic processes usually produce more gasoline having a higher octane number, but with lesser yields of gases and heavy fuel oil. The gases produced by catalytic cracking contain more olefins than those produced by thermal cracking.

In the thermal cracking process, a feedstock is fed to the fractionator with their thermal reactivity to separate gasoline, light, and heavy oil. The light oil is then fed to the heater at 540°C–595°C (1,000°F–1,100°F) within a pressure range on the order of 350–700 psi, and the light oil transforms to the vapor phase and is sent to the soaker. If the feedstock is a viscous feedstock, temperatures on the order of 400°C–480°C (750°F–900°F) are used along with a higher pressure (350–700 psi) to maintain the feedstock in the liquid phase, and then it is fed to the soaker. The liquid- and vapor-phase mix in the soaker and sent to the separator, with the products coming out on the bottom as fuel oil and the light recycle back to the fractionator. Coking in the reactor is the main problem when heavy oil is heated at high temperatures.

The heavier oils produced by cracking are light and heavy gas oils as well as a residual oil which could also be used as heavy fuel oil. Gas oils from catalytic cracking were suitable for domestic and industrial fuel oils or as diesel fuels when blended with straight-run gas oils. The gas oils produced by cracking were also a further important source of gasoline. In a once-through cracking operation, all of the cracked material is separated into products and may be used as such. However, the gas oils produced by cracking (cracked gas oils) are more resistant to cracking (more refractory) than gas oils produced by distillation (straight-run gas oils) but could still be cracked to produce more gasoline. This was achieved using a later innovation (post-1940) involving a recycle operation in which the cracked gas oil was combined with fresh feed for another trip through the cracking unit. The extent to which recycling was carried out affected the yield of gasoline from the process.

The majority of the thermal cracking processes use temperatures on the order of 455°C–540°C (850°F–1005°F) and pressures of 100–1,000 psi. An example is the Dubbs process in which the viscous feedstock is preheated by direct exchange with the cracking products in the fractionating columns. Cracked gasoline and heating oil are removed from the upper section of the column. Light and heavy distillate fractions are removed from the lower section and are pumped to separate heaters. Higher temperatures are used to crack the more refractory light distillate fraction. The streams from the heaters are combined and sent to a soaking chamber where additional time is provided to complete the cracking reactions. The cracked products are then separated in a low-pressure flash chamber where a nonvolatile heavy fuel oil is removed and the volatile cracked products are sent to the fractionating columns.

Mild cracking conditions, with a low conversion per cycle, favor a high yield of gasoline components, with low gas and coke production, but the gasoline quality is not high, whereas more severe conditions give increased gas and coke production and reduced gasoline yield (but of higher quality). With limited conversion per cycle, the heavier residues must be recycled, but these recycled oils become increasingly refractory upon repeated cracking, and if they are not required as a fuel-oil stock, they may be coked to increase gasoline yield or refined by means of a hydrogen process.

The thermal cracking of higher-boiling crude oil fractions to produce gasoline is now virtually obsolete. The antiknock requirements of modern automobile engines together with the different nature of crude oils (compared to those of 50 or more years ago) have reduced the ability of the thermal cracking process to produce gasoline on an economic basis. Very few new units have been installed since the 1960s, and some refineries may still operate the older cracking units.

Finally, another cracking process that cannot be omitted is the steam cracking process that is used to produce olefin derivatives (such as ethylene, $CH_2=CH_2$) for the manufacture of petrochemicals. The feedstocks range from ethane (C_2H_6, CH_3CH_3) to vacuum gas oil, with higher-boiling feedstocks giving higher yields of by-products, such as naphtha. The most common feeds are ethane, butane, and naphtha. The process involves the use of temperature on the order of 815°C–870°C (1,500°F–1,600°F) and at pressures (24–30 psi) slightly above atmospheric pressure. Naphtha produced from steam cracking contains benzene, which is extracted prior to hydrotreating. In addition, the nonvolatile residue from a steam cracking process may be used as a blend stock for the production of heavy fuel oil.

1.2.3　Coking

Coking is a generic term for a series of thermal processes used for the conversion of nonvolatile heavy feedstocks into lighter, distillable products (Parkash, 2003; Gary et al., 2007; Speight, 2014; Hsu and Robinson, 2017; Speight, 2017, 2021). The feedstock is typically a residuum, and the products are gas, naphtha, fuel oil, gas oil, and coke. Gas oil which, depending upon the process parameters, can be the primary product of a coking operation and serves primarily as a feedstock for catalytic cracking units. The coke obtained is usually used as fuel, but specialty uses, such as electrode manufacture and the production of chemicals and metallurgical coke, are also possible, thus increasing the value of the coke.

In contract to the visbreaking process, the coking process is a severe method of thermal cracking used to convert viscous feedstocks into lower-boiling products (i.e., distillates). Coking produces straight-run naphtha (coker naphtha) as well as a variety of other fractions (middle distillates) that

---

TABLE 1.5

Comparison of Visbreaking with Delayed Coking and Fluid Coking

Visbreaking

Purpose: to reduce viscosity of fuel oil to acceptable levels

Mild (470°C–495°C; 880°F–920°F) heating at pressures of 50–200 psi

Reactions quenched and not allowed to proceed to completion

Low conversion (10%) to products boiling less than 220°C (430°F)

Delayed Coking

Purpose: to produce maximum yields of distillate products

Moderate (480°C–515°CC; 900°F–960°F) heating

Pressure on the order of 90 psi

Reactions allowed to proceed to completion

Soak drums (845°F–900°F) used in pairs (one on-stream and one off-stream being de-coked)

Coke yield: 20%–40% w/w of feedstock (dependent upon feedstock)

Yield of distillate <220°C (<430°F): on the order of 30% w/w (feedstock dependent)

Fluid Coking

Purpose: to produce maximum yields of distillate products

Severe (480°C–565°C; 900°F–1,050°F) heating at pressures of 10 psi

Reactions allowed to proceed to completion

Oil contacts refractory coke

Bed fluidized with steam; heat dissipated throughout the fluid bed

Higher yields of light ends (<C_5) than delayed coking

Less coke-make than delayed coking (for the same feedstock)

Source: Speight, J.G. 2014. *The Chemistry and Technology of Petroleum* 5th Edition. CRC
          Press, Taylor & Francis Publishers, Boca Raton, Florida. Table 15.2, page 404.

---

are suitable for use as feedstocks for catalytic cracking units. The two most common processes are delayed coking and continuous (contact or fluid) coking, and three types of coke are obtained: (1) sponge coke, (2) honeycomb coke, and (3) needle coke which depend upon the reaction mechanism, time, temperature, and the crude feedstock.

Thus, a coking process is a thermal process for the continuous conversion of heavy, low-grade oils into lighter products. Unlike visbreaking, coking involved compete thermal conversion of the feedstock into volatile products and coke (Table 1.5).

The feedstock is typically a residuum, and the products are gases, naphtha, fuel oil, gas oil, and coke. The coke obtained is usually used as fuel but specialty uses, such as electrode manufacture, production of chemicals and metallurgical coke are also possible and increases the value of the coke. For these uses, the coke may require treatment to remove sulfur and metal impurities.

After a gap of several years, the recovery of viscous feedstocks either through secondary recovery techniques from reservoirs which contain heavy crude oil or from bituminous deposits which contain extra heavy crude oil or bitumen as well as the increased amounts of residua produced from conventional crude oil formations caused a renewal of interest in these feedstocks in the 1960s and, henceforth, for coking operations. Furthermore, the increasing attention paid to reducing atmospheric pollution has also served to direct some attention to coking, since the process not only concentrates pollutants such as feedstock sulfur in the coke, but also can usually yield volatile products that can be conveniently desulfurized.

The coking processes exist in to configuration, thus: (1) delayed coking, which is a semi-batch process, and (2) fluid coking, which is a continuous process that uses the heated feedstocks and a fluidized bed as the method of conversion to products.

1.2.3.1 Delayed Coking

Delayed coking is a semicontinuous process in which the heated feedstock is transferred to large soaking drum (sometime called the coking, drum), which provides the residence time needed for the cracking reactions to proceed to completion. The process was originally designed to process vacuum residua as well as residua from other thermal processes are also used (Parkash, 2003; Gary et al., 2007; Speight, 2014; Hsu and Robinson, 2017; Speight, 2017, 2021).

In the process, the viscous feedstock is heated and then transferred to large coke drums which provide the long residence time needed to allow the cracking reactions to proceed to completion. Initially, the viscous feedstock is heated to a high temperature (on the order of 480°C–510°C; 900°F–950°F) at low pressure (25–30 psi). The mixture is passed from the heater to one or more coker drums where the hot material is held approximately 24 hours (delayed) at pressures of 25–75 psi, until it cracks into lighter products. Vapors from the drums are returned to a fractionator where gas, naphtha, and gas oils are separated out. The higher-boiling products that are produced in the fractionator are recycled through the furnace. After the coke reaches a predetermined level in one drum, the flow is diverted to another drum to maintain continuous operation. The full drum is steamed to strip out uncracked hydrocarbons, cooled by water injection, and decoked by mechanical or hydraulic methods. The coke is mechanically removed by an auger rising from the bottom of the drum. Hydraulic decoking consists of fracturing the coke bed with high-pressure water ejected from a rotating cutter.

The volatile products include naphtha (often referred to as coker naphtha), light (lighter boiling, lower density) coker gas oil (LCGO), and heavy (higher-boiling, higher-density) coker gas oil (HCGO). The volatile products require further processing (to produce saleable products or suitable blend stock) which may involve hydrotreating or hydrocracking.

In order to provide the continuous operation, two drums are used, and while one drum is on-stream, the one off-stream drum is cleaned, steamed, water-cooled, and decoked in the same time interval. The temperature in the coke drum is on the order of 415°C–450°C (780°F–840°F) with a pressure in the range of 15–90 psi. The volatile (overhead) products go to the fractionator where the naphtha and heating oil fractions are recovered. The nonvolatile material is combined with preheated fresh feed and returned to the reactor. The coke drum is usually, for example, on stream for 24 hours before becoming filled with porous coke, after which the coke is removed hydraulically.

The coke drum is usually on stream for approximately 24 hours before becoming filled with porous coke after which the coke is removed hydraulically and to prepare the coke drum for subsequent use on stream.

Once the coke drum is filled with coke, the feedstock flow is directed to the second drum. To remove the coke from the filled drum, the top and bottom heads of the drum are removed and a rotating cutting tool which uses high-pressure jets of water to drill a hole through the center of the coke from top to bottom is used for the removal operation. In addition to cutting the hole, the water also cools the coke, forming steam as it does so, after which the cutter is raised, step by step, cutting the coke into pieces which can be retrieved from the bottom of the drum.

Typically, the coke drums operate on 18–24-hour cycle which includes preheating the drum, filling the drum with the heated feedstock, allowing coke and liquid products to form, cooling the drum, and the decoking operation. The coke can take several forms (Table 1.6) which can be used accordingly.

1.2.3.2 Fluid Coking

Fluid coking (often referred to as continuous coking or continuous contact coking) uses the fluidized-solid technique to convert atmospheric and vacuum residua to more valuable products. The operating temperature is higher than the temperature used for delayed coking operations.

In the process (a continuous moving-bed process that uses a temperature that is typically higher than the temperature used for the delayed coking process), thermal cracking occurs by using heat transferred from hot, recycled coke particles to feedstock in a radial mixer, called a reactor, at a pressure of 50 psi.

TABLE 1.6

The Forms of Coke Produced by Coking Processes

Type[a]	Properties
Coke	Unspecified coke is a byproduct of petroleum refining, useful in the production of electrodes used as carbon anodes for the aluminum industry, graphite electrodes for steel making, as fuel in the firing of solid fuel boilers used to generate electricity, and as a fuel for cement kilns
	More specific types of coked are described below
Green coke	Obtained in delayed coking units; has a high fixed carbon content comprising hydrocarbons and low levels of inorganic compounds. Typically used for electricity generation and in cement kilns
	Must have sufficiently low metal content to be used as anode material. Green coke with this low metal content is called anode-grade coke
	Green sponge coke (i.e., raw unprocessed coke directly out of the coker) must be calcined before it can be used for anodes. Fuel coke may not require calcination
Needle coke	Named for the needle-like structure, is produced from feedstocks that do not contain any asphaltene constituents – an example of such feedstocks includes hydrotreated decant oil decant oil from a fluid catalytic unit
	Needle coke is a high-value product used to make graphite electrodes for electric-arc furnaces in the steel industry
Shot coke	An undesirable product that is produced when the concentration of asphaltene constituents in the feedstock and/or the temperature in the coke drum is too high. Excessive feedstock oxygen content can also induce the formation of shot coke
	Begins to form as the oil flows into the coke drum. As the volatile products are flashed away, small globules of a heavy tar product remain and form coke rapidly due to the heat produced by reaction of the asphaltene constituents of the feedstock
	Adding aromatic feedstocks, such as decant oil from a fluid catalytic cracking unit, can eliminate shot coke formation. Other methods of eliminating shot coke include (1) decreasing the process temperature, (2) increasing the pressure in the coking drum, (3) increasing the amount of product recycle, and perhaps the least desirable (4) decreasing the yield of liquid products
Sponge coke	Named for the sponge-like appearance and is produced from feedstocks that have low-to-moderate concentration of asphaltene constituents
	Typically, meets the required specifications is used to make carbon anodes for the aluminum industry. Non-specification sponge coke is often used as refinery fuel

[a] Listed alphabetically rather than by properties or use.

In this process, the viscous feedstock is coked by being sprayed into a fluidized bed of hot, fine coke particles, which permits the coking reactions to be conducted at higher temperatures and shorter contact times than can be employed in delayed coking. Moreover, these conditions result in decreased yields of coke; greater quantities of more valuable liquid product are recovered in the fluid coking process. Gases and vapors are taken from the reactor, quenched to stop any further reaction, and fractionated. The reacted coke enters a surge drum and is lifted to a feeder and classifier where the larger coke particles are removed as product. The remaining coke is dropped into the preheater for recycling with feedstock. The process is automatic insofar as there is a continuous flow of the feedstock and the product (coke).

The process employs two vessels: a reactor and a burner; coke particles are circulated between these to transfer heat (generated by burning a portion of the coke) to the reactor. The reactor holds a bed of fluidized coke particles, and steam is introduced at the bottom of the reactor to fluidize the bed.

The volatile products flow from the reactor, are quenched to stop any further reaction, and are fractionated. The coke passes to a surge drum, then to a classifier, where the larger particles are removed as product. The smaller coke particles are recycled to a preheater, where they mix with fresh feed.

1.2.3.3 Flexicoking

Flexicoking is also a continuous process that is a direct descendent (or adaptation) of the fluid coking process. The unit uses the same configuration as the fluid coker but has a gasification section in which excess coke can be gasified to produce refinery fuel gas. The flexicoking process was designed during the late 1960s and 1970s as a means by which excess coke-make could be reduced in view of the gradual incursion of the heavier feedstocks in refinery operations. Such feedstocks are notorious for producing high yields of coke (>15% by weight) in thermal and catalytic operations.

1.3   CATALYTIC PROCESSES

By definition, thermal cracking is a process in which the constituents of crude oil (particularly the hydrocarbon constituents) are subjected to high temperature to decompose (break) the molecular bond in the higher molecular weight constituents (i.e., the long chain, higher-boiling hydrocarbon derivatives) into shorter-chained, lower-boiling hydrocarbon derivatives. Catalytic cracking is also a conversion process that is used in a crude oil refinery but uses a catalyst to increase the rate of chemical reactions by reducing the activation energy required for these reactions. Also, the process can be applied to a variety of feedstocks ranging from gas oil to viscous feedstocks (including atmospheric distillation residua and vacuum distillation residua).

There are many processes in a refinery that employ a catalyst to improve process efficiency (Table 1.7). The original incentive arose from the need to increase gasoline supplies in the 1930s and 1940s, and since cracking could virtually double the volume of gasoline from a barrel of crude oil, cracking was justifiable on this basis alone.

TABLE 1.7
Summary of Catalytic Cracking Processes

Conditions

Solid acidic catalyst (such as silica-alumina or a zeolite)
Temperature: 480°C–540°C (900°F–1,000°F) (solid/vapor contact)
Pressure: 10–20 psi
Provisions needed for continuous catalyst replacement with heavier feedstocks (residua)
Catalyst may be regenerated or replaced

Feedstocks

Gas oils and residua
Residua pretreated to remove salts (metals)
Residua pretreated to remove high molecular weight (asphaltic constituents)

Products

Lower boiling (i.e., molecular weight) than feedstock
Some gases (feedstock and process parameters dependent)
Iso-paraffins in product
Coke deposited on catalyst

Variations

Fixed bed
Moving bed
Fluidized bed

Source: Speight, J.G. 2014. *The Chemistry and Technology of Petroleum* 5th Edition. CRC Press, Taylor & Francis Publishers, Boca Raton, Florida. Table 15.3, page 407.

In the 1930s, thermal cracking units produced approximately 50% of the total gasoline. The octane number of this gasoline was approximately 70 compared to approximately 60 for straight-run (distilled) gasoline. The thermal reforming and polymerization processes that were developed during the 1930s could be expected to further increase the octane number of gasoline to some extent, but an additional innovation was needed to increase the octane number of gasoline to enhance the development of more powerful automobile engines.

1.3.1 CATALYTIC CRACKING

Catalytic cracking is another innovation that truly belongs to the 20th century and is regarded as the modern method for converting high-boiling crude oil fractions, such as gas oil, into gasoline, and other low-boiling fractions. Thus, catalytic cracking in the usual commercial process involves contacting a gas oil faction with an active catalyst under suitable conditions of temperature, pressure, and residence time so that a substantial part (>50%) of the gas oil is converted into gasoline and lower-boiling products, usually in a single-pass operation.

In the reactor, the feedstock is contacted by the catalyst at a high temperature (on the order of 480°C, 900°F), and the cracking reaction takes place. Typically, the reaction is endothermic, and the heat necessary to heat the feedstock to the reaction temperature and to supply the heat of reaction is provided by the heat of combustion from burning the coke that has deposited on the catalyst. In addition, the catalytic cracking produces higher yields a higher of branched-chain hydrocarbon derivatives, unsaturated hydrocarbon derivatives, and aromatic hydrocarbon derivatives compared to the yields from thermal cracking. Also, catalytic cracking is a better-controlled process than thermal cracking, and typically, the products obtained by catalytic cracking have a lower sulfur content.

Catalytic cracking has a number of advantages over thermal cracking: (1) the gasoline produced has a higher octane number, and (2) the catalytically cracked gasoline consists largely of iso-paraffins and aromatics, which have high octane numbers and greater chemical stability than mono-olefins and di-olefins which are present in much greater quantities in thermally cracked naphtha. Substantial quantities of olefin derivatives (which are suitable for the manufacture of polymer gasoline) are produced along with quantities of methane, ethane, and ethylene are produced by catalytic cracking. Sulfur-containing constituents of the feedstock compounds are changed in such a way that the sulfur content of the product (known as catalytically cracked naphtha or catalytically cracked gasoline) is lower than in thermally cracked gasoline. Catalytic cracking produces less heavy residual oil or tar and more useful gas oils than thermal cracking. The process has considerable flexibility, permitting the manufacture of both motor and aviation gasoline and a variation in the gas oil yield to meet changes in the fuel-oil market.

1.3.2 BED TYPES

In the latter part of the 20th century, substantial advances in the development of catalytic processes were achieved which involved not only rapid advances in the chemistry and physics of the catalysts themselves but also major engineering advances in reactor design. For example, the evolution of the design of the catalyst beds from (1) a fixed bed, (2) a moving bed, and (3) a fluidized bed. Catalyst chemistry/physics and bed design have allowed major improvements in process efficiency and product yields.

In the fixed catalytic cracking process, the feedstock is heated and passed over a series of trays containing catalyst. Also, in the fixed-bed designs, the catalyst is not moved, and the equipment consists of a series of reactor vessels containing the catalyst bed. On the other hand, in the moving-bed process, the catalyst is in the form of pellets that are moved continuously to the top of the unit by conveyor or pneumatic lift tubes to a storage hopper, then flow downward by gravity through the reactor, and finally to a regenerator, which is a vessel that is separate from the reactor.

In the fluidized bed designs, an upward current of gas suspends particles which move upward in the center of the bed and fall downward, under gravity, at the walls. In dense beds, a "bubble phase" is often observed which is a pocket of gas moving upward through the bed. In the wake of these pockets is a stream of high-velocity particles. In extreme cases when the bubble diameter approaches that of the catalyst vessel, a phenomenon known as "slugging" may occur which lifts the material in the top sections of the bed causing loss of solid material and low efficiency.

However, whichever unit is employed in the process, carbonaceous material (a product of the cracking process) is deposited on the catalyst, which markedly reduces catalytic activity, and removal of the deposit is necessary. Also, there are several process units that currently employed for catalytic cracking that differ mainly in the method of catalyst handling, although there is overlap with regard to catalyst type and the nature of the products.

The fluid-bed catalytic cracking process differs from the fixed-bed process and the moving-bed process insofar as the powdered catalyst is circulated essentially as a fluid with the feedstock. The several fluid catalytic cracking processes in use differ primarily in mechanical design. For example, the two main mechanical variations are (1) the side-by-side reactor-regenerator option along with (2) the unitary vessel option in which the reactor is either above or below the regenerator.

1.3.3 CATALYSTS

Catalytic cracking has progressively supplanted thermal cracking as the most advantageous means of converting distillate oils into gasoline and a better yield of higher-octane gasoline can be obtained than by any known thermal operation. At the same time, the gas produced consists mostly of propane and butane with less methane and ethane. The production of heavy oils and tars, higher in molecular weight than the charge material, is also minimized, and both the gasoline and the uncracked cycle oil are more saturated than the feedstock.

The acid catalysts first used in catalytic cracking were designated low alumina catalysts; amorphous solids composed of 87% w/w silica, SiO_2, and 13% w/w alumina, Al_2O_3. Later, high alumina catalysts containing 25% w/w alumina and 75% silica w/w were used. However, this type of catalyst has largely been replaced by catalysts containing crystalline aluminosilicates (zeolites) or molecular sieves (Parkash, 2003; Gary et al., 2007; Speight, 2014; Hsu and Robinson, 2017; Speight, 2017, 2021).

The catalyst is employed in bead, pellet, or microspherical form and can be used as a fixed, moving, or fluid bed. The cycle of operations consists of (1) the flow of feedstock through the catalyst bed, (2) the discontinuance of feedstock flow and removal of coke from the catalyst by burning, and (3) the insertion of the reactor back on-stream. On the other hand, the moving-bed process uses a reaction vessel in which cracking takes place, and a kiln in which the spent catalyst is regenerated and catalyst movement between the vessels is provided by various means. The fluid-bed process differs from the fixed-bed and moving-bed processes, insofar as the powdered catalyst is circulated essentially as a fluid with the feedstock.

Natural clays have long been known to exert a catalytic influence on the cracking of oils, but it was not until about 1936 that the process using silica-alumina catalysts was developed sufficiently for commercial use. Since then, catalytic cracking has progressively supplanted thermal cracking as the most advantageous means of converting distillate oils into gasoline. The main reason for the wide adoption of catalytic cracking is the fact that a better yield of higher-octane gasoline can be obtained than by any known thermal operation. At the same time, the gas produced consists mostly of propane and butane with less methane and ethane. The production of heavy oils and tars, higher in molecular weight than the charge material, is also minimized, and both the gasoline and the uncracked material (referred to as cycle oil) are more saturated than the products of the thermal cracking process.

The major innovations of the 20th century lie not only in reactor configuration and efficiency but also in catalyst development. There is probably not an oil company in the United States that does not have some research and development activity related to catalyst development. Much of the work is proprietary and, therefore, can only be addressed here in generalities.

The cracking of crude oil fractions occurs over many types of catalytic materials, but high yields of desirable products are obtained with hydrated aluminum silicates. These may be either activated (acid-treated) natural clays of the bentonite type of synthesized silica-alumina or silica-magnesia preparations. Their activity to yield essentially the same products may be enhanced to some extent by the incorporation of small amounts of other materials such as the oxides of zirconium, boron (which has a tendency to volatilize away on use), and thorium. Natural and synthetic catalysts can be used as pellets or beads and also in the form of powder; in either case, replacements are necessary because of attrition and gradual loss of efficiency. It is essential that they be stable to withstand the physical impact of loading and thermal shocks, and that they withstand the action of carbon dioxide, air, nitrogen compounds, and steam. They should also be resistant to sulfur and nitrogen compounds, and synthetic catalysts, or certain selected clays, appear to be better in this regard than average untreated natural catalysts.

The catalysts are porous and highly adsorptive, and their performance is affected markedly by the method of preparation. Two chemically identical catalysts having pores of different size and distribution may have different activity, selectivity, temperature coefficients of reaction rates, and responses to poisons. The intrinsic chemistry and catalytic action of a surface may be independent of pore size, but small pores produce different effects because of the manner in which hydrocarbon vapors are transported into and out of the pore systems.

1.4 ADDITIONAL REFINERY PROCESSES

Beyond the thermal and catalytic processes, there is the need for other processes to produce the products that are in demand by the market. These processes are given mention here in order to place the thermal cracking processes and the catalytic processes in the perspective of the modern refinery. More detailed information about these processes will be presented in other books in the series. These processes are (1) hydroprocesses, (2) reforming processes, (3) isomerization processes, (4) alkylation processes, (5) polymerization processes, and (6) solvent processes (Parkash, 2003; Hsu and Robinson, 2006; Gary et al., 2007; Speight, 2011a, 2011b, 2014; Hsu and Robinson, 2017; Speight, 2017).

1.4.1 HYDROPROCESSES

Hydroprocesses is a term that covered the processes relating to the processes of hydrocracking and hydrotreating in which hydrogen and a catalyst are used to upgrade feedstocks form, for example, thermal and catalytic cracking processes. Thus, the term "hydrotreating" is employed to indicate the removal of (alphabetically) nitrogen, oxygen, and sulfur as well as metals from crude oil (without cracking) and crude oil products. The term "hydrocracking" has a somewhat different meaning. The main difference is that the term "hydrocracking" refers to the conversion of high molecular weight constituents of the feedstock to low molecular weight products.

Thus, hydrogenation processes for the conversion of crude oil fractions and crude oil products may be classified as destructive and nondestructive. Destructive hydrogenation (hydrogenolysis or hydrocracking) is characterized by the conversion of the higher molecular weight constituents in a feedstock to lower-boiling products. Such treatment requires severe processing conditions and the use of high hydrogen pressures to minimize polymerization and condensation reactions that lead to coke formation.

Nondestructive or simple hydrogenation is generally used for the purpose of improving product quality without appreciable alteration of the boiling range. Mild processing conditions are employed so that only the more unstable materials are attacked. Nitrogen, sulfur, and oxygen compounds undergo reaction with the hydrogen to remove ammonia, hydrogen sulfide, and water, respectively.

Unstable compounds which might lead to the formation of gums, or insoluble materials, are converted to more stable compounds.

Furthermore, the use of hydrogen in thermal processes is perhaps the single most significant advance in refining technology during the 20th century. The process uses the principle that the presence of hydrogen during a thermal reaction of a crude oil feedstock will terminate many of the coke-forming reactions and enhance the yields of the lower-boiling components such as naphtha, kerosene, and light gas oil (Table 1.8) (Parkash, 2003; Hsu and Robinson, 2006; Gary et al., 2007; Speight, 2011a, 2011b, 2014; Hsu and Robinson, 2017; Speight, 2017).

Hydroprocesses (which includes hydrotreating processes and hydrocracking processes) use the principle that the presence of hydrogen during a thermal reaction of a crude oil feedstock will terminate many of the coke-forming reactions and enhance the yields of the lower-boiling components such as gasoline, kerosene, and jet fuel. The outcome is the conversion of a variety of feedstocks to a range of products (Table 1.2) (Parkash, 2003; Hsu and Robinson, 2006; Gary et al., 2007; Speight, 2011a, 2011b, 2014; Hsu and Robinson, 2017; Speight, 2017).

The key difference between the hydrotreating process and the hydrocracking process is that hydrotreaters tend to (1) operate at lower pressure, (2) use different catalysts, and (3) use a higher linear hourly space velocity (LHSV) which is the volume of feedstock on an hourly basis divided by the catalyst volume. The higher linear hourly space velocity means that a given volume of feedstock

---

TABLE 1.8
Summary of Hydrocracking Processes

Conditions

Solid acid catalyst (silica-alumina with rare earth metals, various other options)
Temperature: 260°C–450°C (500°F–845°F) (solid/liquid contact)
Pressure: 1,000–6,000 psi hydrogen
Frequent catalysts renewal for heavier feedstocks
Gas oil: catalyst life up to 3 years
Heavy oil/tar sand bitumen: catalyst life less than 1 year

Feedstocks

Refractory (aromatic) streams
Coker oils
Cycle oils
Gas oils
Residua (as a full hydrocracking or hydrotreating option)
In some cases, asphaltic constituents (S, N, and metals) removed by deasphalting

Products

Lower molecular weight paraffins
Some methane, ethane, propane, and butane
Hydrocarbon distillates (full range depending on the feedstock)
Residual tar (recycle)
Contaminants (asphaltic constituents) deposited on the catalyst as coke or metals

Process Variations

Fixed bed (suitable for liquid feedstocks)
Ebullating bed (suitable for heavy feedstocks)

Source: Speight, J.G. 2014. *The Chemistry and Technology of Petroleum* 5th Edition. CRC Press, Taylor & Francis Publishers, Boca Raton, Florida. Table 15.4, page 409.

---

requires more catalyst. In terms of process conditions and conversion yield, a mild hydrocracking process has parameters that are between hydrotreating process parameter and full-conversion hydrocracking process parameters.

The purpose of hydroprocessing is (1) to improve existing crude oil products or develop new products or uses; (2) to convert inferior or low-grade materials into valuable products; and (3) to transform near-solid residua to liquid fuels. Products are as follows: (1) from naphtha: reformed feedstock and liquefied petroleum gas; (2) from atmospheric gas oil: naphtha, kerosene, and petrochemical feedstock; (3) from vacuum gas oil: liquefied petroleum gas, naphtha, kerosene, and lubricating oil; and (4) from residuum: naphtha, kerosene, catalytic cracking feedstock, and feedstock for a coking unit.

1.4.1.1 Hydrotreating

During the thermal cracking processes (as well as the catalytic cracking processes), there is often a quantity of olefin derivatives (-CH=CH-) in the product mix. Because of the adverse nature that the olefinic derivatives can have on the performance of the derived products, hydrotreating is an essential part of refinery operations. The same is true for the product that contain sulfur and nitrogen-containing derivatives.

Hydrotreating is a catalytic process converts olefins to saturate hydrocarbon derivatives as well as converting sulfur- and/or nitrogen-containing hydrocarbons into low-sulfur low-nitrogen liquids, hydrogen sulfide, and ammonia (Parkash, 2003; Hsu and Robinson, 2006; Gary et al., 2007; Speight, 2011a, 2011b, 2014; Hsu and Robinson, 2017; Speight, 2017). A wide variety of metals are active hydrogenation catalysts; those of most interest are nickel, palladium, platinum, cobalt, and iron. Special preparations of the first three are active at room temperature and atmospheric pressure. The metallic catalysts are easily poisoned by sulfur- or arsenic-containing compounds, and even by other metals. To avoid such poisoning, less effective, but more resistant metal oxides or sulfides are frequently employed, generally those of tungsten, cobalt, or molybdenum. Alternatively, catalyst poisoning can be minimized by mild hydrogenation to remove nitrogen, oxygen, and sulfur from feedstocks in the presence of more resistant catalysts, such as cobalt-molybdenum-alumina ($Co/Mo/Al_2O_3$).

The process temperature affects the rate and the extent of hydrogenation as it does any chemical reaction. In fact, it is also worthy of note that the hydrogenation reaction can be reversed by increasing temperature (referred to as dehydrogenation). If a second functional group is present, a high temperature often leads to the loss of selectivity and, therefore, loss of desired product yield. As a practical measure, hydrogenation is carried out at as low a temperature as possible which is still compatible with a satisfactory reaction rate.

Hydrotreating is carried out by charging the feed to the reactor together with hydrogen at 300°C–345°C (570°F–655°F); the hydrogen pressures are on the order of 500–1,000 psi. The reaction generally takes place in the vapor phase but, depending on the application, can also be a mixed-phase reaction.

After passing through the reactor, the treated oil is cooled and separated from the excess hydrogen recycled through the reactor. The treated oil is pumped to a stripper tower where hydrogen sulfide, formed by the hydrogenation reaction, is removed by steam or by hydrocarbon vapor via reboiling, and the finished product leaves the bottom of the stripper tower. The catalyst can be regenerated in situ and ultimately be replaced after several regenerations.

Distillate hydrotreating is carried out by charging the feed to the reactor, together with hydrogen in the presence of catalysts such as tungsten-nickel sulfide, cobalt-molybdenum-alumina, nickel oxide-silica-alumina, and platinum-alumina (Parkash, 2003; Hsu and Robinson, 2006; Gary et al., 2007; Speight, 2011a, 2011b, 2014; Hsu and Robinson, 2017; Speight, 2017). Most processes employ cobalt-molybdena catalysts which generally contain approximately 10% of molybdenum oxide and less than 1% of cobalt oxide supported on alumina. The temperatures employed are in the range of 260°C–345°C (500°F–655°F), while the hydrogen pressures are approximately 500–1,000 psi.

The reaction generally takes place in the vapor phase but, depending on the application, may be a mixed-phase reaction. Generally, it is more economical to hydrotreat high-sulfur feedstocks prior to catalytic cracking than to hydrotreat the products from catalytic cracking. The advantages are that (1) sulfur is removed from the catalytic cracking feedstock, and corrosion is reduced in the cracking unit; (2) carbon formation during cracking is reduced so that higher conversions result; and (3) the cracking quality of the gas oil fraction is improved.

Hydrofining is a process that first went on-stream in the 1950s and can be applied to lubricating oils, naphtha, and gas oils. The feedstock is heated in a furnace and passed with hydrogen through a reactor containing a suitable metal oxide catalyst, such as cobalt and molybdenum oxides on alumina. Reactor operating conditions range from 205°C to 425°C (400°F to 800°F) and from 50 to 800 psi and depend on the kind of feedstock and the degree of treating required. Higher-boiling feedstocks, high sulfur content, and maximum sulfur removal require higher temperatures and pressures.

After passing through the reactor, the treated oil is cooled and separated from the excess hydrogen which is recycled through the reactor. The treated oil is pumped to a stripper tower where hydrogen sulfide, formed by the hydrogenation reaction, is removed by steam, vacuum, or flue gas, and the finished product leaves the bottom of the stripper tower. The catalyst is not usually regenerated; it is replaced after use for approximately 1 year.

1.4.1.2 Hydrocracking

Hydrocracking is similar to catalytic cracking, with hydrogenation superimposed and with the reactions taking place either simultaneously or sequentially (Parkash, 2003; Hsu and Robinson, 2006; Gary et al., 2007; Speight, 2011a, 2011b, 2014; Hsu and Robinson, 2017; Speight, 2017). Hydrocracking was initially used to upgrade low-value distillate feedstocks, such as cycle oils (high aromatic products from a catalytic cracker which usually are not recycled to extinction for economic reasons), thermal and coker gas oils, and heavy-cracked and straight-run naphtha. These feedstocks are difficult to process by either catalytic cracking or reforming since they are characterized usually by a high polycyclic aromatic content and/or by high concentrations of the two principal catalyst poisons: sulfur and nitrogen compounds.

In the process (>350°C, 660°F) in which hydrogenation accompanies cracking, relatively high pressures (1,000–3,000 psi) are employed, and the overall result is the conversion of the feedstock to lower-boiling products. Another attractive feature of hydrocracking is the low yield of gaseous components, such as methane, ethane, and propane, which are less desirable than the gasoline components. Essentially all the initial reactions of catalytic cracking occur, but some of the secondary reactions are inhibited or stopped by the presence of hydrogen.

In the first, pretreating stage of the process, the main purpose is to convert organic nitrogen compounds and organic sulfur in the feedstock to hydrocarbons and to ammonia and hydrogen sulfide by hydrogenation and mild hydrocracking. The purpose is to reduce the organic nitrogen and sulfur compounds to low levels (<50 ppm). Typical conditions are 340°C–390°C at a pressure on the order of 1,500–2,500 psi, and a catalyst contact time of 0.5–1.5 hours, up to 1.5% w/w hydrogen is absorbed, partly by conversion of the nitrogen compounds, but chiefly by aromatic compounds that are hydrogenated. This first stage is usually carried out with a bifunctional catalyst containing hydrogenation promoters, such as nickel and tungsten or molybdenum sulfides, on an acidic support. The metal sulfides hydrogenate aromatics and nitrogen compounds, and retard deposition of carbonaceous deposits; the acidic support accelerates nitrogen removal as ammonia by breaking carbon-nitrogen bonds. The catalyst is generally used as 0.32–0.32-cm or 0.16–0.32-cm pellets, as well as spheres or other shapes.

Most of the hydrocracking is accomplished in the second stage. Hydrogen sulfide, ammonia, and low-boiling products are usually removed from the first-stage product; the remaining oil, which is low in nitrogen–sulfur compounds, is passed over the second-stage catalyst. Some catalyst systems do not require the removal of adsorbed compounds between stages. In the second stage, typical conditions are 300°C–370°C (570°F to −698°F), at a hydrogen pressure on the order of 1,000–2,500 psi and 0.5–1.5 hours contact time during which time 1%–1.5% w/w hydrogen may be absorbed. Conversion to gasoline or jet fuel is seldom complete in one contact with the catalyst, so the lighter oils are removed by distillation of the products, and the higher-boiling (higher-density) product is combined with fresh feed and recycled over the catalyst until it is completely converted.

The catalyst for the second stage is also a bifunctional catalyst containing hydrogenating and acidic components. Metals such as nickel, molybdenum, tungsten, or palladium are used in various combinations and dispersed on solid acidic supports, such as synthetic amorphous or crystalline silica-alumina (i.e., zeolites). These supports contain strongly acidic sites and sometimes are enhanced by the incorporation of a small amount of fluorine.

A long period of operation (typically in excess of 3 years) between catalyst regeneration is desirable; this is achieved by keeping a low nitrogen content in the feed and avoiding high temperatures and high end-point feedstock, which leads to excess cracking and consequent deposition of coke on the catalyst. Feedstock conversion is the key insofar as the conversion dictates the temperature employed. When activity of the catalyst has decreased, it can often be restored by controlled burning of the coke.

In the destructive hydrogenation process, hydrogenolysis occurs and involves the cleavage of carbon-carbon bonds, carbon-oxygen bonds, carbon-sulfur bonds, and carbon-nitrogen bonds. An example of hydrogenolysis is the hydrodealkylation of toluene to form benzene and methane:

$$C_6H_5CH_3 + H_2 \rightarrow C_6H_6 + CH_4$$

On the other hand, nondestructive or simple hydrogenation is generally used for the purpose of improving product quality without appreciable alteration of the boiling range. Examples are the removal of various sulfur compounds (Table 1.3), which would otherwise have an adverse effect on product quality. Treatment under such mild conditions is often referred to as hydrotreating or hydrofining and is essentially a means of eliminating, in addition to sulfur, nitrogen and oxygen as ammonia and water, respectively.

The older hydrogenolysis type of hydrocracking practiced in Europe during, and after, World War II used tungsten sulfide or molybdenum sulfide as the catalyst and required high reaction temperatures and operating pressures, sometimes in excess of approximately 3,000 psi for continuous operation. The modern hydrocracking processes were initially developed for converting refractory feedstocks (such as gas oils) to gasoline and jet fuel, but process and catalyst improvements and modifications have made it possible to yield products from gases and naphtha to furnace oils and catalytic cracking feedstocks (Parkash, 2003; Hsu and Robinson, 2006; Gary et al., 2007; Speight, 2011a, 2011b, 2014; Hsu and Robinson, 2017; Speight, 2017).

A comparison of hydrocracking with hydrotreating is useful in assessing the parts played by these two processes in refinery operations. Hydrotreating of distillates may be defined simply as the removal of nitrogen-containing compounds, sulfur-containing compounds, and oxygen-containing compounds. The hydrotreating catalysts are usually cobalt plus molybdenum or nickel plus molybdenum (in the sulfide) form impregnated on an alumina base. The hydrotreated operating conditions (370°C, 700°F; 1,000–2,000 psi hydrogen) and are such that appreciable hydrogenation of aromatics will not occur:).

Hydrocracking is an extremely versatile process which can be utilized in many different ways such as conversion of the high-boiling aromatic streams which are produced by catalytic cracking or by coking processes. The commercial processes for treating, or finishing, crude oil fractions with hydrogen all operate in essentially the same manner. The feedstock is heated and passed with hydrogen gas through a tower or reactor filled with catalyst pellets. The reactor is maintained at a temperature of 260°C–425°C (500°F–800°F) at pressures from 100 to 1,000 psi, depending on the particular process, the nature of the feedstock, and the degree of hydrogenation required. After leaving the reactor, any hydrogen is separated from the treated product and recycled through the reactor after removal of any hydrogen sulfide. The liquid product is passed into a stripping tower where steam removes dissolved hydrogen and hydrogen sulfide, and after cooling, the product is taken to product storage or, in the case of feedstock preparation, pumped to the next processing unit.

The manufacture of base stocks for lubricating oil production is an essential part of modern refining. In the past four decades, the majority of the expansion of lubricating oil production is being achieved by production using catalytic hydroprocessing (hydrocracking and hydroisomerization) because of the demand for higher-quality lube base oils. Base oils are subdivided into a number of categories: Groups I, II, II, and IV. Group I base oils are typically conventional solvent-refined products. Groups II and III were added to lubricant classifications in the early 1990s to represent low-sulfur, low-aromatic, and high-viscosity-index (VI) lubricants with good oxidative stability and soot handling. The reduction of wax content in the lubricants also improves the operating range, low-temperature performance via improved pour and cloud point. The first catalytic based plants were introduced in the 1980s, but at that time the catalytic route only produced conventional base oil (Group I). In the 1990s, hydroisomerization was introduced to produce base oils with higher stability. Hydroisomerization has propagated such that a considerable amount of lube base oils is produced in this manner.

1.4.2 Reforming Processes

The term "reforming" as used in the crude oil refining industry is a processing technique by which the molecular structure of a hydrocarbon is rearranged to alter the properties. The process is frequently applied to low-quality naphtha stocks (that are used as a gasoline blend stock) to improve the combustion characteristics in the internal combustion engine.

When the demand for higher-octane gasoline developed during the early 1930s, attention was directed to ways and means of improving the octane number of fractions within the boiling range of gasoline (Parkash, 2003; Hsu and Robinson, 2006; Gary et al., 2007; Speight, 2011a, 2011b, 2014; Hsu and Robinson, 2017; Speight, 2017). Straight-run (distilled) gasoline frequently had very low octane numbers, and any process that would improve the octane numbers would aid in meeting the demand for higher-octane-number gasoline. Such a process (called thermal reforming) was developed and used widely, but to a much lesser extent than thermal cracking. Thermal reforming was a natural development from older thermal cracking processes; cracking converts heavier oils into gasoline, whereas reforming converts (reforms) gasoline into higher-octane gasoline. The equipment for thermal reforming is essentially the same as for thermal cracking, but higher temperatures are used.

Thus, reforming processes are used to change the inherent chemical structures of the hydrocarbons that exist in distillation fractions of crude oil into different compounds. Catalytic reforming is one of the most important processes in a modern refinery, altering straight-run fraction or fractions from a catalytic cracker into new compounds through a combination of heat and pressure in the presence of a catalyst. Reforming processes are particularly important in producing high-quality gasoline fuels. Reforming processes are classified as (1) continuous processes, (2) cyclic processes, or (3) semi-regenerative processes, depending upon the frequency of catalyst regeneration.

1.4.2.1 Thermal Reforming

Thermal reforming was a natural development from older thermal cracking processes; cracking converts heavier oils into gasoline whereas reforming converts (reforms) gasolines into higher-octane gasolines. The equipment for thermal reforming is essentially the same as for thermal cracking, but higher temperatures are used in the former (Parkash, 2003; Hsu and Robinson, 2006; Gary et al., 2007; Speight, 2011a, 2011b, 2014; Hsu and Robinson, 2017; Speight, 2017).

In carrying out thermal reforming, a feedstock such as 205°C (400°F) end-point naphtha or a straight-run gasoline is heated to 510°C–595°C (950°F–1,100°F) in a furnace, much the same as a cracking furnace, with pressures from 400 to 1,000 psi. As the heated naphtha leaves the furnace, it is cooled or quenched (typically) by the addition of cold naphtha. The product then enters a fractional distillation tower where any high-boiling products are separated. The remainder of the reformed material leaves the top of the tower to be separated into gases and reformate. The higher octane number of the reformate is due primarily to the cracking of longer-chain paraffins into higher-octane olefins.

The products of thermal reforming are gases, gasoline, and residual oil or tar, the latter being formed in very small amounts (1%). The amount and quality of the product, known as reformate, are very dependent on the temperature. As a general rule of thumb: the higher the reforming temperature, the higher the octane number, but the lower the yield of reformate.

Thermal reforming is less effective and less economical than catalytic processes and has been largely supplanted. As it used to be practiced, a single-pass operation was employed at temperatures in the range of 540°C–760°C (1,000°F–1,140°F) and pressures of approximately 500–1,000 psi. The degree of improvement of the octane number depended on the extent of conversion but was not directly proportional to the extent of crack per pass. However, at very high conversions, the production of coke and gas became prohibitively high. The gases produced were generally olefinic, and the process required either a separate gas polymerization operation or one in which C3–C4 gases were added back to the reforming system.

More recent modifications of the thermal reforming process due to the inclusion of hydrocarbon gases with the feedstock are known as gas reversion and polyforming. Thus, olefin derivatives that produced by cracking and reforming can be converted into liquid products that boiling in the naphtha range (typically 35°C–200°C, 95°F–390°F) by heating them under high pressure. Since the resulting liquids (polymers) have high octane numbers, they increase the overall quantity and quality of gasoline produced in a refinery.

1.4.2.2 Catalytic Reforming

The catalytic reforming process is a process of the 1950s and usually involves feeding a naphtha (after pretreating with hydrogen if necessary to remove nitrogen and sulfur compounds) and hydrogen mixture to a furnace where the mixture is heated to the desired temperature (450°C–520°C; 844°F–970°F) and then passed through fixed-bed catalytic reactors at hydrogen pressures on the order of 50–400 psi. Typically, several reactors are used in series, and heaters are located between adjoining reactors in order to compensate for the endothermic reactions taking place (Parkash, 2003; Hsu and Robinson, 2006; Gary et al., 2007; Speight, 2011a, 2011b, 2014; Hsu and Robinson, 2017; Speight, 2017).

Thermal reforming, less effective and less economical than catalytic processes, has been largely supplanted. The catalytic reforming process, in a similar manner to the thermal reforming process, converts low-octane naphtha into high-octane naphtha (i.e., reformate). Whereas thermal reforming produces reformate having research octane numbers in the range of 65–80, depending on the yield, catalytic reforming produces reformate having octane numbers on the order of 90–105. Catalytic reforming is conducted in the presence of hydrogen over hydrogenation–dehydrogenation catalysts, such as in the platforming process. Feedstocks for catalytic reforming processes are saturated compounds and do not contain olefin derivatives. Catalytic cracker naphtha and hydrocracker naphtha that contains substantial quantities of naphthenes are also suitable reformer feedstocks.

The dehydrogenation reaction is a major chemical change that occurs in the catalytic reforming process with the production of hydrogen. Using the cyclohexane conversion to benzene as the example:

$$C_6H_{12} \rightarrow C_6H_6 + 3H_2$$

Isomerization is used with the objective of providing additional feedstock consequently produced in large quantities. In order to provide the atmosphere necessary for the chemical reactions and also to prevent carbon from being deposited on the catalyst, thus extending its operating life, hydrogen is recycled through the reactors where the reforming takes place. Because of an excess of hydrogen above, whatever is consumed in the process is produced, catalytic reforming processes are unique in that they are the only crude oil refinery processes to produce hydrogen as a byproduct.

The composition of a reforming catalyst is dictated by the composition of the feedstock and the desired reformate. The catalysts used for the process are typically platinum or platinum-rhenium on an alumina base. The purpose of platinum on the catalyst is to promote dehydrogenation and hydrogenation reactions. Nonplatinum catalysts are used in regenerative processes for feedstocks containing sulfur, although pretreatment (hydrodesulfurization) may permit platinum catalysts to be employed.

The commercial processes available for use can be broadly classified as the moving-bed, fluid-bed, and fixed-bed types. The fluid-bed and moving-bed processes used mixed non-precious metal oxide catalysts in units equipped with separate regeneration facilities. The fixed-bed processes use predominantly platinum-containing catalysts, especially in units that are equipped for cycle with occasional regeneration or even no regeneration.

The feedstocks for catalytic reforming are typically saturated feedstocks (i.e., not olefinic) and in the majority of cases may be straight-run naphtha as well as other by-products such as low-octane naphtha (e.g., coker naphtha) which can be processed after treatment to remove olefins and other contaminants. Hydrocracker naphtha that contains substantial quantities of naphthenes is also a suitable feed.

Catalytic reforming usually is carried out by feeding a naphtha (after pretreating with hydrogen if necessary) and hydrogen mixture to a furnace where the mixture is heated to the desired temperature, 450°C–520°C (840°F–965°F), and then passed through fixed-bed catalytic reactors at hydrogen pressures of 100–1,000 psi (Parkash, 2003; Hsu and Robinson, 2006; Gary et al., 2007; Speight, 2011a, 2011b, 2014; Hsu and Robinson, 2017; Speight, 2017). Generally, reactor pairs are used in series with heaters located between adjoining reactors in order to compensate for the endothermic reactions taking place. In some cases, as many as four or five reactors are kept on stream in series while one or more is being regenerated.

The on-stream cycle of any one reactor may vary from several hours to many days, depending on the feedstock and reaction conditions.

The composition of a reforming catalyst is dictated by the composition of the feedstock and the desired reformate. The catalysts used are principally molybdena-alumina, chromia-alumina, or platinum on a silica-alumina or alumina base. The nonplatinum catalysts are widely used in regenerative process for feeds containing, for example, sulfur, which poisons platinum catalysts, although pretreatment processes (e.g., hydrodesulfurization) may permit platinum catalysts to be employed.

The purpose of platinum on the catalyst is to promote dehydrogenation and hydrogenation reactions, i.e., the production of aromatics, participation in hydrocracking, and rapid hydrogenation of carbon-forming precursors. For the catalyst to have an activity for isomerization of both paraffins and naphthenes – the initial cracking step of hydrocracking – and to participate in paraffin dehydrocyclization, it must have an acid activity. The balance between these two activities is most important in a reforming catalyst. In fact, in the production of aromatics from cyclic saturated materials (naphthenes), it is important that hydrocracking be minimized to avoid loss of the desired product, and thus, the catalytic activity must be moderated relative to the case of gasoline production from a paraffinic feed, where dehydrocyclization and hydrocracking play an important part.

1.4.3 ISOMERIZATION PROCESSES

The term "isomerization" as used in the refinery is a process by which a compound is transformed into any of its isomeric forms (i.e., forms with the same chemical composition but with different structure or configuration) and with different physical and chemical properties. For example, the process is employed for the conversion of n-butane into isobutane to provide additional feedstock for alkylation units (Figure 1.2).

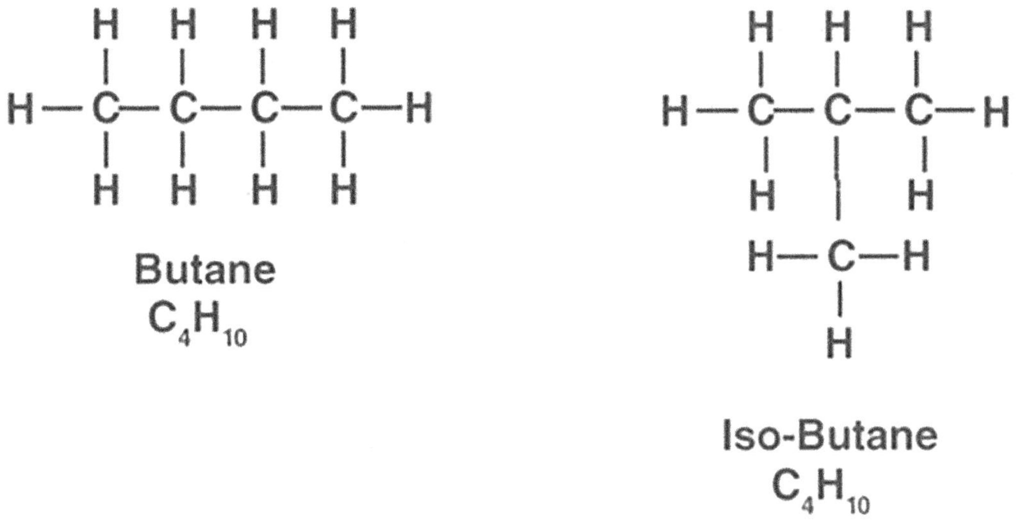

FIGURE 1.2 Butane isomers.

The conversion of n-pentane and n-hexane into their respective branched isomers (for gasoline blending) can be achieved by a similar process.

Catalytic reforming processes provide high-octane constituents in the higher molecular weight constituents of the naphtha (gasoline) fraction, but the normal paraffin components of the lighter gasoline fraction, especially butanes, pentanes and hexanes, have poor octane ratings (Parkash, 2003; Hsu and Robinson, 2006; Gary et al., 2007; Speight, 2011a, 2011b, 2014; Hsu and Robinson, 2017; Speight, 2017). The conversion of these normal paraffin derivatives to the respective isomers (isomerization) yields gasoline components of high octane rating in this lower boiling range. Conversion is obtained in the presence of a catalyst (aluminum chloride activated with hydrochloric acid), and it is essential to inhibit side reactions such as cracking and olefin formation (Parkash, 2003; Hsu and Robinson, 2006; Gary et al., 2007; Speight, 2011a, 2011b, 2014; Hsu and Robinson, 2017; Speight, 2017).

Isomerization – another innovation of the 20th century – found initial commercial applications during World War II for making high-octane aviation gasoline components and additional feed for alkylation units. The lowered alkylate demands in the post–World War II period led to the majority of the butane isomerization units being shut down. In recent years, the greater demand for high-octane motor fuel has resulted in new butane isomerization units being installed.

The earliest process of note was the production of isobutane, which is required as an alkylation feed. The isomerization may take place in the vapor phase, with the activated catalyst supported on a solid phase, or in the liquid phase with a dissolved catalyst (Parkash, 2003; Hsu and Robinson, 2006; Gary et al., 2007; Speight, 2011a, 2011b, 2014; Hsu and Robinson, 2017; Speight, 2017). In the process, pure butane or a mixture of isomeric butanes is mixed with hydrogen (to inhibit olefin formation) and passed to the reactor, at 110°C–170°C (230°F–340°F) and 200–300 psi. The product is cooled, the hydrogen separated, and the cracked gases are then removed in a stabilizer column. The stabilizer bottom product is passed to a superfractionator where the normal butane is separated from the isobutane.

The processes are used to produce feedstocks for alkylation units (isobutane) or high-octane fractions for gasoline blending (pentane and hexane) (Parkash, 2003; Hsu and Robinson, 2006; Gary et al., 2007; Speight, 2011a, 2011b, 2014; Hsu and Robinson, 2017; Speight, 2017). The latter application is useful in the production of reformulated gasoline by increasing the octane number while converting or removing benzene (Parkash, 2003; Hsu and Robinson, 2006; Gary et al., 2007; Speight, 2011a, 2011b, 2014; Hsu and Robinson, 2017; Speight, 2017).

Initially, aluminum chloride was the catalyst used to isomerize butane, pentane, and hexane. Since then, supported metal catalysts have been developed for use in high-temperature processes that operate at 370°C–480°C (700°F–895°F) and a pressure within the range 300–750 psi, whereas aluminum chloride and hydrogen chloride are universally used for the low-temperature processes.

Nonregenerable aluminum chloride catalyst is employed with various carriers in a fixed-bed or liquid contactor. Platinum or other metal catalyst processes that utilize fixed-bed operation can be either regenerable or nonregenerable. The reaction conditions vary widely, between 40°C and 480°C (105°F and 895°F) and 150 and 1,000 psi, depending on the particular process and feedstock.

Present isomerization applications in crude oil refining are used with the objective of providing additional feedstock for alkylation units or high-octane fractions as a blend stock for gasoline production as well as for the production of petrochemicals (Tables 1.9 and 1.10).

Straight-chain paraffins (n-butane, n-pentane, n-hexane) are converted to respective iso-compounds by continuous catalytic (aluminum chloride, noble metals) processes. Natural gasoline or light straight-run gasoline can provide feed by first fractionating as a preparatory step. High volumetric yields (>95%) and 40%–60% conversion per pass are characteristics of the isomerization reaction.

During World War II, aluminum chloride was the catalyst used to isomerize butane, pentane, and hexane. Since then, supported metal catalysts have been developed for use in high-temperature processes which operate in the range of 370°C–480°C (700°F–900°F) and 300–750 psi, while aluminum chloride plus hydrogen chloride is universally used for the low-temperature processes.

Nonregenerable aluminum chloride catalyst is employed with various carriers in a fixed-bed or liquid contactor. Platinum or other metal catalyst processes utilized fixed-bed operation and can be regenerable or nonregenerable. The reaction conditions vary widely depending on the particular process and feedstock: 40°C–480°C (100°F–900°F) and 150–1,000 psi.

TABLE 1.9

Component Streams for Gasoline Production

Stream	Producing Process	Boiling Range	
		°C	°F
	Paraffinic		
Butane	Distillation conversion	0	32
Iso-pentane	Distillation conversion isomerization	27	81
Alkylate	Alkylation	40–150	105–300
Isomerate	Isomerization	40–70	105–160
Naphtha	Distillation	30–100	85–212
Hydrocrackate	Hydrocracking	40–200	105–390
	Olefinic		
Catalytic naphtha	Catalytic cracking	40–200	105–390
Cracked naphtha	Steam cracking	40–200	105–390
Polymer	Polymerization	60–200	140–390
	Aromatic		
Catalytic reformate	Catalytic reforming	40–200	105–390

Source: Speight, J.G. 2014. *The Chemistry and Technology of Petroleum* 5th Edition. CRC Press, Taylor & Francis Publishers, Boca Raton, Florida. Table 15.5, page 416.

TABLE 1.10

Sources of Feedstocks for the Production of Petrochemicals

Hydrocarbon	Source
Methane	Natural gas
Ethane	Natural gas
Ethylene	Cracking processes
Propane	Natural gas
	Cracking processes
	Catalytic reforming processes
Propylene	Cracking processes
Butane	Natural gas
	Cracking processes
Butene(s)	Cracking processes
	Reforming processes
Cyclohexane	Distillation
	Hydrotreating processes
Benzene	Catalytic reforming processes
Toluene	Catalytic reforming processes
Xylene(s)	Catalytic reforming processes
Ethylbenzene	Catalytic reforming processes
Alkylbenzenes	Alkylation processes
$>C_9$	Polymerization processes

Source: Speight, J.G. 2014. *The Chemistry and Technology of Petroleum* 5th Edition. CRC Press, Taylor & Francis Publishers, Boca Raton, Florida. Table 15.7, page 426.

1.4.4 ALKYLATION PROCESSES

The term "alkylation," as used in petroleum refining, is a process in which gaseous hydrocarbon derivatives are combined to produce high-octane components of gasoline. The gaseous hydrocarbon derivatives consist of olefins such as propylene and butylene and iso-paraffin derivatives such as isobutane.

The process is another 20th-century refinery innovation, and developments in crude oil processing in the late 1930s and during World War II were directed toward production of high-octane liquids for aviation gasoline. The sulfuric acid process was introduced in 1938, and hydrogen fluoride alkylation was introduced in 1942. Rapid commercialization took place during the war to supply military needs, but many of these plants were shut down at the end of World War II.

The combination of olefins with paraffins to form higher iso-paraffins is termed alkylation. Since olefins are reactive (unstable) and are responsible for exhaust pollutants, their conversion to high-octane iso-paraffins is desirable when possible. Although alkylation is possible without catalysts, commercial processes use aluminum chloride, sulfuric acid, or hydrogen fluoride as catalysts, when the reactions can take place at low temperatures, minimizing undesirable side reactions, such as polymerization of olefins (Parkash, 2003; Hsu and Robinson, 2006; Gary et al., 2007; Speight, 2011a, 2011b, 2014; Hsu and Robinson, 2017; Speight, 2017).

The alkylate product is composed of a mixture of iso-paraffin derivatives which have octane numbers that vary with the olefin derivatives from which alkylate was produced. Butylene derivatives produce the highest octane numbers, propylene produces alkylates with low-octane numbers, and pentylene derivatives produce alkylates with intermediate values. All alkylates, however, have relatively high octane numbers (>87), which make them particularly valuable.

In the mid-1950s, the demand for aviation fuel (i.e., aviation gasoline) started to decline, but motor-gasoline quality requirements rose sharply. Wherever practical, refiners shifted the use of alkylate to premium motor fuel. To aid in the improvement of the economics of the alkylation process and also the sensitivity of the premium gasoline pool, additional olefins were gradually added to alkylation feed.

The alkylation process as now practiced in a refinery involves the union, through the agency of a catalyst, of an olefin derivative (ethylene, propylene, butylene, and amylene) with isobutane to yield high-octane branched-chain hydrocarbons in the naphtha boiling range that have high octane numbers. The olefin feedstock is derived from the gas produced in a catalytic cracker, while isobutane is recovered by refinery gases or produced by catalytic butane isomerization.

In the process, propylene, butylene derivatives, or amylene derivatives are combined with isobutane in the presence of an acid catalyst, such as sulfuric acid or hydrofluoric acid, at low temperatures (typically 1°C–40°C; 34°F–104°F) and pressures on the order of 15–150 psi. Sulfuric acid and hydrogen fluoride are the catalysts used commercially in refineries. The acid is pumped through the reactor and forms an emulsion with reactants, and the emulsion is maintained at 50% acid. The rate of deactivation varies with the feed and isobutane charge rate. Butene feeds cause less acid consumption than the propylene feeds.

To accomplish this, either ethylene ($CH_2=CH_2$) or propylene ($CH_3CH=CH_2$) is combined with isobutane at 50°C–280°C (125°F–450°F) and 300–1,000 psi in the presence of metal halide catalysts such as aluminum chloride. Conditions are less stringent in catalytic alkylation; olefins (propylene, butylene derivatives, or pentylene derivatives) are combined with isobutane in the presence of an acid catalyst (sulfuric acid or hydrofluoric acid) at low temperatures and pressures (1°C–40°C, 30°F–105°F and 14.8–150 psi) (Parkash, 2003; Hsu and Robinson, 2006; Gary et al., 2007; Speight, 2011a, 2011b, 2014; Hsu and Robinson, 2017; Speight, 2017).

Sulfuric acid, hydrogen fluoride, and aluminum chloride are the general catalysts used commercially. Sulfuric acid is used with propylene and higher-boiling feeds, but not with ethylene, because it reacts to form ethyl hydrogen sulfate. The acid is pumped through the reactor and forms an air emulsion with reactants, and the emulsion is maintained at 50% acid. The rate of deactivation varies with the feed and isobutane charge rate. Butene feeds cause less acid consumption than the propylene feeds.

Aluminum chloride is not widely used as an alkylation catalyst, but when employed, hydrogen chloride is used as a promoter and water is injected to activate the catalyst as an aluminum chloride/hydrocarbon complex. Hydrogen fluoride is used for alkylation of higher-boiling olefins, and the advantage of hydrogen fluoride is that it is more readily separated and recovered from the resulting product.

1.4.5 POLYMERIZATION PROCESSES

The polymerization process, as used in the crude oil industry, is a process by which olefin gases are converted to liquid products which may be suitable for gasoline (polymer gasoline) or other liquid fuels (Parkash, 2003; Hsu and Robinson, 2006; Gary et al., 2007; Speight, 2011a, 2011b, 2014; Hsu and Robinson, 2017; Speight, 2017). The feedstock usually consists of propylene and butylene derivatives from cracking processes or may even be selective olefins for dimer, trimer, or tetramer production (Parkash, 2003; Hsu and Robinson, 2006; Speight, 2011a, 2011b, 2014; Hsu and Robinson, 2017; Speight, 2017). The molecular size of the product is limited insofar as the reaction is terminated at the dimer or trimer stage. Thus the process is more properly termed oligomerization. The 4- to 12-carbon compounds required as the constituents of liquid fuels are the prime products.

Thermal polymerization is not as effective as catalytic polymerization but has the advantage that it can be used to polymerize saturated materials that cannot be induced to react by catalysts. The process consists of the vapor-phase cracking, such as propane and butane, followed by prolonged periods at high temperature (510°C–595°C; 950°F–1,105°F) for the reactions to proceed to near completion. Olefins can also be conveniently polymerized by means of an acid catalyst. Thus, the treated olefin-rich feed stream is contacted with a catalyst, such as sulfuric acid, copper pyrophosphate, or phosphoric acid, at 150°C–220°C (300°F–430°F) and 150–1,200 psi, depending on feedstock and product requirement.

Phosphate derivatives are the principal catalysts used in polymerization units; the commercially used catalysts are liquid phosphoric acid, phosphoric acid on kieselguhr, copper pyrophosphate pellets, and phosphoric acid film on quartz. The last is the least active and has the disadvantage that carbonaceous deposits must occasionally be burned off the support. Compared to other processes, the one using liquid phosphoric acid catalyst is far more responsive to attempts to raise production by increasing temperature.

1.4.6 SOLVENT PROCESSES

Many refineries also incorporate solvent extraction processes (also called solvent refining processes) for manufacturing lubricants and petrochemical units with which to recover propylene, benzene, toluene, and xylenes for further processing into polymers (Parkash, 2003; Gary et al., 2007; Speight, 2014; Hsu and Robinson, 2017; Speight, 2017, 2021). While all solvent processes serve a purpose, the processes included here are (1) solvent deasphalting and (2) solvent dewaxing. In these processes, the feedstock is contacted directly with the solvents in order to disrupt the molecular forces within the feedstock and extract a specific fraction as the desired soluble product or as the raffinate leaving an insoluble product.

Solvent processes can be divided into two main categories which are (1) solvent deasphalting, sometimes referred to as solvent extraction, and (2) solvent dewaxing. The solvent used in the extraction processes includes liquid propane (sometimes liquid butane or propane-butane mixtures) and cresylic acid, 2,20-dichlorodiethyl ether, phenol, furfural, sulfur dioxide, benzene, and nitrobenzene. In the dewaxing process (28), the principal solvents are benzene, methyl ethyl ketone, methyl isobutyl ketone, propane, crude oil naphtha, ethylene dichloride, methylene chloride, sulfur dioxide, and N-methyl pyrrolidinone.

The early developments of solvent processing were concerned with the lubricating oil end of the crude. Solvent extraction processes are applied to many useful separations in the purification of gasoline, kerosene, diesel fuel, and other oils.

1.4.6.1 Deasphalting and Solvent Extraction

Solvent deasphalting processes are a major part of refinery operations and are not often appreciated for the tasks for which they are used (Parkash, 2003; Gary et al., 2007; Speight, 2014; Hsu and Robinson, 2017; Speight, 2017, 2021).

Solvent deasphalting takes advantage of the fact that aromatic pounds are insoluble in paraffins. Propane deasphalting was designed for the treatment of residua but is applicable to a wide range of viscous feedstocks. The deasphalted oil (DAO, sometimes referred to as deasphaltened oil) is sent to hydrotreaters, FCC units, hydrocrackers, or fuel-oil blending. In hydrocrackers and FCC units, the deasphalted oil is easier to process than straight-run residual oils because the asphaltene fraction readily forms coke under thermal conditions and often contains catalyst poisons such as nickel and vanadium, and the asphaltene content of deasphalted oil is (by definition) zero.

In the solvent deasphalting processes, a liquid alkane is injected into the feedstock to disrupt the dispersion of components and causes the polar constituents to precipitate. Propane (or sometimes propane/butane mixtures) is extensively used for deasphalting and produces a deasphalted oil (DAO) and propane deasphalter asphalt (PDA or PD tar) (Dunning and Moore, 1957). Propane has unique solvent properties; at lower temperatures (38°C–60°C; 100°F–140°F), paraffins are very soluble in propane and at higher temperatures (approximately 93°C; 200°F) all hydrocarbons are almost insoluble in propane.

A solvent deasphalting unit processes the residuum from the vacuum distillation unit and produces deasphalted oil (DAO), used as feedstock for a fluid catalytic cracking unit, and the asphaltic residue (deasphalter tar, deasphalter bottoms) which, as a residual fraction, can only be used to produce asphalt or as a blend stock or visbreaker feedstock for low-grade fuel oil (Parkash, 2003; Hsu and Robinson, 2006; Gary et al., 2007; Speight, 2011a, 2011b, 2014; Hsu and Robinson, 2017; Speight, 2017). Solvent deasphalting processes have not realized their maximum potential. With ongoing improvements in energy efficiency, such processes would display its effects in a combination with other processes. Solvent deasphalting allows removal of sulfur and nitrogen compounds as well as metallic constituents by balancing yield with the desired feedstock properties (Ditman, 1973).

The propane deasphalting process is similar to solvent extraction in that a packed or baffled extraction tower or rotating disc contactor is used to mix the feedstock with the solvent. In the tower method, four to eight volumes of propane are fed to the bottom of the tower for every volume of feed flowing down from the top of the tower. The oil, which is more soluble in the propane, dissolves and flows to the top. The asphaltene and resins flow to the bottom of the tower where they are removed in a propane mix. Propane is recovered from the two streams through two-stage flash systems followed by steam stripping in which propane is condensed and removed by cooling at high pressure in the first stage and at low pressure in the second stage. The asphalt recovered can be blended with other asphalts or heavy fuels or can be used as feed to the coker.

The major process variables are temperature, pressure, solvent-to-oil ratio, and solvent type. Pressure and temperature are both variables because the solvent power of light hydrocarbon is approximately proportional to the density of the solvent. Higher temperature always results in decreased yield of deasphalted oil. On the other hand, increasing solvent-to-oil ratio increases the recovery of deasphalted oil with increase in viscosity. However, for the given product quality which can be maintained with change in temperature, solvent-to-oil ratio increases the yield of deasphalted oil. It has been shown that the solvent power of paraffin solvent increases with increase in solvent molecular weight.

Solvent deasphalting processes are a major part of refinery operations (Parkash, 2003; Gary et al., 2007; Speight, 2014; Hsu and Robinson, 2017; Speight, 2017, 2021) and are not often appreciated for the tasks for which they are used. In the solvent deasphalting processes, an alkane is injected into the feedstock to disrupt the dispersion of components and causes the polar constituents to precipitate. Propane (or sometimes propane/butane mixtures) is extensively used for deasphalting and produces a deasphalted oil (DAO) and propane deasphalter asphalt (PDA or PD tar) (Dunning and Moore, 1957). Propane has unique solvent properties; at lower temperatures (38°C–60°C; 100°F–140°F), paraffins are very soluble in propane, and at higher temperatures (approximately 93°C; 200°F), all hydrocarbons are almost insoluble in propane.

The deasphalting process has evolved to the stage where an alternate process concept is offered. This is the ROSE process (residuum oil supercritical extraction process). In this process, the feedstock and solvent are mixed and heated to a temperature in excess of the critical temperature of the solvent at which point the feedstock is almost totally insoluble. The process allows a higher recovery of the deasphalted liquid.

On the other hand, the solvent extraction is used to remove aromatic constituents and other impurities from lubricating oil stocks and grease stocks. In a typical process, the feedstock is dried and then contacted with the solvent in a counter-current or rotating disk extraction unit. The solvent is separated from the product stream by heating, evaporation, or fractionation, and any remaining traces of solvent are removed from the raffinate by steam stripping. In addition, electrostatic precipitators may be used to enhance the separation of any inorganic contaminants after which the solvent is regenerated and recycled.

Commonly used solvents for the process are phenol, furfural, or cresylic acid. Chlorinated ethers and nitrobenzene also have been used for the extraction process. There is a related extraction process (the Edeleanu process) in which the solvent is liquid sulfur dioxide.

1.4.6.2 Dewaxing and Wax Deoiling

Paraffinic crude oils often contain microcrystalline or paraffin waxes, and the solvent dewaxing removes wax (normal paraffins) from deasphalted lubricating oil base stocks.

In the process, the feedstock is mixed with the solvent (commonly used solvents include toluene and methyl ethyl ketone, MEK), and the mixture is cooled to encourage (enable) crystallization of the wax after which the solvent is recovered. Methyl isobutyl ketone (MIBK) is used in a wax deoiling process to prepare food-grade wax.

More specifically, the feedstock is treated with a solvent (MEK) to remove this wax before it is processed (Parkash, 2003; Hsu and Robinson, 2006; Gary et al., 2007; Speight, 2011a, 2011b, 2014; Hsu and Robinson, 2017; Speight, 2017). Solvent dewaxing processes are designed to remove wax from lubricating oils to enable the product to exhibit satisfactory fluidity characteristics at low temperatures (e.g., low pour points) rather than from the whole crude oil. The mechanism of solvent dewaxing involves either the separation of wax as a solid that crystallizes from the oil solution at low temperature or the separation of wax as a liquid that is extracted at temperatures above the melting point of the wax through preferential selectivity of the solvent.

In the 1930s, two types of stocks – paraffinic feedstocks and naphthenic feedstocks – were used for the manufacture of making automobile oil. Both types of feedstock were solvent extracted to improve their quality, but in the high-temperature conditions encountered in service, the naphthenic type could not stand up as well as the paraffinic type. Nevertheless, the naphthenic type was the preferred oil, particularly in cold weather, because of its fluidity at low temperatures. Previous to 1938, the highest-quality lubricating oils were of the naphthenic type and were phenol treated to pour points of −40°C to −7°C (−40°F to 20°F), depending on the viscosity of the oil. Paraffin-type oils were also available and could be treated (with phenol) to produce higher-quality oil but (often) the wax content was so high that the oils were solid at room temperature.

In the solvent dewaxing process, the feedstock is diluted with a solvent that has a high affinity for oil, chilled to precipitate the wax, filtered to remove the wax, stripped of solvent, and dried. The solvents (principally propane, naphtha, MEK) act as diluents for the high molecular weight oil fractions to reduce the viscosity of the mixture and provide sufficient liquid volume to permit pumping and filtering. Wax, produced by the solvent dewaxing process, is used to make (1) paraffin derivatives for the production of candle wax, (2) microwax for the production of cosmetics, and (3) wax for the manufacture of petroleum jelly.

Catalytic dewaxing is a process in which the chemical composition of the feed is changed (Parkash, 2003; Hsu and Robinson, 2006; Gary et al., 2007; Speight, 2011a, 2011b, 2014; Hsu and Robinson, 2017; Speight, 2017). The process involved catalytic cracking of long paraffin chains into shorter chains to remove the wax and produced lower molecular weight products suitable for other uses. As an example, the feedstock is contacted with hydrogen at elevated temperature and pressure over a catalyst (such as a zeolite) that selectively cracks the normal paraffins to methane, ethane, and propane. This process is also known as hydrodewaxing. There are two types of catalytic dewaxing: (1) a single-catalyst process that is used for a reduction in the pour point and to improve the oxygen stability of the product, and (2) a two-catalyst process that uses a fixed-bed reactor and essentially no methane or ethane is formed in the reaction.

The lowest viscosity paraffinic oils were dewaxed by the cold-press method to produce oils with a pour point of 2°C (35°F). The light paraffin distillate oils contained a paraffin wax that crystallized into large crystals when chilled and could thus readily be separated from the oil by the cold-press filtration method. Because the wax could not be removed from intermediate and heavy paraffin distillates, high-quality, high-viscosity lubricating oils in them could not be used except as cracking stock.

Therefore, processes were therefore developed to dewax these high-viscosity paraffinic oils. The processes were similar insofar as the waxy oil was dissolved in a solvent that would keep the oil in solution; the wax separated as crystals when the temperature was lowered. The processes differed chiefly in the use of the solvent. Commercially used solvents were naphtha, propane, sulfur dioxide, acetone-benzene, trichloroethylene, ethylene dichloride-benzene (Barisol), methyl ethyl ketone-benzene (benzol), methyl-n-butyl ketone, and methyl-n-propyl ketone.

The process as now practiced involves mixing the feedstock with one to four times its volume of the ketone (Parkash, 2003; Hsu and Robinson, 2006; Gary et al., 2007; Speight, 2011a, 2011b, 2014;

Hsu and Robinson, 2017; Speight, 2017). The mixture is then heated until the oil is in solution, and the solution is chilled at a slow, controlled rate in double-pipe, scraped-surface exchangers. Cold solvent, such as filtrate from the filters, passes through the 2-in. annular space between the inner and outer pipes and chills the waxy oil solution flowing through the inner 6-in. pipe.

1.4.7 TREATING PROCESSES

The term "treating" as used in the refining industry refers to the removal of contaminants (such as organic compounds containing sulfur, nitrogen, and oxygen; dissolved metals and inorganic salts; and soluble salts dissolved in emulsified water) from crude oil stream or product streams.

There is a variety of treating processes, but the primary purpose of the majority of the processes is the elimination of unwanted sulfur compounds. A variety of intermediate and finished products such as gases, naphtha, and middle distillates are dried and sweetened. The so-called sweetening processes are used predominantly for the treatment of naphtha to remove sulfur compounds (such as hydrogen sulfide, thiophene derivatives, and mercaptan derivative) to improve color, odor, and oxidation stability. Sweetening also reduces concentrations of carbon dioxide.

Treating can be accomplished at an intermediate stage in the refining process, or just before sending the finished product to storage. The choice of a treating method depends on (1) the nature of the crude oil products, (2) the amount and type of impurities in the products to be treated, (3) the extent to which the process removes the impurities, and (4) the specifications of the saleable product. Treating materials include acids, solvents, alkalis, oxidizing, and adsorption agents. Treating materials include alkalis, acids, and adsorption agents (such as clay or lime).

Since the original crude oils contain some sulfur compounds, the resulting products and naphtha fraction (sometimes referred to as gasoline fractions) also contain sulfur compounds, including hydrogen sulfide, mercaptan derivatives, sulfide derivatives, disulfide derivatives, and thiophene derivatives. The processes used to sweeten the products (i.e., desulfurize the products) depend on the type and amount of the sulfur compounds present and the specifications of the finished gasoline or other stocks.

Hydrotreating is the most widely practiced treating process for all types of crude oil products. However, there are other treating processes suitable for the removal of mercaptans and hydrogen sulfide; such processes are necessary and are performed as part of the product improvement and finishing procedures. For example, the removal of mercaptan derivatives (RSH) is achieved by using regenerative solution processes, in which the treatment solutions are regenerated rather than discarded. Mercaptan conversion is essentially a process of oxidation to disulfides, RSSR, by lead sulfide treatment, copper chloride–oxygen treatment, sodium hypochlorite treatment, or oxygen treatment with a chelated cobalt catalyst in either a caustic solution or fixed bed.

$$2RSH + [O] \rightarrow RSSR + H_2O$$

Hydrogen sulfide (H_2S) is removed by a variety of processes, of which one is a regenerative solution process using aqueous solutions of sodium hydroxide (NaOH), calcium hydroxide ($Ca(OH)_2$), sodium phosphate (Na_3PO_4), and sodium carbonate (Na_2CO_3) (Parkash, 2003; Hsu and Robinson, 2006; Gary et al., 2007; Speight, 2011a, 2011b, 2014; Hsu and Robinson, 2017; Speight, 2017).

1.4.7.1 Acid Treatment

The treatment of crude oil products with acids has been in use for a considerable time in the crude oil industry. Various acids, such as hydrofluoric acid, hydrochloric acid, nitric acid, and phosphoric acid, have been used in addition to the most commonly used sulfuric acid, but in most instances, there is little advantage in using any acid other than sulfuric.

Sulfuric acid is the most commonly used acid-treating process and results in partial or complete removal of unsaturated hydrocarbon derivatives, sulfur, nitrogen, and oxygen compounds, as well as resinous and asphaltic compounds. It is used to improve the odor, color, stability, carbon residue, and other properties of the oil.

For example, sulfuric acid also has been employed for refining kerosene distillates and lubricating oil stocks. Although a greater part of the acid-treating processes has been superseded by other processes, acid treating has continued to some extent; it is used for desulfurizing high-boiling fractions of cracked gasoline distillates, for refining paraffinic kerosene, for manufacturing low cost lubricating oils, and for making specialty products such as insecticides, pharmaceutical oils, and insulating oils.

1.4.7.2 Alkali Treatment

Alkali treating (also referred to as caustic treating using with NaOH or KOH) is used to improve odor and color by removing organic acids (such as naphthenic acid derivatives, phenol derivatives) and sulfur compounds (H_2S and RSH) by a caustic wash. When caustic soda (sodium hydroxide) is combined with various solubility promoters (such as methyl alcohol, CH_3OH, and cresol derivatives, $CH_3C_6H_4OH$), there is a near quantitative removal of all mercaptan derivatives as well as oxygen-containing derivatives and nitrogen-containing derivatives can be dissolved and removed from crude oil fractions.

In the caustic washing process, products from crude oil refining are treated with solutions of caustic soda (NaOH). The crude oil fraction is mixed with a water solution of NaOH (often referred to as "lye"). The treatment is carried out as soon as possible after the crude oil fraction is distilled, since contact with air forms free sulfur, which is corrosive and difficult to remove. The lye reacts either with any hydrogen sulfide present to form Na_2S, which is soluble in water, or with mercaptan derivatives (e.g., RSH), followed by oxidation, to form the less nocuous disulfide derivatives (e.g., RSSR).

$$H_2S + 2NaOH \rightarrow Na_2S + 2H_2O$$

$$RSH + NaOH \rightarrow NaSR + H_2O$$

$$4NaSR + O_2 + 2H_2O \rightarrow 2RSSR + 4NaOH$$

Nonregenerative caustic treatment is generally economically applied when the contaminating materials are low in concentration and waste disposal is not a problem. However, the use of nonregenerative systems is on the decline because of the frequently occurring waste disposal problems that arise from environmental considerations and because of the availability of numerous other processes that can effect more complete removal of contaminating materials.

Steam-regenerative caustic treatment is directed toward the removal of mercaptans from such products as gasoline and low-boiling solvents (naphtha). The caustic is regenerated by steam blowing in a stripping tower. The nature and concentration of the mercaptans to be removed dictate the quantity and temperature of the process. However, the caustic gradually deteriorates because of the accumulation of material that cannot be removed by stripping; the caustic quality must be maintained either by continuous discard or by intermittent discard and/or replacement of a minimum amount of the operating solution.

1.4.7.3 Clay Treatment

Treatment with clay (or lime, CaO) of acid-refined oil removes traces of asphaltic materials and other compounds improving product color, odor, and stability.

The original method of clay treating was to percolate a crude oil fraction through a tower containing coarse clay pellets. As the clay adsorbed impurities from the crude oil fraction, the clay became less effective. The activity of the clay was periodically restored by removing it from the tower and burning the adsorbed material under carefully controlled conditions so as not to sinter the clay. The percolation method of clay treating was widely used for lubricating oils but has been largely replaced by clay contacting.

However, this use of clay treating has been superseded by other processes; such as by the use of inhibitors, which, when added in small amounts to gasoline, can prevent gums from forming. Nevertheless, clay treating is still used as a finishing step in the manufacture of lubricating oils and waxes. The clay removes traces of asphaltic materials and other compounds that give oils and waxes unwanted odors and colors.

REFERENCES

Bridjanian, H., and Khadem Samimi, A. 2011. Bottom of the Barrel: An Important Challenge of the Petroleum Refining Industry. *Petroleum & Coal*, 53(1): 13–21.

Casey, D. 2011. Making Lighter Work of Heavier Oil: Customized Solutions Help Refiners to Meet the Residue Upgrading Challenge. Criterion Catalysts and Technologies. http://s05.static-shell.com/content/dam/shell/static/criterion/downloads/pdf/impact/criterion-impactarticleissue12011makinglighterworkofheavieroil.pdf

Colyar, J. 2009. Has the Time for Partial Upgrading of Heavy Oil and Bitumen Arrived? *PTQ*, Q4: 43–55. http://www.digitalrefining.com/article/1000607

Ditman, J.G. 1973. Solvent Deasphalting. *Hydrocarbon Processing*, 52(5): 110.

Dunning, H.N., and Moore, J.W. 1957. Propane Removes Asphalts from Crudes. *Petroleum Refiner*, 36(5): 247–250.

Gary, J.H., Handwerk, G.E., and Kaiser, M.J. 2007. *Petroleum Refining: Technology and Economics*. 5th Edition. CRC Press, Taylor & Francis Group, Boca Raton, Florida.

Hsu, C.S., and Robinson, P.R. (Editors). 2017. *Handbook of Petroleum Technology*. Springer, Cham, Switzerland.

Kressmann, S., Boyer, C., Colyar, J.J., Schweitzer, J.M., and Viguié, J.C. 2000. Improvements of Ebullated-Bed Technology for Upgrading Heavy Oils. *Oil & Gas Science and Technology – Rev. IFP*, 55(4): 397–406.

Larraz, R. 2021. A Brief History of Oil Refining. *Substantia: An International Journal of the History of Chemistry*, 5(2): 129–152. https://riviste.fupress.net/index.php/subs/article/view/1191; https://pdfs.semanticscholar.org/43cf/2e106720b7461bf9f255e74360c9338c2213.pdf?_ga=2.94704988.1370212699.1660937454-706071713.1660937454

Parkash, S. 2003. *Refining Processes Handbook*. Gulf Professional Publishing, Elsevier, Amsterdam, Netherlands.

Phillips, G., and Liu, F. 2002. Advances in Residue Upgrading Technologies Offer Refiners Cost-Effective Options for Zero Fuel Oil Production. Proceedings of the 2002 European Refining Technology Conference. Paris, France. November.

Ross, J., Roux, R., Gauthier, T., and Anderson, L.R. 2005. Fine-tune FCC Operations for Changing Fuels Market. *Hydrocarbon Processing*, 84(9): 65–73.

Schermer, W.E.M., Melein, P.M.J., and Van den Berg, F.G.A. 2004. Simple Techniques for Evaluation of Crude Oil Compatibility. *Petroleum Science and Technology*, 22: 1045–1054.

Speight, J.G., and Moschopedis, S.E. 1979. The Production of Low-Sulfur Liquids and Coke from Athabasca Bitumen. *Fuel Processing Technology*, 2: 295–302.

Speight, J.G. 1987. Initial Reactions in the Coking of Residua. Preprints, *American Chemical Society, Division of Fuel Chemistry*, 32(2): 413–417.

Speight, J.G. 2011a. *An Introduction to Petroleum Technology, Economics, and Politics*. Scrivener Publishing, Salem, Massachusetts.

Speight, J.G. 2011b. *The Refinery of the Future*. Gulf Professional Publishing, Elsevier, Oxford, United Kingdom.

Speight, J.G. 2013. *Heavy and Extra Heavy Oil Upgrading Technologies*. Gulf Professional Publishing, Elsevier, Oxford, United Kingdom.

Speight, J.G. 2014. *The Chemistry and Technology of Petroleum*. 5th Edition. CRC Press, Taylor & Francis Publishers, Boca Raton, Florida.

Speight, J.G. 2016. *Introduction to Enhanced Recovery Methods for Heavy Oil and Tar Sands*. 2nd Edition. Gulf Publishing Company, Elsevier, Waltham, Massachusetts.

Speight, J.G. 2017. *Handbook of Petroleum Refining*. CRC Press, Taylor & Francis Group, Boca Raton, Florida.

Speight, J.G. 2019. *Handbook of Petrochemical Processes*. CRC Press, Taylor & Francis Group, Boca Raton, Florida.

Speight, J.G. 2021. *Refinery Feedstocks*. CRC Press, Taylor & Francis Group, Boca Raton, Florida.

Speight, J.G. 2022. *Dewatering, Desalting, and Distillation in Petroleum Refining*. CRC Routledge, Taylor & Francis Group, New York.

Stark, J.L., and Asomaning, S. 2003. Crude Oil Blending Effects on Asphaltene Stability in Refinery Fouling. *Petroleum Science and Technology*, 21: 569–579.

Stratiev, D., and Petkov, K. 2009. Residue Upgrading: Challenges and Perspectives. *Hydrocarbon Processing*, 88(9): 93–96.

Van den Berg, F.G.A., Kapusta, S.D., Ooms, A.C., and Smith, A.J. 2003. Fouling and Compatibility of Crudes as the Basis for a New Crude Selection Strategy. *Petroleum Science and Technology*, 21: 557–568.

2 Feedstocks and Feedstock Composition

2.1 INTRODUCTION

In the unrefined (raw) state crude oil, heavy crude oil, extra heavy crude oil, and tar sand bitumen have minimal value, but when used as feedstocks for the refinery (Figure 2.1), the result is a variety of high-value liquid fractions that can serve (1) as feedstocks for the production of fuels, solvents, lubricants, or (2) as a sources of many other products (Speight, 2014, 2017, 2021). In fact, the liquid products derived from crude oil contribute approximately one-third to one-half of the total world energy supply and are used not only for transportation fuels (i.e., gasoline, diesel fuel, and aviation fuel, among others) but also to heat buildings.

As a side note, the term black oil is a term that has arisen and used within the past two decades and has seen more common use recently. This term is of unknown scientific and engineering origin (other than color which is reputed to indicate relatively high concentrations of resin constituents and asphaltene constituents in the oil) and is not used in any way in this text.

However, in the context of this chapter, the same is not true for heavy crude oil, extra heavy crude oil, and tar sand bitumen which also include the atmospheric and vacuum residua produced by the distillation of crude oil (Chapter 1) that may not be destined for use as asphalt and are often referred to as viscous feedstocks and require application of more novel recovery methods that are used for conventional crude oil (Speight, 2014, 2016). These feedstocks require additional refining steps such as thermal cracking processes and catalytic processes involving the use of a catalyst – thermal processes involving the use of hydrogen and a catalyst are deliberately excluded from this text and will be subject of a later volume in this series. Thus, the subject of this text is the thermal processes and catalytic processes that are required to produce the distillate products which can then be used for the production of fuels, solvents, lubricants, and other high-value products such as petrochemical products (Speight, 2014, 2017, 2019, 2021).

Furthermore, a major aspect of refinery design is to ensure that the design is adequate to process the feedstocks of different composition. For example, a heavy oil refinery would differ somewhat from a conventional refinery, and a refinery for tar sand bitumen would be significantly different to both (Speight, 2014, 2017, 2020a). Furthermore, the composition of biomass is variable (Speight, 2020b) which is reflected the range of heat value (heat content, calorific value) of biomass, which is somewhat lesser than for coal and much lower than the heat value for crude oil, generally falling in the range of 6,000–8,500 Btu/lb. Moisture content is probably the most important determinant of heating value. Air-dried biomass typically has approximately 15%–20% moisture, whereas the moisture content for oven-dried biomass is around 0%. Moisture content is also an important characteristic of coals, varying in the range of 2%–30%. However, the bulk density (and hence energy density) of most biomass feedstocks is generally low, even after densification, approximately 10% and 40% of the bulk density of most fossil fuels.

Therefore, it is the purpose of this chapter to provide a description of the processes that can be used to convert the viscous feedstocks (i.e., heavy crude oil, extra heavy crude oil, and tar sand bitumen) into more useful distillates from which the high-value products can be produced.

DOI: 10.1201/9781003184904-2

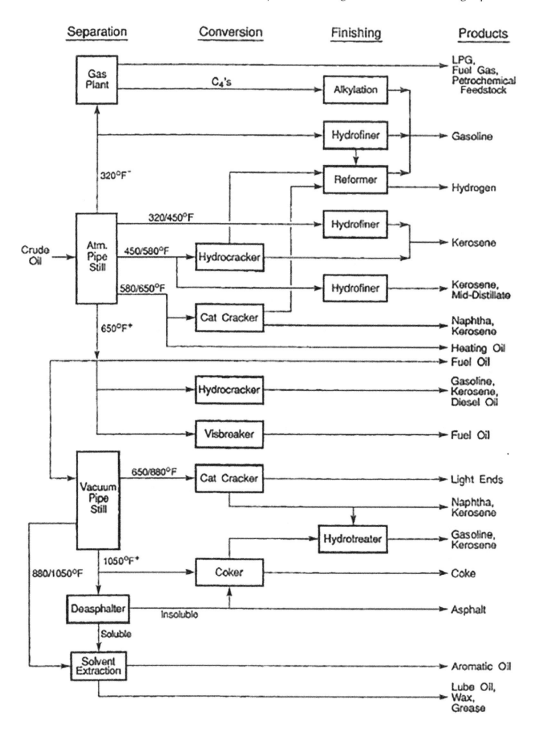

FIGURE 2.1   Schematic representation of a refinery showing the placement of the various conversion units. (Source: Speight, J.G. 2017. *Handbook of Petroleum Refining.* CRC Press, Taylor & Francis Group, Boca Raton, Florida. Figure 8.1, page 293.)

2.2 NON-VISCOUS FEEDSTOCKS

While the focus of this chapter is predominantly on the various viscous refinery feedstocks, the non-viscous fractions of crude oil are also used for the production of products through the application of thermal cracking processes and catalytic cracking processes. These are (1) naphtha, (2) kerosene, and (3) gas oil. These products are typically produced from the conventional refinery feedstocks (i.e., conventional crude oil) but are, nevertheless, worthy of mention as potential feedstocks for thermal; cracking processes and catalytic cracking processes as feedstocks that lead to a range of products that also includes petrochemical products (Parkash, 2003; Gary et al., 2007; Speight, 2014; Hsu and Robinson, 2017; Speight, 2017, 2019, 2021).

2.2.1 NAPHTHA

The term naphtha (often referred to as naft in the older literature) is actually a generic term applied to refined, partly refined, or an unrefined crude oil fraction. In the strictest sense of the term, not less than 10% of the material should distill below 175°C (345°F) and not less than 95% w/w (or v/v) of the material should distill below 240°C (465°F) under standardized distillation conditions.

Generally, naphtha is an unrefined crude oil that distills below 240°C (465°F) and is (after the gases constituents) the most volatile fraction of the crude oil. In fact, in some specifications, not less than 10% of material should distill below approximately 75°C (167°F), but this can be refinery dependent (Pandey et al., 2004). Naphtha resembles gasoline in terms of boiling range and carbon number, being a precursor to gasoline, and it is typically used as a precursor to gasoline or to a variety of solvents. However, the solvents obtained from the petrochemical processes – such as alcohols and ethers – are not included in this chapter.

Naphtha has been available since the early days of the crude oil industry and historically goes back even further into history. Indeed, the infamous Greek fire documented as being used in warfare during the last three millennia is a crude oil derivative. It was produced either by distillation of crude oil isolated from a surface seepage or (more likely) by destructive distillation of the bituminous material obtained from bitumen seepages (of which there are/were many known during the heyday of the civilizations of the Fertile Crescent). As an example, the bitumen obtained from the area of Hit (Tuttul) in Iraq (Mesopotamia) is such an occurrence (Abraham, 1945; Forbes, 1958a, 1958b, 1959). Other crude oil products boiling within the naphtha boiling range include (1) industrial spirit and white spirit (Guthrie, 1960; Speight, 2014).

As an example, of product divergence, the product termed "industrial spirit" comprises liquids distilling between 30°C and 200°C (−1°F and 390°F), with a temperature difference between 5% volume and 90% volume distillation points, including losses, of not more than 60°C (140°F). There are several eight grades of industrial spirit, depending on the distillation range. On the other hand, white spirit is an industrial spirit with a flash point above 30°C (99°F) and has a distillation range from 135°C to 200°C (275°F to 390°F).

2.2.1.1 Manufacture

Typically, naphtha is prepared by any one of several available methods, which include (1) fractionation of straight-run distillate or cracked distillate or reformer distillate or even; (2) solvent extraction; (3) hydrogenation of cracked distillate; (4) polymerization of unsaturated compounds, i.e., olefin derivatives; and (5) alkylation processes. In fact, the naphtha product may be a combination (a blend) of naphtha streams from more than one of these processes.

The more common method of naphtha preparation is distillation of the crude oil – the viscous feedstocks by virtue of their definition and composition contain very little (if any) naphtha constituents. Depending on the design of the distillation unit, either one or two naphtha steams may be produced: (1) a single naphtha with an end point of about 205°C (400°F) and similar to straight-run gasoline, or (2) this same fraction divided into a low-boiling (low-density) naphtha and a high-boiling (high-density) naphtha. The end point of the low-boiling (low-density) naphtha is varied to suit the subsequent subdivision of the naphtha into narrower boiling fractions and may be of the order of 120°C (250°F).

Before the naphtha is redistilled into a number of fractions with boiling ranges suitable for aliphatic solvents, the naphtha are usually treated to remove sulfur compounds, as well as aromatic hydrocarbon derivatives, which are present in sufficient quantity to cause an odor. Aliphatic solvents that are specially treated to remove aromatic hydrocarbon derivatives are known as deodorized solvents. The term "odorless solvent" is the name given to high-boiling (high-density) alkylate used as an aliphatic solvent, which is a by-product in the manufacture of aviation alkylate (Parkash, 2003; Gary et al., 2007; Speight, 2014; Hsu and Robinson, 2017; Speight, 2017).

Sulfur compounds are most commonly removed or converted to a harmless form by chemical treatment with lye, doctor solution, copper chloride, or similar treating agents (Chapter 24). Hydrorefining processes are also often used in place of chemical treatment to reduce the sulfur content to an acceptable level (Parkash, 2003; Gary et al., 2007; Speight, 2014; Hsu and Robinson, 2017; Speight, 2017). Also, naphtha with a low content of aromatic constituents (increase the solvent power of the naphtha) has a slight odor, but there is no need to remove the aromatic constituents unless an odor-free product is specified.

Furthermore, naphtha that is either naturally sweet (no odor) or has been treated until sweet and is subdivided into several fractions in efficient fractional distillation towers frequently called pipe stills, columns, and column steam stills (Parkash, 2003; Gary et al., 2007; Speight, 2014; Hsu and Robinson, 2017; Speight, 2017). A typical arrangement consists of primary and secondary fractional distillation towers and a stripper. High-boiling (high-density) naphtha, for example, is heated by a steam heater and passed into the primary tower, which is usually operated under vacuum. The low pressure in the unit allows vaporization of the naphtha at the temperatures obtainable from the steam heater.

The primary tower separates the naphtha into three parts, which are (1) a high-boiling or high-density product that is removed as a bottom product and sent to a cracking unit; (2) a side stream product of narrow boiling range that, after passing through the stripper, may be suitable for use as an the aliphatic solvent; and (3) an overhead product – a lower-boiling or lower-density product that is sent the secondary tower (i.e., the vacuum) tower where the higher-boiling product from the primary tower is divided into an overhead product and a bottom product due to the partial vacuum with steam injected into the bottom of the tower to assist in the fractionation. The overhead product and the bottom (lower-volatile) product are finished aliphatic solvents, or if the feed to the primary tower is low-boiling (low-density) naphtha instead of high-boiling (high-density) naphtha, other aliphatic solvents of different boiling ranges are produced.

In terms of further purification of the naphtha, several methods, involving solvent extraction or destructive hydrogenation (hydrocracking), can accomplish the removal of aromatic hydrocarbon derivatives from naphtha. By this latter method, aromatic hydrocarbon constituents are converted into odorless, straight-chain paraffin hydrocarbon derivatives that are required in aliphatic solvents (Parkash, 2003; Gary et al., 2007; Speight, 2014; Hsu and Robinson, 2017; Speight, 2017, 2019).

The Edeleanu extraction process was originally developed to improve the burning characteristics of kerosene by extraction of the smoke-forming aromatic compounds. The process is an extraction process in which liquid sulfur dioxide (SO_2) is used to selectively extract (dissolves) the aromatic derivatives from kerosene-type liquids to produce the insoluble low aromatic product. Thus it is not surprising that its use has been extended to the improvement of other products as well as to the segregation of aromatic hydrocarbon derivatives for use as solvents. Naphtha fractions rich in aromatic derivatives may be treated by the Edeleanu process for the purpose of recovering the aromatic derivatives, or the product stream from a catalytic reformer unit – particularly when the unit is operated to product maximum aromatic derivatives – may be treated using the Edeleanu process to recover the aromatic derivatives. The other most widely used processes for this purpose are the extractive distillation process and the Udex processes. Processes such as the Arosorb process and cyclic adsorption processes are used to a lesser extent.

Extractive distillation (Chapter 17) is used to recover aromatic hydrocarbon derivatives from, say, reformate fractions in the following manner. By means of preliminary distillation in a 65-tray pre-fractionator, a fraction containing a single aromatic can be separated from reformate, and this aromatic concentrate is then pumped to an extraction distillation tower near the top and aromatic concentrate enters near the bottom. A reboiler in the extractive distillation tower induces the aromatic concentrate to ascend the tower, where it contacts the descending solvent.

The solvent removes the aromatic constituents and accumulates at the bosom of the tower; the non-aromatic portion of the concentrate leaves the top of the tower and may contain about 1% of the aromatic derivatives. The solvent and dissolved aromatic derivatives are conveyed from the bottom of the extractive distillation tower to a solvent stripper, where fractional distillation separates the aromatic derivatives from the solvent as an overhead product. The solvent is recirculated to the extractive distillation tower, whereas the aromatic stream is treated with sulfuric acid and clay to yield a finished product of high purity.

The Udex process (Chapter 24) is also employed to recover aromatic streams from reformate fractions. This process uses a mixture of water and diethylene glycol to extract aromatic derivatives. Unlike extractive distillation, an aromatic concentrate is not required and the solvent removes all of the aromatic derivatives which are then separated from one another by subsequent fractional distillation.

The reformate is pumped into the base of an extractor tower. The feed rises in the tower countercurrent to the descending diethylene glycol-water solution, which extracts the aromatic derivatives from the feed. The non-aromatic portion of the feed leaves the top of the tower, and the aromatic-rich solvent leaves the bottom of the tower. Distillation in a solvent stripper separates the solvent from the aromatic derivatives, which are sulfuric acid and clay treated and then separated into individual aromatic derivatives by fractional distillation.

Silica gel (SiO_2) is an adsorbent for aromatic derivatives and has found use in extracting aromatic derivatives from refinery streams (Arosorb and cyclic adsorption processes) (Speight, 2014, 2015). Silica gel is manufactured amorphous silica that is extremely porous and has the property of selectively removing and specific chemical compounds from mixtures. For example, silica gel selectively removes aromatic derivatives from a crude oil fraction, and after the non-aromatic portion of the fraction is drained from the silica gel, the adsorbed aromatic derivatives are washed from the silica gel by a stripped (or desorbent). Depending on the kind of feedstock, xylene, kerosene, or pentane may be used as the desorbent.

However, silica gel can be poisoned by contaminants, and the feedstock must be treated to remove water as well as nitrogen, oxygen, and sulfur-containing compounds by passing the feedstock through beds of alumina and/or other materials that remove impurities. The treated feedstock is then sent to one of several silica gel cases (silica gel columns) where the aromatic derivatives are removed by adsorption. The time period required for adsorption depends on the nature of the feedstock; for example, reformate product streams have been known to require substantially less treatment time than kerosene fractions.

2.2.1.2 Composition

Naphtha contains varying amounts of paraffin derivatives, olefin derivatives, naphthene derivatives, and aromatic derivatives and olefin derivatives in different proportions in addition to potential isomers of paraffin that exist in naphtha boiling range. As a result, naphtha is divided predominantly into two main types: (1) aliphatic naphtha and (2) aromatic (naphtha). The two types differ in two ways in the types of the hydrocarbon derivatives that are present and in the methods used for their manufacture.

For example, the aliphatic solvents are composed of paraffin hydrocarbon derivatives and cyclo-paraffin derivatives (naphthene derivatives) and may be obtained directly from conventional crude oil by distillation. On the other hand, an aromatic naphtha contains higher amounts of aromatic derivatives – typically alkyl-substituted benzene derivatives – and is very rarely, if at all, obtained from crude oil as a distillation (straight-run) fraction.

Stoddard solvent is a crude oil distillate widely used as a dry cleaning solvent and as a general cleaner and degreaser. It may also be used as paint thinner, as a solvent in some types of photocopier toners, in some types of printing inks, and in some adhesives. Stoddard solvent is considered to be a form of mineral spirits, white spirits, and naphtha but not all forms of mineral spirits, white spirits, and naphtha are considered to be Stoddard solvent. Stoddard solvent consists of linear alkane derivatives (30%–50% v/v), branched alkane derivatives (20%–40% v/v), cycloalkane derivatives (30%–40% v/v), and aromatic hydrocarbon derivatives (10%–20% v/v). The typical hydrocarbon chain ranges from C_7 through C_{12} in length.

2.2.1.3 Properties and Uses

Naphtha is required to have a low level of odor to meet the specifications for use (Pandey et al., 2004), which is related to the chemical composition – generally paraffin hydrocarbon derivatives possess the mildest odor, and the aromatic hydrocarbon derivatives have a much stronger odor. Naphtha containing higher proportions of aromatic constituents may be pale yellow – usually, naphtha is colorless (water white) and can be tested for the level of contaminants.

Naphtha is used as automotive fuel, engine fuel, and jet-B (naphtha type). More generally, naphtha is classified as low-boiling (low-density) naphtha and high-boiling (high-density) naphtha. Low-boiling (low-density) naphtha is used as rubber solvent, lacquer diluent, while high-boiling (high-density) naphtha finds its application as varnish solvent, dyer's naphtha used in the dyestuffs industry and in the cleaning industry.

More specifically, naphtha is valuable as for solvents because of good dissolving power. The wide range of naphtha available, from the ordinary paraffin straight-run to the highly aromatic types, and the varying degree of volatility possible offer products suitable for many uses (Parkash, 2003; Gary et al., 2007; Speight, 2014; Hsu and Robinson, 2017; Speight, 2017, 2021).

The main uses of naphtha fall into the general areas of (1) solvents (diluents) for paints, for example (2) dry cleaning solvents; (3) solvents for cutback asphalt; (4) solvents in the rubber industry; and (5) solvents for industrial extraction processes.

The boiling ranges of fractions that evaporate at rates permitting the deposition of good films have been fairly well established. Depending on conditions, products are referred to as "light" (i.e., low boiling, low density) as those fractions that boil from 38°C to 150°C (100°F to 300°F) and as "heavy" (i.e., high boiling, high density) as those boiling between 150°C and 230°C (300°F and 450°F). The latter are used mainly in the manufacture of backed and forced-drying products.

In addition to solvent power (the ability of the solvent to dissolve a variety of solutes) and correct evaporation rate, a paint thinner should also be resistant to oxidation, i.e., the thinner should not develop bad color and odor during use. The thinner should be free of corrosive impurities and reactive materials, such as certain types of sulfur compounds, when employed with paints containing lead and similar metals. The requirements are best met by straight-run distillates from paraffinic crude oils that boil from 120°C to 205°C (250°F to 400°F). The components of enamels, vanishes, nitrocellulose lacquers, and synthetic resin finishes are not as soluble in paraffinic naphtha as the materials in conventional paints, and hence naphthenic and aromatic naphtha are favored for such uses.

Dry cleaning is a well-established industry, and the standardized requirements for the solvent are usually met by straight-run naphtha from a low-sulfur, suitably refined paraffinic crude oil. An aromatic hydrocarbon content is not desirable, since it may cause removal of dyes from fabrics or too efficient removal of natural oils from wool, for example. Such a product is usually high boiling and, hence, safe from fire risks, as well as stable enough for extensive reuse and reclaiming. It is especially important that dry cleaning solvents leave no odor on the cloth, and for this reason (coupled with reuse and reclaiming), the solvents cannot be treated with sulfuric acid. The acid treatment leaves the oil with very small quantities of sulfonated hydrocarbon derivatives, which leave a residual odor on the cloth and render the solvent unstable when exposed to distillation temperatures.

Cutback asphalt is asphalt cement diluted with a crude oil distillate to make it suitable for direct application to road surfaces with little or no heating. Asphalt cement, in turn, is a combination of hard asphalt with a heavy distillate (a such as a vacuum gas oil distillate) or with a viscous residuum obtained from the distillation of an asphaltic crude oil. The products are classified as rapid, medium, and slow curing, depending on the rate of evaporation of the solvent. A rapid-curing product may contain 40%–50% of material distilling up to 360°C (680°F); a slow-curing mixture may have only 25% of such material. Gasoline naphtha, kerosene, and low-boiling (low-density) fuel oils boiling from 38°C to 330°C (100°F to 30°F) are used in different products and for different purposes; the use may also dictate the nature of the asphaltic residuum that can be used for the asphalt.

One of the grades of naphtha is used in the rubber industry for dampening the play and tread stocks of automobile tires during manufacture to obtain better adhesion between the units of the tire. Naphtha is also consumed extensively in making rubber cements (adhesives) or are employed in the fabrication of rubberized cloth, hot-water bottles, bathing caps, gloves, overshoes, and toys. These cements are solutions of rubber and were formerly made with benzene, but crude oil naphtha is now preferred because of the less toxic character.

Crude oil distillates are also added in amounts up to 25% and higher at various stages in the polymerization of butadiene-styrene to synthetic rubber. The distillates employed for use in oil extended rubber are of the aromatic type. These distillates are generally high-boiling fractions (such as gas oil) and preferably contain no wax, boil from 425°C to 510°C (800°F to 950°F), have characterization factors of 10.5–11.6, a viscosity index lower than 0, bromine numbers of 6–30, and API gravity of 3–24.

Naphtha is used for extraction on a fairly wide scale. They are supplied in extracting residual oil from castor beans, soybeans, cottonseed, and wheat germ and in the recovery of grease from mixed garbage and refuse. The solvent employed in these cases is a hexane cut, boiling from about 65°C to 120°C (150°F to 250°F).

The recovery of wood resin by naphtha extraction of the resinous portions of dead trees of the resin-bearing varieties or stumps, for example, is also used in the wood industry. The chipped wood is steamed to distill out the resinous products recoverable in this way and then extracted with a naphtha solvent, usually a well-refined, low-sulfur, paraffinic product boiling from, say, 95°C to 150°C (200°F to 300°F).

Crude oil distillates of various compositions and volatility are also employed as solvents in the manufacture of printing inks, leather coatings, diluents for dyes, and degreasing of wool fibers, polishes, and waxes, as well as rust proofing and water proofing compositions, mildew-proofing compositions, insecticides, and wood preservatives.

2.2.2 MIDDLE DISTILLATES

The middle distillate fraction of crude oil (sometime referred to as kerosene and also often referred to as paraffin oil or liquid paraffin, boiling range: 205°C–260°C, 400–500°F) is typically a collection of more specific boiling range distillates that are obtained in the so-called "middle" boiling range on the order of 180°C–360°C (355°F–680°F) during the crude oil distillation process and are removed at mid-height in the distillation tower during the multi-stage process of thermal separation.

Using kerosene as the example, it is a flammable pale yellow or colorless oily liquid with a characteristic odor and is intermediate in volatility between gasoline and gas/diesel oil. It is a medium oil distilling between 150°C and 300°C (300°F and 570°F) and has a flash point about 25°C (77°F) and is suitable for use as an illuminant when burned in a wide lamp (Parkash, 2003; Gary et al., 2007; Speight, 2014; Hsu and Robinson, 2017; Speight, 2017, 2019, 2021). The term kerosene is also often incorrectly applied to various types of fuel oil, but a fuel oil is actually any liquid or liquid crude oil product that produces heat when burned in a suitable container or that produces power when burned in an engine.

Kerosene was the major refinery product before the onset of the automobile age, but now kerosene can be termed one of several secondary crude oil products after the primary refinery product – gasoline. Kerosene originated as a straight-run crude oil fraction that boiled between approximately 205°C and 260°C (400°F and 500°F) (Parkash, 2003; Gary et al., 2007; Speight, 2014; Hsu and Robinson, 2017; Speight, 2017, 2019, 2021). Some crude oils, for example those from the Pennsylvania oil fields, contain kerosene fractions of very high quality, but other crude oils, such as those having an asphalt base, must be thoroughly refined to remove aromatic derivatives and sulfur compounds before a satisfactory kerosene fraction can be obtained.

Jet fuel comprises both gasoline and kerosene-type jet fuels meeting specifications for use in aviation turbine power units and is often referred to as gasoline-type jet fuel and kerosene-type jet fuel.

Jet fuel is a low-boiling (low-density) crude oil distillate that is available in several forms suitable for use in various types of jet engines. The major jet fuels used by the military are JP-4, JP-5, JP-6, JP-7, and JP-8. Briefly, JP-4 is a wide-cut fuel developed for broad availability. JP-6 is a higher cut than JP-4 and is characterized by fewer impurities. JP-5 is specially blended kerosene, and JP-7 is high flash point special kerosene used in advanced supersonic aircraft. JP-8 is kerosene modeled on Jet A-1 fuel (used in civilian aircraft). From what data are available, typical hydrocarbon chain lengths characterizing JP-4 range from C_4 to C_{16}. Aromatic hydrocarbon derivatives are limited to 20%–25% of the total mixture because they produce smoke when burned. A maximum of 5% alkenes is specified for JP-4, and the distribution of the constituents by chemical class is on the order (1) straight-chain alkane derivatives, 32% v/v; (2) branched alkane derivatives, 31% v/v; (3) cycloalkane derivatives, 16% v/v; and (4) aromatic hydrocarbon derivatives, 21% v/v.

Gasoline-type jet fuel includes all light hydrocarbon oils for use in aviation turbine power units that distill between 100°C and 250°C (212°F and 480°F). It is obtained by blending kerosene and gasoline or naphtha in such a way that the aromatic content does not exceed 25% v/v. Additives can be included to improve fuel stability and combustibility.

Kerosene-type jet fuel is a medium distillate product that is used for aviation turbine power units. It has the same distillation characteristics and flash point as kerosene (between 150°C and 300°C, 300°F and 570°F, but not generally above 250°C, 480°F). In addition, kerosene has particular specifications (such as freezing point) which are established by the International Air Transport Association (IATA).

2.2.2.1 Manufacture

Kerosene was first manufactured in the 1850s from coal tar, hence the name coal oil as often applied to kerosene, but crude oil became the major source after 1859. From that time, the kerosene fraction is, and has remained, a distillation fraction of crude oil. However, the quantity and quality vary with the type of crude oil, and although some crude oils yield high-quality kerosene, other crude oils produce kerosene that requires substantial refining. Kerosene is now largely produced by cracking the less volatile portion of crude oil at atmospheric pressure and elevated temperatures (Chapter 4).

In the early days, the poorer quality kerosene was treated with large quantities of sulfuric acid to convert them to marketable products. However, this treatment resulted in high acid and kerosene losses, but the later development of the Edeleanu process (Chapter 24) overcame these problems.

2.2.2.2 Composition

Chemically, kerosene is a mixture of hydrocarbon derivatives; the chemical composition depends on its source, but it usually consists of about 10 different hydrocarbon derivatives, each containing from 10 to 16 carbon atoms per molecule; the constituents include n-dodecane (n-$C_{12}H_{26}$), alkyl benzenes, and naphthalene and its derivatives. Kerosene is less volatile than gasoline; it boils between about 140°C (285°F) and 320°C (610°F).

Kerosene, because of its use as a burning oil, must be free of aromatic and unsaturated hydrocarbon derivatives, as well as free of the more obnoxious sulfur compounds. The desirable constituents of kerosene are saturated hydrocarbon derivatives, and it is for this reason that kerosene is manufactured as a straight-run fraction, not by a cracking process.

Although the kerosene constituents are predominantly saturated materials, there is evidence for the presence of substituted tetrahydronaphthalene. Dicycloparaffin derivatives also occur in substantial amounts in kerosene. Other hydrocarbon derivatives with both aromatic and cycloparaffin rings in the same molecule, such as substituted indan, also occur in kerosene. The predominant structure of the di-nuclear aromatic derivatives appears to be that in which the aromatic rings are condensed, such as naphthalene, whereas the isolated two-ring compounds, such as biphenyl, are only present in traces, if at all.

2.2.2.3 Properties and Uses

Kerosene is by nature a fraction distilled from crude oil that has been used as a fuel oil from the beginning of the crude oil-refining industry. As such, low proportions of aromatic and unsaturated hydrocarbon derivatives are desirable to maintain the lowest possible level of smoke during burning. Although some aromatic derivatives may occur within the boiling range assigned to kerosene, excessive amounts can be remove by extraction; that kerosene is not usually prepared from cracked products almost certainly excludes the presence of unsaturated hydrocarbon derivatives.

The essential properties of kerosene are flash point, fire point, distillation range, burning, sulfur content, color, and cloud point (Speight, 2014, 2015). In the case of the flash point, the minimum flash temperature is generally placed above the prevailing ambient temperature; the fire point determines the fire hazard associated with its handling and use.

The boiling range is of less importance for kerosene than for gasoline, but it can be taken as an indication of the viscosity of the product, for which there is no requirement for kerosene. The ability of kerosene to burn steadily and cleanly over an extended period is an important property and gives some indication of the purity or composition of the product.

The significance of the total sulfur content of a fuel oil varies greatly with the type of oil and the use to which it is put. Sulfur content is of great importance when the oil to be burned produces sulfur oxides that contaminate the surroundings. The color of kerosene is of little significance, but a product darker than usual may have resulted from contamination or aging, and in fact a color darker than specified may be considered by some users as unsatisfactory. Finally, the cloud point of kerosene gives an indication of the temperature at which the wick may become coated with wax particles, thus lowering the burning qualities of the oil.

2.2.3 GAS OIL

Gas oil is a catch-all phrase that is used to describe a liquid distillation product that is obtained from the distillation section of the refinery that boils between kerosene and lubricating oil. However, this definition can vary from refinery to refinery and, in some refineries, may even include kerosene (at the lower end of the boiling range) and liberating oil (at the upper end of the boiling range).

For the purpose of this text, gas oil occurs in the refinery in as two fractions which are (1) atmospheric gas oil, often referred to as AGO, and (2) vacuum gas oil, often referred to as VGO (Parkash, 2003; Gary et al., 2007; Speight, 2014; Hsu and Robinson, 2017; Speight, 2017, 2019, 2021). Typically, the atmospheric gas oil has a boiling range higher than the kerosene boiling range and is on the order of 300°C (570°F) to 360°C (680°F), which is the temperature just before the distillation is terminated. On the other hand, the vacuum gas oil is the first fraction that is obtained in the vacuum distillation process and has a boiling range on the order of 360°C to (950°F) which is the temperature at which the vacuum distillation process is terminated to produce the vacuum residuum. In some refineries, the vacuum distillation process may be terminated at 565°C (1,050°F).

The gas oil fractions vary in composition depending upon the source and character of the crude oil. Generally, the gas oil fractions may be used as a source of fuel oil or fuel oil blending stock or as a source of lubricating oil.

2.2.3.1 Manufacture

There are two types of gas oil fractions that are produced in a refinery and are: (1) atmospheric gas oil and (2) vacuum gas oil (Table 2.1). The former – the atmospheric gas oil – boils above the middle distillate fractions (at approximately 260°C–425°C, 400°F–800°F). The latter – the vacuum gas oil – is the fraction that is isolated during the initial stages of the vacuum distillation process (boils at approximately 425°C–600°C, 800°F–1,100°F). These data are subject to change depending upon the refinery type of nomenclature.

2.2.3.2 Composition

Typically, a gas oil fraction consists of complex mixtures of aliphatic and aromatic hydrocarbon derivatives, the relative amounts depending on the source and properties of the gas oil. The aliphatic alkane derivatives (paraffin derivatives) and cycloalkane derivatives (naphthene derivatives) are hydrogen saturated and compose as much as 75% v/v (or more) of the gas oil. Aromatic constituents (e.g., benzene, naphthalene, and phenanthrene derivatives) compose up to 25% v/v of the gas oil. In addition, the sulfur content of (atmospheric or vacuum) gas oil can vary up to 5% w/w.

The saturate constituents contribute less to the vacuum gas oil (VGO) than the aromatic constituents but more than the polar constituents that are now present at percentage rather than trace levels. Vacuum gas oil is occasionally used as a heating oil, but most commonly it is processed by catalytic cracking to produce naphtha or extraction to yield lubricating oil.

Within the vacuum gas oil, saturate derivatives, distribution of paraffin derivatives, iso-paraffin derivatives, and naphthene derivatives is highly dependent upon the crude oil source. Generally, the naphthene constituents account for approximately two-thirds 60% of the saturate constituents, but the overall range of variation is from <20% to >80%. In most samples, the n-paraffin derivatives

TABLE 2.1
Boiling Fractions of Conventional Crude Oil

Fraction	Boiling Range[a]	
	°C	°F
Low-boiling naphtha	30–150	30–300
High-boiling naphtha	150–180	300–400
Middle distillates[b]	180–290	400–500
Kerosene	180–260	355–500
Fuel oil	205–290	400–550
Atmospheric gas oil	260–315	500–800
Vacuum gas oil	425–600	800–1100
Light-vacuum gas oil	315–425	600–800
Heavy-vacuum gas oil	425–600	425–600
Residuum	>510	>950

[a] For convenience, boiling ranges which can vary from refinery to refinery are approximate and, for convenience, are converted to the nearest 5°C.

[b] Obtained in the "middle" boiling range which is on the order of 180°C–260°C (355°F–500°F) during the crude oil distillation process. The middles distillate is so named because the fraction is removed at mid-height in the distillation tower during the multi-stage process of thermal separation.

from C_{20} to C_{44} are still present in sufficient quantity to be detected as distinct peaks in gas chromatographic analysis.

The bulk of the saturated constituents in vacuum gas oil consist of iso-paraffin derivatives and especially naphthene species although isoprenoid compounds, such as squalane (C_{30}) and lycopane (C_{40}), have been detected. Analytical techniques show that the naphthene derivatives contain from one to more than six fused rings accompanied by alkyl substitution. For mono- and di-aromatic derivatives, the alkyl substitution typically involves several methyl and ethyl substituents. Hopanes and steranes have also been identified and are also used as internal markers for estimating biodegradation of crude oils during bioremediation processes (Prince et al., 1994).

The aromatic constituents in vacuum gas oil may contain one to six fused aromatic rings that may bear additional naphthene rings and alkyl substituents in keeping with their boiling range. Mono- and di-aromatic derivatives account for approximately 50% v/v of the aromatic derivatives in crude oil vacuum gas oil samples. Analytical data show the presence of up to four fused naphthenic rings on some aromatic compounds. This is consistent with the suggestion that these species originate from the aromatization of steroids. Although present at lower concentration, alkyl benzenes and naphthalene derivatives show one long side chain and multiple short side chains.

The fused ring aromatic compounds (having three or more rings) in crude oil include phenanthrene, chrysene, and picene as well as fluoranthene, pyrene, benzo(a)pyrene, and benzo(ghi)perylene. The most abundant reported individual phenanthrene compounds appear to be the 3-derivatives. In addition, phenanthrene derivatives outnumber anthracene derivatives by as much as 100:1. In addition, chrysene derivatives are favored over pyrene derivative.

Heterocyclic constituents are significant contributors to the vacuum gas oil fraction. With respect to sulfur-containing compounds, thiophene and thiacyclane sulfur predominate over sulfide-type sulfur. Some constituents may even contain more than one sulfur atom. The benzothiophene derivatives and dibenzothiophene derivatives are the prevalent thiophene forms of sulfur. In the vacuum gas oil range, the nitrogen-containing compounds include higher molecular weight pyridine derivatives, quinoline derivatives, benzoquinoline derivatives, amide derivatives, indole derivatives, carbazole derivative, and molecules with two nitrogen atoms (diaza-compounds) with three and four aromatic rings are especially prevalent (Green et al., 1989). Typically, approximately one-third of the compounds are basic, i.e., pyridine and its benzo-derivatives, while the remainder are present as neutral species (amide derivatives and carbazole derivatives). Although benzo- and dibenzo-quinoline derivatives found in crude oil are rich in sterically hindered structures, hindered and unhindered structures have been found to be present at equivalent concentrations in source rocks. This has been rationalized as geo-chromatography in which the less polar (hindered) structures moved more readily to the reservoir and are not adsorbed on any intervening rocks structures.

Oxygen levels in the vacuum gas oil parallel the nitrogen content. Thus, the most commonly identified oxygen compounds are the carboxylic acids and phenols, collectively called naphthenic acids (Seifert and Teeter, 1970).

Lubricating oil – a valuable product that falls within the boiling range of gas oil – has a high boiling point (>400°C, >750°F), boiling point, and high viscosity. The gas oil fraction that is considered suitable for the production of lubricating oil is comprised principally of hydrocarbon derivatives containing from 25 to 35 or even 40 carbon atoms per molecule, whereas residual stocks may contain hydrocarbon derivatives with 50 or more carbon atoms per molecule (in fact, as many as 80 or more carbon atoms per molecule). The composition of lubricating oil may be substantially different from the lubricant fraction from which it was derived since wax (normal paraffin derivatives) is removed by distillation or refining by solvent extraction and adsorption preferentially removes non-hydrocarbon constituents as well as polynuclear aromatic compounds and the multi-ring cycloparaffin derivatives.

When distillate production is to be maximized, the amount of gas oil allowed to remain in the nonvolatile distillation residue bottoms stream must be minimized.

2.2.3.3 Properties and Uses

The gas oil fractions contain mostly high-molecular-weight hydrocarbon derivatives and can be employed to produce lubricating oil and other high-value oils for a variety of uses. In many cases, the gas oil fractions are used as feedstocks for catalytic cracking processes and for hydrocracking processes to produce a variety of products. After hydrotreatment, gas oil can be blended and/or sold as fuel oil.

2.2.4 WAX

The use of paraffin wax in a historical sense is varied but for the purpose of this chapter can be taken to the 18th century (Burke, 1996).

At that time, documents were written or drawn on damp paper with special ink that included gum Arabic, which stayed moist for 24 hours, during which copies could be made by pressing another smooth white sheet against the original and transferring the ink marks to the new sheet. Initially, the copier was not a success. Banks were opposed because they thought it would encourage forgery. Counting houses argued that it would be inconvenient when they were rushed, or working by candlelight. But by the end of the first year, Watt had sold 200 examples and had made a great impression with a demonstration at the houses of Parliament, causing such a stir that members had to be reminded they were in session. By 1785, the copier was in common use.

Then in 1823 Cyrus P. Dalkin of Concord, Massachusetts, improved on the technique by using two different materials whose effect on history was to be startling. By rolling a mixture of carbon black and hot paraffin wax onto the back of a sheet of paper, Dalkin invented carbon copies. The development lay relatively unnoticed until the 1868 balloon ascent by Lebbeus H. Rogers, the 21-year-old partner in a biscuit-and-greengrocery firm. His aerial event was being covered by the Associated Press and in the local newspaper office after the flight. Rogers was interviewed by a reporter who was using the carbon paper developed by Dalkin. Impressed by what he saw, Rogers terminated his ballooning and biscuits efforts and started a business producing carbon paper for use in order books, receipt books, invoices, etc. In 1873, he conducted a demonstration for the Remington typewriter company, and the new carbon paper became an instant success.

The paraffin wax used, and which was therefore half-responsible (together with carbon black) for changing the world of business, had originally been produced from oil shale rocks. After the discovery of crude oil in Pennsylvania, in 1857 (Chapter 1), paraffin oil was produced by distillation and was used primarily as an illuminant to make up for the dwindling supply of sperm-whale oil in a rapidly growing lamp market. Chilled-down paraffin solidified into paraffin wax. Apart from its use in lighting, the wax was also used to preserve the crumbling Cleopatra's Needle obelisk in New York's Central Park.

Crude oil wax is of two general types: (1) paraffin wax in crude oil distillates and (2) microcrystalline wax in crude oil residua. However, the melting point of crude oil-derived wax is not directly related to the boiling point because wax contains hydrocarbon derivatives of different chemical structure. Nevertheless, wax is graded according to their melting point and oil content.

Microcrystalline waxes form approximately 1% to 2% w/w of crude oil and are a valuable product having numerous applications. These waxes are usually obtained from high-boiling (high-density) lube distillates by solvent dewaxing and from tank bottom sludge by acid clay treatment (Agrawal et al., 1997). However, these crude waxes usually contain appreciable quantity (10%–20% w/w) of residual oil and, as such, are not suitable for many applications such as paper coating, electrical insulation, textile printing, and polishes.

2.2.4.1 Manufacture

Paraffin wax from a solvent dewaxing operation is commonly known as slack wax, and the processes employed for the production of waxes are aimed at de-oiling the slack wax (crude oil wax concentrate) (Parkash, 2003; Gary et al., 2007; Speight, 2014; Hsu and Robinson, 2017; Speight, 2017, 2019, 2021).

Wax sweating was originally used in Scotland to separate wax fractions with various melting points from the wax obtained from shale oil. The wax sweating process is still used to some extent but is being replaced by the more convenient was recrystallization process. In wax sweating, a cake of slack wax is slowly warmed to a temperature at which the oil in the wax and the lower melting waxes become fluid and drip (or sweat) from the bottom of the cake, leaving a residue of higher melting wax. However, wax sweating can be carried out only when the residual wax consists of large crystals that have spaces between them, through which the oil and lower melting waxes can percolate; it is therefore limited to wax obtained from low-boiling (low density) paraffin distillate.

The amount of oil separated by sweating is now much smaller than it used to be owing to the development of highly efficient solvent dewaxing techniques (Chapter 20). In fact, wax sweating is now more concerned with the separation of slack wax into fractions with different melting points. A wax sweater consists of a series of about nine shallow pans arranged one above the other in a sweater house or oven, and each pan is divided horizontally by a wire screen. The pan is filled to the level of the screen with cold water. Molten wax is then introduced and allowed to solidify, and the water is then drained from the pan leaving the wax cake supported on the screen.

A single sweater oven may contain more than 600 barrels of wax, and steam coils arranged on the walls of the oven slowly heat the wax cakes, allowing oil and the lower melting waxes to sweat from the cakes and drip into the pans. The first liquid removed from the pans is called foots oil, which melts at 38°C (100°F) or lower, followed by interfoots oil, which melts in the range of 38°C–44°C (100°F–112°F). Crude scale wax next drips from the wax cake and consists of wax fractions with melting points over 44°C (112°F).

When oil removal was an important function of sweating, the sweating operation was continued until the residual wax cake on the screen was free of oil. When the melting point of the wax on the screen has increased to the required level, allowing the oven to cool terminates sweating. The wax on the screen is a sweated wax with the melting point of a commercial grade of paraffin wax, which after a finished treatment becomes refined paraffinic wax. The crude scale wax obtained in the sweating operation may be recovered as such or treated to improve the color, in which case it is white crude scale wax. The crude scale wax and interfoots, however, are the sources of more waxes with lower melting points. The crude scale wax and interfoots are re-sweated several times to yield sweated waxes, which are treated to produce a series of refined paraffin waxes with melting points ranging from about 50°C to 65°C (125°F to 150°F).

Sweated waxes generally contain small amounts of unsaturated aromatic and sulfur compounds, which are the source of unwanted color, odor, and taste that reduce the ability of the wax to resist oxidation; the commonly used method of removing these impurities is clay treatment of the molten wax.

The wax recrystallization process, like wax sweating, is employed to separate slack wax into fractions, but instead of using the differences in melting points, it makes use of the different solubility of the wax fractions in a solvent, such as the ketone used in the dewaxing process (Speight, 2014). When a mixture of ketone and slack wax is heated, the slack wax usually dissolves completely, and if the solution is cooled slowly, a temperature is reached at which a crop of wax crystals is formed. These crystals will all be of the same melting point, and if they are removed by filtration, a wax fraction with a specific melting point is obtained. If the clear filtrate is further cooled, a second crop of wax crystals with a lower melting point is obtained. Thus by alternate cooling and filtration, the slack wax can be subdivided into a number of fractions, each with a different melting point.

This method of producing wax fractions is much faster and more convenient than sweating and results in a much more complete separation of the various fractions. Furthermore, recrystallization can also be applied to the microcrystalline waxes obtained from intermediate and high-boiling (high-density) paraffin distillates, which cannot be sweated. Indeed, the microcrystalline waxes have higher melting points and differ in their properties from the paraffin waxes obtained from low-boiling (low-density) paraffin distillates; thus wax recrystallization makes new kinds of waxes available.

The physical properties of microcrystalline wax is affected significantly by the oil content (Kumar et al., 2007), and by achieving the desired level of oil content, wax of the desired physical properties and specifications can be produced. Deep de-oiling of microcrystalline wax is comparatively difficult compared to de-oiling paraffin because the oil remains occluded in the microcrystalline wax and is difficult to separate by sweating. Also since wax and residual oil have similar boiling ranges, separation by distillation is difficult. However, these waxes can be de-oiled by treatment with solvents at lower temperature that have high oil miscibility and poor wax solubility, and these have been used extensively to separate.

These conventional solvent de-oiling processes to upgrade the quality of high oil content microcrystalline waxes involve agitation followed by extractive crystallization at low temperatures with solvents such as methyl iso-butyl ketone, methyl ethyl ketone-toluene mixtures, and dichloroethane – typically using drum filters.

2.2.4.2 Composition

Paraffin wax is a solid crystalline mixture of straight-chain (normal) hydrocarbon derivatives ranging from C_{20} to C_{30} and possibly higher, that is, $CH_3 (CH_2)_n CH_3$ where $n \geq 18$. This type of wax is distinguished by its solid state at ordinary temperatures (25°C, 77°F) and low viscosity (35–45 SUS at 99°C, 210°F) when melted. However, in contrast to crude oil wax, crude oil jelly (also known as petrolatum), although typically solid at ambient temperature, does in fact contain both solid and liquid hydrocarbon derivatives and is, in fact, essentially a low-melting, ductile, microcrystalline wax.

2.2.4.3 Properties and Uses

The physical properties of microcrystalline waxes are greatly affected by the oil content (Kumar et al., 2007) and hence by achieving the desired level of oil content waxes of desired physical properties can be obtained.

The melting point of paraffin wax has both direct and indirect significance in most wax utilization. All wax grades are commercially indicated in a range of melting temperatures rather than at a single value, and a range of 1°C (2°F) usually indicates a good degree of refinement. Other common physical properties that help to illustrate the degree of refinement of the wax are color (ASTM D-156), oil content, API gravity, flash point (ASTM D-92), and viscosity, although the last three properties are not usually given by the producer unless specifically requested.

Crude oil waxes (and petrolatum) find many uses in pharmaceuticals, cosmetics, paper manufacturing, candle making, electrical goods, rubber compounding, textiles, and many more too numerous to mention here – for additional information, more specific texts on crude oil waxes should be consulted.

2.2.5 BIOMASS AND BIO-OIL

Biomass is a term used to describe any material of recent biological origin, including plant materials such as trees, grasses, agricultural crops, and even animal manure. Thus, biomass (also referred to as bio-feedstock) refers to living and recently dead biological material which can be used as fuel or for industrial production of chemicals (Lee, 1996; Wright et al., 2006; Lorenzini et al., 2010; Nersesian, 2010; Speight, 2020b).

In terms of the chemical composition, biomass is a mixture of complex organic compounds that containing, for the most part carbon, hydrogen, and oxygen, with small amounts of nitrogen and sulfur, as well as with traces of other elements including metals. In the most cases, the biomass composition is approximately carbon 47%–53% w/w, hydrogen 5.9%–6.1% w/w, and oxygen 41%–45% w/w. The presence of a large amount of oxygen in biomass makes a significant difference with fossil-derived hydrocarbon derivatives. When used as fuel this is less efficient, but more is proving to be more suited for producing higher value bio-products, which contain functional entities within the constituent molecules and a variety of petrochemical products (Parkash, 2003; Gary et al., 2007; Speight, 2014; Hsu and Robinson, 2017; Speight, 2017, 2019).

The biomass used as industrial feedstock can be supplied by agriculture, forestry, and aquaculture, as well as resulting from various waste materials. The biomass can be classified as follows: (1) agricultural feedstocks, such as sugarcane, sugar-beet, and cassava; (2) starch feedstocks, such as wheat, maize, and potatoes; (3) oil feedstocks, such as rapeseed and soy; (4) dedicated energy crops, such as short rotation coppice which includes poplar, willow, and eucalyptus; (5) high-yield perennial grass, such as miscanthus and switchgrass; (6) non-edible oil plants, such as jatropha, camellia, sorghum; and (7) lignocellulosic waste material, which includes forestry wood, straw, corn stover, bagasse, paper pulp, and algal crops from land farming.

The utilization of biomass through the adoption of the conventional crude oil refinery systems and infrastructure to produce substitutes for fuels and other chemicals currently derived from conventional fuels (coal, oil, natural gas) is one of the most favored methods to combat fossil fuel depletion as the 21st century matures. In a biorefinery, a solid biomass feedstock is converted through either a thermochemical process (such as gasification, pyrolysis) or a biochemical process (such as hydrolysis, fermentation) into a mixture of organic (such as hydrocarbon derivatives, alcohol derivatives, and ester derivatives) and inorganic compounds (such as carbon monoxide and hydrogen) that can be upgraded through catalytic reactions to high-value fuels or chemicals (Speight, 2014, 2017, 2019, 2020a).

Chemically, biomass is carbonaceous feedstock that is composed of a variety of organic constituents that contain carbon, hydrogen, oxygen, often nitrogen, and also small quantities of other atoms, including alkali metals, alkaline earth metals, and heavy metals. The alkali metals consist of the chemical elements lithium (Li), sodium (Na), potassium (K), rubidium (Rb), cesium (Cs), and francium (Fr). Together with hydrogen, they make up Group I of the Periodic Table (Figure 2.2).

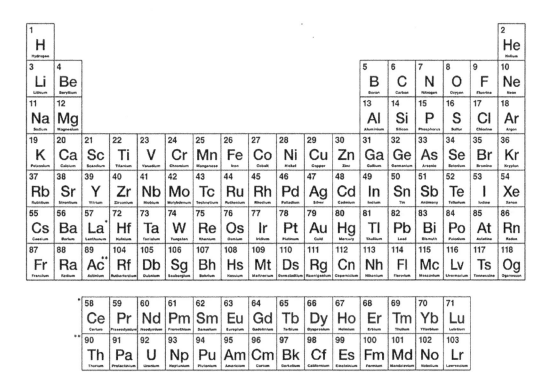

FIGURE 2.2 Periodic table of the elements. *Lanthanide series of elements. **Actinide series of elements.

On the other hand, the alkaline earth metals are the six chemical elements in Group 2 of the Periodic Table and are beryllium (be), magnesium (Mg), calcium (Ca), strontium (Sr), barium (Ba), and radium (Ra) (Figure 2.2). These elements have very similar properties – they are shiny, silvery-white, and somewhat reactive at standard temperature and pressure. Finally, the heavy metals are less easy to define but are generally recognized as metals with relatively high density, atomic weight, or atomic number. The common transition metals such as copper (Cu), lead (Pb), and zinc (Zn) are often classed as heavy metals, but the criteria used for the definition and whether metalloids (types of chemical elements which have properties in between, or that are a mixture of, those of metals and nonmetals) are included vary depending on the context. These metals are often found in functional molecules such as the porphyrin molecules which include chlorophyll and which contains magnesium.

Biomass feedstocks and fuels exhibit a wide range of physical, chemical, and agricultural/process engineering properties and is subdivided into three different grades (or types) and the feedstock origin determines the so-called biomass generation. In some cases, the third-generation biomass may also include high-yield algal crops which can be fed directly with concentrated carbon dioxide streams resulting from industrial processes, as from coal power plants and from fermentation of sugars – algal cultures can produce various hydrocarbon derivatives as well as various volatile olefin derivatives (Dimian, 2015).

In addition to using biomass for energy for power and heat generation by means of co-firing and gasification, woody crops with their high hemicellulose and cellulose content are well suited for biorefining to yield liquid fuels such as methanol, ethanol, and distillable oil (sometimes referred to as pyrolysis oil or bio-oil), as well as other products, such as specialty chemicals (Ben et al., 2019; Speight, 2019).

The thermal decomposition of biomass (typically referred to as pyrolysis) gives usually rise to three phases: (1) gases, (2) condensable liquids, and (3) char/coke. However, there are various types of related kinetic pathways ranging from very simple paths to more complex paths, and all usually include several elementary processes occurring in series or in competition. As anticipated, the kinetic paths are different for cellulose, lignin, and hemicelluloses (biomass main basic components) and also for usual biomasses according to their origin, composition, and inorganic contents (Speight, 2020b).

Thus, the thermal decomposition of biomass offers a flexible and attractive way of converting solid biomass into an easily stored and transported liquid (bio-oil), which can be successfully used for the production of chemicals in any one of several crude oil refinery-type scenarios. The pyrolysis processes can be categorized as (1) slow pyrolysis, which can take several hours to complete and results in biochar as the main products, or (2) fast pyrolysis, which is completed in a matter of seconds and is currently the most widely used pyrolysis system and can produce a high of bio-oil – on the order of 60% w/w bio-oil along with synthesis gas (20% w/w) and biochar (20% w/w) (Table 2.2).

Fast pyrolysis processes include ablative fast pyrolysis, cyclonic fast pyrolysis, open-core fixed bed pyrolysis, and rotating core fast pyrolysis systems. The essential features of a fast pyrolysis process are (1) a high heating, which require a finely ground feedstock; (2) a high rate of heat transfer

TABLE 2.2

Biomass Products by Thermal Decomposition

Biomass Pyrolysis
(550°C, 1020°F, No Air)

Gases	H_2, CO, CO_2, C_nH_{2n+2}
Liquids	C_nH_{2n+2}
Char	C_n

rates, which also requires a finely ground feedstock; (3) a carefully controlled reaction temperature on the order of 500°C/930°F; (4) a residence time of the pyrolysis vapors in the reactor of less than 1 second; and (5) quenching – i.e., rapid cooling – of the pyrolysis vapors to condense the give the bio-oil.

Bio-oil (sometimes known as pyrolysis oil or bio-crude) is a synthetic oil that is obtained by the pyrolysis of dried biomass in the absence of oxygen at a temperature on the order of about 500°C (900°F) with subsequent cooling. The oil may appear as a light-to-dark brown liquid or a fluid type of tar that typically contains levels of oxygen too high to consider the oil as a pure hydrocarbon product. The high oxygen content results in nonvolatility, corrosiveness, immiscibility with fossil fuels, thermal instability, and a tendency to polymerize (with an accompanying increase in viscosity) when exposed to air and allowed to oxidize.

Bio-oil is a promising alternative refinery feedstocks which can be produced from various types of biomass. Typically, bio-oil produced from by pyrolysis of biomass is a complex mixture containing significant quantities of nitrogen and oxygen with high total acid number (TAN). Thus, further treatment, such as denitrogenation and deoxygenation, is required before further use.

2.2.5.1 Manufacture

Bio-oil is produced by pyrolysis (fast pyrolysis, flash pyrolysis) of biomass, and the process occurs when solid fuels are heated at temperatures between 350°C and 500°C (570°F and 930°F) for a very short period of time (<2 seconds) and the temperatures are attainable within an active refinery (Table 2.3).

In another process, the feedstock is fed into a fluidized bed (at 450°C–500°C, 840°F–930°F) and the feedstock flashes and vaporizes. The resulting vapors pass into a cyclone where solid particles, char, are extracted. The gas from the cyclone enters a quench tower where they are quickly cooled by heat transfer using bio-oil already made in the process. The bio-oil condenses into a product

---

TABLE 2.3

Examples of the Stages That Occur During Pyrolysis

Temperature	Event
<100°C (<212°F)	Volatiles, including some water, evaporate.
	Heat-sensitive substances may partially decompose.
100°C (212°F)	Water remaining absorbed in the material is evolved.
	Water trapped in crystal structure of hydrates may be evolved at somewhat higher temperatures.
	Some solid substances, like fats, sugars, and waxes, may melt.
100°C–500°C (212°F–930°F)	Organic molecules break down.
	Most sugar derivatives start decomposing at 160°C–180°C.
	Cellulose decomposes at approximately 350°C.
	Lignin commences the starts decomposing process at approximately 350°C and continues releasing volatile products up to 500°C. The decomposition products usually include water, carbon monoxide, and/or carbon dioxide as well as a large number of organic compounds.
	The nonvolatile residues typically become richer in carbon and form large, disordered molecules, with colors ranging between brown and black.
200°C–300°C (390°F–570°F)	When oxygen has not been excluded, the carbonaceous residue may start to burn in a highly exothermic reaction releasing carbon dioxide and/or monoxide.
	At this stage, some of the nitrogen still remaining in the residue may be oxidized into nitrogen oxides (such as NO_2 and N_2O).
	Sulfur and other elements (such as chlorine and arsenic) may be oxidized and volatilized.

receiver and any non-condensable gases are returned to the reactor to maintain process heating. The entire reaction from injection to quenching takes only 2 seconds.

In the hydrothermal liquefaction process, wet biomass is converted to bio-oil under a moderate temperature and high pressure (350°C, 660°F, and 3,000 psi). The process differs from the pyrolysis (in which the biomass must be dried first) process insofar as wet biomass can be processed and produce a bio-oil. The presence of water in pyrolysis drastically increases the heat of vaporization of the organic material, increasing the energy required to decompose the biomass. Typical pyrolysis processes require a water content of less than 40% w/w to suitably convert the biomass to bio-oil. This requires considerable pretreatment of wet biomass such as tropical grasses, which can contain a water content as high as 80%–85% w/w. The properties of the resulting bio-oil are affected by temperature, reaction time, biomass type, reaction atmosphere, and catalysts.

Processing bio-oil requires larger efforts for commercial development because of the overall poor quality of the bio-oils, and the conventional hydrotreating catalysts are expected to have a considerably lower catalyst life in bio-oil upgrading operations than that observed with a crude oil-based feedstock. While the current generation commercial catalysts are excellent hydroprocessing catalysts, they are optimized for crude oil-based feedstock, and since the bio-oils have significantly different properties than crude oil feedstock, it would be worthwhile to dedicate efforts to developing catalysts specifically designed for upgrading bio-oil.

2.2.5.2 Composition

Bio-oil – depending on the source – can be divided into three categories which are (1) the liquid products derived from the destructive distillation of biomass, which is called wood tar that is similar to coal tar, and consists of polycyclic aromatic hydrocarbon and asphaltene; (2) the oil derived from rapid pyrolysis or liquidation whose major component is liquid fuel which, when compared with conventional crude oil, has some disadvantages, such as unsteady physical and chemical properties, component changes with time and temperature, high viscosity, and high oxygen and water content; and (3) the tar produced from the gasification of biomass which is an inevitable by-product that will affect the operation of systems once condensed and coked.

The oxygen content of bio-oil is typically on the order of 35%–40% w/w and is contained in oxygenated organic compounds making up bio-oil which typically contains a high proportion of organic acid derivatives, such as acetic acid (CH_3CO_2H) and formic acid (HCO_2H), which lead to an acidic liquid with a typical pH value of 2–3. Due to this inherent acidity, bio-oil is corrosive, which requires specific construction materials being used for storage vessels or subsequent upgrading processes.

The viscosity of bio-oil can vary greatly depending on (1) the feedstock used to produce the oil, (2) the parameters of the pyrolysis process, (3) content of low-boiling constituents, (4) the temperature, and (5) the time of storage time. When bio-oil is stored, there is an aging process which leads to an increase in viscosity that is caused by condensation reactions that occur between the constituents of the oil.

The majority of mineral matter (inorganic matter) contained in biomass is concentrated in char, but small amounts of fine char can be entrained in bio-oil. Also, alkali metals derivatives within the ash can lead to cracking reactions within the bio-oil.

2.2.5.3 Properties and Uses

Compounds that occur in bio-oil fall into the following five broad categories (Piskorz et al., 1988): (1) hydroxy-aldehyde derivatives, (2) hydroxy-ketone derivatives, (3) sugars and dehydrosugar derivatives, (4) carboxylic acid derivatives, and (5) phenolic derivatives.

The composition and physical and chemical properties of bio-oil are affected by many factors, such as (1) the source of the raw material sources, (2) the water content, (3) the reactor type, (4) the reaction parameters, and (5) the production collection methods. However, bio-oils derived from different routes still have commonalities, such as high oxygen content (35%–40% w/w), high corrosive ability (pH 2–3), poor thermal stability (the tendency for polymerization of the constituents at temperatures in excess of 80°C, >176°F), and a relatively high viscosity compared to similar hydrocarbon liquids. The quality of bio-oil can be improved through catalytic reforming, which is similar to that of crude oil. Bio-oil can be gasified or converted into manufactured gas.

2.3 VISCOUS FEEDSTOCKS

While the terminology used for the various refinery feedstocks has been described in detail elsewhere (Chapter 1) (Speight, 2021), for the sake of completeness and to mitigate the potential for any confusion or misunderstanding, a short description of the various viscous feedstocks is worthy of inclusion at this point.

The definition of crude oil has been varied, unsystematic, diverse, and often archaic (Speight, 2014, 2017, 2021) because the terminology used to describe crude oil is subject to many conventions of which a major ones are chemistry-aligned definitions, company-aligned definitions, engineering-aligned definitions, and definitions aligned with the geological sciences. Thus the long-established use of an expression, however inadequate it may be, is altered with difficulty, and a new term, however precise, is at best adopted only slowly.

Of the many forms of terminology that have been used not all have survived, and the more commonly used are illustrated here but, as a note, particularly troublesome, and more confusing, are those terms that are applied to the more viscous materials, for example, the use of the terms bitumen and asphalt. This part of the text attempts to alleviate much of the confusion that exists, but it must be remembered that the terminology of crude oil is still open to personal choice and historical usage.

By way of introduction, when crude oil occurs in a reservoir that allows the crude material to be recovered by pumping operations as a free-flowing dark to light-colored liquid, it is often referred to as conventional crude oil. Heavy crude oils are categorized as types of crude oil that are different from conventional crude oil insofar as they are much more difficult to recover from the subsurface reservoir. Arbitrarily, the definition of heavy crude oil is often based on the API gravity or viscosity, although there have been arbitrary attempts to rationalize the definition based upon viscosity, API gravity, and density when it is preferable to use the method of recovery of the material (Speight, 2014, 2017, 2021).

Viscous feedstocks are refinery feedstocks such as (1) crude oil residua, (2) heavy crude oil, (3) extra heavy crude oil, (4) tar sand bitumen, and (5) distillation residua that have a low volatility, high density, and high viscosity. Although other feedstocks (Section 2.2.5) such as naphtha, kerosene, and gas oil are also used as feedstocks for thermal processes and catalytic processes in the refinery, the focus of this book remains on the viscous feedstocks.

2.3.1 HEAVY CRUDE OIL

Heavy crude oil is typically categorized as a type of crude oil but is different to the other types of crude oil insofar as heavy crude oil differs conventional crude oil insofar as the heavy crude oil is more difficult to recover from the subsurface reservoir. The heavy crude oil has a higher viscosity (and lower API gravity) than conventional crude oil, and recovery of heavy crude oil usually requires thermal stimulation of the reservoir leading to application of various thermal methods (such as coking) in the refinery for suitable conversion to low-boiling distillates.

However, heavy crude oil is more difficult to recover from the subsurface reservoir than conventional crude oil (often referred to as light oil, i.e., low boiling (low density). There has been the continuing tendency to define heavy crude oil using the API gravity or viscosity, and the definition is quite arbitrary although there have been attempts to rationalize the definition based upon viscosity, API gravity, and density. For example, heavy crude oil is considered to be the type of crude oils that had gravity somewhat less than 20° API with many heavy crude oils falling into the API gravity range 10°–15°. For example, Cold Lake heavy crude oil (Alberta, Canada) has an API gravity equal on the order of 12°, and tar sand bitumen (Athabasca, Fort McMurray, Alberta, Canada) usually has an API gravity on the order of 5°–10° (Athabasca bitumen=8° API). Using this scale, crude oil residua vary depending upon the temperature at which distillation was terminated, but atmospheric residua are typically in the range of 10°–15° API, while vacuum residua are in the range of 2°–8° API (Parkash, 2003; Gary et al., 2007; Speight, 2014; Hsu and Robinson, 2017; Speight, 2017, 2021).

In the simplest terms, heavy crude oil is a type of crude oil which is very viscous and does not flow easily. The common characteristic properties (relative to conventional crude oil) are high specific gravity, low hydrogen-to-carbon ratios, high carbon residues, and high contents of asphaltenes, heavy metals (i.e., metallic chemical elements that have a relatively high density and are toxic or poisonous at low concentrations; examples of heavy metals include (alphabetically): arsenic (As), cadmium (Cd), chromium (Cr), mercury (Hg), lead (Pb), and thallium (Tl)), sulfur, and nitrogen. Specialized refining processes are required to produce more useful fractions, such as naphtha, kerosene, gas oil, and wax.

Thus, the term heavy crude oil has also been arbitrarily used to describe the viscous crude oil that requires a thermal stimulation method to recover the oil from the reservoir.

2.3.2 Extra Heavy Crude Oil

Extra heavy oil is a non-descript term (related to viscosity) of little scientific meaning, which is usually applied to tar sand bitumen, which is generally incapable of free flow under reservoir conditions. Thus, the term extra heavy oil is used to describe materials that occur in the solid or near-solid state in the deposit or reservoir and are generally incapable of free flow under ambient conditions. Whether or not such a material exists in the near-solid or solid state in the reservoir can be determined from the pour point and the reservoir temperature.

The general difference is that extra heavy oil, which may have properties similar to tar sand bitumen in the laboratory but, unlike tar sand bitumen in the deposit, has some degree of mobility in the reservoir or deposit (Table 2.4) (Delbianco and Montanari, 2009; Speight, 2014). Extra heavy oils can flow at reservoir temperature and can be produced economically, without additional viscosity-reduction techniques, through variants of conventional processes such as long horizontal wells or multilaterals.

This is the case, for instance, in the Orinoco basin (Venezuela) or in offshore reservoirs of the coast of Brazil but, once outside of the influence of the high reservoir temperature, these oils are too viscous at surface to be transported through conventional pipelines and require heated pipelines for transportation. Alternatively, the oil must be partially upgraded or fully upgraded or diluted with a low-boiling (low-density) hydrocarbon (such as aromatic naphtha) to create a mix that is suitable for transportation (Speight, 2014).

Thus, extra heavy oil is a material that occurs in the solid or near-solid state and is generally has mobility under reservoir conditions. While this type of oil may resemble tar sand bitumen and does not flow easily, extra heavy oil is generally recognized as having mobility in the reservoir compared to tar sand bitumen, which is typically incapable of mobility (free flow) under reservoir conditions. For example, the tar sand bitumen located in Alberta, Canada, is not mobile in the deposit and requires extreme methods of recovery to recover the bitumen. On the other hand, much of the extra heavy oil located in the Orinoco basin of Venezuela requires recovery methods that are less extreme because of the mobility of the material in the reservoir. Whether the mobility of extra heavy oil is due to a high reservoir temperature (that is, higher than the pour point of the extra heavy oil) or due to other factors is variable and subject to localized conditions in the reservoir. This may also be reflected in the choice of suitable extra heavy oil or bitumen conversion processes in the refinery.

In the context of this book, the methods outlined in this book for refining viscous feedstocks focus on heavy oil with an API gravity of less than 20 with a variable sulfur content (which includes atmospheric residua and vacuum residua) (Parkash, 2003; Gary et al., 2007; Speight, 2014; Hsu and Robinson, 2006; Speight, 2017). However, it must be recognized that some of the heavy oil sufficiently liquid to be recovered by pumping operations and are already being recovered by this

TABLE 2.4

Simplified Differentiation between Conventional Crude Oil, Tight Oil, Heavy Crude Oil, Extra Heavy Crude Oil, and Tar Sand Bitumen[a]

Conventional Crude Oil

Mobile in the reservoir; API gravity: >25°
Primary recovery methods
Secondary recovery methods

Heavy Crude Oil

More viscous than conventional crude oil; API gravity: 10°–20°
Mobile in the reservoir
Secondary recovery methods
Tertiary recovery (enhanced oil recovery – EOR, e.g., steam stimulation)

Extra Heavy Crude Oil

Similar properties to the properties of tar sand bitumen; API gravity: <10°
Temperature of the reservoir/deposit higher than pout point of the oil
Mobile in the reservoir
Secondary recovery methods
Tertiary recovery methods (enhanced oil recovery – EOR, e.g., steam stimulation)

Tar Sand Bitumen

Immobile in the deposit; API gravity: <10°
Temperature of the reservoir/deposit lower than pout point of the oil
Immobile in the deposit
Mining (often preceded by explosive fracturing)
Extreme heating methods
Innovative methods[b]

Residuum

Semi-mobile to immobile at ambient conditions.
Typically, API gravity: 5°–15°; sulfur: 2.0%–5.0% w/w; metals: 100–1,000 ppm
Does not occur naturally; manufactured product produced by refinery distillation
Atmospheric residuum: source of heavy gas oil under vacuum distillation
Vacuum residuum: nonvolatile under vacuum distillation conditions
Typically immobile under ambient conditions

[a] This list is not intended for use as a means of classification.

[b] Innovative methods exclude tertiary recovery methods and methods such as steam-assisted gravity drainage (SAGD) and vapor-assisted extraction (VAPEX) methods but does include variants or hybrids thereof (Speight, 2016).

method. Refining depends on the characteristics (properties) since these heavy oils fall into a range of high viscosity and the viscosity is subject to temperature effects (Speight, 2014, 2015), which is the reason for the application of thermal conditions or dilution with a suitable solvent (such as aromatic naphtha) to enable heavy oil to flow in pipeline or within the refinery system.

2.3.3 TAR SAND BITUMEN

In addition to heavy crude oil and the so-called extra heavy crude oil, there remains an even more viscous material that offers some relief to the potential shortfalls in supply (Meyer and De Witt,

1990; Speight, 2014, 2016). This is the bituminous material that occurs found in tar sand (oil sand) deposits (referred to as oil sand deposits in Canada). However, many of these reserves are only available with some difficulty, and optional refinery scenarios will be necessary for conversion of these materials to liquid products because of the substantial differences in character between conventional crude oil and tar sand bitumen (Speight, 2000, 2014, 2016). Tar sands, also variously called oil sands or bituminous sands, are a loose-to-consolidated sandstone or a porous carbonate rock, impregnated with bitumen, a heavy asphaltic crude oil with an extremely high viscosity under reservoir conditions.

The term tar sand bitumen (also, on occasion, incorrectly referred to as native asphalt) includes a wide variety of near-black to black materials that occur in the form of near solid to solid in the deposit either (1) with no mineral impurity or (2) with a mineral matter content that may exceed 50% w/w. Bitumen is frequently found filling pores and crevices of sandstone, limestone, or argillaceous sediments, in which case the organic and associated mineral matrix is known as rock asphalt. Even though the term tar sand is incorrectly since tar is a product of the thermal processing of coal (Speight, 2013a), the term has remained in use. Also, it is incorrect to refer to native bituminous materials as tar or pitch even though the word "tar" is descriptive of the black, nonvolatile bituminous material, it is best to avoid its use with respect to natural materials and to restrict the meaning to the volatile or near-volatile products produced in the destructive distillation of such organic substances as coal (Speight, 2013a). In the simplest sense, pitch is the distillation residue of the various types of tar. Thus, alternative names, such as bituminous sand or oil sand, are gradually finding usage, with the former name (bituminous sands) more technically correct. The term oil sand is also used in the same way as the term tar sand, and these terms are used interchangeably throughout this text.

The permeability of a tar sand deposit low and passage of fluids through the deposit can only be achieved by prior application of fracturing techniques. Alternatively, bitumen recovery can be achieved by conversion of the bitumen to a product in situ (in situ upgrading) followed by product recovery from the deposit (Speight, 2013b, 2013c, 2014, 2016). Tar sand bitumen is a high-boiling material with little, if any, material boiling below 350°C (660°F), and the boiling range approximates the boiling range of an atmospheric residuum.

There have been many attempts to define tar sand deposits and the bitumen contained therein. In order to define conventional crude oil, heavy oil, and bitumen, the use of a single physical parameter such as viscosity is not sufficient. Other properties such as API gravity, elemental analysis, composition, and, most of all, the properties of the bulk deposit must also be included in any definition of these materials. In fact, the most appropriate definition of tar sands is found in the writings of the United States government (Speight, 2014, 2016):

> Tar sands are the several rock types that contain an extremely viscous hydrocarbon which is not recoverable in its natural state by conventional oil well production methods including currently used enhanced recovery techniques. The hydrocarbon-bearing rocks are variously known as bitumen-rocks oil, impregnated rocks, oil sands, and rock asphalt.

This definition refers to the character of the bitumen through the method of recovery. Mining methods match the requirements of this definition (since mining is not one of the specified recovery methods), and the bitumen can be recovered by alteration of its natural state such as thermal conversion to a product that is then recovered. In this sense, changing the natural state (the chemical composition) as occurs during several thermal processes (such as some in situ combustion processes) also matches the requirements of the definition.

By inference and by omission, conventional crude oil and heavy crude oil are also included in this definition. Crude oil is the material that can be recovered by conventional oil well production methods whereas heavy oil is the material that can be recovered by enhanced recovery methods. Tar sand currently recovered by a mining process followed by separation of the bitumen by the hot-water process. The bitumen is then used to produce hydrocarbon derivatives by a conversion process.

The only commercial operations for the recovery of bitumen from tar sand and its subsequent conversion to liquid fuels exist in the Canadian Province of Alberta where Suncor (initially as Great Canadian Oil Sands before the name change) went on-stream in 1967 and Syncrude (a consortium of several companies) went on-steam in 1977. Thus, throughout this text, frequent reference is made to tar sand bitumen, but because commercial operations have been in place for approximately 50 years it is not surprising that more is known about the Alberta (Canada) tar sand reserves than any other reserves in the world. Therefore, when discussion is made of tar sand deposits, reference is made to the relevant deposit, but when the information is not available, the Alberta material is used for the purposes of the discussion.

Refining heavy crude oil, extra heavy crude oil, tar sand bitumen, and residua depends to a large degree on the composition of the feedstock (Parkash, 2003; Gary et al., 2007; Speight, 2014; Hsu and Robinson, 2006; Speight, 2017). Generally, the bitumen found in tar sand deposits is an extremely viscous material that is immobile and cannot be refined by the application of conventional refinery. On the other hand, extra heavy oil, which is often likened to tar sand bitumen because of similarities in the properties of the two, has a degree of mobility under reservoir or deposit conditions but typically suffers from the same drawbacks as the immobile bitumen in refinery operations.

Thus, alternative names, such as bituminous sand or oil sand, are gradually finding usage, with the former name (bituminous sands) more technically correct. The term oil sand is also used in the same way as the term tar sand, and these terms are used interchangeably throughout this text.

Bituminous rock and bituminous sand are those formations in which the bituminous material is found as a filling in veins and fissures in fractured rocks or impregnating relatively shallow sand, sandstone, and limestone strata. The deposits contain as much as 20% bituminous material, and if the organic material in the rock matrix is bitumen, it is usual (although chemically incorrect) to refer to the deposit as rock asphalt to distinguish it from bitumen that is relatively mineral free. A standard test method (ASTM D4) is available for determining the bitumen content of various mixtures with inorganic materials, although the use of word bitumen as applied in this test might be questioned and it might be more appropriate to use the term organic residues to include tar and pitch. If the material is the asphaltite-type or asphaltoid-type, the corresponding terms should be used: rock asphaltite or rock asphaltoid.

Bituminous rocks generally have a coarse, porous structure, with the bituminous material in the voids. A much more common situation is that in which the organic material is present as an inherent part of the rock composition insofar as it is a diagenetic residue of the organic material detritus that was deposited with the sediment. The organic components of such rocks are usually refractory and are only slightly affected by most organic solvents.

Tar sand deposits occur throughout the world, and the largest deposits occur in Alberta, Canada (the Athabasca, Wabasca, Cold Lake, and Peace River areas), and in Venezuela. Smaller deposits occur in the United States, with the larger individual deposits in Utah, California, New Mexico, and Kentucky. The term tar sand, also known as oil sand (in Canada), or bituminous sand, commonly describes sandstones or friable sand (quartz) impregnated with a viscous organic material known as bitumen (a hydrocarbonaceous material that is soluble in carbon disulfide). Significant amounts of fine material, usually largely or completely clay, are also present. The degree of porosity in tar sand varies from deposit to deposit and is an important characteristic in terms of recovery processes. The bitumen makes up the desirable fraction of the tar sands from which liquid fuels can be derived. However, the bitumen is usually not recoverable by conventional crude oil production techniques (Speight, 2013b, 2013c, 2014, 2016).

The properties and composition of the tar sand deposits and the bitumen contained therein significantly influence the selection of recovery and the bitumen conversion processes and vary among the bitumen from different deposits. In the so-called wet sands or water-wet sands of the Athabasca deposit, a layer of water surrounds the sand grain, and the bitumen partially fills the voids between the wet grains. Utah tar sands lack the water layer; the bitumen is directly in contact with the sand grains without any intervening water; such tar sands are sometimes referred to as oil-wet sands.

Typically, more than 99% w/w of mineral matter is composed of quartz and clays – the latter are detrimental to most refining processes. The general composition of typical deposits at the P.R. Spring Special Tar Sand Area showed a porosity of 8.4 vol% with the solid/liquid fraction being 90.5% w/w sand, 1.5% w/w fines, 7.5% w/w bitumen, and 0.5% w/w water. High concentrations of heteroatoms (nitrogen, oxygen, sulfur, and metals) tend to increase viscosity, increase the bonding of bitumen with minerals, reduce yields, and make processing more difficult.

2.3.4 RESIDUUM

In contract to heavy crude oil, extra heavy crude oil, and tar sand bitumen, a residuum (pl. residua, also shortened to resid, pl. resids) are manufactured materials insofar as a residuum is produced as a result of crude oil refining (Parkash, 2003; Gary et al., 2007; Speight, 2014; Hsu and Robinson, 2006; Speight, 2017). In many cases, residua are subject to further refining after production and are worthy of mention here as optional refinery viscous feedstocks either further refined as single feedstocks or as blends with other feedstocks, such as gas oil, as well as conventional whole crude oil, heavy crude oil, extra heavy crude oil, and tar sand bitumen.

Typically, a residuum is the residue obtained from crude oil after non-destructive distillation has removed all the volatile materials (Figure 2.3).

The temperature of the distillation is usually maintained below 350°C (660°F) since the rate of thermal decomposition of crude oil constituents is minimal below this temperature, but the rate of thermal decomposition of crude oil constituents is substantial above 350°C (660°F). If the temperature of the distillation unit rises above 350°C (660°F) as happens in certain units where

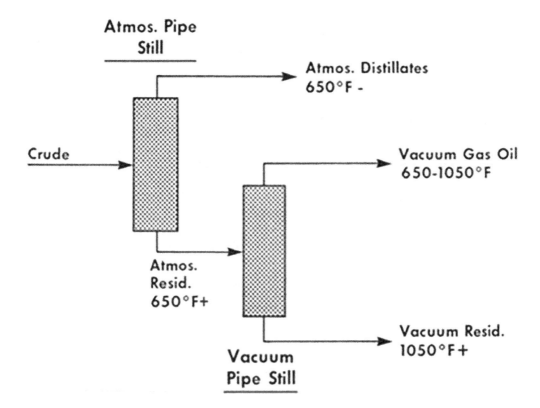

FIGURE 2.3 Crude oil distillation scheme. (Source: Speight, J.G. 2017. *Handbook of Petroleum Refining*. CRC Press, Taylor & Francis Group, Boca Raton, Florida. Figure 1.4, page 34.)

temperatures up to 395°C (740°F) are known to occur, cracking can be controlled by adjustment of the residence time.

Residua (pl: residuum, often shortened to "resid") are black, viscous materials that are obtained by distillation of a crude oil under atmospheric pressure (atmospheric residuum) or under reduced pressure (vacuum residuum). More typically, resids are black, viscous materials that may be near liquid at room temperature (typically, an atmospheric residuum) or solid (typically, a vacuum residuum) depending upon the properties of the crude oil source and are obtained by distillation of a crude oil under atmospheric pressure (atmospheric residuum) or under reduced pressure (vacuum residuum).

When a residuum is obtained from crude oil and thermal decomposition has commenced, it is more usual to refer to this product as cracked residuum (sometime referred to as pitch, a name derived from the coal industry). The differences between conventional crude oil and the related residua are due to the relative amounts of various constituents present, which are removed or retained by according to the relative volatility of the constituents.

The vacuum residuum (*vacuum bottoms*, typically 950°F⁺ or 1050°F⁺) is the most complex of crude oil and, in many cases, may even resemble heavy oil or extra heavy oil or tar sand bitumen in composition. Vacuum residua contain the majority of the heteroatoms originally in the crude oil, and molecular weight of the constituents range up to several thousand (as near as can be determined but subject to method dependence). The fraction is so complex that the characterization of individual species is virtually impossible, no matter what claims have been made or will be made. Separation of vacuum residua by group type can be difficult because of the multi-substitution of aromatic and naphthenic species as well as by the presence of multiple functionalities in single molecules.

Classically, n-pentane or n-heptane precipitation is used as the initial step for the characterization of vacuum residuum. The insoluble fraction, the pentane- or heptane-asphaltenes, may be as much as 50% by weight of a vacuum residuum. The pentane- or heptane-soluble portion (maltene) constituents of the residuum is then fractionated chromatographically into several solubility or adsorption classes for characterization. However, in spite of claims to the contrary, the method is not a separation by chemical type. Kit is a separation by solubility and adsorption. However, the separation of the asphaltene constituents does, however, provide a simple way to remove some of the highest molecular weight and most polar components, but the asphaltene fraction is so complex that compositional detail based on average parameters is of questionable value.

For the 565°C⁺ (1,050°F⁺) fractions of crude oil, the levels of nitrogen and oxygen may begin to approach the concentration of sulfur. These elements consistently concentrate in the most polar fractions to the extent that every molecule contains more than one heteroatom. At this point, structural identification is somewhat fruitless and characterization techniques are used to confirm the presence of the functionalities found in lower-boiling fractions such as, for example, acids, phenols, non-basic (carbazole-type) nitrogen, and basic (quinoline-type) nitrogen.

The nickel and vanadium that are concentrated into the vacuum residuum appear to occur in two forms: (1) porphyrins and (2) non-porphyrins (Reynolds, 1998). Because the metalloporphyrins can provide insights into crude oil maturation processes, they have been studied extensively and several families of related structures have been identified. On the other hand, the non-porphyrin metals remain not clearly identified although some studies suggest that some of the metals in these compounds still exist in a tetra-pyrrole (porphyrin-type) environment (Pearson and Green, 1993).

It is more than likely that, in a specific residuum molecule, the heteroatoms are arranged in different functionalities, making an incredibly complex molecule. Considering how many different combinations are possible, the chances of determining every structure in a residuum are very low. Because of this seemingly insurmountable task, it may be better to determine ways of utilizing the residuum rather attempting to determine (at best questionable) molecular structures.

In summary, the chemical composition of a residuum is complex. Physical methods of fractionation usually indicate high proportions of asphaltene constituents and resin constituents, even in amounts up to 50% (or higher) of the residuum (Parkash, 2003; Gary et al., 2007; Speight, 2014; Hsu and Robinson, 2006; Speight, 2017). In addition, the presence of ash-forming metallic constituents, including such organometallic compounds as those of vanadium and nickel, is also a distinguishing feature of residua and the heavier oils. Furthermore, the deeper the cut into the crude oil, the greater is the concentration of sulfur and metals in the residuum and the greater the deterioration in physical properties.

2.4 FEEDSTOCK COMPOSITION

Like crude oil in the natural state, heavy crude oil, extra heavy crude oil, and tar sand bitumen are not homogeneous materials and the physical characteristics differ depending on where the material was produced. This is due to the fact that any of these feedstocks from different geographical locations will naturally has unique properties. In its natural, unrefined state, crude oil, heavy oil, extra heavy oil, and tar sand bitumen range in density and consistency from very thin, lightweight, and volatile fluidity to an extremely thick, semisolid (Speight, 2014). There is also a tremendous gradation in the color that ranges from a light, golden yellow (conventional crude oil) to black Tar sand bitumen.

Thus, heavy crude oil, extra heavy crude oil, and tar sand bitumen are not (within the individual categories) uniform materials, and the chemical and physical (fractional) composition of each of these refinery feedstocks can vary not only with the location and age of the reservoir or deposit but also with the depth of the individual well within the reservoir or deposit. On a molecular basis, the three feedstocks are complex mixtures containing (depending upon the feedstock) hydrocarbon derivatives with varying amounts of hydrocarbonaceous constituents that contain sulfur, oxygen, and nitrogen as well as constituents containing metallic constituents, particularly those containing vanadium nickel, iron, and copper. The hydrocarbon content may be as high as 97% w/w, for example, in a conventional (light) crude oil or less than 50% w/w in heavy crude oil, extra heavy crude oil, and tar sand bitumen (Speight, 2014, 2017).

Heavy crude oil, extra heavy crude oil, tar sand bitumen, and residua (as a result of the concentration effect of distillation) contain more heteroatomic species and less hydrocarbon constituents than conventional crude oil. Thus, to obtain more gasoline and other liquid fuels, there have been different approaches to refining the heavier feedstocks as well as the recognition that knowledge of the constituents of these higher-boiling feedstocks is also of some importance. The problems encountered in processing the heavier feedstocks can be equated to the chemical character and the amount of complex, higher-boiling constituents in the feedstock. Refining these materials is not just a matter of applying know-how derived from refining conventional crude oils but requires knowledge of the chemical structure and chemical behavior of these more complex constituents.

2.5 ELEMENTAL COMPOSITION

The analysis of feedstocks for the percentages by weight of carbon, hydrogen, nitrogen, oxygen, and sulfur (elemental composition, ultimate composition) is perhaps the first method used to examine the general nature, and perform an evaluation, of a feedstock. The atomic ratios of the various elements to carbon (i.e., H/C, N/C, O/C, and S/C) are frequently used for indications of the overall character of the feedstock. It is also of value to determine the amounts of trace elements, such as vanadium and nickel, in a feedstock since these materials can have serious deleterious effects on catalyst performance during refining by catalytic processes.

For example, carbon content can be determined by the method designated for coal and coke (ASTM D3178) or by the method designated for municipal solid waste (ASTM E777). There are also methods designated for: (1) hydrogen content – ASTM D1018, ASTM D3178, ASTM D3343, ASTM D3701, and ASTM E777; (2) nitrogen content – ASTM D3179, ASTM D3228, ASTM E258, and ASTM E778; (3) oxygen content – ASTM E385; and (4) sulfur content – ASTM D1266, ASTM D1552, ASTM D1757, ASTM D2622, ASTM D3177, ASTM D4045, and ASTM D4294. For all feedstocks, the higher the atomic hydrogen-carbon ratio, the higher is its value as refinery feedstock because of the lower hydrogen requirements for upgrading. Similarly, the lower the heteroatom content, the lower the hydrogen requirements for upgrading. Thus, inspection of the elemental composition of feedstocks is an initial indication of the quality of the feedstock and, with the molecular weight, indicates the molar hydrogen requirements for upgrading (Parkash, 2003; Gary et al., 2007; Speight, 2014; Hsu and Robinson, 2017; Speight, 2017).

However, it has become apparent, with the introduction of the viscous feedstocks into refinery operations, that these ratios are not the only requirement for predicting feedstock character before refining. The use of more complex feedstocks (in terms of chemical composition) has added a new dimension to refining operations. Moreover, although atomic ratios, as determined by elemental analyses, may be used on a comparative basis between feedstocks, there is now no guarantee that a particular feedstock will behave as predicted from these data. Product slates cannot be predicted accurately, if at all, from these ratios. Additional knowledge such as defining the various chemical reactions of the constituents as well as the reactions of these constituents with each other also play a role in determining the processability of a feedstock.

The elemental analysis of heavy crude oil, extra heavy crude oil, and tar sand bitumen has also been widely reported (Speight, 1990, 2009), but the data suffer from the disadvantage that identification of the source is too general (i.e., Athabasca bitumen which covers several deposits) and is often not site specific. In addition, the analysis is quoted for separated bitumen, which may have been obtained by any one of several procedures and may therefore not be representative of the total bitumen on the sand. However, recent efforts have focused on a program to produce sound, reproducible data from samples for which the origin is carefully identified (Wallace et al., 1988).

The data that are available the elemental composition of oil sand bitumen is generally constant and, like the data for crude oil, falls into a narrow range:

Carbon: 83.4%–0.5%
Hydrogen: 10.4%–0.2%
Nitrogen: 0.4%–0.2%
Oxygen: 1.0%–0.2%
Sulfur: 5.0%–0.5%
Metals (Ni and V): >1,000 ppm

The major exception to these narrow limits is the oxygen content that can vary from as little as 0.2% to as high as 4.5%. This is not surprising, since when oxygen is estimated by difference the analysis is subject to the accumulation of all of the errors in the other elemental data. In addition, bitumen is susceptible to aerial oxygen, and the oxygen content is very dependent on sample history.

The viscosity of a viscous feedstock is related to the hydrogen-to-carbon atomic ratio and hence the required supplementary heat energy for thermal extraction processes. It also affects the bitumen's distillation curve or thermodynamic characteristics, its gravity, and its pour point. Atomic hydrogen-to-carbon ratios as low as 1.3 have been observed for tar sand bitumen although an atomic hydrogen-to-carbon ratio of 1.5 is more typical. The higher the hydrogen-carbon ratio of bitumen, the higher is its value as refinery feedstock because of the lower hydrogen requirements. Elements related to the hydrogen-carbon ratio are distillation curve, bitumen gravity, pour point, and bitumen viscosity.

The occurrence of sulfur in a viscous feedstock as organic or elemental sulfur or in produced gas as compounds of oxygen and hydrogen is an expensive nuisance. It must be removed from the bitumen at some point in the upgrading and refining process. Sulfur contents of some tar sand bitumen can exceed 10% w/w. Elements related to sulfur content are hydrogen content, hydrogen-carbon ratio, nitrogen content, distillation curve, and viscosity.

The nitrogen content of a viscous feedstock can be as high as 1.3% by weight and nitrogen-containing constituents complicate the refining process by poisoning the catalysts employed in the refining process. Elements related to nitrogen content are sulfur content, hydrogen content, hydrogen-carbon ratio, bitumen viscosity, distillation profile, and viscosity.

Furthermore, heteroatoms in viscous feedstocks affect every aspect of refining. The occurrence of *sulfur* in feedstocks as organic or elemental sulfur or in produced gas as compounds of oxygen (SO_x) and hydrogen (H_2S) is an expensive aspect of refining. It must be removed at some point in the upgrading and refining process. Sulfur contents of many crude oil are on the order of 1% by weight, whereas the sulfur content of tar sand bitumen can exceed 5% or even 10% by weight. Of all of the heteroelements, sulfur is usually the easiest to remove, and many commercial catalysts are available that routinely remove 90% of the sulfur from a feedstock (Speight, 2000).

Metals (particularly vanadium and nickel) are found in most viscous feedstocks which contain relatively high proportions of metals either in the form of salts or as organometallic constituents (such as the metallo-porphyrins), which are extremely difficult to remove from the feedstock. Indeed, the nature of the process by which residua are produced virtually dictates that all the metals in the original crude oil are concentrated in the residuum (Speight, 2014, 2017). The metallic constituents that may actually *volatilize* under the distillation conditions and appear in the higher-boiling distillates are the exceptions here.

Metal constituents of viscous feedstocks cause problems by poisoning the catalysts used for sulfur and nitrogen removal as well as the catalysts used in other processes such as catalytic cracking. Thus, serious attempts are being made to develop catalysts that can tolerate a high concentration of metals without serious loss of catalyst activity or catalyst life.

A variety of tests have been designated for the determination of metals refinery feedstocks in crude oil products (ASTM D1318, ASTM D3340, ASTM D3341, ASTM D3605). Determination of metals in whole feeds can be accomplished by combustion of the sample so that only inorganic ash remains. The ash can then be digested with an acid and the solution examined for metal species by atomic absorption (AA) spectroscopy or by inductively coupled argon plasma (ICP) spectrometry.

2.6   CHEMICAL COMPOSITION

Heavy crude oil, extra heavy crude oil, and tar sand bitumen contain an extreme range of organic functionality and molecular size. In fact, the variety is so great that it is unlikely that a complete compound-by-compound description for even a single crude oil would not be possible. As already noted, the composition of crude oil can vary with the location and age of the field in addition to any variations that occur with the depth of the individual well. Two adjacent wells are more than likely to produce crude oil with very different characteristics.

In very general terms (and as observed from elemental analyses), the viscous feedstocks (including residua) are a complex composition of: (1) hydrocarbon derivatives, (2) nitrogen-containing derivatives, (3) oxygen-containing derivatives, (4) sulfur-containing derivatives, and (e) metal-containing derivatives. However, this general definition is not adequate to describe the composition of the various viscous feedstocks as it relates to the behavior of these feedstocks. Indeed, the consideration of hydrogen-to-carbon atomic ratio, sulfur content, and API gravity are no longer adequate to the task of determining the refining behavior of these feedstocks.

Furthermore, the molecular composition of the feedstocks can be described in terms of three classes of compounds: saturate derivatives, aromatic derivatives, and compounds bearing heteroatoms (sulfur, oxygen, or nitrogen). Within each class, there are several families of related compounds which are (1) saturated constituents include normal alkane derivatives, branched alkane derivatives, and cycloalkane derivatives (paraffin derivative derivatives, iso-paraffin derivative derivatives, and naphthene derivatives, in crude oil terms); (2) alkene constituents (olefin derivatives) are rare to the extent of being considered an oddity; (3) monoaromatic constituents range from benzene to multiple fused ring analogs (naphthalene, phenanthrene, etc.); (4) thiol constituents – mercaptan) constituents – contain sulfur as do thioether derivatives and thiophene derivatives; and (5) nitrogen-containing and oxygen-containing constituents are more likely to be found in polar forms (pyridines, pyrroles, phenols, carboxylic acids, amides, etc.) than in non-polar forms (such as ethers). The distribution and characteristics of these molecular species account for the rich variety of crude oils.

In fact, the behavior of a feedstock behavior during refining is better addressed through consideration of the molecular make-up of the feedstock (perhaps, by analogy, just as genetic make-up dictates human behavior). The occurrence of amphoteric species (i.e., compounds having a mixed acid/base nature) is rarely addressed nor is the phenomenon of molecular size or the occurrence of specific functional types which can play a major role in the interactions between the constituents of a feedstock (Speight, 2014). All of these items are important in determining feedstock behavior during refining operations.

An understanding of the chemical types (or composition) of any feedstock can lead to an understanding of the chemical aspects of processing the feedstock. Processability is not only a matter of knowing the elemental composition of a feedstock; it is also a matter of understanding the bulk properties as they relate to the chemical or physical composition of the material. For example, it is difficult to understand, a priori, the process chemistry of various feedstocks from the elemental composition alone. From such data, it might be surmised that the major difference between a heavy crude oil and a more conventional material is the H/C atomic ratio alone. This property indicates that a heavy crude oil (having a lower H/C atomic ratio and being more aromatic in character) would require more hydrogen for upgrading to liquid fuels. This is, indeed, true but much more information is necessary to understand the processability of the feedstock.

With the necessity of processing the viscous feedstocks to obtain more saleable products, there has been the recognition that knowledge of the constituents of these higher-boiling feedstocks is also of some importance. Indeed, the problems encountered in processing the heavier feedstocks can be equated to the chemical character and the amount of complex, higher-boiling constituents in the feedstock. Refining these materials is not just a matter of applying know-how derived from refining conventional crude oils but requires knowledge of the chemical structure and chemical behavior of these more complex constituents.

However, heavy crude oil, extra heavy crude oil, tar sand bitumen, and residua are extremely complex and very little direct information can be obtained by distillation. It is not possible to isolate and identify the constituents of the heavier feedstocks (using analytical techniques that rely upon volatility). Other methods of identifying the chemical constituents must be employed. Such techniques include a myriad of fractionation procedures as well methods designed to draw inferences related to the hydrocarbon skeletal structures and the nature of the heteroatomic functions.

Processability is not only a matter of knowing the elemental composition of a feedstock; it is also a matter of understanding the bulk properties as they relate to the chemical or physical composition of the material. Understanding of the chemical types (or composition) of any feedstock can lead to a better understanding of the chemical aspects of processing. For example, it is difficult to understand, a priori, the process chemistry of various feedstocks from the elemental composition alone. From such data, it might be surmised that the major difference between a heavy crude oil and a more conventional material is, for example, the H/C atomic ratio. This property indicates that a heavy crude oil (having a lower H/C atomic ratio and being more aromatic in character) would require more hydrogen for upgrading to liquid fuels. This is, indeed, true but much more information is necessary to understand the processability of the feedstock.

The hydrocarbon content of a viscous feedstock may be as low (if not lower than) as 50% w/w. Nevertheless, feedstocks with as little as 50% w/w hydrocarbon constituents can, in some cases, be assumed to retain some (if not most) of the essential characteristics of the hydrocarbon derivatives. It is, nevertheless, the non-hydrocarbon (sulfur, oxygen, nitrogen, and metal) constituents that play a large part in determining the processability of the crude oil and will determine the processability of crude oil, heavy oil, and tar sand bitumen in the future (Speight, 2020a). But there is more to the composition of crude oil than the hydrocarbon content.

The inclusion of organic compounds of sulfur, nitrogen, and oxygen serves only to present crude oils as even more complex mixtures, and the appearance of appreciable amounts of these non-hydrocarbon compounds causes some concern in the refining of crude oils. Even though the concentration of non-hydrocarbon constituents (i.e., those organic compounds containing one or more sulfur, oxygen, or nitrogen atoms) in certain fractions may be quite small, they tend to concentrate in the higher-boiling fractions of crude oil. Indeed, their influence on the processability of the crude oil is important irrespective of their molecular size and the fraction in which they occur. It is, nevertheless, the non-hydrocarbon (sulfur, oxygen, nitrogen, and metal) constituents that play a large part in determining the processability of the crude oil, and their influence on the processability of the crude oil is important irrespective of their molecular size (Green et al., 1989; Speight, 2000, 2014, 2015). The occurrence of organic compounds of sulfur, nitrogen, and oxygen serves only to present crude oil as even more complex mixture, and the appearance of appreciable amounts of these non-hydrocarbon compounds causes some concern in crude oil refining. The non-hydrocarbon constituents (i.e., those organic compounds containing one or more sulfur, oxygen, or nitrogen atoms) tend to concentrate in the higher-boiling fractions of crude oil (Speight, 2000). In addition, as the feedstock series progresses to higher molecular weight feedstocks from crude oil to heavy crude oil to tar sand bitumen, not only does the number of the constituents increase but the molecular complexity of the constituents also increases (Speight, 2014, 2015).

The presence of traces of non-hydrocarbon compounds can impart objectionable characteristics to finished products, leading to discoloration and/or lack of stability during storage. On the other hand, catalyst poisoning and corrosion are the most noticeable effects during refining sequences when these compounds are present. It is therefore not surprising that considerable attention must be given to the non-hydrocarbon constituents of crude oil as the trend in the refining industry, of late, has been to process more heavy crude oil as well as residua that contain substantial proportions of these non-hydrocarbon materials.

2.6.1 HYDROCARBON CONSTITUENTS

The isolation of pure compounds from any refinery feedstock is an exceedingly difficult task, and the overwhelming complexity of the hydrocarbon constituents of the higher molecular weight fractions and the presence of compounds of sulfur, oxygen, and nitrogen are the main causes for the difficulties encountered. It is difficult on the basis of the data obtained from synthesized hydrocarbon derivatives to determine the identity or even the similarity of the synthetic hydrocarbon derivatives to those that constitute many of the higher-boiling fractions of crude oil. Nevertheless, it has been well established that the hydrocarbon components of crude oil are composed of paraffinic, naphthenic, and aromatic groups (Table 2.5). Olefin groups are not usually found in crude oils, and acetylene-type hydrocarbon derivatives are very rare indeed.

It is therefore convenient to divide the hydrocarbon components of crude oil into the following three classes:

1. *Paraffin derivatives*, which are saturated hydrocarbon derivatives with straight or branched chains, but without any ring structure.
2. *Naphthene derivatives*, which are saturated hydrocarbon derivatives containing one or more rings, each of which may have one or more paraffinic side chains (more correctly known as *alicyclic hydrocarbon derivatives*).
3. *Aromatic derivatives*, which are hydrocarbon derivatives containing one or more aromatic nuclei, such as benzene, naphthalene, and phenanthrene ring systems, which may be linked up with (substituted) naphthene rings and/or paraffinic side chains.

2.6.1.1 Paraffin Derivatives

The proportion of paraffin derivatives in a feedstock varies with the type of feedstock, but within any feedstock the proportion of paraffinic hydrocarbon derivatives usually decreases with increasing molecular weight and there is a concomitant increase in aromaticity and the relative proportion of heteroatoms (nitrogen, oxygen, and sulfur) (Speight, 2014, 2015).

TABLE 2.5

Hydrocarbon and Heteroatom Types in Crude Oil, Heavy Oil, and Tar Sand Bitumen

Class	Compound Types
Saturated hydrocarbons	*n*-Paraffins
	iso-Paraffins and other branched paraffins
	Cycloparaffins (naphthenes)
	Condensed cycloparaffins (including steranes, hopanes)
	Alkyl side chains on ring systems
Unsaturated hydrocarbons	Olefins: non-indigenous;
	Present in products of thermal reactions
Aromatic hydrocarbons	Benzene systems
	Condensed aromatic systems
	Condensed naphthene-aromatic systems
	Alkyl side chains on ring systems
Saturated heteroatomic systems	Alkyl sulfides
	Cycloalkyl sulfides
Sulfides	Alkyl side chains on ring systems
Aromatic heteroatomic systems	Furans (single-ring and multi-ring systems)
	Thiophenes (single-ring and multi-ring systems)
	Pyrroles (single-ring and multi-ring systems)
	Pyridines (single-ring and multi-ring systems)
	Mixed heteroatomic systems
	Amphoteric (acid-base systems)
	Alkyl side chains on ring systems

Source: Speight, J.G. 2014. *The Chemistry and Technology of Petroleum.* 5th Edition. CRC Press, Taylor & Francis Publishers, Boca Raton Florida. Table 8.1, page 191.

The abundance of the different members of the same homologous series varies considerably in absolute and relative values. However, in any particular crude oil or crude oil fraction, there may be a small number of constituents forming the greater part of the fraction, and these have been referred to as the predominant constituents (Bestougeff, 1961). This generality may also apply to other constituents and is very dependent upon the nature of the source material as well as the relative amounts of the individual source materials prevailing during maturation conditions (Speight, 2014).

Normal paraffin hydrocarbon derivatives (such as n-paraffin derivatives, straight-chain paraffin derivatives) occur in varying proportions in most crude oils. In fact, paraffinic crude oil may contain up to 20%–50% w/w n-paraffin derivatives in the gas oil fraction. However, naphthenic crude oil or asphaltic crude oil sometimes contain only very small amounts of normal paraffin derivatives.

Considerable quantities of iso-paraffin derivatives have been noted to be present in the straight-run naphtha fraction of crude oil. The 2- and 3-methyl derivatives are the most abundant, and the 4-methyl derivative is present in small amounts. The proportion tends to decrease with increasing boiling point; it appears that if the iso-paraffin derivatives are present in lubricating oils their amount is too small to have any significant influence on the physical properties of the lubricating oils.

As the molecular weight (or boiling point) of the crude oil fraction increases, there is a concomitant decrease in the amount of free paraffin derivatives in the fraction. In certain types of crude oil, there may be no paraffin derivatives at all in the vacuum gas oil fraction. For example, in the paraffinic crude oils, free paraffin derivatives will separate as a part of the asphaltene fraction but in the naphthenic crude oils, free paraffin derivatives are not expected in the gas oil and asphaltene fractions. The vestiges of paraffin derivatives in the asphaltenes fractions occur as alkyl side chains on aromatic systems and on naphthenic systems, and furthermore, these alkyl chains can contain 20 or more carbon atoms (Speight, 1994, 2014).

2.6.1.2 Cycloparaffin Derivatives

Cycloparaffin derivatives (such as cyclohexane derivatives, cyclopentane derivatives, and decahydronaphthalene also known as decalin) are largely represented in a variety of refinery feedstocks. Crude oil also contains polycyclic naphthene derivatives, such as terpenes, and such molecules (often designated bridge-ring hydrocarbon derivatives), occur even in the heavy gasoline fractions (boiling point 150°C–200°C, 300°F–390°F). Naphthene rings may be built up of a varying number of carbon atoms, and among the synthesized hydrocarbon derivatives, there are individual constituents with rings of the three, four, five, six, seven, and eight carbon atoms. Only naphthene derivatives with five- and six-membered rings have been isolated from the lower-boiling fractions. Thermodynamic studies show that naphthene rings with five and six carbon atoms are the most stable. The naphthenic acid derivatives contain predominantly cyclopentane ring systems as well as cyclohexane ring systems.

Cycloparaffin derivatives (naphthene derivatives) are represented in all fractions in which the constituent molecules contain more than five carbon atoms. The content of cycloparaffin derivatives in crude oil varies up to 60% w/w of the total hydrocarbon derivatives. Nevertheless, the structure of these constituents may change within the same crude oil, as a function of the molecular weight or boiling range of the individual fractions as well as from one crude oil to another.

The principal structural variation of the naphthene derivatives is the number of rings present in the molecule. The monocyclic and bicyclic naphthene derivatives are generally the major types of cycloparaffin derivatives in the lower-boiling factions of crude oil, with boiling point or molecular weight increased by the presence of alkyl chains. The higher-boiling-point fractions, such as the lubricating oils, may contain two to six rings per molecule. As the molecular weight (or boiling point) of the fraction increases, there is a concomitant increase the amount of cycloparaffin (naphthene) species in the fraction. In the asphaltic (naphthenic) crude oils, the gas oil fraction can contain considerable amounts of naphthenic ring systems that increase even more in consideration of the molecular types in the asphaltenes.

There is also the premise that the naphthene ring systems carry alkyl chains that are generally shorter than the alkyl substituents carried by aromatic systems. There are indications from spectroscopic studies that the short chains (methyl and ethyl) appear to be characteristic substituents of the aromatic portion of the molecule, whereas a limited number (one or two) of longer chains may be attached to the cycloparaffin rings. The total number of chains, which is in general four to six, as well as their length, increases according to the molecular weight of the naphthene-aromatic compounds.

In the asphaltene constituents, free condensed naphthenic ring systems may occur but general observations favor the occurrence of combined aromatic-naphthenic systems that are variously substituted by alkyl systems. There is also general evidence that the aromatic systems are responsible for the polarity of the asphaltenes constituents (Speight, 1994, 2014).

2.6.1.3 Aromatic Derivatives

The concept of the occurrence of identifiable aromatic systems in nature is a reality, and the occurrence of monocyclic and polycyclic aromatic systems in natural product chemicals is well documented (Sakarnen and Ludwig, 1971; Durand, 1980; Weiss and Edwards, 1980). However, one source of aromatic systems that is often ignored is crude oil (Eglinton and Murphy, 1969; Tissot and Welte, 1978; Speight, 1981; Brooks and Welte, 1984). Therefore, attempts to identify such systems in the nonvolatile constituents of crude oil should be an integral part of the repertoire of the crude oil chemist as well as the domain of the natural product chemist.

Crude oil is a mixture of compounds and aromatic compounds are common to all crude oil, and it is the difference in extent that becomes evident upon examination of a series of crude oil. For the majority of these aromatic derivatives contain paraffinic chains, naphthene rings, and aromatic rings (often co-joined) often occur side by side.

There is a general increase in the proportion of aromatic hydrocarbon derivatives with increasing molecular weight. However, aromatic hydrocarbon derivatives without the accompanying naphthene rings or alkyl-substituted derivatives are also present in the lower fractions of crude oil.

All known aromatic derivatives are present in naphtha fractions, but the benzene content is usually low compared to the benzene homologues, such as toluene and the xylene isomer. In addition to the 1- and 2-methylnaphthalenes, other simple alkyl naphthalene derivatives have also been isolated from crude oil. Aromatic derivatives without naphthene rings appear to be relatively rare in the higher-boiling fractions of crude oil. In the higher molecular weight fractions, the rings are usually condensed together. Thus components with two aromatic rings are presumed to be naphthalene derivatives, and those with three aromatic rings may be phenanthrene derivatives. Currently, and because of the consideration of the natural product origins of crude oil, phenanthrene derivatives are favored over anthracene derivatives.

In summation, all hydrocarbon compounds that have aromatic rings, in addition to the presence of alkyl chains and naphthenic rings within the same molecule, are classified as aromatic compounds. Many separation procedures that have been applied to crude oil (Speight, 2001, 2014, 2015) result in the isolation of a compound as an *aromatic* even if there is only one such ring (i.e., six carbon atoms) that is substituted by many more than six non-aromatic carbon atoms.

It should also be emphasized that in the higher-boiling-point crude oil fractions, many polycyclic structures occur in naphthene-aromatic systems. The naphthene-aromatic hydrocarbon derivatives, together with the naphthenic hydrocarbon series, form the major content of higher-boiling-point crude oil fractions. Usually the different naphthene-aromatic components are classified according to the number of aromatic rings in their molecules. The first to be distinguished is the series with an equal number of aromatic and naphthenic rings, and the first members of this bicyclic series C_9–C_{11} are the simplest, such as the 1-methyl-, 2-methyl, and 4-methyl indane derivatives and 2-methyl- and 7-methyl tetralin derivative. Tetralin and methyl-, dimethyl-, methyl ethyl-, and tetramethyl tetralin have been found in several crude oils, particularly in the heavier, naphthenic, crude oils, and there are valid reasons to believe that this increase in the number of rings, and side-chain complexity continues into the heavy oil and bitumen feedstocks.

Of special interest in the present context are the aromatic systems that occur in the nonvolatile asphaltene fraction (Speight, 1994). These polycyclic aromatic systems are complex molecules that fall into a molecular weight and boiling range where very little is known about model compounds (Speight, 1994, 2014). There has not been much success in determining the nature of such systems in the higher-boiling constituents of crude oil, i.e., the residua or nonvolatile constituents. In fact, it has been generally assumed that as the boiling point of a crude oil fraction increases, so does the number of condensed rings in a polycyclic aromatic system. To an extent, this is true but the simplicities of such assumptions cause an omission of other important structural constituents of the crude oil matrix, the alkyl substituents, the heteroatoms, and any polycyclic systems that are linked by alkyl chains or by heteroatoms.

The active principle is that crude oil is a continuum (Chapters 12 and 13) and has natural product origins (Long, 1979; Speight, 1980; Long, 1981; Speight, 1981, 1994, 2014). As such, it might be anticipated that there is a continuum of aromatic systems throughout crude oil that might differ from volatile to nonvolatile fractions but which, in fact, are based on natural product systems. It might also be argued that substitution patterns of the aromatic nucleus that are identified in the volatile fractions, or in any natural product counterparts, also apply to the nonvolatile fractions.

The application of thermal techniques to study the nature of the volatile thermal fragments from crude oil asphaltenes has produced some interesting data relating to the nature of the aromatic systems and the alkyl side chins in crude oil, heavy oil, and bitumen (Speight, 1971; Schucker and Keweshan, 1980; Gallegos, 1981). These thermal techniques have produced strong evidence for the presence of small (1–4 ring) aromatic systems (Speight and Pancirov, 1984; Speight, 1987). There was a preponderance of single-ring (cycloparaffin and alkylbenzene) species as well as the domination of saturated material over aromatic material.

Further studies using pyrolysis/gas chromatography/mass spectrometry (py/gc/ms) showed that different constituents of the asphaltene fraction produce the same type of polycyclic aromatic systems in the volatile matter, but the distribution was not constant (Speight and Pancirov, 1984). It was also possible to compute the hydrocarbon distribution from which a noteworthy point here is preponderance of single-ring (cycloparaffin and alkylbenzene) species as well as the domination of saturated material over aromatic material. The emphasis on low-molecular-weight material in the volatile products is to be anticipated on the basis that more complex systems remain as nonvolatile material and, in fact, are converted to coke.

One other point worthy of note is that the py/gc/ms program does not accommodate nitrogen and oxygen species whether or not they be associated with aromatic systems. This matter is resolved, in part, not only by the concentration of nitrogen and oxygen in the nonvolatile material (coke) but also by the overall low proportions of these heteroatoms originally present in the asphaltenes (Speight, 1970; Speight and Pancirov, 1984). The major drawback to the use of the py/gc/ms technique to study of the aromatic systems in asphaltenes is the amount of material that remains as a nonvolatile residue.

Of all of the methods applied to determining the types of aromatic systems in crude oil asphaltenes, one with considerable potential, but given the least attention, is ultraviolet spectroscopy. Typically, the ultraviolet spectrum of an asphaltene shows two major regions with very little fine structure. Interpretation of such a spectrum can only be made in general terms. However, the technique can add valuable information about the degree of condensation of polycyclic aromatic ring systems through the auspices of high performance liquid chromatography (*hplc*) (Lee et al., 1981; Bjorseth, 1983; Monin and Pelet, 1983; Felix et al., 1985; Killops and Readman, 1985; Speight, 1986). Indeed, when this approach is taken, the technique not only confirms the complex nature of the asphaltene fraction but also allows further detailed identifications to be made of the individual functional constituents of asphaltenes. The constituents of the fraction produce a multi-component chromatogram, but sub-fractions produce a less complex and much narrower chromatograph that may even approximate a single peak which may prove much more difficult to separate by a detector.

These data provide strong indications of the ring-size distribution of the polycyclic aromatic systems in crude oil asphaltenes (Speight, 1994, 2014). On the other hand, acid sub-fractions (phenolic/carboxylic functions) and neutral polar sub-fractions (amides/imino functions) contain few if any polycyclic aromatic systems having more than three rings per system. Moreover, the differences in the functionality of the sub-fractions result in substantial differences in thermal and catalytic reactivity that can lead to unanticipated phase separation and, subsequently, coke formation in a thermal reactor as well as structural orientation on, and blocking of, the active sites on a catalyst. This is especially the case when the behavior of the functional types that occur in the various high-boiling fractions of heavy oil and tar sand bitumen are considered (Speight, 1994, 2014).

In all cases examined, the evidence favored the preponderance of the smaller (one-to-four) ring systems, but perhaps what is more important about these investigations is that the data show that the asphaltene fraction is a complex mixture of compound types which confirms fractionation studies and cannot be adequately represented by any particular formula that is construed to be average (Speight, 1986). Therefore, the concept of a large polycyclic aromatic ring system as the central feature of the asphaltene fraction must be abandoned for lack of evidence.

In summary, the premise is that crude oil is a natural product and that the aromatic systems are based on identifiable structural systems that are derived from natural product precursors.

2.6.1.4 Unsaturated Hydrocarbon Derivatives

The presence of olefin derivatives ($RCH=CHR^1$) in crude oil has been under dispute for many years because there are investigators who claim that olefin derivatives are actually present. In fact, these claims usually refer to distilled fractions, and it is very difficult to entirely avoid cracking during the distillation process. Nevertheless, evidence for the presence of considerable proportions of olefin derivatives in Pennsylvanian crude oils has been obtained; spectroscopic and chemical methods showed that the crude oils, as well as all distillate fractions, contained up to 3% w/w olefin derivatives. Hence, although the opinion that crude oil does not contain olefin derivatives requires some revision, it is perhaps reasonable to assume that the Pennsylvania crude oils may hold an exceptional position and that olefin derivatives are present in crude oil in only a few special cases. The presence of diene derivatives (RCH=CH=CHR') and acetylene derivatives (RC≡CR') is considered to be extremely unlikely.

In summary, a variety of hydrocarbon compounds occur throughout the various feedstocks. Although the amount of any particular hydrocarbon varies from one crude oil to another, the family from which that hydrocarbon arises is well represented.

2.6.2 NON-HYDROCARBON CONSTITUENTS

The previous sections present some indication of the types and nomenclature of the organic hydrocarbon derivatives that occur in various crude oils. Thus it is not surprising that crude oil which contains only hydrocarbon derivatives is, in fact, an extremely complex mixture. The phenomenal increase in the number of possible isomers for the higher-boiling (higher-molecular-weight) hydrocarbon derivatives makes it very difficult, if not impossible in most cases, to isolate individual members of any one series having more than, say, 12 carbon atoms.

Inclusion of organic compounds of nitrogen, oxygen, and sulfur serves only to present crude oil as an even more complex mixture than was originally conceived. Nevertheless, considerable progress has been made in the isolation and/or identification of the lower molecular weight hydrocarbon derivatives, as well as accurate estimations of the overall proportions of the hydrocarbon types present in crude oil. Indeed, it has been established that, as the boiling point of the crude oil fraction increases, not only does the number of the constituents but the molecular complexity of the constituents also increases (Speight, 2014, 2015).

Crude oils contain appreciable amounts of organic non-hydrocarbon constituents, mainly sulfur-, nitrogen-, and oxygen-containing compounds and, in smaller amounts, organometallic compounds in solution and inorganic salts in colloidal suspension. These constituents appear throughout the entire boiling range of the crude oil but tend to concentrate mainly in the heavier fractions and in the nonvolatile residues.

Although their concentration in certain fractions may be quite small, their influence is important. For example, the decomposition of inorganic salts suspended in the crude can cause serious breakdowns in refinery operations; the thermal decomposition of deposited inorganic chlorides with evolution of free hydrochloric acid can give rise to serious corrosion problems in the distillation equipment. The presence of organic acid components, such as mercaptan derivatives and acid derivatives, can also promote metallic corrosion. In catalytic operations, passivation and/or poisoning of the catalyst can be caused by deposition of traces of metals (vanadium and nickel) or by chemisorption of nitrogen-containing compounds on the catalyst, thus necessitating the frequent regeneration of the catalyst or its expensive replacement.

The presence of traces of non-hydrocarbon derivatives may impart objectionable characteristics in finished products, such as discoloration, lack of stability on storage, or a reduction in the effectiveness of organic lead antiknock additives. It is thus obvious that a more extensive knowledge of these compounds and of their characteristics could result in improved refining methods and even in finished products of better quality. The non-hydrocarbon compounds, particularly the porphyrins and related compounds, are also of fundamental interest in the elucidation of the origin and nature of crude oils.

Although sulfur is the most important (certainly the most abundant) heteroatom (i.e., non-hydrocarbon) present in crude oil with respect to the current context, other non-hydrocarbon atoms can exert a substantial influence not only on the nature and properties of the products but also on the nature and efficiency of the process. Such atoms are nitrogen, oxygen, and metals, and because of their influence on the process, some discussion of each is warranted here. Furthermore, a knowledge of their surface-active characteristics is of help in understanding problems related to the migration of oil from the source rocks to the actual reservoirs.

2.6.2.1 Sulfur Compounds

Although the concentration of heteroatom constituents in certain fractions may be quite small, their influence is important. For example, the decomposition of inorganic salts suspended in the crude can cause serious breakdowns in refinery operations; the thermal decomposition of deposited inorganic chlorides with evolution of free hydrochloric acid can give rise to serious corrosion problems in the distillation equipment. The presence of organic acid components, such as mercaptan derivatives and acid derivatives, can also promote metallic corrosion. In catalytic operations, passivation and/or poisoning of the catalyst can be caused by deposition of traces of metals (vanadium and nickel) or by chemisorption of nitrogen-containing compounds on the catalyst, thus necessitating the frequent regeneration of the catalyst or its expensive replacement.

Sulfur compounds are among the most important heteroatomic constituents of crude oil, and although there are many varieties of sulfur compounds (Speight, 2000, 2014), the prevailing conditions during the formation, maturation, and even in situ alteration may dictate that only preferred types exist in any particular crude oil. Nevertheless, sulfur compounds of various types are present in all crude oils (Thompson et al., 1976). In general, the higher the density of the crude oil for the lower the API gravity of the crude oil, the higher the sulfur content, and the total sulfur in the crude oil can vary from approximately 0.04% w/w for light crude oil to approximately 5.0% for heavy crude oil and tar sand bitumen. However, the sulfur content of crude oils produced from broad geographic regions varies with time, depending on the composition of newly discovered fields, particularly those in different geological environments.

The presence of sulfur compounds in finished crude oil products often produces harmful effects. For example, in the naphtha fraction, the sulfur-containing constituents are believed to promote corrosion of engine parts, especially under winter conditions, when water containing sulfur dioxide from the combustion may accumulate in the crankcase. In addition, mercaptan derivatives in hydrocarbon solution cause the corrosion of copper and brass in the presence of air and also affect lead susceptibility and color stability. Free sulfur is also corrosive, as are sulfide derivatives, disulfide derivatives, and thiophene derivatives, which are detrimental to the octane number response to tetraethyllead. However, gasoline with a sulfur content between 0.2% and 0.5% has been used without obvious harmful effect. In diesel fuels, sulfur compounds increase wear and can contribute to the formation of engine deposits. Although a high sulfur content can sometimes be tolerated in industrial fuel oils, the situation for lubricating oils is that a high content of sulfur compounds in lubricating oils seems to lower resistance to oxidation and increases the deposition of solids.

Although it is generally true that the proportion of sulfur increases with the boiling point during distillation (Speight, 2000, 2014), the middle fractions may actually contain more sulfur than higher-boiling fractions as a result of decomposition of the higher molecular weight compounds during the distillation. High sulfur content is generally considered harmful in most crude oil products, and the removal of sulfur compounds or their conversion to less deleterious types is an important part of refinery practice. The distribution of the various types of sulfur compounds varies markedly among crude oils of diverse origin as well as between the various heavy feedstocks, but fortunately some of the sulfur compounds in crude oil undergo thermal reactions at relatively low temperatures. If elemental sulfur is present in the oil, a reaction, with the evolution of hydrogen sulfide, begins at approximately 150°C (300°F) and is very rapid at 220°C (430°F), but organically bound sulfur compounds do not yield hydrogen sulfide until higher temperatures are reached. Hydrogen sulfide is, however, a common constituent of many crude oils, and some crude oils with >1% w/w sulfur are often accompanied by a gas having substantial properties of hydrogen sulfide.

Various thiophene derivatives have also been isolated from a variety of crude oils; benzothiophene derivatives are usually present in the higher-boiling crude oil fractions. On the other hand, disulfides are not regarded as true constituents of crude oil but are generally formed by oxidation of thiols during processing:

$$2R\text{-}SH + [O] \rightarrow R\text{-}S\text{-}S\text{-}R + H_2O$$

2.6.2.2 Nitrogen Compounds

Nitrogen in a feedstock may be classified arbitrarily as basic and non-basic (Speight, 2014). The basic nitrogen compounds, which are composed mainly of pyridine homologues and occur throughout the boiling ranges, have a decided tendency to exist in the higher-boiling fractions and residua. The non-basic nitrogen compounds, which are usually of the pyrrole, indole, and carbazole types, also occur in the higher-boiling fractions and residua.

In general, the nitrogen content of crude oil is low and generally falls within the range 0.1%–0.9% w/w, although early work indicates that some crude oil may contain up to 2% nitrogen. However, crude oils with no detectable nitrogen or even trace amounts are not uncommon, but in general the more asphaltic the oil, the higher its nitrogen content. Insofar as an approximate correlation exists between the sulfur content and API gravity of crude oils (Speight, 2000, 2014), there also exists a correlation between nitrogen content and the API gravity of crude oil. It also follows that there is an approximate correlation between the nitrogen content and the carbon residue: the higher the nitrogen content, the higher the carbon residue. The presence of nitrogen in crude oil is of much greater significance in refinery operations than might be expected from the small amounts present. Nitrogen compounds can be responsible for the poisoning of cracking catalysts, and they also contribute to gum formation in such products as domestic fuel oil. The trend in recent years toward cutting deeper into the crude to obtain stocks for catalytic cracking has accentuated the harmful effects of the nitrogen compounds, which are concentrated largely in the higher-boiling portions.

Basic nitrogen compounds with a relatively low molecular weight can be extracted with dilute mineral acids; equally strong bases of higher molecular weight remain unextracted because of unfavorable partitioning between the oil and aqueous phases. A method has been developed in which the nitrogen compounds are classified as basic or non-basic, depending on whether they can be titrated with perchloric acid in a 50:50 solution of glacial acetic acid and benzene. Application of this method has shown that the ratio of basic to total nitrogen is approximately constant (0 to 30 0.05) irrespective of the source of the crude. Indeed, the ratio of basic to total nitrogen was found to be approximately constant throughout the entire range of distillate and residual fractions. Nitrogen compounds extractable with dilute mineral acids from crude oil distillates were found to consist of alkyl pyridine derivatives, alkyl quinoline derivatives, and alkyl iso-quinoline derivatives carrying alkyl substituents, as well as pyridine derivatives in which the substituent was a cyclopentyl or cyclohexyl group. The compounds that cannot be extracted with dilute mineral acids contain the greater part of the nitrogen in crude oil and are generally of the carbazole, indole, and pyrrole types.

2.6.2.3 Oxygen Compounds

Oxygen in organic compounds can occur in a variety of forms (ROH, ArOH, ROR', RCO_2H, $ArCO_2H$, RCO_2R, $ArCO_2R$, $R_2C=O$ as well as the cyclic furan derivatives, where R and R' are alkyl groups and Ar is an aromatic group) in nature so it is not surprising that the more common oxygen-containing compounds occur in crude oil (Speight, 2014). The total oxygen content of crude oil is usually less than 2% w/w, although larger amounts have been reported, but when the oxygen content is phenomenally high it may be that the oil has suffered prolonged exposure to the atmosphere either during or after production. However, the oxygen content of crude oil increases with the boiling point of the fractions examined; in fact, the nonvolatile residua may have oxygen contents up to 8% w/w. Although these high-molecular-weight compounds contain most of the oxygen in crude oil, little is known concerning their structure, but those of lower molecular weight have been investigated with considerably more success and have been shown to contain carboxylic acids and phenols.

It has generally been concluded that the carboxylic acids in crude oil with fewer than eight carbon atoms per molecule are almost entirely aliphatic in nature; monocyclic acids begin at C_6 and predominate above C_{14}. This indicates that the structures of the carboxylic acids correspond with those of the hydrocarbon derivatives with which they are associated in the crude oil. In the range in which paraffin derivatives are the prevailing type of hydrocarbon, the aliphatic acids may be expected to predominate. Similarly, in the ranges in which mono-cycloparaffin derivatives and di-cycloparaffin derivatives prevail, one may expect to find principally monocyclic and di-cyclic acid derivatives, respectively.

In addition to the carboxylic acids and phenolic compounds, the presence of ketones, esters, ethers, and anhydrides has been claimed for a variety of crude oils. However, the precise identification of these compounds is difficult because most of them occur in the higher molecular weight nonvolatile residua and are claimed (with some justification) to be the products of the air blowing of the residua.

Although comparisons are frequently made between the sulfur and nitrogen contents and such physical properties as the API gravity, it is not the same with the oxygen contents of crude oils. It is possible to postulate, and show, that such relationships exist. However, the ease with which some of the crude oil constituents can react with oxygen (aerial or dissolved) to incorporate oxygen functions into their molecular structure often renders the exercise somewhat futile if meaningful deductions are to be made.

Carboxylic acid derivatives (RCO_2H) may be less detrimental than other heteroatom constituents because there is the high potential for decarboxylation to a hydrocarbon and carbon dioxide at the temperatures (>340°C, >645°F) used during distillation of flashing (Speight and Francisco, 1990):

$$RCO_2H \rightarrow RH + CO_2$$

In addition to the carboxylic acids and phenolic compounds (ArOH, where Ar is an aromatic moiety), the presence of ketones (>C=O), esters [>C(=O)-OR], ethers (R-O-R), and anhydrides [>C(=O)-O-(O=)C<] has been claimed for a variety of crude oils. However, the precise identification of these compounds is difficult because most of them occur in the higher molecular weight nonvolatile residua.

2.6.2.4 Metal-Containing Constituents

Metallic constituents are found in every crude oil, and the concentrations have to be reduced to convert the oil to transportation fuel. Metals affect many upgrading processes and cause particular problems because they poison catalysts used for sulfur and nitrogen removal as well as other processes such as catalytic cracking. The trace metals Ni and V are generally orders of magnitude higher than other metals in crude oil, except when contaminated with co-produced brine salts (Na, Mg, Ca, Cl) or corrosion products gathered in transportation (Fe).

The occurrence of metallic constituents in crude oil is of considerably greater interest to the crude oil industry than might be expected from the very small amounts present. Even minute amounts of iron, copper, and particularly nickel and vanadium in the charging stocks for catalytic cracking affect the activity of the catalyst and result in increased gas and coke formation and reduced yields of gasoline. In high-temperature power generators, such as oil-fired gas turbines, the presence of metallic constituents, particularly vanadium in the fuel, may lead to ash deposits on the turbine rotors, thus reducing clearances and disturbing their balance. More particularly, damage by corrosion may be very severe. The ash resulting from the combustion of fuels containing sodium and especially vanadium reacts with refractory furnace linings to lower their fusion points and so cause their deterioration.

Thus, the ash residue left after burning of a crude oil is due to the presence of these metallic constituents, part of which occur as inorganic water-soluble salts (mainly chlorides and sulfates of sodium, potassium, magnesium, and calcium) in the water phase of crude oil emulsions (Abdel-Aal et al., 2016). These are removed in the desalting operations, either by evaporation of the water and subsequent water washing or by breaking the emulsion, thereby causing the original mineral content of the crude to be substantially reduced. Other metals are present in the form of oil-soluble organo-metallic compounds as complexes, metallic soaps, or in the form of colloidal suspensions, and the total ash from desalted crude oils is of the order of 0.1–100 mg/L. Metals are generally found only in the nonvolatile portion of crude oil (Altgelt and Boduszynski, 1994; Reynolds, 1998).

Two groups of elements that are associated with well-defined types of compounds occur in significant concentrations in the original crude oil. Zinc, titanium, calcium, and magnesium appear in the form of organometallic soaps with surface-active properties adsorbed in the water/oil interfaces and act as emulsion stabilizers. However, vanadium, copper, nickel, and part of the iron found in crude oils seem to be in a different class and are present as oil-soluble compounds. These metals are capable of complexing with pyrrole pigment compounds derived from chlorophyll and hemoglobin and are almost certain to have been present in plant and animal source materials. It is easy to surmise that the metals in question are present in such form, ending in the ash content. Evidence for the presence of several other metals in oil-soluble form has been produced, and thus zinc, titanium, calcium, and magnesium compounds have been identified in addition to vanadium, nickel, iron, and copper. Examination of the analyses of a number of crude oil for iron, nickel, vanadium, and copper indicates a relatively high vanadium content, which usually exceeds that of nickel, although the reverse can also occur.

Distillation concentrates the metallic constituents in the residues (Reynolds, 1998), although some can appear in the higher-boiling distillates, but the latter may be due in part to entrainment. Nevertheless, there is evidence that a portion of the metallic constituents may occur in the distillates by volatilization of the organometallic compounds present in the crude oil. In fact, as the percentage of overhead obtained by vacuum distillation of a reduced crude is increased, the amount of metallic constituents in the overhead oil is also increased. The majority of the vanadium, nickel, iron, and copper in residual stocks may be precipitated along with the asphaltenes by hydrocarbon solvents. Thus, removal of the asphaltenes with n-pentane reduces the vanadium content of the oil by up to 95% with substantial reductions in the amounts of iron and nickel.

2.6.2.5 Porphyrin Derivatives

Porphyrin derivatives are a naturally occurring chemical species that exist in crude oil and usually occur in the non-basic portion of the nitrogen-containing concentrate (Bonnett, 1978; Reynolds, 1998). They are not usually considered among the usual nitrogen-containing constituents of crude oil, nor are they considered a metallo-containing organic material that also occurs in some crude oils. As a result of these early investigations there arose the concept of porphyrins as biomarkers that could establish a link between compounds found in the geosphere and their corresponding biological precursors.

Porphyrins are derivatives of porphine that consists of four pyrrole molecules joined by methine (–CH=) bridges (Figure 2.4). The methine bridges establish conjugated linkages between the component pyrrole nuclei, forming a more extended resonance system.

Although the resulting structure retains much of the inherent character of the pyrrole components, the larger conjugated system gives increased aromatic character to the porphine molecule. Furthermore, the imino functions (–NH–) in the porphine system allow metals such as nickel to be included into the molecule through chelation (Figure 2.5).

A large number of different porphyrin compounds exist in nature or have been synthesized. Most of these compounds have substituents other than hydrogen on many of the ring carbons. The nature of the substituents on porphyrin rings determines the classification of a specific porphyrin compound into one of various types according to one common system of nomenclature (Bonnet, 1978). Porphyrin derivatives also have well-known trivial names or acronyms that are often in more common usage than the formal system of nomenclature. When one or two double bonds of a

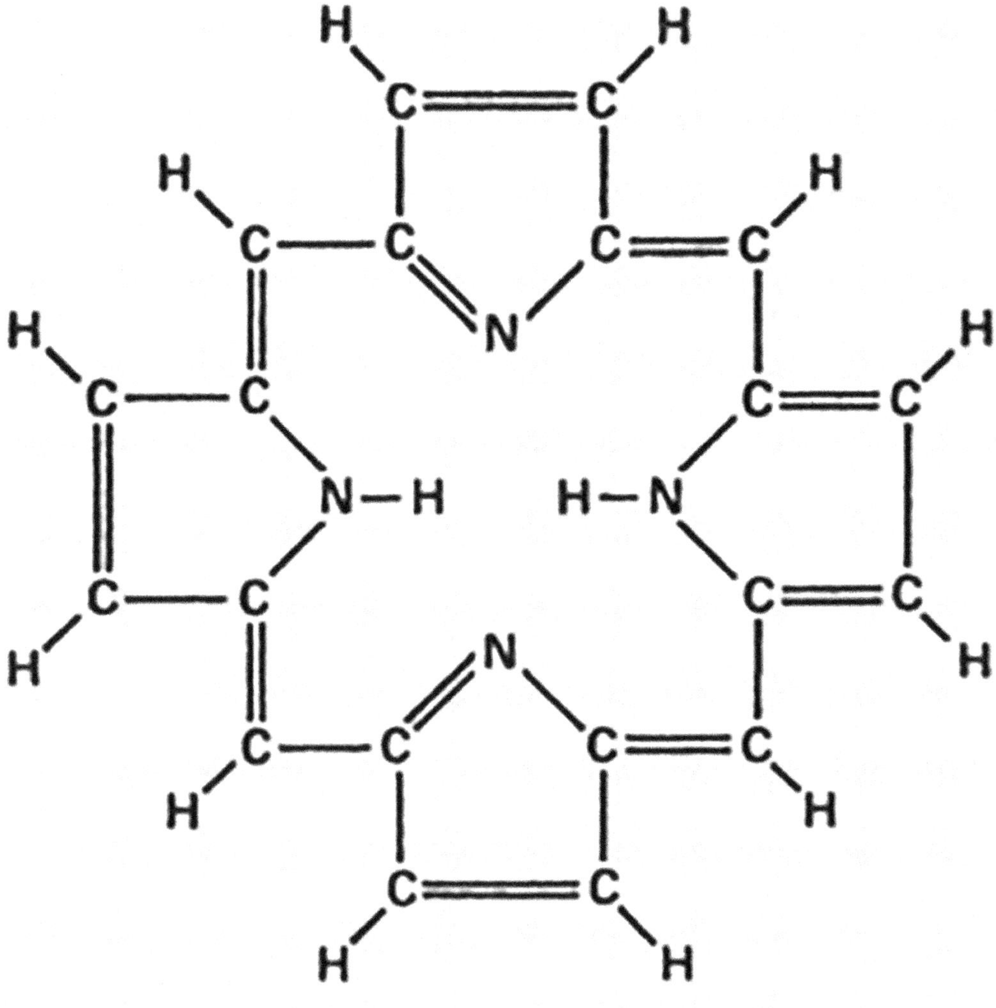

FIGURE 2.4 Porphine – the basic structure of poyphyrins. (Source: Speight, J.G. 2014. *The Chemistry and Technology of Petroleum.* 5th Edition. CRC Press, Taylor & Francis Publishers, Boca Raton Florida. Figure 8.3, page 202.)

porphyrin are hydrogenated, a chlorin or a phlorin is the result. Chlorin derivatives are components of chlorophyll and possess an iso-cyclic ring formed by two methylene groups bridging a pyrrole-type carbon to a methine carbon. Geological porphyrins that contain this structural feature are assumed to be derived from various chlorophyll derivatives.

Almost all crude oil, heavy oil, and bitumen contain detectable amounts of vanadium and nickel porphyrin derivatives. More mature, lighter crude oils usually contain only small amounts of these compounds. Heavy oils may contain large amounts of vanadium and nickel porphyrin derivatives. Vanadium concentrations of over 1,000 ppm are known for some crude oil, and a substantial amount of the vanadium in these crude oils is chelated with porphyrins. In high-sulfur crude oil of marine origin, vanadium porphyrin derivatives are more abundant than nickel porphyrin derivatives. Low-sulfur crude oils of lacustrine origin usually contain more nickel porphyrin derivatives than vanadium porphyrin derivatives.

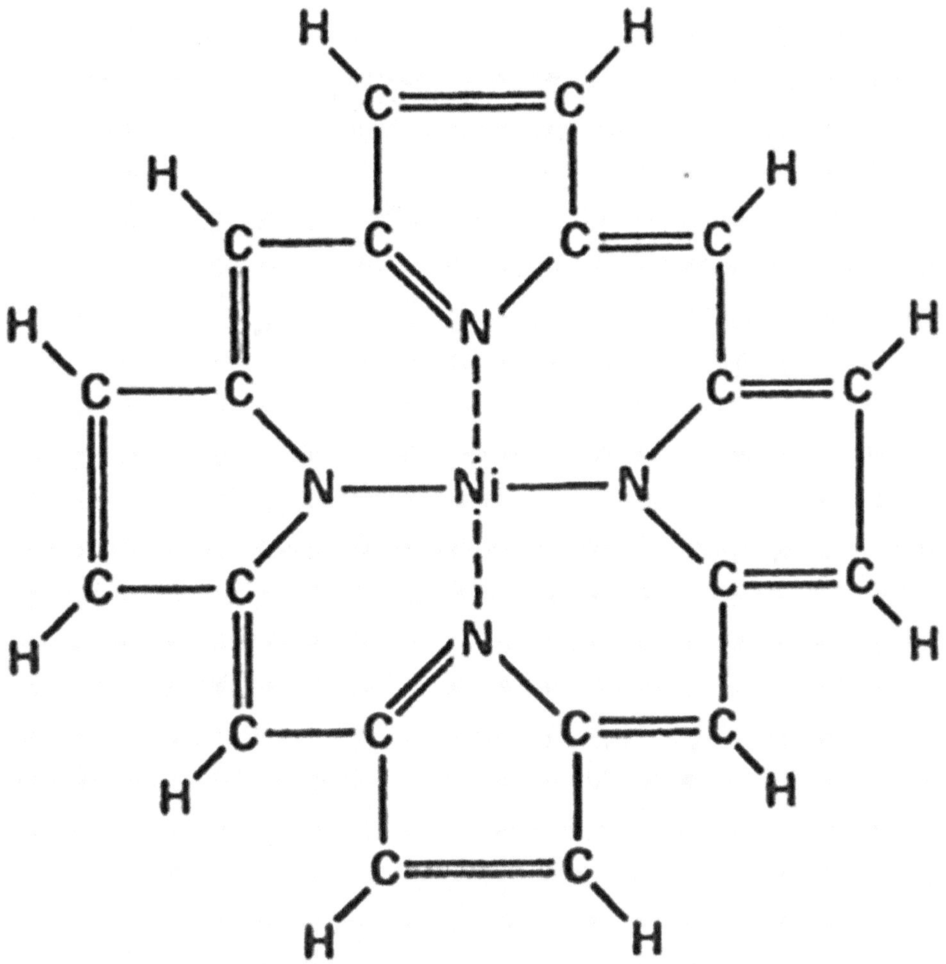

FIGURE 2.5 Nickel chelate of porphine. (Source: Speight, J.G. 2014. *The Chemistry and Technology of Petroleum*. 5th Edition. CRC Press, Taylor & Francis Publishers, Boca Raton Florida. Figure 8.4, page 202.)

Of all the metals in the periodic table, only vanadium and nickel have been proven definitely to exist as chelates in significant amounts in a large number of crude oils and tar sand bitumen.

If the vanadium and nickel contents of crude oils are measured and compared with porphyrin concentrations, it is usually found that not all the metal content can be accounted for as porphyrin constituents (Reynolds, 1998). In some crude oils, as little as 10% w/w of total metals appears to be chelated with porphyrins. Only rarely can all measured nickel and vanadium in a crude oil be accounted for as porphyrin-type. Currently, some investigators believe that part of the vanadium and nickel in crude oils is chelated with ligands that are not porphyrins. These metal chelates are referred to as non-porphyrin metal chelates or complexes (Reynolds et al., 1987).

Finally, during the fractionation of crude oil the metallic constituents (metalloporphyrins and non-porphyrin metal chelates) are concentrated in the asphaltene fraction. The deasphaltened oil contains smaller concentrations of porphyrins than the parent materials and usually very small concentrations of non-porphyrin metals.

2.7   FRACTIONAL COMPOSITION

Refining any feedstock (viscous or non-viscous) involves subjecting the feedstock to a series of integrated physical and chemical unit processes (Figure 2.1) as a result of which a variety of products are generated. In some of the processes, e.g., distillation, the constituents of the feedstock are isolated unchanged, whereas in other processes (such as cracking) considerable changes are occur in the constituents.

Thus, feedstocks can be defined (on a *relative* or *standard* basis) in terms of three or four general fractions: asphaltenes, resins, saturate derivatives, and aromatic derivatives (Figure 2.6).

For the fractionation of the viscous feedstocks, it is typically necessary to blend the feedstock with an equal volume of a solvent in order to enable the diffusion of the paraffinic precipitating liquid into the feedstocks. In such a case, the amount of paraffin (pentane or heptane) added must be 40 time the volume of the feedstock plus the volume of the solvent.

Thus, it is possible to compare inter-laboratory investigations and thence to apply the concept of predictability to refining sequences and potential products. Recognition that refinery behavior is related to the composition of the feedstock has led to a multiplicity of attempts to establish crude oil and its fractions as compositions of matter. As a result, various analytical techniques have been developed for the identification and quantification of *every molecule* in the lower-boiling fractions of crude oil. It is now generally recognized that the name *crude oil* does not describe a composition of matter but rather a mixture of various organic compounds that includes a wide range of molecular weights and molecular types that exist in balance with each other (Speight, 1994; Long and Speight, 1998). There must also be some questions of the advisability (perhaps *futility* is a better word) of attempting to describe *every molecule* in crude oil. The true focus should be to what ends these molecules can be used.

The fractionation methods available to the crude oil industry allow a reasonably effective degree of separation of hydrocarbon mixtures (Speight, 2014, 2015). However, the problems are separating

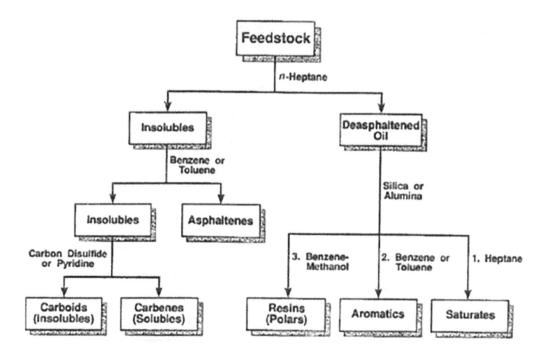

FIGURE 2.6   Feedstock fractionation scheme. (Source: Speight, J.G. 2014. *The Chemistry and Technology of Petroleum*. 5th Edition. CRC Press, Taylor & Francis Publishers, Boca Raton Florida. Figure 9.1, page 212.)

the crude oil constituents without alteration of their molecular structure and obtaining these constituents in a substantially pure state. Thus, the general procedure is to employ techniques that segregate the constituents according to molecular size and molecular type.

It is more generally true, however, that the success of any attempted fractionation procedure involves not only the application of one particular technique but also the utilization of several integrated techniques, especially those techniques involving the use of chemical and physical properties to differentiate among the various constituents. For example, the standard processes of physical fractionation used in the crude oil industry are those of distillation and solvent treatment, as well as adsorption by surface-active materials. Chemical procedures depend on specific reactions, such as the interaction of olefin derivatives with sulfuric acid or the various classes of adduct formation. Chemical fractionation is not always successful because of the complex nature of crude oil. This may result in unprovoked chemical reactions that have an adverse effect on the fractionation and the resulting data. Indeed, caution is advised when using methods that involve chemical separation of the constituents.

The order in which the several fractionation methods are used is determined not only by the nature and/or composition of the crude oil but also by the effectiveness of a particular process and its compatibility with the other separation procedures to be employed. Thus, although there are wide variations in the nature of refinery feedstocks, there have been many attempts to devise standard methods of crude oil fractionation. However, various laboratories are inclined to adhere firmly to, and promote, their own particular methods. Recognition that no one particular method may satisfy all the requirements of crude oil fractionation is the first step in any fractionation study. This is due, in the main part, to the complexity of crude oil not only from the distribution of the hydrocarbon species but also from the distribution of the heteroatom (nitrogen, oxygen, and sulfur) species.

2.7.1 SOLVENT METHODS

Fractionation of crude oil by distillation is an excellent means by which the volatile constituents can be isolated and studied. However, the nonvolatile residuum, which may actually constitute from 1% to 60% of the crude oil, cannot be fractionated by distillation without the possibility of thermal decomposition, and as a result, alternative methods of fractionation have been developed.

The distillation process separates light (lower-molecular-weight, lower-boiling) and heavy (higher-molecular-weight, higher-boiling) constituents by virtue of their volatility and involves the participation of a vapor phase and a liquid phase. These are, however, physical processes that involve the use of two liquid phases, usually a solvent phase and an oil phase.

Solvent methods have also been applied to crude oil fractionation on the basis of molecular weight. The major molecular weight separation process used in the laboratory as well as in the refinery is solvent precipitation. Solvent precipitation occurs in a refinery in a deasphalting unit and is essentially an extension of the procedure for separation by molecular weight, although some separation by polarity might also be operative. The deasphalting process is usually applied to the higher molecular weight fractions of crude oil such as atmospheric and vacuum residua and, depending upon the composition, can be applied to other viscous feedstocks (Parkash, 2003; Gary et al., 2007; Speight, 2014; Hsu and Robinson, 2017; Speight, 2017).

These fractionation techniques can also be applied to cracked residua, asphalt, bitumen, and even to virgin crude oil, but in the last case the possibility of losses of the lower-boiling constituents is apparent; hence the recommended procedure for virgin crude oil is, first, distillation followed by fractionation of the residua.

The simplest application of solvent extraction consists of mixing crude oil with another liquid, which results in the formation of two phases. This causes distribution of the crude oil constituents over the two phases which are (1) the dissolved portion, also known as the extract and (2) the non-dissolved portion, also known as the raffinate.

The ratio of the concentration of any particular component in the two phases is known as the distribution coefficient K:

$$K = C_1 / C_2$$

C_1 and C_2 are the concentrations of the components in the various phases. The distribution coefficient is usually constant and may vary only slightly, if at all, with the concentration of the other components. In fact, the distribution coefficients may differ for the various components of the mixture to such an extent that the ratio of the concentrations of the various components in the solvent phase differs from that in the original crude oil; this is the basis for solvent extraction procedures.

It is generally molecular type, not molecular size, which is responsible for the solubility of species in various solvents. Thus, solvent extraction separates crude oil fractions according to type, although within any particular series there is a separation according to molecular size. Lower molecular weight hydrocarbon derivatives of a series (the light fraction) may well be separated from their higher molecular weight homologues (the heavy fraction) by solvent extraction procedures.

In general, it is advisable that selective extraction be employed with fairly narrow boiling range fractions. However, the separation achieved after one treatment with the solvent is rarely complete, and several repetitions of the treatment are required. Such repetitious treatments are normally carried out by movement of the liquids countercurrently through the extraction equipment (countercurrent extraction), which affords better yields of the extractable materials.

The list of compounds that have been suggested as selective solvents for the preferential extraction fractionation of crude oil contains a large selection of different functional types (Speight, 2014). However, before any extraction process is attempted, it is necessary to consider the following criteria: (1) the differences in the solubility of the crude oil constituents in the solvent should be substantial, (2) the solvent should be significantly less dense or denser than the crude oil (product) to be separated to allow easier countercurrent flow of the two phases, and (3) separation of the solvent from the extracted material should be relatively easy.

It may also be advantageous to consider other properties, such as viscosity, surface tension, and the like, as well as the optimal temperature for the extraction process. Thus, aromatic derivatives can be separated from naphthene and paraffinic hydrocarbon derivatives by the use of selective solvents. Furthermore, aromatic derivatives with differing numbers of aromatic rings that may exist in various narrow boiling fractions can also be effectively separated by solvent treatment.

The separation of crude oil into two fractions – the asphaltene fraction and (2) the maltene fraction – is conveniently brought about by use of low-molecular-weight paraffin hydrocarbon derivatives, which were recognized to have selective solvency for hydrocarbon derivatives, and simple relatively low-molecular-weight hydrocarbon derivatives. The more complex, higher molecular weight compounds are precipitated particularly well by addition of 40 volumes of n-pentane or n-heptane in the methods generally preferred at present (Speight et al., 1984; Speight, 1994) although hexane is used on occasion (Yan et al., 1997). It is no doubt a separation of the chemical components with the most complex structures from the mixture, and this fraction, which should correctly be called n-pentane asphaltene fraction or the n-heptane asphaltene fraction, is qualitatively and quantitatively reproducible (Speight, 2014, 2015).

Variation in the solvent type also causes significant changes in asphaltene yield. For example, in the ease of a Western Canadian bitumen and, indeed, for conventional crude oils, branched chain paraffin derivatives or terminal olefin derivatives do not precipitate the same amount of asphaltenes as do the corresponding normal paraffin derivatives (Mitchell and Speight, 1973). Cycloparaffin derivatives (naphthene derivatives) have a remarkable effect on asphaltene yield and give results totally unrelated to those from any other non-aromatic solvent. For example, when cyclopentane, cyclohexane, or their methyl derivatives are employed as precipitating media, only approximately 1% w/w of the material remains insoluble.

To explain those differences, it was necessary to consider the solvent power of the precipitating liquid, which can be related to molecular properties (Hildebrand et al., 1970). The solvent power of non-polar solvents has been expressed as a solubility parameter (δ) and equated to the internal pressure of the solvent, that is, the ratio between the surface tension (γ) and the cubic root of the molar volume (V):

$$\delta_1 = \sqrt[3]{V}$$

Alternatively, the solubility parameter of non-polar solvents can be related to the energy or vaporization ΔR^v and the molar volume:

$$\delta_2 = (\Delta E^v / V)^{1/2}$$

Also:

$$\delta_2 = (\Delta H^v - RT/V)^{1/2}$$

In these equations, ΔH^v is the heat of vaporization, R is the gas constant, and T is the absolute temperature.

The introduction of a polar group (heteroatom function) into the molecule of the solvent has significant effects on the quantity of precipitate. For example, treatment of a residuum with a variety of ethers or treatment of asphaltenes with a variety of solvents illustrates this point (Speight, 1979). In the latter instance, it was not possible to obtain data from addition of the solvent to the whole feedstock *per se* since the majority of the non-hydrocarbon materials were not miscible with the feedstock. It is nevertheless interesting that, as with the hydrocarbon derivatives, the amount of precipitate, or asphaltene solubility, can be related to the solubility parameter.

The solubility parameter allows an explanation of certain apparent anomalies, for example, the insolubility of asphaltenes in pentane and the near complete solubility of the materials in cyclopentane. Moreover, the solvent power of various solvents is in agreement with the derivation of the solubility parameter; for any one series of solvents, the relationship between amount of precipitate (or asphaltene solubility) and the solubility parameter δ is quite regular.

In any method used to isolate asphaltenes as a separate fraction, standardization of the technique is essential. For many years, the method of asphaltenes separation was not standardized, and even now, it remains subject to the preferences of the standards organizations of different countries. The use of both n-pentane and n-heptane has been widely advocated, and although n-heptane is becoming the deasphalting liquid of choice, this is by no means a hard-and-fast rule. And it must be recognized that large volumes of solvent may be required to effect a reproducible separation, similar to amounts required for consistent asphaltene separation. It is also preferable that the solvents be of sufficiently low boiling point that complete removal of the solvent from the fraction can be effected, and most important, the solvent must not react with the feedstock. Hence, the preference for hydrocarbon liquids, although the several standard methods that have been used are not unanimous in the ratio of hydrocarbon liquid to feedstock. For example:

Method	Deasphalting Liquid	Volume, ml/gm
ASTM D893	n-Pentane	10
ASTM D2007	n-Pentane	10
ASTM D3279	n-Heptane	100
ASTM D4124	n-Heptane	100

However, it must be recognized that some of these methods were developed for use with feedstocks other than heavy oil and adjustments are necessary.

Although n-pentane and n-heptane are the solvents of choice in the laboratory (other solvents can be used (Speight, 1979) and cause the separation of asphaltenes as brown-to-black powdery materials. In the refinery, supercritical low-molecular-weight hydrocarbon derivatives (e.g., liquid propane, liquid butane, or mixtures of both) are the solvents of choice, and the product is a semisolid (tacky) to solid asphalt. The amount of asphalt that settles out of the paraffin/residuum mixture depends on the size of the paraffin, the temperature, and the paraffin-to-feedstock ratio (Corbett and Petrossi, 1978; Speight et al., 1984; Speight, 2014, 2015).

Insofar as industrial solvents are very rarely one compound, it was also of interest to note that the physical characteristics of two different solvent types, in this case benzene and n-pentane, are additive on a mole-fraction basis (Mitchell and Speight, 1973) and also explain the variation of solubility with temperature. The data also show the effects of blending a solvent with the bitumen itself and allowing the resulting solvent-heavy oil blend to control the degree of bitumen solubility. Varying proportions of the hydrocarbon alter the physical characteristics of the oil to such an extent that the amount of precipitate (asphaltenes) can be varied accordingly within a certain range.

At constant temperature, the quantity of precipitate first increases with increasing ratio of solvent to feedstock and then reaches a maximum (Speight et al., 1984). In fact, there are indications that when the proportion of solvent in the mix is <35% little or no asphaltenes are precipitated. In addition, when pentane and the lower molecular weight hydrocarbon solvents are used in large excess, the quantity of precipitate and the composition of the precipitate change with increasing temperature (Mitchell and Speight, 1973).

Contact time between the hydrocarbon and the feedstock also plays an important role in asphaltene separation (Speight et al., 1984). Yields of the asphaltenes reach a maximum after approximately 8 hours, which may be ascribed to the time required for the asphaltene particles to agglomerate into particles of a *filterable size* as well as the diffusion-controlled nature of the process. Heavier feedstocks also need time for the hydrocarbon to penetrate their mass.

After removal of the asphaltene fraction, further fractionation of crude oil is also possible by variation of the hydrocarbon solvent. For example, liquefied gases, such as propane and butane, precipitate as much as 50% by weight of the residuum or bitumen, and the black, tacky, semisolid precipitate is in direct contrast to the pentane-precipitated or heptane-precipitated asphaltene fraction, which is typically a brown, amorphous solid. Treatment of the propane precipitate with pentane then yields the insoluble brown, amorphous asphaltenes and soluble, near-black, semisolid resins, which are, as near as can be determined, equivalent to the resins isolated by adsorption techniques (Speight, 2014, 2015).

2.7.2 ADSORPTION METHODS

The separation of, for example, the maltene fraction by adsorption chromatography essentially commences with the preparation of a porous bed of finely divided solid, the adsorbent ("Porous Bed – an overview I ScienceDirect Topics"). The adsorbent may be contained in an open tube (often referred to as column chromatography), and the sample is introduced at one end of the adsorbent bed and induced to flow through the bed by means of a suitable solvent.

As the sample moves through the bed, various components are held (adsorbed) to a greater or lesser extent depending on the chemical nature of the component. Thus, those molecules that are strongly adsorbed spend considerable time on the adsorbent surface rather than in the moving (solvent) phase, but components that are slightly adsorbed move through the bed comparatively rapidly.

It is essential that, before application of the adsorption technique to the crude oil, the asphaltenes first be completely removed, for example by any of the methods outlined in the previous section. The prior removal of the asphaltenes is essential insofar as they are usually difficult to remove from the earth or clay and may actually be irreversibly adsorbed on the adsorbent.

By definition, the saturate fraction consists of paraffin derivatives and cycloparaffin derivatives (naphthene derivatives). The single-ring naphthene derivatives, or cycloparaffin derivatives, present in crude oil are primarily alkyl-substituted cyclopentane and cyclohexane rings. The alkyl groups are usually quite short, with methyl, ethyl, and isopropyl groups the predominant substituents. As the molecular weight of the naphthene derivatives increases, the naphthene fraction contains more condensed rings with six-membered rings predominating. However, five-membered rings are still present in the complex higher molecular weight molecules.

The aromatic fraction consists of those compounds containing an aromatic ring and varies from monoaromatic derivatives (containing one benzene ring in a molecule) to di-aromatic derivatives (substituted naphthalene) to tri-aromatic derivatives (substituted phenanthrene). Higher condensed ring systems (tetra-aromatic derivatives, penta-aromatic derivatives) are also known but are somewhat less prevalent than the lower ring systems, and each aromatic type will have increasing amounts of condensed ring naphthene attached to the aromatic ring as molecular weight is increased.

However, depending upon the adsorbent employed for the separation, a compound having an aromatic ring (i.e., six aromatic carbon atoms) carrying side chains consisting in toto of more than six carbon atoms (i.e., more than six non-aromatic carbon atoms) will appear in the aromatic fraction.

Careful monitoring of the experimental procedures and the nature of the adsorbent has been responsible for the successes achieved with this particular technique. Early procedures consisted of warming solutions of the crude oil fraction with the adsorbent and subsequent filtration. This procedure has continued to the present day, and separation by adsorption is used commercially in plant operations in the form of clay treatment of crude oil fractions and products. In addition, the proportions of each fraction are subject to the ratio of adsorbent to deasphalted oil.

It is also advisable, once a procedure using a specific adsorbent has been established, that the same type of adsorbent be employed for future fractionation since the ratio of the product fractions varies from adsorbent to adsorbent. It is also very necessary that the procedure be used with caution and that the method not only be reproducible but quantitative recoveries be guaranteed; reproducibility with only, say, 85% of the material recoverable is not a criterion of success.

There are two procedures that have received considerable attention over the years, and these are (1) the United States Bureau of Mines-American Crude oil Institute (USBM-API) method and (2) the saturate derivatives-aromatic derivatives-resin constituents-asphaltene constituents (SARA) method. This latter method is often also called the saturate derivatives-aromatic derivatives-polar constituents-asphaltene constituents (SAPA) method. These two methods are used as representing the standard methods of crude oil fractionation. Other methods are also noted, especially when the method has added further meaningful knowledge to compositional studies (Speight, 2014, 2015).

However, there are precautions that must be taken when attempting to separate heavy feedstocks (heavy oil, tar sand bitumen) or polar feedstocks into constituent fractions. The disadvantages in using ill-defined adsorbents are that adsorbent performance differs with the same feed and, in certain instances, may even cause chemical and physical modification of the feed constituents. The use of a chemical reactant like sulfuric acid should only be advocated with caution since feedstocks react differently and may even cause irreversible chemical changes and/or emulsion formation (Abdel-Aal et al., 2016). These advantages may be of little consequence when it is not, for various reasons, the intention to recover the various product fractions in toto or in the original state, but in terms of the compositional evaluation of different feedstocks the disadvantages are very real.

In summary, the terminology used for the identification of the various methods might differ. However, in general terms, group-type analysis of crude oil is often identified by the acronyms for the names: PONA (paraffin derivatives, olefin derivatives, naphthene derivatives, and aromatic derivatives), PIONA (paraffin derivatives, iso-paraffin derivatives, olefin derivatives, naphthene derivatives, and aromatic derivatives), PNA (paraffin derivatives, naphthene derivatives, and aromatic derivatives), PINA (paraffin derivatives, iso-paraffin derivatives, naphthene derivatives, and aromatic derivatives), or SARA (saturate derivatives, aromatic derivatives, resins, and asphaltenes). However, it must be recognized that the fractions produced by the use of different adsorbents will differ in content and will also be different from fractions produced by solvent separation techniques.

The variety of fractions isolated by these methods and the potential for the differences in composition of the fractions make it even more essential that the method is described accurately and that it reproducible not only in any one laboratory but also between various laboratories.

2.7.3 CHEMICAL METHODS

Methods of fractionation using chemical reactants are entirely different in nature from the methods described in the preceding sections. Although several methods using chemical reactants have been applied to fractionation, method such as adsorption, solvent treatment, and treatment with alkali (Speight, 2014, 2015), these methods are often applied to product purification as well as separation.

The method of chemical separation commonly applied to separate crude oil into various fractions is treatment with sulfuric acid and, since this method has also been applied in the refinery but with limited success in the fractionation of heavy oil and/or bitumen due to the formation of complex sulfates and difficult-to-break emulsions (Speight, 2014, 2015). Obviously, the success of this fractionation method is feedstock dependent, and in conclusion, it would appear that the test be left more as a method of product cleaning for which it was originally designed rather than a method of separation of the various fractions.

2.8 USE OF THE DATA

In the simplest sense, any viscous feedstock (in fact, any unrefined crude oil-related feedstock crude oil is a composite of four major fractions that are defined by the method of separation but, more important, the behavior and properties of any feedstock are dictated by composition (Speight, 2014, 2015). Although early studies were primarily focused on the composition and behavior of asphalt, the techniques developed for those investigations have provided an excellent means of studying heavy feedstocks (Tissot, 1984). Later studies have focused not only on the composition of crude oil and its major operational fractions but on further fractionation that allows different feedstocks to be compared on a relative basis and to provide a very simple but convenient feedstock *map.*

Such a map does not give any indication of the complex interrelationships of the various fractions (Koots and Speight, 1975), although predictions of feedstock behavior are possible using such data. It is necessary to take the composition studies one step further using sub-fractionation of the major fractions to obtain a more representative indication of crude oil composition.

Furthermore, crude oil can be viewed as consisting of two continuous distributions: one of molecular weight and the other of molecular type. Using data from molecular weight studies and elemental analyses, the number of nitrogen and sulfur atoms in the aromatic fraction and in the polar aromatic fraction can also be also exhibited. These data showed that not only can every molecule in the resins and asphaltenes have more than one sulfur atom or more than one nitrogen atom but also some molecules probably contain both sulfur and nitrogen. As the molecular weight of the aromatic fraction decreases, the sulfur and nitrogen contents of the fractions also decrease. In contrast to the sulfur-containing molecules, which appear in both the naphthene-aromatic derivatives and the polar aromatic fractions, the oxygen compounds present in the heavy fractions of crude oil are normally found in the polar aromatic derivatives fraction.

More recent work (Long and Speight, 1989) involved in the development of a different type of compositional map using the molecular weight distribution and the molecular type distribution as coordinates. The separation involved the use of an adsorbent such as clay, and the fractions were characterized by *solubility parameter* as a measure of the polarity of the molecular types. The molecular weight distribution can be determined by gel permeation chromatography. Using these two distributions, a map of composition can be prepared using molecular weight and solubility parameter as the coordinates for plotting the two distributions. Such a composition map can provide insights into many separation and conversion processes used in crude oil refining.

The molecular type was characterized by the polarity of the molecules, as measured by the increasing adsorption strength on an adsorbent. At the time of the original concept, it was unclear how to characterize the continuum in molecular type or polarity. For this reason, the molecular type coordinate of their first maps was the yield of the molecular types ranked in order of increasing polarity. However, this type of map can be somewhat misleading because the areas are not related to the amounts of material in a given type. The horizontal distance on the plot is a measure of the yield, and there is not a continuous variation in polarity for the horizontal coordinate. It was suggested that the solubility parameter of the different fractions could be used to characterize of both polarity and adsorption strength.

In order to attempt to remove some of these potential ambiguities, more recent developments of this concept have focused on the solubility parameter, estimated by the values for the eluting solvents that remove the fractions from the adsorbent. The simplest maps that can be derived using the solubility parameter are produced with the solubility parameters of the solvents used in solvent separation procedures, and equating these parameters to the various fractions (Long and Speight, 1998; Speight, 2014).

Thus, a composition map can be used to show where a particular physical or chemical property tends to concentrate on the map. For example, the *coke-forming propensity*, i.e., the amount of the carbon residue, is shown for various regions on the map for a sample of atmospheric residuum (Long and Speight, 1998). The plot shows molecular weight plotted against weight percent yield in order of increasing polarity. The dashed line is the envelope of composition of the total sample. The slanted lines show the boundaries of solvent-precipitated fractions, and the vertical lines show the boundaries of the fractions obtained by clay adsorption of the pentane-deasphalted oil.

A composition map can be very useful for predicting the effectiveness of various types of separations or conversions of crude oil (Long and Speight, 1998). These processes are adsorption, distillation, solvent precipitation with relatively non-polar solvents, and solvent extraction with polar solvents. The vertical lines show the cut points between saturate derivatives, aromatic derivatives, and polar aromatic derivatives as determined by clay chromatography. The slanted lines show how distillation, extraction, and solvent precipitation can divide the composition map. The line for distillation divides the map into distillate, which lies below the dividing line, and bottoms, which lies above the line. As the boiling point of the distillate is raised, the line moves upward, including higher molecular weight materials and more of the polar species in the distillate and rejecting lower molecular weight materials from the bottoms. As more of the polar species are included in the distillate, the carbon residue of the distillate rises. In contrast to the cut lines generated by separation processes, conversion processes move materials in the composition from one molecular type to another (Long and Speight, 1998).

The ultimate decision in the choice of any particular fractionation technique must be influenced by the need for the data. For example, there are those needs that require only that the crude oil be separated into four bulk fractions. On the other hand, there may be the need to separate the crude oil into many sub-fractions in order to define specific compound types (Green et al., 1988; Vogh and Reynolds, 1988). Neither method is incorrect; each method is merely being used to answer the relevant questions about the character of the crude oil.

REFERENCES

Abdel-Aal, H.K., Aggour, M.A., and Fahim, M.A. 2016. *Petroleum and Gas Filed Processing*. CRC Press, Taylor & Francis Publishers, Boca Raton, Florida.

Abraham, H. 1945. *Asphalt and Allied Substances*. 5th Edition. Van Nostrand Inc., New York. Volume I. p. 1.

Agrawal, K.M., Prakash, J., Gomkale, A.V., Kumar, S., and Rawat, B.S. 1997. Production of Microcrystalline Waxes from Crude Oil Tank Sludge: Techno-Economic Aspects. Proceedings of the 2nd International Petroleum Conference and Exhibition. PETROTECH-97. New Delhi. Page 59–62.

Altgelt, K.H., and Boduszynski, M.M. 1994. *Compositional Analysis of Heavy Petroleum Fractions*. Marcel Dekker Inc., New York.

ASTM D4. 2015. *Test Method for Bitumen Content. Annual Book of Standards.* ASTM International, West Conshohocken, Pennsylvania.

ASTM D893. 2015. *Standard Test Method for Insolubles in Used Lubricating Oils. Annual Book of Standards.* ASTM International, West Conshohocken, Pennsylvania.

ASTM D1018. 2015. *Standard Test Method for Hydrogen in Petroleum Fractions. Annual Book of Standards.* ASTM International, West Conshohocken, Pennsylvania.

ASTM D1266. 2015. *Standard Test Method for Sulfur in Petroleum Products (Lamp Method). Annual Book of Standards.* ASTM International, West Conshohocken, Pennsylvania.

ASTM D1318. 2015. *Standard Test Method for Sodium in Residual Fuel Oil (Flame Photometric Method). Annual Book of Standards.* ASTM International, West Conshohocken, Pennsylvania.

ASTM D1552. 2015. *Standard Test Method for Sulfur in Petroleum Products (High-Temperature Method). Annual Book of Standards.* ASTM International, West Conshohocken, Pennsylvania.

ASTM D1757. 2015. *Standard Test Method for Sulfur in Ash from Coal and Coke. Annual Book of Standards.* ASTM International, West Conshohocken, Pennsylvania.

ASTM D2007. 2015. *Standard Test Method for Characteristic Groups in Rubber Extender and Processing Oils and Other Petroleum-Derived Oils by the Clay-Gel Absorption Chromatographic Method. Annual Book of Standards.* ASTM International, West Conshohocken, Pennsylvania.

ASTM D2622. 2015. Standard Test Method for Sulfur in Petroleum Products by Wavelength Dispersive X-ray Fluorescence Spectrometry. Annual Book of Standards. ASTM International, West Conshohocken, Pennsylvania.

ASTM D3177. 2015. *Standard Test Methods for Total Sulfur in the Analysis Sample of Coal and Coke. Annual Book of Standards.* ASTM International, West Conshohocken, Pennsylvania.

ASTM D3178. 2015. *Standard Test Methods for Carbon and Hydrogen in the Analysis Sample of Coal and Coke. Annual Book of Standards.* ASTM International, West Conshohocken, Pennsylvania.

ASTM D3179. 2015. *Standard Test Methods for Nitrogen in the Analysis Sample of Coal and Coke. Annual Book of Standards.* ASTM International, West Conshohocken, Pennsylvania.

ASTM D3228. 2015. *Standard Test Method for Total Nitrogen in Lubricating Oils and Fuel Oils by Modified Kjeldahl Method. Annual Book of Standards.* ASTM International, West Conshohocken, Pennsylvania.

ASTM D3279. 2015. *Standard Test Method for n-Heptane Insolubles. Annual Book of Standards.* ASTM International, West Conshohocken, Pennsylvania.

ASTM D3340. 2015. *Standard Test Method for Lithium and Sodium in Lubricating Greases by Flame Photometer. Annual Book of Standards.* ASTM International, West Conshohocken, Pennsylvania.

ASTM D3341. 2015. *Standard Test Method for Lead in Gasoline—Iodine Monochloride Method. Annual Book of Standards.* ASTM International, West Conshohocken, Pennsylvania.

ASTM D3343. 2015. *Standard Test Method for Estimation of Hydrogen Content of Aviation Fuels. Annual Book of Standards.* ASTM International, West Conshohocken, Pennsylvania.

ASTM D3605. 2015. *Standard Test Method for Trace Metals in Gas Turbine Fuels by Atomic Absorption and Flame Emission Spectroscopy.* Annual Book of Standards. ASTM International, West Conshohocken, Pennsylvania.

ASTM D3701. 2015. *Standard Test Method for Hydrogen Content of Aviation Turbine Fuels by Low Resolution Nuclear Magnetic Resonance Spectrometry. Annual Book of Standards.* ASTM International, West Conshohocken, Pennsylvania.

ASTM D4045. 2015. *Standard Test Method for Sulfur in Petroleum Products by Hydrogenolysis and Rateometric Colorimetry. Annual Book of Standards.* ASTM International, West Conshohocken, Pennsylvania.

ASTM D4124. 2015. *Standard Test Method for Separation of Asphalt into Four Fractions. Annual Book of Standards.* ASTM International, West Conshohocken, Pennsylvania.

ASTM D4294. 2015. *Standard Test Method for Sulfur in Petroleum and Petroleum Products by Energy Dispersive X-ray Fluorescence Spectrometry. Annual Book of Standards.* ASTM International, West Conshohocken, Pennsylvania.

ASTM E258. 2015. *Standard Test Method for Total Nitrogen in Organic Materials by Modified Kjeldahl Method. Annual Book of Standards.* ASTM International, West Conshohocken, Pennsylvania.

ASTM E385. 2015. *Standard Test Method for Oxygen Content Using a 14-MeV Neutron Activation and Direct-Counting Technique. Annual Book of Standards.* ASTM International, West Conshohocken, Pennsylvania.

ASTM E777. 2015. *Standard Test Method for Carbon and Hydrogen in the Analysis Sample of Refuse-Derived Fuel. Annual Book of Standards.* ASTM International, West Conshohocken, Pennsylvania.

ASTM E778. 2015. *Standard Test Methods for Nitrogen in Refuse-Derived Fuel Analysis Samples. Annual Book of Standards.* ASTM International, West Conshohocken, Pennsylvania.

Bestougeff, M.A. 1961. J. Études Methodes Separation Immediate Chromatogr. Comptes Réndus (Paris). Page 55.

Bjorseth, A. 1983. *Handbook of Polycyclic Aromatic Hydrocarbons*. Marcel Dekker Inc., New York.

Bonnett, R. 1978. *The Porphyrins. Volume I. Structure and Synthesis. Part A*. Academic Press, New York.

Brooks, J., and Welte, D.H. 1984. *Advances in Petroleum Geochemistry*. Academic Press Inc., New York.

Burke, J. 1996. *The Pinball Effect*. Little, Brown and Company, New York. Page 25 and 26.

Corbett, L.W., and Petrossi, U. 1978. Differences in Distillation and Solvent Separated Asphalt Residua. *Industrial & Engineering Chemistry Product Research and Development*, 17: 342.

Delbianco, A., and Montanari, R. 2009. *Encyclopedia of Hydrocarbons, Volume III/New Developments: Energy, Transport, Sustainability*. Eni S.p.A., Rome, Italy.

Durand, B. 1980. *Kerogen: Insoluble Organic Matter from Sedimentary Rocks*. Editions Technip, Paris, France.

Eglinton, G., and Murphy, B. 1969. *Organic Geochemistry: Methods and Results*. Springer-Verlag, New York.

Felix, G., Bertrand, C., and Van Gastel, F. 1985. Hydroprocessing of Heavy Oils and Residua. *Chromatographia*, 20: 155–160.

Forbes, R.J. 1958a. *A History of Technology*. Oxford University Press, Oxford, United Kingdom. Volume V. p. 102.

Forbes, R.J. 1958b. *Studies in Early Petroleum Chemistry*. E. J. Brill, Leiden, The Netherlands.

Forbes, R.J. 1959. *More Studies in Early Petroleum Chemistry*. E.J. Brill, Leiden, The Netherlands.

Gallegos, E.J.J. 1981. Alkylbenzenes Derived from Carotenes in Coals by GC/MS. *Journal of Chromatographic Science*, 19: 177–182.

Gary, J.G., Handwerk, G.E., and Kaiser, M.J. 2007. *Petroleum Refining: Technology and Economics*. 5th Edition. CRC Press, Taylor & Francis Group, Boca Raton, Florida.

Green, J.B., Grizzle, P.L., Thomson, P.S., Shay, J.Y., Diehl, B.H., Hornung, K. W., and Sanchez, V. 1988. Report No. DE88 001235. Contract FC22-83F460149. United States Department of Energy, Washington, DC.

Green, J.A., Green, J.B., Grigsby, R.D., Pearson, C.D., Reynolds, J.W., Sbay, I.Y., Sturm, O.P. Jr., Thomson, J.S., Vogh, J.W., Vrana, R.P., Yu, S.K.Y., Diem, B.H., Grizzle, P.L., Hirsch, D.E., Hornung, K.W., Tang, S.Y., Carbognani, L., Hazos, M., and Sanchez, V. 1989. Analysis of Heavy Oils: Method Development and Application to Cerro Negro Heavy Petroleum, NIPER-452 (DE90000200). Volumes I and II. Research Institute, National Institute for Petroleum and Energy Research (NIPER), Bartlesville, Oklahoma.

Guthrie, V.B. 1960. *The Petroleum Products Handbook*. McGraw-Hill Education, New York.

Hildebrand, J.H., Prausnitz, J.M., and Scott, R.L. 1970. *Regular Solutions*. Van Nostrand-Reinhold, New York.

Hsu, C.S., and Robinson, P.R. (Editors). 2017. *Handbook of Petroleum Technology*. Springer, Cham, Switzerland.

Killops, S.D., and Readman, J.W. 1985. HPLC Fractionation and GC-MS Determination of Aromatic Hydrocarbons for Oils and Sediments. *Organic Geochemistry*, 8: 247–257.

Koots, J.A., and Speight, J.G. 1975. The Relationship of Petroleum Resins to Asphaltenes. *Fuel*. 54: 179.

Kumar, S., Nautiyal, S.P., and Agrawal, K.M. 2007. Physical Properties of Petroleum Waxes 1: Effect of Oil Content. *Petroleum Science and Technology*, 25: 1531–1537.

Lee, M.L., Novotny, M.S., and Bartle, K.D. 1981. *Analytical Chemistry of Polycyclic Aromatic Compounds*. Academic Press Inc., New York.

Long, R.B. 1979. The Concept of Asphaltenes. Preprints. *American Chemical Society, Division of Petroleum Chemistry*, 24(4): 891.

Long, R.B. 1981. The Concept of Asphaltenes. In: *The Chemistry of Asphaltenes*. J.W. Bunger and N. Li (Editors). Advances in Chemistry Series No. 195. American Chemical Society, Washington, DC.

Long, R.B., and Speight, J.G. 1989. Studies in Petroleum Composition. I: Development of a Compositional Map for Various Feedstocks. *Revue de l'Institut Français du Pétrole*, 44: 205.

Long, R.B., and Speight, J.G. 1998. The Composition of Petroleum. In: *Petroleum Chemistry and Refining*. J.G. Speight (Editor). Taylor & Francis Publishers, Washington, DC. Chapter 1.

Meyer, R.F., and De Witt, W. Jr. 1990. Definition and World Resources of Natural Bitumens. Bulletin No. 1944. US Geological Survey, Reston, Virginia.

Mitchell, D.L., and Speight, J.G. 1973. The Solubility of Asphaltenes in Hydrocarbon Solvents. *Fuel*, 52: 149.

Monin, J.C., and Pelet, R. 1983. In *Advances in Organic Geochemistry*. M. Bjorev (Editor). John Wiley & Sons Inc., New York.

Pandey, S.C., Ralli, D.K., Saxena, A.K., and Alamkhan, W.K. 2004. Physicochemical Characterization and Application of Naphtha. *Journal of Scientific and Industrial Research*, 63(3): 276–282.

Parkash, S. 2003. *Refining Processes Handbook*. Gulf Professional Publishing, Elsevier, Amsterdam, Netherlands.

Pearson, C.D., and Green, J.B. 1993. Vanadium and Nickel Complexes in Petroleum Resid Acid, Base, and Neutral Fractions. *Energy & Fuels*, 7: 338–346.

Prince, R.O. Elmendoff, D.L., Lute, B.R., Hsu, C.S., Hath, C.., Sunnis, B.P., Decherd, G., Douglas, D., and Butler, E. 1994. *Environmental Science and Technology*, 28: 142.

Sakarnen, K.V., and Ludwig, C.H. 1971. *Lignins: Occurrence, Formation, Structure and Reactions.* John Wiley & Sons Inc., New York.

Schucker, R.C., and Keweshan, C.F. 1980. Reactivity of Cold Lake Asphaltenes. Preprints. *Division of Fuel Chemistry American Chemical Society*, 25: 155.

Seifert, W.K., and Teeter, R.M. 1970. Identification of Polycyclic Aromatic and Heterocyclic Crude. Oil Carboxylic Acids. *Analytical Chemistry*, 42; 750–758.

Speight, J.G. 1971. Thermal Cracking of Athabasca Bitumen, Athabasca Asphaltenes, and Athabasca Deasphalted Heavy Oil. *Fuel*, 49: 134.

Speight, J.G. 1981. Asphaltenes as an Organic Natural Product and Their Influence on Crude Oil Properties. Proceedings. Division of Geochemistry. American Chemical Society. . New York Meeting.

Speight, J.G., and Pancirov, R.J. 1984. Structural Types in Asphaltenes as Deduced from Pyrolysis-Gas Chromatography-Mass Spectrometry. *Liquid Fuels Technology*, 2: 287.

Speight, J.G. 1986. Polynuclear Aromatic Systems in Petroleum. Preprints. *American Chemical Society, Division of Petroleum Chemistry*, 31(4): 818.

Speight, J.G. 1987. Initial Reactions in the Coking of Residua. Preprints. *American Chemical Society, Division of Petroleum Chemistry*, 32(2): 413.

Speight, J.G. 1990. Tar Sands. In: *Fuel Science and Technology Handbook.* J.G. Speight (Editor). Marcel Dekker Inc., New York. Chapter 12.

Speight, J.G., and Francisco, M.A. 1990. Studies in Petroleum Composition IV: Changes in the Nature of Chemical Constituents during Crude Oil Distillation. *Revue de l'Institut Français du Pétrole*, 45: 733.

Speight, J.G. 1994. Chemical and Physical Studies of Petroleum Asphaltenes. In: *Asphaltenes and Asphalts. I. Developments in Petroleum Science*, 40. T.F. Yen and G.V. Chilingarian (Editors). Elsevier, Amsterdam, Netherlands. Chapter 2.

Speight, J.G. 2000. *The Desulfurization of Heavy Oils and Residua.* 2nd Edition. Marcel Dekker, New York.

Speight, J.G. 2001. *Handbook of Petroleum Analysis.* John Wiley & Sons Inc., Hoboken, New Jersey.

Speight, J.G. 2009. *Enhanced Recovery Methods for Heavy Oil and Tar Sands.* Gulf Publishing Company, Houston, Texas.

Speight, J.G. 2013a. *The Chemistry and Technology of Coal.* 3rd Edition. CRC Press, Taylor & Francis Group, Boca Raton, Florida.

Speight, J.G. 2013b *Heavy Oil Production Processes.* Gulf Professional Publishing, Elsevier, Oxford, United Kingdom.

Speight, J.G. 2013c. *Oil Sand Production Processes.* Gulf Professional Publishing, Elsevier, Oxford, United Kingdom.

Speight, J.G. 2014. *The Chemistry and Technology of Petroleum.* 5th Edition. CRC Press, Taylor & Francis Group, Boca Raton, Florida.

Speight, J.G. 2015. *Handbook of Petroleum Product Analysis.* 2nd Edition. John Wiley & Sons Inc., Hoboken, New Jersey.

Speight, J.G. 2016. *Introduction to Enhanced Recovery Methods for Heavy Oil and Tar Sands.* 2nd Edition. Gulf Professional Publishing, Elsevier, Oxford, United Kingdom.

Speight, J.G. 2017. *Handbook of Petroleum Refining.* CRC Press, Taylor & Francis Group, Boca Raton, Florida.

Speight, J.G. 2019. *Handbook of Petrochemical Processes.* CRC Press, Taylor & Francis Group, Boca Raton, Florida.

Speight, J.G. 2020a. *The Refinery of the Future.* 2nd Edition. Gulf Professional Publishing, Elsevier, Oxford, United Kingdom.

Speight, J.G. 2020b. *Synthetic Fuels Handbook: Properties, Processes, and Performance.* 2nd Edition. McGraw-Hill, New York.

Speight, J.G. 2021. *Refinery Feedstocks.* CRC Press, Taylor & Francis Group, Boca Raton, Florida.

Thompson, C.J., Ward, C.C., and Ball, J.S. 1976. Characteristics of World's Crude Oils and Results of API Research Project 60. Report BERC/RI-76/8. Bartlesville Energy Technology Center, Bartlesville, Oklahoma.

Tissot, B.P., and Welte, D.H. 1978. *Petroleum Formation and Occurrence.* Springer-Verlag, New York.

Tissot, B.P. (Editor). 1984. *Characterization of Heavy Crude Oils and Petroleum Residues.* Editions Technip, Paris.

Vogh, J.W., and Reynolds, J.W. 1988. Report No. DE88 001242. Contract FC22-83FE60149. United States Department of Energy, Washington, DC.

Wallace, D., Starr, J., Thomas, K.P., and Dorrence. S.M. 1988. *Characterization of Oil Sands Resources.* Alberta Oil Sands Technology and Research Authority, Edmonton, Alberta, Canada.

Weiss, V., and Edwards, J.M. 1980. *The Biosynthesis of Aromatic Compounds.* John Wiley & Sons, Inc., New York.

Yan, J., Plancher, H., and Morrow, N.R. 1997. Wettability Changes Induced by Adsorption of Asphaltenes. Paper No. SPE 37232. Proceedings of the SPE International Symposium on Oilfield Chemistry. Houston, Texas. Society of Petroleum Engineers, Richardson, Texas.

3 Feedstock Evaluation

3.1 INTRODUCTION

Just as the type of feedstock controls the processes used in a crude oil refinery, the pre-refining evaluation of the feedstock is a necessary aspect of refinery practice. Evaluation, in this context, is the determination of the physical and chemical characteristics of crude oil, heavy oil, extra heavy oil, and tar sand bitumen since the yields and properties of products or factions produced from these feedstocks vary considerably and are dependent on the concentration of the various types of hydrocarbon derivatives as well as the amounts of the heteroatom compounds (i.e., molecular constituents contacting nitrogen and/or oxygen and/or sulfur and/or metals). Moreover, the physical and chemical characteristics of the non-viscous feedstocks and the viscous feedstocks and the yields and properties of products or factions prepared from them vary considerably and are dependent on the concentration of the various types of hydrocarbon derivatives and minor constituents present.

The evaluation of the suitability of crude oil feedstocks and viscous feedstocks as refinery feedstocks is determined by using a series of standard test methods that are set by organizations varying from country to country.

Since refinery feedstocks exhibit a wide range of physical properties, it is not surprising the behavior of various feedstocks in these refinery operations is not simple. The atomic ratios from ultimate analysis can give an indication of the nature of a feedstock and the generic hydrogen requirements to satisfy the refining chemistry, but it is not possible to predict with any degree of certainty how the feedstock will behave during refining. In addition, the chemical composition of a feedstock is also an indicator of refining behavior insofar as the composition can be a pointer to the manner which a particular feedstock should be processed (Parkash, 2003; Gary et al., 2007; Speight, 2014; Hsu and Robinson, 2017; Speight, 2017).

Evaluation of a viscous feedstock usually involves an examination of one or more of the physical properties of the material (Table 3.1).

By this means, a set of basic characteristics can be obtained that can be correlated with utility. The physical properties of crude oil and crude oil products are often equated with those of the hydrocarbon derivatives for although crude oil is indeed a very complex mixture, there is gasoline, produced by nondestructive distillation, in which very few hydrocarbon derivatives make up at least 50% of the material (Speight, 2014, 2015).

To satisfy specific needs with regard to the type of crude oil to be processed, as well as to the nature of the product, most refiners have, through time, developed their own methods of crude oil analysis and evaluation. Consequently, various standards organizations, such as the ASTM International in North America, have devoted considerable time and effort to the correlation and standardization of methods for the inspection and evaluation of crude oil and crude oil products.

For the purposes of this book, the United States typically uses the ASTM International (formerly the American Society for Testing and Materials ASTM) standard test methods and the standard test methods of the American Petroleum Institute (API), which have established the standard test methods for determining the properties of refinery feedstocks (and, hence, the refinery processes to be applied to the feedstocks) and also standard test methods' application to refinery products which, in this case of refinery products, contribute to the formulation of product specifications. Other countries also have institutions that formulate the standard test methods. On the other hand, many crude oil companies, while not using the ASTM standard test methods and/or the API standard test methods, may have formulated in-company analytical test methods for crude oil and crude

DOI: 10.1201/9781003184904-3

TABLE 3.1

Simplified Differentiation between Conventional Crude Oil, Tight Oil, Heavy Crude Oil, Extra Heavy Crude Oil, and Tar Sand Bitumen

Conventional Crude Oil

Mobile in the reservoir; API gravity: >25°
Primary recovery methods
Secondary recovery methods

Heavy Crude Oil

More viscous than conventional crude oil; API gravity: 10°–20°
Mobile in the reservoir
Secondary recovery methods
Tertiary recovery (enhanced oil recovery – EOR, e.g., steam stimulation)

Extra Heavy Crude Oil

Similar properties to the properties of tar sand bitumen; API gravity: <10°
Mobile in the deposit due to deposit temperature higher than pour point of the oil
Secondary recovery methods
Tertiary recovery methods (enhanced oil recovery – EOR, e.g., steam stimulation)

Tar Sand Bitumen

Similar properties the properties of extra heavy oil; API gravity: <10°
Immobile in the deposit due to deposit temperature lower than pour point of the oil
Mining (often preceded by explosive fracturing)
Extreme heating methods

Note: This list is not intended for use as a means of classification of the feedstocks (Speight, 2021).

oil products, but these individual in-company standard test methods remain proprietary and very rarely published.

In relation to the ASTM and API standard test methods, three frequently specified properties are density–specific gravity–API gravity, characterization factor, and sulfur content (Speight, 2014, 2015). In addition, the API gravity is a measure of density or specific gravity:

$$°API = (141:5 \text{ / specific gravity}) - 131.5$$

The specific gravity is the ratio of the weight of a given volume of oil to the weight of the same volume of water at a standard temperature, usually 60°F (15.6°C). This method of measuring density and gravity first arose as a result of the need to define the character of products in more detail; it was natural to extend the measure to crude oils in general.

A complete discussion of the large number of routine tests available for crude oil fills an entire book (Speight, 2015; ASTM, 2021). However, it seems appropriate that in any discussion of the physical properties of crude oil and crude oil products reference be made to the corresponding test, and accordingly, the various test numbers have been included in the text.

It is the purpose of this chapter to present an outline of the tests that may be applied to crude oil, heavy crude oil, extra heavy crude oil, and tar sand bitumen as well as to their respective products as a pointer to the chemical properties and physical properties from which a feedstock or product can be evaluated (Speight, 2013b, c, d, 2014, 2015, 2016). For this purpose, data relating to various chemical-physical properties have been included as illustrative examples, but theoretical discussions of the physical properties of hydrocarbon derivatives were deemed irrelevant and are omitted.

3.2 FEEDSTOCK EVALUATION

The evaluation of refinery feedstocks is a necessary series of actions because of the variety of materials that are accepted as feedstocks for the modern refinery. Also, the time when one crude oil was the sole feedstock for a refinery is limited (if not non-existent) in the modern refinery because of the use of a blend of two or more crude oils as the refinery feedstock. This evaluation of the blend (in fact, each of the components of the blend) is necessary to allow an assessment of the product slate and the processes that are necessary to create the saleable products. Thus, feedstock evaluation is the necessary application of a series of test methods in order to assess the chemical and physical properties of each of the constituents of the blend as well as the blend itself against a specific set of refinery-developed standards. A comprehensive feedstock evaluation is based on a detailed analysis is necessary to determine the value of these crude oils to the refinery.

The series of test methods are often referred to as the feedstock assay which consists of determination of (1) the feedstock assay, (2) the physical properties, (3) the thermal properties, (4) the chromatographic properties, and, when deemed appropriate, and (5) the spectroscopic properties which offer insights into the chemical constituents in the feedstock and the potential for reaction of these chemical constituents with other constituents of the blend and with catalysts used in catalytic processes.

Thus, analyses are performed (by a series of standard test methods) to determine whether each batch of crude oil or the viscous feedstock received at the refinery is not only suitable for refining purposes but also to determine the preferential process options and the order in which the processes should be applied. To obtain the necessary information, two different analytical schemes are commonly used and these are: (1) an inspection assay and (2) a comprehensive assay (Table 3.2). The

---

TABLE 3.2
Recommended Inspection Data Required for Crude Oil and Viscous Feedstocks[a]

Crude Oil	Viscous Feedstocks
Density, specific gravity	Density, specific gravity
API gravity	API gravity
Carbon, wt. %	Carbon, wt. %
Hydrogen, wt. %	Hydrogen, wt. %
Nitrogen, wt. %	Nitrogen, wt. %
Sulfur, wt. %	Sulfur, wt. %
	Nickel, ppm
	Vanadium, ppm
	Iron, ppm
Pour point	Pour point
Wax content	
Wax appearance temperature	
Viscosity (various temperatures)	Viscosity (various temperatures)
Carbon residue of residuum	Carbon residue[a]
	Ash, wt. %
Distillation profile:	Fractional composition:
All fractions plus vacuum residue	Asphaltenes, wt. %
	Resin constituents, wt. %
	Aromatics, wt. %
	Saturates, wt. %

Source: Speight, J.G. 2014. *The Chemistry and Technology of Petroleum.* 4th Edition. CRC Press, Taylor & Francis Publishers, Boca Raton Florida. Table 10.1, page 244.

[a] Includes heavy crude oil, extra heavy crude oil, tar sand bitumen, and residua.

test applied to the feedstock to obtain the assay data are, for the purposes of this text, the standard methods as described by the ASTM International (formerly the American Society for Testing and Materials) (ASTM, 2021).

An inspection assay (which is routinely performed on all feedstocks received at a refinery) typically involves determination of several bulk properties of the feedstock (such as API gravity, sulfur content, pour point, distillation range, viscosity, salt content, water, and sediment content, and trace metals) as a means to determine if any major changes in characteristics have occurred since the last comprehensive assay was performed within the refinery. Other properties that are determined on an as-needed basis include, but are not limited to, the following: (1) vapor pressure (Reid method, ASTM D323) and (2) total acid number (ASTM D664). The results from these tests with the archived data from a comprehensive assay provide an estimate of any changes that have occurred in the crude oil that may be critical to refinery operations. The aforementioned inspection assay test methods are not exhaustive but are the ones most commonly used and provide data on the impurities present as well as a general idea of the products that may be recoverable (Speight, 2014, 2015).

On the other hand, a full assay (also known as a comprehensive assay) is more complex and is usually performed when the feedstocks from a new field (in the case of crude oil and heavy crude oil) or a new deposit (in the case of extra heavy oil and tar sand bitumen) comes on stream, or when the inspection assay indicates that significant changes in the composition of the feedstock have occurred. Except for these circumstances, a comprehensive assay of a particular feedstock may not (unfortunately) be updated for several years.

The data derived from the test assay can be used to assess the quality (and potential behavior) of the feedstock in the refinery and, therefore, indicate a degree of predictability of feedstocks performance during refining. However, knowledge of the basic concepts of refining will help the analyst understand the production and, to a large extent, the anticipated properties of the product, which in turn is related to storage, sampling, and handling the products.

3.2.1 Physical Properties

For the purposes of this text, a physical property is any property that is measurable and the value of which describes the physical state of crude oil that does not change the chemical nature of the crude oil.

Before any volatility tests are carried out, it must be recognized that the presence of more than 0.5% water in test samples of crude can cause several problems during various test procedures and produce erroneous results. For example, during various thermal tests, water (which has a high heat of vaporization) requires the application of additional thermal energy to the distillation flask. In addition, water is relatively easily superheated, and therefore, excessive bumping can occur, leading to erroneous readings and the potential for destruction of the glass equipment is real. Steam formed during distillation can act as a carrier gas, and high-boiling-point components may end up in the distillate (often referred to as steam distillation).

Removal of water (and sediment) can be achieved by centrifugation if the sample is not a tight (almost unbreakable) emulsion. Other methods that are used to remove water include: (1) heating in a pressure vessel to control loss of light ends, (2) addition of calcium chloride as recommended in ASTM D1160, (3) addition of an azeotroping agent such as iso-propanol or n-butanol, (4) removal of water in a preliminary low-efficiency or flash distillation followed by re-blending the hydrocarbon which co-distills with the water into the sample, and (5) separation of the water from the hydrocarbon distillate by freezing.

Thus, the standard test methods described below in the various sections (presented in alphabetical order rather than order of preference) are based on the assumption that any water in the sample has been reduced to an acceptable level is usually defined by each standard test.

3.2.1.1   Acid Number

Another characteristic of crude oil is the total acid number (TAN), which represents a composite of acids present in the crude oil (ASTM D664) and is often also expressed as the neutralization number. Crude oils having a high acid number (high TAN number) account for an increasing percentage of the global crude oil market, and, in fact, the increase in world production of heavy, sour, and high-TAN crude oils will impact many world (especially US) refineries (Shafizadeh et al., 2003; Sheridan, 2006).

High acid crude oils are considered to be those with an acid content >1.0 mg KOH/g sample (many refiners consider TAN number greater than 0.5 mg KOH/g to be high), and refiners looking for discounted crude supplies will import and use greater volumes of high total acid number (TAN) crude oils.

In the United States, using California as the example, Wilmington crude oil and Kern crude oil have a TAN ranging from 2.2 to 3.2 mg, respectively. However, some acids are relatively inert and the TAN number does not always give a true reflection of the corrosive properties of the crude oil. Furthermore, different naphthenic acids (a broad group of organic acids in crude oil) will react at different temperatures – making it difficult to pinpoint the processing units within the refinery that will be affected by a particular high-TAN crude oil. However, the impact of corrosive, high-TAN, crude oils can be overcome by blending higher- and lower-TAN crude oils, installing or retrofitting equipment with anticorrosive materials, or by developing low-temperature catalytic decarboxylation processes using metal catalysts such as copper.

In the test method, the sample normally dissolved in toluene/iso-propyl alcohol/water is titrated with potassium hydroxide, and the results are expressed as milligrams of potassium hydroxide per gram of sample (mg KOH/g). Crude oils having high acid numbers have a high potential to cause corrosion problems in the refineries, especially in the atmospheric and vacuum distillation units where the hot crude oil first comes into contact with hot metal surfaces. Crude oil typically has a total acid number (TAN value) on the order of 0.05–6.0 mg KOH/g of sample.

Current methods for the determination of the acid content of hydrocarbon compositions are well established (ASTM D664), which include potentiometric titration in non-aqueous conditions to clearly defined endpoints as detected by changes in millivolts readings versus volume of titrant used. A color indicator method (ASTM D974) is also available.

In addition, the total acid number (TAN) values as conventionally analyzed in accordance with standard test methods (ASTM D664) do not correlate at all with their risk of forming naphthenates or other soaps during production in oilfields. The TAN of oil has frequently been used to quantify the presence of naphthenic acids because the carboxylic acid components of oils are believed to be largely responsible for oil acidity. However, more recent research has begun to highlight deficiencies in relying upon this method for such a direct correlation, and the total acid number is no longer considered to be such a reliable indicator. Furthermore, the ASTM D974 test method is an older method and used for distillates, while the ASTM D664 test method is more accurate but measures acid gases and hydrolyzable salts in addition to organic acids. These differences are important on crude oils but less significant on distillates (Haynes, 2006).

3.2.1.2   Elemental Analysis

The elemental analysis (ultimate analysis) of crude oil for the percentages of carbon, hydrogen, nitrogen, oxygen, and sulfur is perhaps the first method used to examine the general nature and perform an evaluation of a feedstock. The atomic ratios of the various elements to carbon (i.e., H/C, N/C, O/C, and S/C) are frequently used for indications of the overall character of the feedstock. It is also of value to determine the amounts of trace elements, such as vanadium, nickel, and other metals, in a feedstock since these materials can have serious deleterious effects on catalyst performance during refining by catalytic processes.

The ultimate analysis (elemental composition) of crude oil is not reported to the same extent as for coal (Speight, 2013a, 2014, 2015). Nevertheless, there are ASTM standard test methods that may have been designed for other materials and can be adapted for application to refinery feedstocks (Speight, 2015). For example, the carbon content of a sample can be determined by the method designated for coal and coke (ASTM D3178) or by the method designated for municipal solid waste (ASTM E777).

For example, carbon content can be determined by the method designated for coal and coke (ASTM D3178) or by the method designated for municipal solid waste (ASTM E777). There are also methods designated for:

1. Carbon and hydrogen content (ASTM D1018, ASTM D3178, and ASTM D5291).
2. Nitrogen content (ASTM D3179, ASTM E258, ASTM D5291, and ASTM E778).
3. Oxygen content (ASTM E385).
4. Sulfur content (ASTM D129 and ASTM D139, ASTM D1266, ASTM D1552, ASTM D1757, ASTM D2622, ASTM D3177, ASTM D4045, and ASTM D4294).
5. Metals content (ASTM D482, ASTM D1318, ASTM D3340, and ASTM D3605).

From the available data, the proportions of the elements in crude oil vary only slightly over narrow limits:

Carbon	83.0%–87.0%
Hydrogen	10.0%–14.0%
Nitrogen	0.1%–2.0%
Oxygen	0.05%–1.5%
Sulfur	0.05%–6.0%
Metals (Ni and V):	<1,000 ppm

Nevertheless, there is a wide variation in physical properties from the lighter more mobile crude oil at one extreme to the viscous asphaltic crude oils at the other extreme. The majority of the more aromatic species and the heteroatoms occur in the higher boiling fractions of feedstocks. The viscous feedstocks are relatively rich in these higher boiling fractions (Speight, 2014, 2015).

Of the ultimate analytical data, more has been made of the sulfur content than any other property. For example, the sulfur content (ASTM D1552 and ASTM D4294) and the API gravity represent the two properties that have, in the past, had the greatest influence on determining the value of crude oil as a feedstock. The sulfur content varies from approximately 0.1% w/w to approximately 3% w/w for the more conventional crude oils to as much as 5%–6% for heavy crude oil, extra heavy crude oil, and tar sand bitumen.

The sulfur content varies from approximately 0.1% w/w to approximately 3% w/w for the more conventional crude oils to as much as 5%–6% w/w for the more viscous feedstocks and, in fact, a residuum, may be of the same order or even have a substantially higher sulfur content depending on the sulfur content of the crude oil from which the residuum was produced.

3.2.1.3 Density and Specific Gravity

The density and specific gravity of refinery feedstocks (ASTM D70, ASTM D71, ASTM D287, ASTM D1217, ASTM D1298, ASTM D1480, ASTM D1481, and ASTM D4052) are two properties that have found wide use in the industry for initial assessment of the character and quality of crude oil.

Briefly, by definition, density is the mass of a unit volume of material at a specified temperature and has the dimensions of grams per cubic centimeter (a close approximation to grams per milliliter). On the other hand, the specific gravity of a feedstock is the ratio of the mass of a volume of the substance to the mass of the same volume of water and is dependent on two temperatures, those at which the masses of the sample and the water are measured. More specifically, when the water temperature is 4°C (39°F), the specific gravity is equal to the density in the centimeter-gram-second (cgs) system, since the volume of 1 g of water at that temperature is 1 mL. Thus the density of water, for example, varies with temperature, and its specific gravity at equal temperatures is always unity. The standard temperatures for a specific gravity in the crude oil industry in North America are 60/60°F (15.6/15.6°C).

In the early years of the crude oil industry, density was the principal specification for crude oil and refinery products; it was used to give an estimation of the gasoline and, more particularly, the kerosene present in the crude oil. However, the derived relationships between the density of crude oil and its fractional composition were valid only if they were applied to a certain type of crude oil and lost some of their significance when applied to different types of crude oil. Nevertheless, density is still used to give a rough estimation of the nature of crude oil and crude oil products. Although density and specific gravity are used extensively throughout the science and engineering fields, in the crude oil industry the API (American Petroleum Institute) gravity is the most noteworthy:

$$\text{Degrees API} = [141.5 / (\text{sp gr} @ 60 / 60°F)] - 131.5$$

Density and specific gravity are influenced by the chemical composition of a crude oil (or, for that matter, any refinery feedstock), and it is generally acknowledged that increased amounts of aromatic derivatives result in an increase in density, whereas an increase in alkane derivatives would result in a decrease in density. It is also possible to recognize certain preferred trends between the density of crude oil and one or another of the physical properties. For example, an approximate correlation exists between the density (API gravity) and sulfur content, Conradson carbon residue, viscosity, and nitrogen content (Speight, 2000, 2015).

The density or specific gravity of a feedstock (heavy crude oil, extra heavy crude oil, and tar sand bitumen may as well as refinery products) is typically determined by means of a hydrometer (ASTM D287 and ASTM D1298), a pycnometer (ASTM D70, ASTM D1217, ASTM D1480, and ASTM D1481) or by means of a digital density meter (ASTM D4052) and a digital density analyzer (ASTM D5002). While all of these methods may not be suitable for measuring the density of a viscous feedstock, some methods lend themselves to adaptation (Speight, 2014, 2015). Other methods such as using the hydrometer (ASTM D1298), the density meter (ASTM D4052), and the digital density analyzer (ASTM D5002) are often preferred. However, surface tension effects can affect the displacement method, and the density meter method loses some of its advantage when measuring the density of heavy oil and bitumen.

The density of refinery feedstocks ranges from approximately 0.8 (45.3° API) for the lighter crude oils to over 1.0 (less than 10° API) for heavy crude oil and bitumen. The variation of density with temperature, effectively the coefficient of expansion, is a property of great technical importance, since most crude oil products are sold by volume and specific gravity is usually determined at the prevailing temperature (21°C, 70°F) rather than at the standard temperature (15.6°C, 60°F).

However, not all of these methods are suitable for measuring the density or specific gravity of heavy oil and bitumen although some methods lend themselves to adaptation.

The API gravity of a feedstock (ASTM D287), which is calculated directly from the specific gravity, shows a fairly wide range of variation. The largest degree of variation is usually due to local reservoir or deposit conditions that affect material close to the faces, or exposures, occurring in surface oil sand beds. The range of specific gravity of heavy crude oil, extra heavy crude oil, and tar sand bitumen usually varies over the range of the order of 0.995–1.04.

It is worthy of note at this point is the variation in density (specific gravity) of the Alberta (Athabasca) tar sand bitumen with temperature. Over the temperature range of 30°C–130°C (85°F–265°F), the bitumen is lighter than water, and flotation of the bitumen (with aeration) on the surface of water is facilitated, hence the logic of the hot water separation process (Speight, 2005, 2009, 2014).

3.2.1.4 Metals Content

Heteroatoms (nitrogen, oxygen, sulfur, and metals) are found in every refinery feedstock and affect every aspect of refining. Sulfur is usually the most concentrated element, and many commercial catalysts are available to the refinery that can routinely effect the removal of 90% w/w of the sulfur in the feedstock. Nitrogen is more difficult to remove than sulfur, and there are fewer catalysts that are specific for nitrogen. Metals cause particular problems because they poison catalysts used for sulfur and nitrogen removal as well as other processes such as catalytic cracking (Speight, 2014, 2017).

Metal derivatives in refinery feedstocks cause problems during refinery operations because they (the metal derivatives) poison the catalysts that are employed to remove sulfur and nitrogen as well as catalysts and other processes such as catalytic cracking. The viscous feedstocks and residua contain relatively high proportions of metals either in the form of salts or as organometallic constituents (such as the metallo-porphyrin derivatives), which are extremely difficult to remove from the feedstock. Indeed, the concentration effect of the distillation process dictates that all the metals in the original feedstock crude oil (unless there is an entrainment effect) are concentrated in the residuum (Speight, 2014, 2015, 2017). In the case of an entrainment effect, some of the metallic constituents may actually "volatilize" under the distillation conditions and appear in the higher boiling distillates, but there are the exception rather than the rule and the refinery must, of necessity, be aware of the potential for entrainment. The deleterious effect of metallic constituents on the catalyst is known, and serious attempts have been made to develop catalysts that can tolerate a high concentration of metals without serious loss of catalyst activity or catalyst life.

A variety of tests have been designated for the determination of metals in refinery products (ASTM D1318, ASTM D3340, and ASTM D3605), some of which can be adapted to whole feedstocks (Speight, 2014, 2015). One method for the determination of metal derivatives in feedstocks is to combust the sample so that only inorganic ash remains which can then be digested with an acid and the solution examined for metal species by atomic absorption (AA) spectroscopy or by inductively coupled argon plasma (ICP) spectrometry.

The viscous feedstocks (including distillation residua) contain relatively high proportions of metals either in the form of salts or as organometallic constituents (such as, for example, the metallo-porphyrin derivatives), which are extremely difficult to remove from the feedstock.

3.2.1.5 Viscosity

Briefly, the viscosity (of a refinery feedstock or product) is the force in dynes required to move a plane of $1\,cm^2$ area at a distance of $1\,cm$ from another plane of $1\,cm^2$ area through a distance of $1\,cm$ in 1 second. In the centimeter-gram-second (cgs) system, the unit of viscosity is the poise or centipoise (0.01 P). Two other terms in common use are kinematic viscosity and fluidity. In addition, the kinematic viscosity is the viscosity in centipoises divided by the specific gravity, and the unit is the stoke (cm^2/s), although centistokes (0.01 cSt) is in more common usage; fluidity is simply the reciprocal of viscosity. The viscosity (ASTM D445, ASTM D88, ASTM D2161, ASTM D341, and ASTM D2270) of crude oils varies markedly over a very wide range with values varying from less than 10 cP at room temperature to many thousands of centipoises at the same temperature.

Although many types of instruments and test methods have been proposed for the determination of viscosity of refinery feedstocks and products, the simplest and most widely used is the capillary tube method (ASTM D445), and the viscosity is derived from the equation:

$$\mu = \pi r^4 P/8nl$$

In this equation, r is the tube radius, l the tube length, P the pressure difference between the ends of a capillary, n the coefficient of viscosity, and the quantity discharged in unit time. Not only are such capillary instruments the most simple, but when designed in accordance with known principle and used with known necessary correction factors, they are probably the most accurate viscometers available. It is usually more convenient, however, to use relative measurements, and for this purpose, the instrument is calibrated with an appropriate standard liquid of known viscosity.

Another, but somewhat less used, measurement of viscosity is the Saybolt universal viscosity (SUS) (ASTM D88) which is the time in seconds required for the flow of 60 mL of feedstock (or product) from a container, at constant temperature, through a calibrated orifice. The Saybolt furol viscosity (SFS) (ASTM D88) is determined in a similar manner except that a larger orifice is employed in the test method.

As a result of the various methods for viscosity determination, it is not surprising that much effort has been spent on interconversion of the several scales, especially converting Saybolt to kinematic viscosity (ASTM D2161),

$$\text{Kinematic viscosity} = a \times \text{Saybolt s} + b / \text{Saybolt s}$$

In this equation, a and b are constants. Factors for the conversion factors of kinematic viscosity from 2 to 70 cSt at 38°C (100°F) and 99°C (210°F) to the equivalent Saybolt universal viscosity in seconds and appropriate multipliers are listed to convert kinematic viscosity that is higher than 70 cSt.

If the kinematic viscosity has been determined at any other temperature, the equivalent Saybolt universal value can be calculated by use of the Saybolt equivalent at 38°C (100°F) and a multiplier that varies with the temperature:

$$\text{Saybolt s at } 100°F \, (38°C) = cSt \times 4.635$$

$$\text{Saybolt s at } 210°F \, (99°C) = cSt \times 4.667$$

Various studies have also been made on the effect of temperature on viscosity since the viscosity of a feedstock (or product) decreases as the temperature increases. The rate of change of the viscosity depends primarily on the nature or composition of the crude oil (especially if the constituents interact with each other under the conditions of the test method), but other factors, such as volatility, may also have an effect. The effect of temperature on viscosity is generally represented by the equation:

$$\log \log(n + c) = A + B \log T$$

In this equation, n is the absolute viscosity, T is the temperature, and A and B are constants. The constants A and B vary widely with different feedstocks, but c remains fixed at 0.6 for all feedstocks having a viscosity in excess of 1.5 cSt but does increase slightly at lower viscosity (0.75 at 0.5 cSt). The viscosity-temperature characteristics of any oil, so plotted, thus creates a straight line, and the parameters A and B are equivalent to the intercept and slope of the line. To express the viscosity and viscosity-temperature characteristics of an oil, the slope and the viscosity at one temperature must be known; the usual practice is to select 38°C (100°F) and 99°C (210°F) as the observation temperatures.

Suitable conversion tables are available (ASTM D341), and each table or chart is constructed in such a way that for any given crude oil or crude oil product the viscosity-temperature points result in a straight line over the applicable temperature range. Thus, only two viscosity measurements need to be made at temperatures far enough apart to determine a line on the appropriate chart from which the approximate viscosity at any other temperature can be read. The charts can be applicable only to measurements made in the temperature range in which a given feedstock is a Newtonian liquid. However, the charts do not give accurate results when either the cloud point or boiling point is approached, but they are useful over the Newtonian range for estimating the temperature at which oil attains a desired viscosity.

In some instances, the viscosity-temperature coefficient of the sample (especially for lubricating oil) is an important expression of its suitability, a convenient number to express this property is very useful, and hence, a viscosity index (ASTM D2270) has been derived. Thus:

$$\text{Viscosity index} = (L - U) / (L - H \times 100)$$

In this equation, L and H are the viscosities of the zero and 100 index reference oils, both having the same viscosity at 99°C (210°F), and U is that of the unknown, all at 38°C (100°F). Originally the viscosity index was calculated from Saybolt viscosity data, but subsequently data were provided for kinematic viscosity.

3.2.2 THERMAL PROPERTIES

The thermal properties of refinery feedstocks crude oil are the properties (or characteristics) that determine how the feedstock will react (or behave) when it is subjected to excessive heat or variations with time when the heat is applied. As with all properties of crude oil, a collection of standard test methods is instrumental in the evaluation and assessment of the thermal properties (Speight, 2015). These standards allow crude oil refineries and other geological and chemical processing plants to appropriately examine and process crude oil in a safe and efficient manner.

3.2.2.1 Carbon Residue

Crude oil products are mixtures of many compounds that differ widely in their physical and chemical properties. Some of them may be vaporized in the absence of air at atmospheric pressure without leaving an appreciable residue. Other nonvolatile constituents may result in the formation (deposition) of a carbonaceous residue when destructively distilled under such conditions. This residue is referred to as the carbon residue when determined in accordance with prescribed procedure.

The carbon residue is a property that can be correlated with several other properties of crude oil (Speight, 2000, 2014); hence, it also presents indications of the volatility of the crude oil and the coke-forming (or gasoline-producing) propensity. However, tests for carbon residue are sometimes used to evaluate the propensity for the formation of a carbonaceous deposit when the sample is subject to heat.

The mechanical design and operating conditions of such equipment have such a profound influence on carbon deposition during service that the comparison of carbon residues between oils should be considered as giving only a rough approximation of relative deposit-forming tendencies. A more precise relationship between carbon residue and hydrogen content, H/C atomic ratio, nitrogen content, and sulfur content has been shown to exist. These data can provide more precise information about the anticipated behavior of a variety of feedstocks in thermal processes.

Because of the extremely small values of carbon residue obtained by the Conradson and Ramsbottom methods when applied to the lighter distillate fuel oils, it is customary to distill such products to 10% residual oil and determine the carbon residue thereof. Such values may be used directly in comparing fuel oils, as long as it is kept in mind that the value is for a residuum oil, and are not to be compared with the carbon residue of the whole feedstock.

There are two commonly used methods for determining the carbon residue of a feedstock or a feedstock product, and they are (1) the Conradson method (ASTM D189) and (2) the Ramsbottom method (ASTM D524) (Speight, 2014, 2015). Both methods are applicable to the relatively nonvolatile portion of feedstocks and feedstock products, which partially decompose when distilled at a pressure of 1 atmosphere. However, a feedstock that contains ash-forming constituents will have an erroneously high-carbon residue by either method unless the ash is first removed from the oil; the degree of error is proportional to the amount of ash.

An alternate method, which involves micro-pyrolysis of the sample, is also available as a standard test method (ASTM D4530), which requires smaller amounts of the sample amount and is, on occasion, referred to as a thermogravimetric method. The carbon residue produced by this method is often referred to as the microcarbon residue (MCR). Agreements between the data from the three test methods are good, making it possible to interrelate all of the data from carbon residue tests (Long and Speight, 1989, 1998; Speight, 2015). However, there has been a tendency to advocate use of the more expedient microcarbon method because of the lesser amounts required in this method.

The carbon residue is a property that can be correlated with several other properties of crude oil. Hence it also presents indications of the volatility of the crude oil and the coke-forming (or gasoline-producing) propensity. However, tests for carbon residue are sometimes used to evaluate the carbonaceous depositing characteristics of fuels used in certain types of oil-burning equipment and internal combustion engines.

The mechanical design and operating conditions of such equipment have such a profound influence on carbon deposition during service that comparison of carbon residues between oils should be considered as giving only a rough approximation of relative deposit-forming tendencies. Recent work has focused on the carbon residue of the different fractions of crude oils, especially the asphaltene constituents. A more precise relationship between carbon residue and hydrogen content, H/C atomic ratio, nitrogen content, and sulfur content has been shown to exist. These data can provide valuable information about the anticipated behavior of a variety of feedstocks in thermal processes, and in fact, there is a fairly universal linear correlation between the carbon residue (Conradson) and the H/C ratio:

$$H/C = 171 - 0.0115 CR \text{ (Conradson)}.$$

This equation holds within two limits; at H/C values $=171$, where the carbon residue is zero (no coke formation), and H/C$=0.5$, where the carbon residue is 100 (all the material converts to coke under test conditions).

Because of the extremely small values of carbon residue obtained by the Conradson and Ramsbottom methods when applied to the lighter distillate fuel oils, it is customary to distill such products to 10% residual oil and determine the carbon residue thereof. Such values may be used directly in comparing fuel oils, as long as it is kept in mind that the values are carbon residues on 10% residual oil and are not to be compared with straight carbon residues.

3.2.2.2 Critical Properties

A study of the pressure, volume, and temperature relationships of a pure component reveals a particular unique state where the properties of a liquid and vapor become indistinguishable from each other. At that state, the latent heat of vaporization becomes zero and no volume change occurs when the liquid is vaporized. This state is called the critical state and the appropriate parameters of state are termed the critical pressure (P_C), critical volume (V_C), and critical temperature (T_C). It is an important characteristic of the critical state for a pure component that with values of P or T greater than either P_C or T_C the vapor and liquid states cannot coexist at equilibrium, and thus P_C and T_C represent the maximum values of P and T at which phase separation can occur.

Since the critical state of a component is unique, it is perhaps not surprising that knowledge of P_C, T_C, and V_C allows many predictions to be made concerning the physical properties of substances. These predictions are based on the Law of Corresponding States that states that substances behave in the same way when they are in the same state with reference to the critical state. The particular corresponding state is characterized by its reduced properties, i.e., $T_r = T/T_C$, $Pr = P/P_C$, and $Vr = V/V_C$.

The use of this concept permits generalized plots in terms of reduced properties to be drawn that are then applicable to all substances (which obey the law) and can be of great value in determining thermodynamic relationships. It is rare in crude oil engineering to have to deal with pure substances, and unfortunately, the application of the Law of Corresponding States to mixtures is complicated by the fact that use of the true critical point for a mixture does not yield correct values of reduced properties for accurate prediction from generalized charts. For a mixture, the critical state no longer represents the maximum temperature and pressure at which a liquid and vapor phase can coexist, and phase separation can occur under retrograde conditions.

For engineering purposes, this difficulty is resolved by the use of pseudo-critical conditions, which are based on the molal average critical temperatures and pressures of the compounds of the mixture. Although the use of pseudo-reduced conditions for mixtures of hydrocarbon derivatives is generally satisfactory, this is not true for states near the true critical, nor, in general, for mixtures of vapor and liquid.

The temperature, pressure, and volume at the critical state are of considerable interest in crude oil physics, particularly in connection with modern high-pressure, high-temperature refinery operations and in correlating pressure-temperature-volume relationships for other states. Critical data are known for most of the lower molecular weight pure hydrocarbon derivatives, and standard methods are generally used for such determinations.

The critical point of a pure compound is the equilibrium state in which its gaseous and liquid phases are indistinguishable and coexistent; they have the same intensive properties. However, localized variations in these phase properties may be evident experimentally. The definition of the critical point of a mixture is the same, but, in fact, mixtures generally have a maximum temperature or pressure at other than the true critical point; maximum here denotes the greatest value at which two phases can coexist in equilibrium.

Thus, when a pure compound is heated at atmospheric pressure, it eventually reaches its boiling point and is completely vaporized at a constant temperature unless the pressure is increased. If the pressure is increased, the compound is completely condensed and cannot be vaporized again unless the temperature is also increased. This mechanism, alternately increasing the pressure and temperature, functions until at some high temperature and pressure it is found that the material cannot be condensed regardless of the amount of pressure applied. This point is called the critical point, and in addition, the temperature and pressure at the critical point are referred to as the critical temperature and the critical pressure, respectively.

The liquid phase and vapor phase merge at the critical point so that one phase cannot be distinguished from the other. No volume change occurs when a liquid is vaporized at the critical point, and no heat is required for vaporization but the coefficient of expansion has become large.

Limited information concerning the behavior of complex mixtures has required that the pseudo-critical temperature and pseudo-critical pressure be used for many crude oil fractions and products. The pseudo-critical point is defined as the molal average critical temperature and pressure of the several constituents that make up a mixture. It may be used as the critical point of a mixture in computing reduced temperatures and pressures. However, in computing the pressure-volume-temperature relations of mixtures by use of the pseudo-critical point, it must he recognized that the values are not accurate in the region of the critical point and that it cannot be applied to mixtures of gas and liquid.

In the correlation of many properties, reduced properties are useful. Reduced properties are defined as the ratio of the actual value of the property to its critical value. Thus for volume, temperature, or pressure the relationships are as follows:

$$\text{Reduced volume, } V_R = V / V_c$$

where V is the volume at specified conditions and V_c is the volume at the critical point. Similarly,

$$\text{Reduced temperature } T_R = T / T_c$$
$$\text{Reduced volume } P_R = P / P_c$$

where T and P are the temperature and volume, respectively, at specified conditions and T_c and V_c are the temperature and volume, respectively, at the critical point

3.2.2.3　Heat of Combustion

The heat of combustion (ASTM D240) is a direct measure of fuel energy content and is determined as the quantity of heat liberated by the combustion of a quantity of fuel with oxygen in a standard bomb calorimeter.

Chemically, the heat of combustion is the energy (heat) released when an organic compound is burned to produce water (H_2O liquid,) carbon dioxide (CO_2 gas), sulfuric acid (H_2SO_4 liquid), and nitric acid (HNO_3 liquid). The value can be calculated using a theoretical equation based upon the elemental composition of the feedstock:

$$H_g / 4.187 = 8,400C + 27,765H + 1,500N + 2,500S - 2,650O$$

H_g is given in kilo joules per kilogram (1.0 kJ/kg=0.43 Btu/lb); C, H, N, S, and O are the normalized weight fractions for these elements in the sample.

The gross heat of combustion for a feedstock and a feedstock product can be calculated (with an acceptable degree of accuracy) by the equation:

$$Q = 12,400 - 2,100d^2$$

In this equation, d is the 60/60°F specific gravity. The ranges for crude oil is 10,000–11,600 calories/g, and the heat of combustion of heavy crude oil, extra heavy crude oil, and tar sand bitumen is considerably higher.

For gasoline, the heat of combustion is 11,000–11,500 calories/g, and for kerosene (and diesel fuel), it falls in the range of 10,500–11,200 calories/g. Finally, the heat of combustion for fuel oil is on the order of 9,500–11,200 calories/g. Heats of combustion of crude oil gases may be calculated from the analysis and data for the pure compounds. Experimental values for gaseous fuels may be obtained by measurement in a water flow calorimeter, and heats of combustion of liquids are usually measured in a bomb calorimeter.

For the thermodynamic calculation of equilibria useful in hydrocarbon research, combustion data of extreme accuracy are required because the heats of formation of water and carbon dioxide are large in comparison with those in the hydrocarbon derivatives.

3.2.2.4 Liquefaction and Solidification

Crude oil and the majority of crude oil products are liquids at ambient temperature, while the viscous feedstocks vary from semi-solid to solid at ambient temperature leading to problems that may arise from solidification during normal use.

The melting point is a test (ASTM D87 and ASTM D127) that is widely used by suppliers of wax and by the wax consumers; it is particularly applied to the highly paraffinic or crystalline waxes. Quantitative prediction of the melting point of pure hydrocarbon derivatives is difficult, but the melting point tends to increase qualitatively with the molecular weight and with symmetry of the molecule.

Unsubstituted and symmetrically substituted compounds (e.g., benzene, cyclohexane, p-xylene, and naphthalene) melt at higher temperatures relative to the paraffin compounds of similar molecular weight: the unsymmetrical isomers generally melt at lower temperatures than the aliphatic hydrocarbon derivatives of the same molecular weight.

Unsaturation affects the melting point principally by its alteration of symmetry; thus ethane (−172°C, −278°F) and ethylene (−169.5°C, −273°F) differ only slightly, but the melting points of cyclohexane (6.2°C, 21°F) and cyclohexane (−104°C, −155°F) contrast strongly. All types of highly unsymmetrical hydrocarbon derivatives are difficult to crystallize; asymmetrically branched aliphatic hydrocarbon derivatives as low as octane and most substituted cyclic hydrocarbon derivatives comprise the greater part of the lubricating fractions of crude oil, crystallize slowly, if at all, and on cooling merely take the form of glasslike solids.

Although the melting points of viscous feedstocks are of limited usefulness, except to estimate the purity or perhaps the composition of waxes, the reverse process, solidification, has received attention in crude oil chemistry. In fact, solidification of crude oil and crude oil products has been differentiated into four categories, namely, freezing point, congealing point, cloud point, and pour point.

Crude oil becomes more or less a plastic solid when cooled to sufficiently low temperatures due to the congealing of the various hydrocarbon derivatives that constitute the oil. The cloud point of a crude oil is the temperature at which paraffin wax or other solidifiable compounds present in the oil appear as a haze when the oil is chilled under definitely prescribed conditions (ASTM D2500 and ASTM D3117). As cooling is continued, all crude oil oils become more and more viscous and flow becomes slower and slower. Thus, the pour point of a crude oil is the lowest temperature at which the oil pours or flows under definitely prescribed conditions when it is chilled without disturbance at a standard rate (ASTM D97).

The solidification characteristics of a crude oil product depend on its grade or kind. For a viscous feedstock, the temperature of interest is that at which fluidity occurs which is often commonly referred to as the dropping point. At this point, the sample passes from a plastic solid to a liquid state and begins to flow under the conditions of the test (ASTM D566 and ASTM D2265).

The melting point of a sample (using wax as the example) is the temperature at which the sample becomes sufficiently fluid to drop from the thermometer. The congealing point is the temperature at which melted sample ceases to flow when allowed to cool under definitely prescribed conditions (ASTM D938).

For paraffin wax, the solidification temperature is of interest. For such purposes, the melting point is the temperature at which the melted paraffin wax begins to solidify, as shown by the minimum rate of temperature change, when cooled under prescribed conditions.

The relationship of cloud point, pour point, melting point, and freezing point to one another varies widely from one crude oil product to another. Hence, their significance for different types of product also varies. In general, cloud, melting, and freezing points are of more limited value, and each is of narrower range of application than the pour point.

The cloud point of crude oil or a crude oil product is the temperature at which a solid (such as paraffin wax in the case of crude oil or crude oil products) or other solidifiable compounds present in the sample appear as a haze when the sample is chilled under definitely prescribed conditions (ASTM D2500 and ASTM D3117).

To determine the cloud point and the pour point (ASTM D97, ASTM D5853, ASTM D5949, ASTM D5950, and ASTM D5985), the oil is contained in a glass test tube fitted with a thermometer and immersed in one of three baths containing coolants. The sample is dehydrated and filtered at a temperature 25°C (45°F) higher than the anticipated cloud point. It is then placed in a test tube and cooled progressively in coolants held at −1°C to +2°C (30°F–35°F), −18°C to −20°C (−4°F to 0°F), and −32°C to −35°C (−26°F to −31°F), respectively. The sample is inspected for cloudiness at temperature intervals of 1°C (2°F). If conditions or oil properties are such that reduced temperatures are required to determine the pour point, alternate tests are available that accommodate the various types of samples. Related to the cloud point, the wax appearance temperature or wax appearance point is also determined (ASTM D3117).

The pour point of crude oil or a crude oil product is determined using this same technique (ASTM D97), and it is the lowest temperature at which the oil pours or flows. It is actually 2°C (3°F) above the temperature at which the oil ceases to flow under these definitely prescribed conditions when it is chilled without disturbance at a standard rate. To determine the pour point, the sample is first heated to 46°C (115°F) and cooled in air to 32°C (90°F) before the tube is immersed in the same series of coolants as used for the determination of the cloud point. The sample is inspected at temperature intervals of 2°C (3°F) by withdrawal and holding horizontal for 5 seconds until no flow is observed during this time interval.

Cloud and pour points are useful for predicting the temperature at which the observed viscosity of oil deviates from the true (Newtonian) viscosity in the low-temperature range. They are also useful for identification of oils or when planning the storage of oil supplies, as low temperatures may cause handling difficulties with some oils.

The pour point of a crude oil was originally applied to crude oil that had a high wax content. More recently, the pour point, like the viscosity, is determined principally for use in pumping arid pipeline design calculations. Difficulty occurs in these determinations with waxy crude oils that begin to exhibit irregular flow behavior when wax begins to separate. These crude oils possess viscosity relationships that are difficult to predict in pipeline operation. In addition, some waxy crude oils are sensitive to heat treatment that can also affect their viscosity characteristics. This complex behavior limits the value of viscosity and pour point tests on waxy crude oils. At the present time, long crude oil pipelines and the increasing production of waxy crude oils make an assessment of the pumpability of a wax-containing crude oil through a given system a matter of some difficulty that can often only be resolved after field trials. Consequently, considerable work is in progress to develop a suitable laboratory pumpability test that gives an estimate of minimum handling temperature and minimum line or storage temperature.

3.2.2.5 Pressure-Volume-Temperature Relationships

Hydrocarbon vapors, like other gases, follow the ideal gas law (i.e., $PV = RT$) but only at relatively low pressures and high temperatures, that is, far from the critical state. Several more empirical equations have been proposed to represent the gas laws more accurately, such as the well-known van der Waals equation, but they are either inconvenient for calculation or require the experimental determination of several constants. A more useful device is to use the simple gas law and to induce a correction, termed the compressibility factor, μ, so that the equation takes the form:

$$PV = \mu RT$$

For hydrocarbon derivatives, the compressibility factor is very nearly a function only of the reduced variables of state, that is, a function of the pressure and temperature divided by the respective critical values.

3.2.2.6 Volatility

The volatility of a sample is the tendency of the sample to vaporize, that is the temperature at which the sample changes from the liquid state to the vapor state or to the gaseous state. Because one of the three essentials for combustion in a flame is that the fuel be in the gaseous state, volatility is a primary characteristic of liquid fuels. A test (ASTM D6) also exists for determining the loss of material when crude oil and asphaltic compounds are heated. Another test (ASTM D20) is a method for the distillation of road tars that might also be applied to estimating the volatility of high-molecular-weight residues.

The flash point of crude oil or a crude oil product is the temperature to which the product must be heated under specified conditions to give of sufficient vapor to form a mixture with air that can be ignited momentarily by a specified flame (ASTM D56, ASTM D92, and ASTM D93). The fire point is the temperature to which the product must be heated under the prescribed conditions of the method to burn continuously when the mixture of vapor and air is ignited by a specified flame (ASTM D92).

From the viewpoint of safety, information about the flash point is of most significance at or slightly above the maximum temperatures (30°C–60°C, 86°F–140°F) that may be encountered in storage, transportation, and use of liquid crude oil products, in either closed or open containers. In this temperature range, the relative fire and explosion hazard can be estimated from the flash point. For example, samples with a flash point below 40°C (104°F) require special precautions for safe handling while samples with a flash point above 60°C (140°F) do not require the same degree of caution.

A further aspect of volatility that receives considerable attention is the vapor pressure of a sample which is the force exerted on the walls of a closed container by the vaporized portion of a liquid. Conversely, the vapor pressure is also the force that must be exerted on the liquid to prevent it from vaporizing further (ASTM D323). The vapor pressure increases with temperature for any given product. The temperature at which the vapor pressure of a liquid, or a pure compound of a mixture of many compounds, equals 1 atmosphere (14.7 psi, absolute) is designated as the boiling point of the liquid.

In each homologous series of hydrocarbon derivatives, the boiling points increase with molecular weight and structure also has a marked influence since it is a general rule that branched paraffin isomers have lower boiling points than the corresponding n-alkane. In any given series, steric effects notwithstanding, there is an increase in boiling point with an increase in carbon number of the alkyl side chain. This particularly applies to alkyl aromatic compounds where alkyl-substituted aromatic compounds can have higher boiling points than polycondensed aromatic systems. And this fact is very meaningful when attempts are made to develop hypothetical structures for asphaltene constituents (Speight, 1994, 2014).

The boiling points of crude oil fractions are rarely, if ever, distinct temperatures, and it is, in fact, more correct (even necessary) to refer to the boiling ranges of the various fractions. To determine these ranges, the crude oil is tested in various methods of distillation, either at atmospheric pressure or at reduced pressure. In general, the limiting molecular weight range for distillation at atmospheric pressure without thermal degradation is 200–250, whereas the limiting molecular weight range for conventional vacuum distillation is 500–600.

As an early part of characterization studies, a correlation was observed between the quality of crude oil products and their hydrogen content since gasoline, kerosene, diesel fuel, and lubricating oil are made up of hydrocarbon constituents containing high proportions of hydrogen. Thus, it is not surprising that test to determine the volatility of crude oil and crude oil products were among the first to be defined. Indeed, volatility is one of the major tests for crude oil products, and it is inevitable that all products will, at some stage of their history, be tested for volatility characteristics.

Distillation involves the general procedure of vaporizing the crude oil liquid in a suitable flask either at atmospheric pressure (ASTM D86) or at reduced pressure (ASTM D1160), and the data are reported in terms of one or more of the following seven items:

1. Initial boiling point is the thermometer reading in the neck of the distillation flask when the first drop of distillate leaves the tip of the condenser tube. This reading is materially affected by a number of test conditions, namely, room temperature, rate of heating, and condenser temperature.

2. Distillation temperatures are usually observed when the level of the distillate reaches each 10% mark on the graduated receiver, with the temperatures for the 5% and 95% marks often included. Conversely, the volume of the distillate in the receiver, that is, the percentage recovered, is often observed at specified thermometer readings.

3. End point or maximum temperature is the highest thermometer reading observed during distillation. In most cases, it is reached when the entire sample has been vaporized. If a liquid residue remains in the flask after the maximum permissible adjustments are made in heating rate, this is recorded as indicative of the presence of very high-boiling compounds.

4. Dry point is the thermometer reading at the instant the flask becomes dry and is for special purposes, such as for solvents and for relatively pure hydrocarbon derivatives. For these purposes, dry point is considered more indicative of the final boiling point than end point or maximum temperature.

5. Recovery is the total volume of distillate recovered in the graduated receiver and residue is the liquid material, mostly condensed vapors, left in the flask after it has been allowed to cool at the end of distillation. The residue is measured by transferring it to an appropriate graduated cylinder. Low or abnormally high residues indicate the absence or presence, respectively, of high-boiling components.

6. Total recovery is the sum of the liquid recovery and residue; distillation loss is determined by subtracting the total recovery from 100% and is a measure of the portion of the vaporized sample that does not condense under the conditions of the test. Like the initial boiling point, distillation loss is affected materially by a number of test conditions, namely, condenser temperature, sampling and receiving temperatures, barometric pressure, heating rate in the early part of the distillation, and others. Provisions are made for correcting

high distillation losses for the effect of low barometric pressure because of the practice of including distillation loss as one of the items in some specifications for motor gasoline.

7. Percentage evaporated is the percentage recovered at a specific thermometer reading or other distillation temperatures, or the converse. The amounts that have been evaporated are usually obtained by plotting observed thermometer readings against the corresponding observed recoveries plus, in each case, the distillation loss. The initial boiling point is plotted with the distillation loss as the percentage evaporated. Distillation data are considerably reproducible, particularly for the more volatile products.

One of the main properties of crude oil that serves to indicate the comparative ease with which the material can be refined is the volatility (Speight, 2015). Investigation of the volatility of crude oil is usually carried out under standard conditions, thereby allowing comparisons to be made between data obtained from various laboratories. Thus, nondestructive distillation data (US Bureau of Mines method) show that, not surprisingly, bitumen is a higher boiling material than the more conventional crude oils. There is usually little, or no, gasoline (naphtha) fraction in bitumen, and the majority of the distillate falls in the gas oil-lubrication distillate range (>260°C, >500°F). In excess of 50% of each bitumen is non-distillable under the conditions of the test, and the yield of the nonvolatile material corresponds very closely to the asphaltic content (asphaltene constituents plus resin content of each feedstock). In fact, detailed fractionation of the sample might be of secondary importance. Thus, it must be recognized that the general shape of a one-plate distillation curve is often adequate for making engineering calculations, correlating with other physical properties, and predicting the product slate.

There is also another method that is increasing in popularity for application to a variety of feedstocks, and this method is commonly known as simulated distillation (ASTM D2887) (Carbognani et al., 2012). The method has been well researched in terms of method development and application (Romanowski and Thomas, 1985; MacAllister and DeRuiter, 1987; Schwartz et al., 1987; Neer and Deo, 1995). The benefits of the technique include good comparisons with other ASTM distillation data as well as the application to higher boiling fractions of crude oil. In fact, data output include the provision of the corresponding Engler profile (ASTM D86) as well as the prediction of other properties such as vapor pressure and flash point. When it is necessary to monitor product properties, as is often the case during refining operations, such data provide a valuable aid to process control and on-line product testing.

For a more detailed distillation analysis of feedstocks and products, a low-resolution, temperature-programmed gas chromatographic analysis has been developed to simulate the time-consuming true boiling point distillation. The method relies on the general observation that hydrocarbon derivatives are eluted from a non-polar adsorbent in the order of their boiling points. The regularity of the elution order of the hydrocarbon components allows the retention times to be equated to distillation temperatures, and the term simulated distillation by gas chromatography (or simdis) is used throughout the industry to refer to this technique.

Simulated distillation by gas chromatography is often applied in the crude oil industry to obtain true boiling point data for distillates and crude oils (Speight, 2001). The ASTM D2887 method utilizes non-polar, packed gas chromatographic columns in conjunction with flame ionization detection. The upper limit of the boiling range covered by this method is to approximately 540°C (1,000°F) atmospheric equivalent boiling point. Recent efforts in which high-temperature gas chromatography were used have focused on extending the scope of the ASTM D2887 method for the more viscous materials to 800°C (1,470°F) atmospheric equivalent boiling point.

3.2.3 Chromatographic Properties

Typically, the term chromatography properties is the collective term for a collection of laboratory techniques for the separation of mixtures.

The method involves the dissolution of a mixture in a fluid (mobile phase) which carries the mixture it through a structure holding another material (stationary phase). The various constituents of the mixture travel at different speeds, causing them to separate which is based on the differential partitioning of the constituents of the mixture between the mobile phase and the stationary phase.

A chromatographic technique may be preparative or analytical. The purpose of preparative chromatography is to separate the components of a mixture for more advanced use (and is thus a form of purification). Analytical chromatography generally requires smaller amounts of material and is for measuring the relative proportions of analytes in a mixture. The two are not mutually exclusive.

3.2.3.1 Adsorption Chromatography

Adsorption chromatography is a technique that is employed to characterize the group composition of crude oils, heavy crude oils, extra heavy crude oils, tar sand bitumen, and distillation residua. The type and relative amount of certain hydrocarbon classes in the matrix can have a profound effect on the quality and performance of the hydrocarbon product, and two standard test methods have been used predominantly over the years (ASTM D2007 and ASTM D4124) (Suatoni and Garber, 1980; Miller et al., 1983; Norris and Rawdon, 1984).

The segregation of individual components from a mixture can be achieved by application of adsorption chromatography in which the adsorbent is either packed in an open tube (column chromatography) or shaped in the form of a sheet (thin-layer chromatography, TLC). A suitable solvent is used to elute from the bed of the adsorbent. Chromatographic separations are usually performed for the purpose of determining the composition of a sample. Even with such complex samples as crude oil, some information about the chemical structure of a fraction can be gained from the separation data.

In the present context, the challenge is the nature of the heteroatomic species in the viscous feedstocks because it is these heteroatom constituents that are largely responsible for coke formation and catalyst deactivation during refining operations. Therefore, it is these constituents that are the focus of much of the study. An ideal integrated separation scheme for the analysis of the heteroatomic constituents should therefore meet several criteria:

1. The various compound types should be concentrated into a reasonable number of discrete fractions, and each fraction should contain specific types of the heteroatomic compounds. It is also necessary that most of the heteroatom constituents be separated from the hydrocarbon derivatives and sulfur compounds that may constitute the bulk of the sample.
2. Perhaps most important, the separation should be reproducible insofar as the yields of the various fractions, and the distribution of the compound types among the fractions should be constant within the limits of experimental error.
3. The separation scheme should be applicable to high-boiling distillates and heavy feedstocks such as residua since heteroatomic compounds often predominate in these feedstocks.
4. The separation procedures should be relatively simple to perform and free of complexity.
5. Finally, the overall separation procedure should yield quantitative or, at worst, near-quantitative recovery of the various heteroatomic species present in the feedstock. There should be no significant loss of these species to the adsorbent or, perhaps more important, any chemical alteration of these compounds. Should chemical alteration occur, it will give misleading data that could have serious effects on refining predictions or on geochemical observations.

Group-type analysis by means of chromatography has been applied to a wide variety of crude oil types and products (Speight, 2014, 2015). These types of analysis are often abbreviated by the names PONA (paraffin derivatives, olefin derivatives, naphthene derivatives, and aromatic derivatives), PIONA (paraffin derivatives, iso-paraffin derivatives, olefin derivatives, naphthene derivatives, and aromatic derivatives), PNA (paraffin derivatives, naphthene derivatives, and aromatic derivatives), PINA (paraffin derivatives, iso-paraffin derivatives, naphthene derivatives, and aromatic derivatives), or SARA (saturate derivatives, aromatic derivatives, resin constituents, and asphaltene constituents).

The USBM-API (US Bureau of Mines-American Petroleum Institute) method allows the fractionation of crude oil samples into acids, bases, neutral nitrogen compounds, saturates, and mono-, di-, and polyaromatic compounds. Multidimensional techniques, that is, the combination of two or more chromatographic techniques, can be very useful to gain further information about the individual components of chemical groups. Compounds can be isolated and identified from complex matrices, and detailed fingerprinting of crude oil constituents is feasible (Altgelt and Gouw, 1979).

3.2.3.2 Gas Chromatography

Gas-liquid chromatography (GLC) is a method for separating the volatile components of various mixtures and is, in fact, a highly efficient fractionating technique, and it is ideally suited to the quantitative analysis of mixtures when the possible components are known, and the interest lies only in determining the amounts of each present. In this type of application, gas chromatography has taken over much of the work previously done by the other techniques; it is now the preferred technique for the analysis of hydrocarbon gases, and gas chromatographic in-line monitors are having increasing application in refinery plant control.

The evolution of gas-liquid chromatography has been a major factor in the successful identification of crude oil constituents. It is, however, almost impossible to apply this technique to the higher boiling crude oil constituents because of the comparatively low volatility. It is this comparative lack of volatility in the higher molecular weight, asphaltic constituents of crude oil that brought about another type of identification procedure, namely, carbon-type analysis.

The technique has proved to be an exceptional and versatile instrumental tool for analyzing compounds that are of low molecular weight and that can be volatilized without decomposition and is also useful for the component analysis of viscous feedstocks (Speight, 2014, 2015).

For example, the number of possible components of a certain molecular weight range increases markedly with increasing molecular weight. Furthermore, there is a corresponding sharp decrease in physical property differences between isometric structures as the molecular weight increases. Thus, it is very difficult, and on occasion almost impossible, to separate and identify single components in the heavier fractions of crude oil by gas chromatography. Indeed, the molecular weights of the constituents dictate that long residence times are necessary. This is inevitably accompanied by the requirement of increased column temperature, which decreases the residence time on the column but, at the same time, increases the possibility of thermal decomposition.

The instrumentation for gas-liquid chromatography is fairly straightforward and involves passing a carrier gas passes through a controller to the column (packed with an adsorbent) at the opening of which is a sample injector. The carrier gas then elutes the components of the mixture through the column to the detector at the end of which may be another gas flow monitor. Any gas, such as helium, argon, nitrogen, or hydrogen that is easily distinguishable from the components in the mixture, may be used as the carrier gas.

Column dimensions vary, but for analytic purposes a packed column may be 6 ft (2 m) long by 3 in. (6 mm) in diameter. It is also necessary to use a dissolving liquid as part of the column substance. This remains stationary on the adsorbent and affects partition of the components of the mixture. The solid support is usually a porous material that allows passage of the gas. For example, kieselguhr (diatomaceous earth), which can absorb up to 40% by weight of a liquid without appearing to be overly moist, is commonly used. The supporting material should not adsorb any of the components of the mixture and must therefore be inert.

Individual components of mixtures are usually identified by their respective retention times, that is, the time required for the component to traverse through the column under the specified conditions. Although tables for retention time data are available, it is more common in practice to determine the retention times of the pure compounds. The retention time of any component is itself a function of the many variables of column operation, such as the flow rate of the carrier gas and column temperature, and exact duplication of other operator's conditions may be difficult, if not impossible.

The sample size used in gas chromatography may vary upward from a microliter, and there is no theoretical upper limit to the size of the sample that may be handled if the equipment is built to accommodate it. The technique can be used for the analysis of mixtures of volatile vaporizable compounds boiling at any temperature between absolute zero (−273°C, −459°F) and 450°C (840°F). Identification of any substance that can be heated sufficiently without decomposing to give a vapor pressure of a few millimeters mercury is also possible.

The use of gas-liquid chromatography for direct component analysis in the higher boiling fractions of crude oil, such as residua, is beset by many problems, not the least of which is the low volatility and tendency for adsorption on solids by the higher molecular weight constituents. The number of possible components in any given molecular weight range increases markedly with the molecular weight (Speight, 2015), and there is a significant drop in the differences in physical properties among similar structural entities. This limits the ability of gas-liquid chromatography, and unless the sample has been fractionated by other techniques to reduce the complexity, complete component analysis is difficult, if not impossible.

The mass spectrometer identifies chemical compounds principally in terms of molecular type and molecular weight, and for many problems, therefore, it becomes necessary to use additional means of identification; the integrated gas-liquid chromatography infrared system is a very valuable complement to the mass spectrometer technique.

Considerable attention has also been given to trapping devices to collect gas chromatographic fractions for examination by one or more of the spectroscopic techniques. At the same time, developments in preparative gas-liquid chromatography have contributed even more to the compositional studies of crude oil and its products. With column size of 4–6 in. in diameter and capable of dealing with sample sizes of 200 mL or more, there is every possibility that gas-liquid chromatography will replace distillation in such areas as standard crude oil assay work.

Gas-liquid chromatography also provides a simple and convenient method for determining n-paraffin distribution throughout the crude oil distillate range. In this method, the n-paraffins are first separated by activated chemical destruction of the sieve with hydrofluoric acid, and the identity of the individual paraffins is determined chromatographically. This allows n-paraffin distribution throughout the boiling range 170°C–500°C (340°F–930°F) to be determined.

Gas chromatographic process analyzers have become very important in crude oil refineries. In some refineries, more samples are analyzed automatically by process chromatographs than are analyzed with laboratory instruments. These chromatographs are usually fully automatic. In some cases, after an analysis the instrument even makes automatic adjustments to the refinery unit. The chromatographs usually determine from 1 to 10 components, and the analyses are repeated at short intervals (15–20 minutes) over 24 hours.

A more recent, very important development in gas chromatography is its combination with a mass spectrometer as the detector. The technique in which gas chromatography is combined with spectrometry (GC/MS) has proved to be a powerful tool for identifying many compounds at very low levels in a wide range of boiling matrix. By the combination of the two techniques in one instrument, the need to trap the fractions from the gas chromatographic column is avoided and higher sensitivities can be attained. In passing through the gas chromatographic column, the sample is separated more or less according to its boiling point.

In view of the molecular characterizing nature of spectrometric techniques, it is not surprising that considerable attention has been given to the combined use of gas-liquid chromatography and these techniques. In recent years, the use of the mass spectrometer to monitor continuously the effluent of a chromatographic column has been reported, and considerable progress has been made in the development of rapid scan infrared spectrometers for this purpose. The mass spectrometer, however, has the advantage that the quantity of material required for the production of a spectrum is considerably less than that necessary to produce an infrared spectrum.

Although insufficient component resolution is observed in most cases, the eluting compounds at any time are usually closely related to each other in boiling point and molecular weight or both and are free from interfering lower and higher molecular weight species. Because of the reduced complexity of the gas chromatographic fractions, mass spectrometric scans carried out at regular intervals yield simpler spectra from which compound classes can more easily be determined.

Pyrolysis gas chromatography can be used for information on the gross composition of heavy crude oil fractions. In this technique, the sample under investigation is pyrolyzed, and the products are introduced into a gas chromatography system for analysis.

In the technique of inverse gas-liquid chromatography, the sample under study is used as the stationary phase and a number of volatile test compounds are chromatographed on this column. The interaction coefficient determined for these compounds is a measure of certain qualities of the liquid phase. The coefficient is therefore indicative of the chemical interaction of the solute with the stationary phase. The technique has been used largely for studies of asphalt.

3.2.3.3 Gel Permeation Chromatography

There are two additional techniques that have evolved from the more recent development of chromatographic methods and are (1) gel filtration chromatography (GFC), which was developed using soft, cross-linked dextran beads, and (2) gel permeation chromatography (GPC), which employs semi-rigid, cross-linked polystyrene beads. In either technique, the packing particles swell in the chromatographic solvent and forming a porous gel structure.

Briefly, dextran is a polymer of anhydro-glucose that is composed of approximately 95% alpha-D-(1–6) linkages. The remaining a(1–3) linkages account for the branching of dextran.

The distinction between the methods is based on the degree of swelling of the packing. The dextran swells to a much greater extent than the polystyrene. Subsequent developments of rigid porous packings of glass, silica, and silica gel have led to their use and classification as packings for gel permeation chromatography.

Gel permeation chromatography, also called size exclusion chromatography (SEC), which in its simplest representation, consists of employing column(s) packed with gels of varying pore sizes in a liquid chromatograph (Carbognani, 1997). Under conditions of constant flow, the solutes are injected onto the top of the column, whereupon they appear at the detector in order of decreasing molecular weight. The separation is based on the principle that the larger solute molecules cannot be accommodated within the pore system of the gel beads and, thus, are eluted first. On the other hand, the smaller solute molecules have increasing volume within the beads, depending upon their relative size, and require more time to elute.

Thus it is possible, with careful flow control, calibration, injection, and detection (usually by refractive index or UV absorption), to obtain an accurate chromatographic representation of the molecular weight distribution of the solute (Carbognani, 1997). This must of course assume that there is no chemical or physical interaction between the solute and the gel that negates the concept of solute size and pore size. For example, highly polar, small molecules that could associate in solution and are difficult to dissociate could conceivably appear in the incorrect molecular weight range.

In theory, gel permeation chromatography is an attractive technique for the determination of the number of average molecular weight (M_n) distribution of crude oil fractions. However, it is imperative to recognize that crude oil contains constituents of widely differing polarity, including non-polar paraffin derivatives and naphthene derivatives (alicyclic compounds), moderately polar aromatics (mononuclear and condensed), and polar nitrogen, oxygen, and sulfur species. Each particular compound type interacts with the gel surface to a different degree. The strength of the interaction increases with increasing polarity of the constituents and with decreasing polarity of the solvent. It must therefore be anticipated that the ideal linear relationship of log M_n against elution volume V_e that may be operative for non-polar hydrocarbon species cannot be expected to remain in operation. It must also be recognized that the lack of realistic standards of known number average molecular weight distribution and of chemical nature similar to that of the constituents of crude oil for calibration purposes may also be an issue. However, gel permeation chromatography has been employed in the study of crude oil constituents, especially the heavier constituents, and has yielded valuable data (Baltus and Anderson, 1984; Reynolds and Biggs, 1988; Speight, 2001, 2015).

The adoption of gel permeation chromatography represents a novel approach to the identification of the constituents since the method is not limited by the vapor pressure of the constituents. However, the situation is different with heavy crude oil samples. These are not homologous mixtures differing only in molecular weight. In any particular crude oil, a large variety of molecular species, varying from paraffinic molecules to the polynuclear aromatic ring systems, may not follow the assumed physical relationships that the method dictates from use with polymers.

Gel permeation chromatography is the separation method that comes closest to differentiating by molecular weight only and is almost unaffected by chemical composition (hence the alternate name size exclusion chromatography). The method actually separates by molecular size and has been used to measure molecular weights (Altgelt and Guow, 1979) although there is some question about the value of the data when the method is applied to asphaltene constituents (Speight et al., 1985).

Size exclusion chromatography is usually practiced with refractive index detection and yields a mass profile (concentration vs. time or elution volume) that can be converted to a mass vs. molecular weight plot by means of a calibration curve. The combination of size exclusion chromatography with element specific detection has widened this concept to provide the distribution of heteroatom constituents in the sample as a function of elution volume and molecular weight.

The use of size exclusion chromatography with reverse-phase high-performance liquid chromatography (HPLC) with a graphite furnace atomic absorption (GFAA) detector has been described for measuring the distribution of vanadium and nickel in high-molecular-weight crude oil fractions, including the asphaltene fraction. Using variants of this technique, inductively coupled and direct current plasma atomic emission spectroscopy (ICP and DCP), the method was extended and improved the former size exclusion chromatography-graphite furnace atomic absorption (SEC-GFAA) method allowing the separation to be continuously monitored.

The combination of gel permeation chromatography with another separation technique also allows the fractionation of a sample separately by molecular weight and by chemical structure. This is particularly advantageous for the characterization of the heavier fractions of crude oil materials because there are limitations to the use of other methods. Thus, it is possible to obtain a matrix of fractions differing in molecular weight and in chemical structure. It is also considered advisable to first fractionate a feedstock by gel permeation chromatography to avoid overlap of the functionality that might occur in different molecular weight species in the separation by other chromatographic methods.

The combination of gel permeation chromatography with another separation technique also allows the fractionation of a sample separately by molecular weight and by chemical structure. This is particularly advantageous for the characterization of the heavier fractions of crude oil materials because there are limitations to the use of other methods. Thus, it is possible to obtain a matrix of fractions differing in molecular weight and in chemical structure. It is also considered advisable to first fractionate a feedstock by gel permeation chromatography to avoid overlap of the functionality that might occur in different molecular weight species in the separation by other chromatographic methods.

In short, the gel permeation chromatographic technique concentrates all of a specific functional type into one fraction, recognizing that there will be a wide range of molecular weight species in that fraction. This is especially true when the chromatographic feedstock is a whole feed rather than a distillate fraction.

3.2.3.4   High-Performance Liquid Chromatography

High-performance liquid chromatography (HPLC), particularly in the normal phase mode, has found great utility in separating different hydrocarbon group types and identifying specific constituent types (Colin and Vion, 1983; Miller et al., 1983). Of particular interest is the application of the HPLC technique to the identification of the molecular types in nonvolatile feedstocks such as residua. The molecular species in the asphaltene fraction have been of particular interest (Colin and Vion, 1983; Felix et al., 1985) leading to identification of the size of polynuclear aromatic systems in the asphaltene constituents (Speight, 1986, 2014).

Several recent high-performance liquid chromatographic separation schemes are particularly interesting since they also incorporate detectors not usually associated with conventional hydrocarbon group types of analyses (Matsushita et al., 1981; Miller et al., 1983; Norris and Rawdon, 1984; Schwartz and Brownlee, 1986).

In general, the amount of information that can be derived from any chromatographic separation, however effective, depends on the detectors. As the field of application for HPLC has increased, the limitations of commercially available conventional detectors, such as ultraviolet/visible absorption (UV/VIS) and refractive index (RI), have become increasingly restrictive to the growth of the technique.

The general advantages of high-performance liquid chromatography method are (1) each sample is analyzed as received; (2) the boiling range of the sample is generally immaterial; (3) the total time per analysis is usually of the order of minutes; and (4) the method can be adapted for on-stream analysis.

3.2.3.5 Ion-Exchange Chromatography

Ion-exchange chromatography is widely used in the analyses of crude oil fractions for the isolation and preliminary separation of acid and basic components (Speight, 2001, 2015).

Ion-exchange resin constituents are prepared from aluminum silicates, synthetic resin constituents, and polysaccharides. The most widely used resin constituents have a skeletal structure of polystyrene cross-linked with varying amounts of divinylbenzene derivatives which have a loose gel structure of cross-linked polymer chains through which the sample ions must diffuse to reach most of the exchange sites. Since ion-exchange resin constituents are usually prepared as beads that are several hundred micrometers in diameter, most of the exchange sites are located at points quite distant from the surface. Because of the polyelectrolyte nature of these organic resin constituents, they can absorb large amounts of water or solvents and swell to volumes considerably larger than the dried gel. The size of the species that can diffuse through the particle is determined by the intermolecular spacing between the polymeric chains of the three-dimensional polyelectrolyte resin.

Cation-exchange chromatography is now used primarily to isolate the nitrogen constituents in a crude oil fraction. The relative importance of these compounds in crude oil has arisen because of their deleterious effects in many crude oil refining processes. They reduce the activity of cracking and hydrocracking catalysts and contribute to gum formation, color, odor, and poor storage properties of the product. Anion-exchange chromatography is used to isolate the acid components (such as carboxylic acids and phenols) from crude oil fractions.

3.2.3.6 Simulated Distillation

Gas-liquid chromatography has also been found useful for the preparation of simulated distillation curves. By integrating increments of the total area of the chromatogram and relating these to the boiling points of the components within each increment, which are calculated from the known boiling points of the easily recognizable n-paraffins, simulated boiling point data are produced.

Distillation is the most widely used separation process in the crude oil industry (Parkash, 2003; Gary et al., 2007; Speight, 2014; Hsu and Robinson, 2017; Speight, 2017). In fact, knowledge of the boiling range of crude feedstocks and finished products has been an essential part of the determination of feedstock quality since the start of the refining industry. The technique has been used for the control of plant and refinery processes as well as for predicting product slates. Thus it is not surprising that routine laboratory scale distillation tests have been widely used for determining the boiling ranges of crude feedstocks and a whole slate of refinery products (Speight, 2015).

There are some limitations to the routine distillation tests. For example, although heavy crude oils contain volatile constituents, it is not always advisable to use distillation for identification of these volatile constituents. Thermal decomposition of the constituents of crude oil is known to occur at approximately 350°C (660°F). Thermal decomposition of the constituents of the heavier, but immature, crude oil has been known to commence at temperatures as low as 200°C (390°F), however. Thus, thermal alteration of the constituents and erroneous identification of the decomposition products as natural constituents is always a possibility.

On the other hand, the limitations to the use of distillation as an identification technique may be economic, and detailed fractionation of the sample may also be of secondary importance. There have been attempts to combat these limitations, but it must be recognized that the general shape of a one-plate distillation curve is often adequate for making engineering calculations, correlating with other physical properties, and predicting the product slate.

However, a low-resolution, temperature-programmed gas chromatographic analysis has been developed to simulate the time-consuming true boiling point distillation (ASTM D2887). The method relies on the general observation that hydrocarbon derivatives are eluted from a non-polar adsorbent in the order of their boiling points. The method has been well researched in terms of method development and application (Romanowski and Thomas, 1985; MacAllister and DeRuiter, 1985; Schwartz et al., 1987). The benefits of the technique include good comparisons with other ASTM distillation data as well as application to higher boiling fractions of crude oil (Speight, 2001, 2015).

3.2.3.7   Supercritical Fluid Chromatography

By definition, a supercritical fluid is defined as a substance above its critical temperature that has properties not usually found at ambient temperatures and pressures. Moreover, use of a fluid under supercritical conditions conveys upon the fluid extraction capabilities that allows the opportunity to improve the recovery of the solute.

In supercritical fluid chromatography, the mobile phase is a substance maintained at a temperature a few degrees above its critical point. The physical properties of this substance are intermediate to those of a liquid and of a gas at ambient conditions. Hence, it is preferable to designate this condition as the supercritical phase.

In a chromatographic column, the supercritical fluid usually has a density approximately one-third to one-fourth of that of the corresponding liquid when used as the mobile phase; the diffusivity is approximately 1/100 that of a gas and approximately 200 times that of the liquid. The viscosity is of the same order of magnitude as that of the gas. Thus, for chromatographic purposes, such a fluid has more desirable transport properties than a liquid. In addition, the high density of the fluid results in a 1,000-fold better solvency than that of a gas. This is especially valuable for analyzing high-molecular-weight compounds.

A primary advantage of chromatography using supercritical mobile phases results from the mass transfer characteristics of the solute. The increased diffusion coefficients of supercritical fluids compared to non-supercritical liquids can lead to greater efficiency in separation methods or greater resolution in the analysis of complex mixtures. Another advantage of supercritical fluids compared with gases is that they can dissolve thermally labile and nonvolatile solutes and, upon expansion (decompression) of this solution, introduce the solute into the vapor phase for detection. Although supercritical fluids are sometimes considered to have superior solvating power, they usually do not provide any advantages in solvating power over liquids given a similar temperature constraint. In fact, many unique capabilities of supercritical fluids can be attributed to the poor solvent properties obtained at lower fluid densities. This dissolution phenomenon is increased by the variability of the solvent power of the fluid with density as the pressure or temperature changes.

The solvent properties that are most relevant for supercritical fluid chromatography are the critical temperature, polarity, and any specific solute-solvent intermolecular interactions (such as hydrogen bonding) that can enhance solubility and selectivity in a separation. Non-polar or low-polarity solvents with moderate critical temperatures (e.g., nitrous oxide, carbon dioxide, ethane, propane, pentane, xenon, sulfur hexafluoride, and various Freon derivatives) have been well explored for use in supercritical fluid chromatography. Carbon dioxide has been the fluid of choice in many super-critical fluid chromatography applications because of its low critical temperature (31°C, 88°F), non-toxic nature, and lack of interference with most detection methods.

3.2.4 SPECTROSCOPIC PROPERTIES

Spectroscopic studies have played an important role in the evaluation of crude oil and of crude oil products for the last three decades, and many of the methods are now used as standard methods of analysis for refinery feedstocks and products. Application of these methods to feedstocks and products is a natural consequence for the refiner.

Ultraviolet spectroscopy is employed for the evaluation of mineral oils (ASTM D2269), and flame photometry has been employed as a means of measuring the lithium/sodium content of lubri-cating greases (ASTM D3340) and the sodium content of residual fuel oil (ASTM D1318).

The contribution of spectroscopic studies has been effective in the application to the delineation of structural types in the viscous feedstocks because of the unknown nature of these feedstocks. One particular example is the ndM method (ASTM D3238), which is designed for the carbon distri-bution and structural group analysis of crude oil oils. Later investigators have taken structural group analysis several steps further than the ndM method.

It is also appropriate at this point to give a brief description of other methods that are used for the identification of the constituents of crude oil (Yen, 1984).

It is not intended to convey here that any one of these methods can be used for identification purposes. However, although these methods may fall short of complete acceptability as methods for the characterization of individual constituents of feedstocks, they can be used as methods by which an overall evaluation of the feedstock may be obtained in terms of molecular types.

3.2.4.1 Infrared Spectroscopy

Conventional infrared spectroscopy yields information about the functional features of various crude oil constituents. For example, infrared spectroscopy will aid in the identification of N-H and O-H functions, the nature of polymethylene chains, the C-H out-of-place bending frequencies, and the nature of any polynuclear aromatic systems.

With the progress of Fourier transform infrared (FTIR) spectroscopy during the past several decades, quantitative estimates of the various functional groups can also be made which is par-ticularly important for application to the higher molecular weight solid constituents of crude oil (i.e., the asphaltene fraction). It is also possible to derive structural parameters from infrared spectro-scopic data, and these are (1) saturated hydrogen to saturated carbon ratio; (2) paraffinic character; (3) naphthenic character; (4) methyl group content; and (5) paraffin chain length.

In conjunction with proton magnetic resonance (see the next section), structural parameters such as the fraction of paraffinic methyl groups to aromatic methyl groups can be obtained.

3.2.4.2 Mass Spectrometry

Mass spectrometry can play a key role in the identification of the constituents of feedstocks and products. The principal advantages of mass spectrometric methods are (1) high reproducibility of quantitative analyses; (2) the potential for obtaining detailed data on the individual components and/or carbon number homologs in complex mixtures; and (3) a minimal sample size is required for analysis. The ability of mass spectrometry to identify individual components in complex mixtures is unmatched by any modern analytical technique. Perhaps the exception is gas chromatography.

However, there are disadvantages arising from the use of mass spectrometry and these are: (1) the limitation of the method to organic materials that are volatile and stable at temperatures up to 300°C (570°F); and (2) the difficulty of separating isomers for absolute identification (Speight, 2014, 2015).

Thus, mass spectrometry does furnish useful information about the composition of feedstocks and products although there may be structural similarities that can hinder the identification of individual components. Consequently, identification by type or by homolog will be more meaningful since similar structural types may be presumed to behave similarly in processing situations.

Mass spectrometry should be used discriminately where a maximum amount of information can be expected. The heavier nonvolatile feedstocks are for practical purposes, beyond the useful range of routine mass spectrometry. At the elevated temperatures necessary to encourage volatility, thermal decomposition will occur in the inlet, and any subsequent analysis would be biased to the low molecular weight end and to the lower molecular products produced by the thermal decomposition.

On the other hand, the occurrence of high-molecular-weight hydrocarbon derivatives in ozocerite and a waxy yellow crude oil from the Uinta Basin (Utah) has been studied successfully by field ionization-mass spectrometry (FIMS) (Del Río and Philp, 1999). The spectra consisted predominantly of molecular ions ranging up to near mass 2,000 and correspond to several series of hydrocarbon derivatives ranging up to C_{110}. The use of method permitted the range of hydrocarbon derivatives identified in geological materials to be extended far beyond that identified by the usual chromatographic techniques. Moreover, from the spectra, it was possible to extract the molecular ions corresponding to series of hydrocarbon derivatives with different degree of unsaturation or ring closures. The ozocerite solid bitumen consisted mainly of a series of branched alkanes (C_nH_{2n+2}) and cyclic alkanes (C_nH_{2n} and C_nH_{2n-2}) up to C_{110} with a predominance of monocyclic alkanes in the high-molecular-weight region (above C_{40}). The waxy yellow crude oil, on the other hand, contained only acyclic compounds, mainly n-alkanes, ranging up to C_{100}.

3.2.4.3 Nuclear Magnetic Resonance Spectroscopy

Nuclear magnetic resonance spectroscopy has frequently been employed for general studies and for the structural studies of crude oil constituents (Bouquet and Bailleul, 1982; Hasan et al., 1989). In fact, proton magnetic resonance (PMR) studies (along with infrared spectroscopic studies) were, perhaps, the first studies of the modern era that allowed structural inferences to be made about the polynuclear aromatic systems that occur in the high-molecular-weight constituents of crude oil.

In general, the proton (hydrogen) types in crude oil fractions can be subdivided into five types which subdivides the hydrogen distribution into (1) aromatic hydrogen; (2) substituted hydrogen next to an aromatic ring; (3) naphthenic hydrogen; (4) methylene hydrogen; and (5) terminal methyl hydrogen remote from an aromatic ring. Other ratios are also derived from which a series of structural parameters can be calculated. However, it must be remembered that the structural details of structural entities obtained by the use of physical techniques are, in many cases, derived by inference, and it must be recognized that some signals can be obscured by intermolecular interactions. This, of course, can cause errors in deduction reasoning which can have a substantial influence on the outcome of the calculations (Ebert et al., 1984, 1987; Ebert, 1990; Speight, 1994, 2014, 2015).

It is in this regard that carbon-13 (^{13}C) magnetic resonance (CMR) can play a useful role, and because carbon magnetic resonance deals with analyzing the carbon distribution types, the obvious structural parameter to be determined is the aromaticity, f_a. A direct determination from the various carbon-type environments is one of the better methods for the determination of aromaticity (Snape et al., 1979). Thus, through a combination of proton and carbon magnetic resonance techniques, refinements can be made on the structural parameters and for the solid-state high-resolution CMR technique additional structural parameters can be obtained (Weinberg et al., 1981).

3.2.5 Molecular Weight

The molecular weight (formula weight) of a compound is the sum of the atomic weights of all the atoms in a molecule and can be determined by a variety of methods (Cooper, 1989). Crude oil, being a complex mixture of (at least) several thousand constituents, requires qualification of the molecular weight as either (1) number average molecular weight or (2) weight average molecular weight.

The number average molecular weight is the ordinary arithmetic mean or average of the molecular weights of the individual constituents and is determined by measuring the molecular weight of n molecules, summing the weights, and dividing by n (Speight, 2014, 2015).

The weight average molecular weight is a way of describing the molecular weight of a complex mixture such as crude oil even if the molecular constituents are not of the same type and exist in different sizes.

Even though refining produces, in general, lower molecular weight species than those originally in the feedstock, there is still the need to determine the molecular weight of the original constituents as well as the molecular weights of the products as a means of understanding the process. For those original constituents and products, e.g., resin constituents and asphaltene constituents that have little or no volatility, vapor pressure osmometry (VPO) has been proven to be of considerable value (Blondel-Telouk et al., 1995).

A particularly appropriate method involves the use of different solvents (at least two), and the data are then extrapolated to infinite dilution. There has also been the use of different temperatures for a particular solvent after which the data are extrapolated to room temperature (Speight et al., 1985; Speight, 1987). In this manner, different solvents are employed, and the molecular weight of a crude oil fraction (particularly the asphaltene constituents) can be determined for which it can be assumed that there is little or no influence from any intermolecular forces. In summary, the molecular weight may be as close to the real value as possible.

In fact, it is strongly recommended that to negate concentration effects and temperature effects the molecular weight determination be carried out at three different concentrations at three different temperatures.

The data for each temperature are then extrapolated to zero concentration, and the zero concentration data at each temperature are then extrapolated to room temperature (Speight, 1987, 2014, 2015).

3.3 BIOMASS FEEDSTOCKS

As the modern refinery evolves, the availability of the fossil fuel feedstocks declines as well as the increased use of bio-based feedstocks (such as biomass and the biomass-derived bio-oil). In the case of biomass (and, hence bio-oil), it will be necessary to include other forms of analyses to determine the properties and potential reliability of the bio-based feedstocks (Chapter 2) (Speight, 2011, 2022). Almost all biomass, whether grown for chemicals, food, or animal feed result in the generation of an organic residue after the primary use has been satisfied and which can be used for the production of chemicals by the direct method (i.e., extraction) or by the indirect method (i.e., gasification).

One extremely important aspect of biomass use as a process feedstock is the preparation of the biomass (also referred to as biomass cleaning or biomass pretreatment) which is the removal of any contaminants that could have (1) an adverse effect of the process, (2) an adverse effect on the yields and quality of the products, and (3) an adverse effect on any catalyst used in the process. Thus, feedstock properties and preparation are, essentially, the pretreatment of the biomass feedstock to assist in the efficiency of the conversion process.

3.3.1 PROPERTIES

In any biorefinery (where bio-oil is often produced from the various types of biomass), there is the need for feedstock evaluation which must include a description of each of the individual feedstocks that are accepted by the biorefinery. This would typically involve an examination of one or more of the physical properties of the potential feedstock, and by this means, a set of basic characteristics can be obtained that can be correlated with the processing options.

Evaluation, in this context of this book, is the determination of the physical and chemical characteristics of the biomass insofar as the selection of a suitable processing option as well as the yields and properties of products produced from the feedstock will vary considerably and are dependent on the concentration of the various constituents of the feedstock.

There are several methods by which the properties of biomass (and its suitability for conversion processes) can be determined, and these are common ways to measure the properties of any carbon product, which will also be used for biomass. Details of the various analytical methods can be found in the list of standard test methods published by ASTM International, formerly the American Society for Testing and Materials (ASTM, 2021). By way of clarification, ASTM International is an international organization headquartered in the United States that provides standard test methods that are used to assert the quality of products (including materials, processes, and services) and personnel for industries that desire an independent third-party demonstration of compliance to standards and/or are facing regulatory pressures to prove compliance to standards.

For the purposes of this text, examples of the properties that might be determined prior to the biomass being used in the biorefinery are (1) proximate analysis, (2) ultimate analysis, (3) heat of combustion, (4) mineral matter analysis, commonly performed by analysis of the combustion ash, and (5) the bulk density.

3.3.1.1 Proximate Analysis

Proximate analysis of a carbonaceous feedstock such as biomass coal is an assay of the feedstock while the ultimate analysis is a determination of the individual elements in the feedstock. The various parameter studied are percentages of moisture (M), volatile matter (VM), ash (A), and fixed carbon (FC).

By way of clarification, feedstock does not contain ash but do contain mineral matter that is reflected in the test for the yield of ash produced by a combustion procedure. The mineral matter is changed in the test method from the form in the feedstock (such as a chloride salt, e.g., NaCl) to the various metal oxides.

The proximate analysis of a feedstock system is both comparative and predictive in nature and allows comparisons of feedstocks on the basis of specific constituents or the yield of theses constituents. This makes it possible to know how much better one process is more suitable for conversion of the feedstock to products.

The properties examined during the proximate analysis tests are determined on a mass basis. In the test method, moisture (M) is driven off at approximately $105°C–110°C$ ($220°F–230°F$) and the data represent only the water in the sample that is physically bound. Volatile matter (VM) is driven off in an inert atmosphere at $950°C$ ($1,740°F$), using a slow heating rate. The ash yield is determined by combusting the remaining material (after loss of any volatile matter) and at a temperature in excess of $700°C$ ($1,290°F$) in an oxygen atmosphere. The fixed carbon is then determined by difference:

$$FC = 1 - M - Ash - VM$$

3.3.1.2 Ultimate Analysis

The ultimate analysis of a feedstock provides the elemental composition of the feedstock. For this analysis, a sample of the feedstock is combusted in an ultimate analyzer, which measures the weight percent of carbon, hydrogen, nitrogen, sulfur, and ash in the sample.

The ultimate analysis of a feedstock is more specific than the proximate analysis (initially derived from the term approximate analysis) insofar as the test methods that are part of the ultimate analysis produce data about the elemental composition of the organic portion of the feedstock. The amounts (% w/w) of carbon (C), hydrogen (H), nitrogen (N), sulfur (S), and oxygen (O) are determined on a mass percent (% w/w) basis, and can be converted to an atomic basis. In some cases, chlorine (Cl) and other elements (depending upon the feedstock history) will also need to be analyzed. Oxygen can be determined directly but is often determined by difference. Water (in the feedstock or produced during the analysis) can interfere with the data for the hydrogen content of the feedstock and must be given consideration.

3.3.1.3 Heat of Combustion

The heat of combustion of a feedstock is the heat released under standard conditions during the combustion process in which the feedstock is combusted in the presence of oxygen to produce carbon dioxide and water.

The heat of combustion of a biomass feedstock (or any carbonaceous feedstock, for that matter) can be measured directly using a bomb calorimeter. This instrument is used to measure the calorific value per mass (in calories/gram or Btu/lb) and he heat of combustion can also be estimated using different formulas that calculate it based on either ultimate analysis or proximate analysis.

In the test method, a sample of the feedstock is placed in a crucible that is put inside of a reactor with high-pressure oxygen after which the sample is ignited. The temperature of the water in the container is measured before ignition and after ignition from which the specific heat of water and the change in temperature are used to determine the heat of combustion.

Heating values are reported on both the wet sample and dry sample bases. For the high heating value (HHV), the value can be determined bon a moisture-free basis while for the low heating value (LHV), a portion of the heat of combustion is recognized as being used to evaporate the moisture.

3.3.1.4 Ash Analysis

The determination of the content of the mineral matter in a biomass feedstock (which is manifested as combustion ash using an analysis often referred to as ashing) is the process of determining the concentration of trace elements prior to a chemical analysis such as (in the case of the content of the mineral matter of the feedstock) or optical analysis, such as the use of a spectroscopic method.

In general, the chemical composition of the ash is typically an examination of the elemental composition of the ash. A common assumption in most of these analyses has been that the minerals present in the ash are oxides of different elements originally in the biomass but this does not give the necessary information related to the thermal stability of the chemical stability of the actual mineral constituents of the feedstock and the effect of these constituents on the process (Misra et al., 1993).

More specifically, information about the following can help to alleviate problem that can occur during the processing (1) the fly ash which can include the solids escaping via the flue during combustion, (2) the combustion temperature which produces two direct effects: dissociation which is the conversion of carbonate derivatives to oxide derivatives and volatilization which leads to the escape of the inorganic products in the flue gas, (3) other chemical reactions such occur when some constituents of the ash are exposed to the environment between combustion and the analysis, oxides may convert back to carbonates by reacting with carbon dioxide in the air, and (4) the type, age, and growing environment of the feedstock affect the composition of the wood, and thus the ash.

In the test method, which must be performed und standard conditions (Misra et al., 1993), the minerals in the feedstock are converted to mineral ash (the metal oxides and any non-metal oxides) during the combustion after which the ash is analyzed for specific constituents that will contain oxygen, such as CaO, K_2O, Na_2O, MgO, SiO_2, Fe_2O_3, P_2O_5, SO_3, and Cl. Once the mineral or ash is isolated, it often must be dissolved in various acids and then analyzed.

Determination of the types of minerals in biomass is an important aspect of biomass composition because of the potential of the mineral mater to have a beneficial effect or an adverse effect on the conversion process. In some cases, de-ashing (or mineral matter removal) before thermal processing can increase the yield of volatile productions as well as affecting the initial decomposition temperature and rate of pyrolysis.

3.3.1.5 Bulk Density

The bulk density (also called apparent density or volumetric density) of a sample is a property of powders, granules, and other solids and is used in reference to other masses of particulate matter (such as mineral components and any chemical substances) in the sample.

While the density of a substance is a ratio of the mass of the substance to the volume of the substance (i.e., the mass per unit volume), the bulk density is used in situations where the particles of the sample (sometimes referred to as chunks of the sample) are loosely packed with space for air between the particles. The bulk density is calculated (in the present context) as the dry weight of biomass divided by the volume of the biomass. This volume includes the volume of any of the particles as well as the volume of the pores between the particles and is typically expressed in rams per cubic meter (g/cm^3). Also, the specific gravity of a substance is the ratio of the weight of the substance in air of a given volume of the substance at a stated temperature to the weight in air of an equal volume of distilled water at that temperature; the bulk density includes the pore volume of the substance.

The bulk density is typically determined by measuring the weight of material per unit volume. The data are usually determined on a dry weight basis (moisture free) or on an as-received basis with moisture content available. The heating value and bulk density are used to determine the energy density.

3.3.2 Pretreatment

The pretreatment of biomass is considered one of the most important steps in the overall processing in a biomass-to-chemicals program, and the method chosen depends very much upon the properties of the biomass. Pretreatment can occur using acidic or alkaline reagents as well as using a variety of physical methods, and the method of choice depends much upon the process needs and the process parameters (Trajano and Wyman, 2013; Modenbach and Nokes, 2014). With the strong advancement in developing lignocellulose biomass-based refinery and algal biomass-based biorefinery (Chapter 8), the major focus is the development of pretreatment processes that are technically feasible and serve the purpose (Pandey et al., 2015).

Typically, the fundamental steps in the pretreatment of biomass involves processes such as (1) washing/separation, which involves removal of inorganic matter such as stones and pebbles – the inherent inorganic constituents of the biomass may require more drastic treatment depending upon the character and properties of these constituents; (2) size reduction, which involves grinding, milling, and crushing; and (3) separation of soluble matter (Speight, 2011, 2022). Also, the pretreatment process that is selected, depending upon the character of the biomass and the process parameters, is likely to be different for diverse types of biomass and production of the planned for (desired) products.

Moisture in the biomass is another consideration for feedstock preparation because moisture in the feedstock will simply vaporize during the process and then re-condense with, say, the bio-oil product which has an adverse impact on the resulting quality of the bio-oil. It should also be noted that water is formed as part of the thermochemical reactions that occur in the reactor (such as the pyrolysis reactor). For example, if dry biomass is subjected to the thermal requirements for fast pyrolysis, the resulting bio-oil will still contain water (as much as 12%–15% w/w). This water originates during the process and is the result of the thermal dehydration of carbohydrate derivatives (e.g., $C_6H_{12}O_6$) in the feedstocks as well as the result of reactions occurring between the hydrogen (H_2) and oxygen (O_2) at the elevated temperature (such as 500°C, 930°F) in the reactor. Thus, using the following equations as simple examples:

$$C_6H_{12}O_6 \rightarrow C + 6H_2O$$

$$2H_2 + O_2 \rightarrow 2H_2O$$

Furthermore, moisture (intrinsic moisture or extrinsic moisture) in the feedstock acts as a heat sink and can have an adverse effect on the amount of heat available to complete the pyrolysis reaction. Ideally, it would be desirable to have little or no moisture in the feedstock but practical considerations make this unrealistic (or, more; likely, uneconomical), and moisture levels on the order of 5%–10% w/w are generally considered acceptable for the pyrolysis process to proceed efficiently. As with the particle size (which should be sufficiently small to allow efficient heat transfer in the reactor), the moisture levels in the feedstock biomass are a trade-off between the cost of drying and the heating value penalty that is due to the presence of moisture in the feedstock.

If the moisture content in biomass feedstock is too high, any bio-oil that is produced exhibits a high moisture content which eventually reduces the value of the bio-oil that is designated for further processing. Therefore, it is preferable that the original biomass feedstock should undergo a pretreatment (drying) process to reduce the water content before being submitted as a feedstock to a pyrolysis process (Dobele et al., 2007). In contrast, the use of an elevated temperature for the drying process could be a critical issue (with some biomass feedstocks) because of the possibility of producing thermal-oxidative reactions, thereby inducing cross-linking reactions and producing a condensation of the components resulting in a complex product that has a higher thermal stability of the biomass complex than the original feedstock. The overall result is a char-like product that is resistant to further conversion.

To achieve high yields of the desirable products (whether the products are gases, liquids, or solids), it is also necessary to prepare the solid biomass feedstock in such a manner that the required heat transfer rates in the pyrolysis process are facilitated and not prolonged. The three primary heat transfer mechanisms available to engineers in designing reaction vessels are (1) convection, (2) conduction, and (3) radiation, and to adequately exploit one or more of these mechanisms as applied to biomass pyrolysis, it is necessary to have a relatively small particle size in the reaction vessel. The small particle size ensures a high surface area per unit volume of particle, and the particle achieves the desired temperature in a short residence time. Another reason for the conversion of biomass feedstock to small particles is the physical transition of feedstock as it undergoes pyrolysis. At this point of the reaction, char can develop on the surface of the particle, and this char can act as an insulator which impedes the heat transfer into the center of the particle, which is counter to the requirements needed for pyrolysis.

One aspect of feedstock preparation in the light of the processes in which the biomass is to be used and converted is the concept of torrefaction which is used as a pretreatment step for biomass conversion techniques, such when the feedstock is used as a feedstock for the gasification process (Prins et al., 2006a, b, c). Typically, torrefaction commences when the temperature of the feedstock reaches 200°C (390°F). During the process, the biomass feedstock is partly devolatilized leading to a decrease in mass, but the initial energy content of the torrefed biomass is preserved in the solid product, thereby creating a feedstock with higher energy per unit weight or higher energy per unit volume.

3.4   USE OF THE DATA

The data derived from the evaluation techniques described here can be employed to give the refiner an indication of the means by which the crude feedstock should be processed as well as for the prediction of product properties (Dolbear et al., 1987; Wallace, 1988; Wallace and Carrigy, 1988; Speight, 2014, 2015). Other properties (Table 3.2) may also be required for further feedstock evaluation, or, more likely, for comparison between feedstocks even though they may not play any role in dictating which refinery operations are necessary. An example of such an application is the calculation of product yields for delayed coking operations by using the carbon residue and the API gravity of the feedstock.

Nevertheless, it must be emphasized that to proceed from the raw evaluation data to full-scale production is not the preferred step; further evaluation of the processability of the feedstock is usually necessary through the use of a pilot-scale operation. To take the evaluation of a feedstock one step further, it may then be possible to develop correlations between the data obtained from the actual plant operations (as well as the pilot plant data) with one or more of the physical properties determined as part of the initial feedstock evaluation (Speight, 2014, 2015).

Even proceeding from the raw evaluation data to full-scale production is not always the preferred step. Further evaluation of the processability of the feedstock is usually necessary through the use of a pilot-scale operation. To take the evaluation of a feedstock one step further, it may then be possible to develop correlations between the data obtained from the actual plant operations (as well as the pilot plant data) with one or more of the physical properties determined as part of the initial feedstock evaluation.

Evaluation of crude oil from known physical properties may also be achieved by use of the refractivity intercept. Thus, if refractive indices of hydrocarbon derivatives are plotted against the respective densities, straight lines of constant slope are obtained, one for each homologous series; the intercepts of these lines with the ordinate of the plot are characteristic, and the refractivity intercept is derived from the formula:

$$\text{Refractivity intercept} = n - d \,/\, 2$$

The intercept cannot differentiate accurately among all series, which restricts the number of different types of compounds that can be recognized in a sample. The technique has been applied to non-aromatic olefin-free materials in the gasoline range by assuming additivity of the constant on a volume basis.

Following this, an equation has been devised that is applicable to straight-run lubricating distillates if the material contains between 25% and 75% of the carbon present in naphthenic rings:

$$\text{Refractivity intercept} = 1.0502 - 0.00020\% C_N$$

Although not specifically addressed in this chapter, the fractionation of crude oil (Speight, 2014, 2015) also plays a role, along with the physical testing methods, of evaluating crude oil as a refinery feedstock. For example, by careful selection of an appropriate technique, it is possible to obtain a detailed overview of feedstock or product composition that can be used for process predictions. Using the adsorbent separation as an example, it becomes possible to develop one or more crude oil maps and determine how a crude oil might behave under specified process conditions.

This concept has been developed to the point where various physical parameters as the ordinates and abscissa. However, it must be recognized that such maps do not give any indication of the complex interactions that occur between, for example, such fractions as the asphaltene constituents and resin constituents (Koots and Speight, 1975; Speight, 1994), but it does allow predictions of feedstock behavior. It must also be recognized that such a representation varies for different feedstocks.

In summary, evaluation of feedstock behavior from test data is not only possible but has been practiced for decades. And such evaluations will continue for decades to come. However, it is essential to recognize that the derivation of an equation for predictability of behavior will not suffice (with a reasonable degree of accuracy) for all feedstocks. Many of the data are feedstock dependent because they incorporate the complex reactions of the feedstock constituents with each other. Careful testing and evaluation of the behavior of each feedstock and blend of feedstocks are recommended. If this is not done, incompatibility or instability (Speight, 2014, 2015) can result leading to higher-than-predicted yields of thermal or catalytic coke.

REFERENCES

Altgelt, K.H., and Gouw, T.H. 1979. *Chromatography in Petroleum Analysis*. Marcel Dekker Inc., New York.

ASTM. 2021. *Annual Book of Standards*. ASTM International, West Conshohocken, Pennsylvania.

ASTM D6. 2021. *Standard Test Method for Loss on Heating of Oil and Asphaltic Compounds. Annual Book of Standards*. ASTM International, West Conshohocken, Pennsylvania.

ASTM D20. 2021. *Standard Test Method for Distillation of Road Tars. Annual Book of Standards*. ASTM International, West Conshohocken, Pennsylvania.

ASTM D56. 2021. *Standard Test Method for Flash Point by Tag Closed Cup Tester. Annual Book of Standards.* ASTM International, West Conshohocken, Pennsylvania.

ASTM D70. 2021. *Standard Test Method for Density of Semi-Solid Bituminous Materials (Pycnometer Method). Annual Book of Standards.* ASTM International, West Conshohocken, Pennsylvania.

ASTM D71. 2025. *Standard Test Method for Relative Density of Solid Pitch and Asphalt (Displacement Method). Annual Book of Standards.* ASTM International, West Conshohocken, Pennsylvania.

ASTM D86. 2021. *Standard Test Method for Distillation of Petroleum Products at Atmospheric Pressure. Annual Book of Standards.* ASTM International, West Conshohocken, Pennsylvania.

ASTM D87. 2021. *Standard Test Method for Melting Point of Petroleum Wax (Cooling Curve). Annual Book of Standards.* ASTM International, West Conshohocken, Pennsylvania.

ASTM D88. 2021. *Standard Test Method for Saybolt Viscosity. Annual Book of Standards.* ASTM International, West Conshohocken, Pennsylvania.

ASTM D92. 2021. *Standard Test Method for Flash and Fire Points by Cleveland Open Cup Tester. Annual Book of Standards.* ASTM International, West Conshohocken, Pennsylvania.

ASTM D93. 2021. *Standard Test Methods for Flash Point by Pensky-Martens Closed Cup Tester. Annual Book of Standards.* ASTM International, West Conshohocken, Pennsylvania.

ASTM D97. 2021. *Standard Test Method for Pour Point of Petroleum Products. Annual Book of Standards.* ASTM International, West Conshohocken, Pennsylvania.

ASTM D127. 2021. *Standard Test Method for Drop Melting Point of Petroleum Wax, Including Petrolatum. Annual Book of Standards.* ASTM International, West Conshohocken, Pennsylvania.

ASTM D129. 2021. *Standard Test Method for Sulfur in Petroleum Products (General High Pressure Decomposition Device Method). Annual Book of Standards.* ASTM International, West Conshohocken, Pennsylvania.

ASTM D139. 2021. *Standard Test Method for Float Test for Bituminous Materials. Annual Book of Standards.* ASTM International, West Conshohocken, Pennsylvania.

ASTM D189. 2021. *Standard Test Method for Conradson Carbon Residue of Petroleum Products. Annual Book of Standards.* ASTM International, West Conshohocken, Pennsylvania.

ASTM D240. 2021. *Standard Test Method for Heat of Combustion of Liquid Hydrocarbon Fuels by Bomb Calorimeter. Annual Book of Standards.* ASTM International, West Conshohocken, Pennsylvania.

ASTM D287. 2021. *Standard Test Method for API Gravity of Crude Petroleum and Petroleum Products (Hydrometer Method). Annual Book of Standards.* ASTM International, West Conshohocken, Pennsylvania.

ASTM D323. 2021. *Standard Test Method for Vapor Pressure of Petroleum Products (Reid Method). Annual Book of Standards.* ASTM International, West Conshohocken, Pennsylvania.

ASTM D341. 2021. *Standard Practice for Viscosity-Temperature Charts for Liquid Petroleum Products. Annual Book of Standards.* ASTM International, West Conshohocken, Pennsylvania.

ASTM D445. 2021. *Standard Test Method for Kinematic Viscosity of Transparent and Opaque Liquids (and Calculation of Dynamic Viscosity). Annual Book of Standards.* ASTM International, West Conshohocken, Pennsylvania.

ASTM D482. 2021. *Standard Test Method for Ash from Petroleum Products. Annual Book of Standards.* ASTM International, West Conshohocken, Pennsylvania.

ASTM D524. 2021. *Standard Test Method for Ramsbottom Carbon Residue of Petroleum Products. Annual Book of Standards.* ASTM International, West Conshohocken, Pennsylvania.

ASTM D566. 2021. *Standard Test Method for Dropping Point of Lubricating Grease. Annual Book of Standards.* ASTM International, West Conshohocken, Pennsylvania.

ASTM D664. 2021. *Standard Test Method for Acid Number of Petroleum Products by Potentiometric Titration. Annual Book of Standards.* ASTM International, West Conshohocken, Pennsylvania.

ASTM D938. 2021. *Standard Test Method for Congealing Point of Petroleum Waxes, Including Petrolatum. Annual Book of Standards.* ASTM International, West Conshohocken, Pennsylvania.

ASTM D974. 2015. *Standard Test Method for Acid and Base Number by Color-Indicator Titration. Annual Book of Standards.* ASTM International, West Conshohocken, Pennsylvania.

ASTM D1018. 2021. *Standard Test Method for Hydrogen in Petroleum Fractions. Annual Book of Standards.* ASTM International, West Conshohocken, Pennsylvania.

ASTM D1160. 2021. *Standard Test Method for Distillation of Petroleum Products at Reduced Pressure. Annual Book of Standards.* ASTM International, West Conshohocken, Pennsylvania.

ASTM D1217. 2021. *Standard Test Method for Density and Relative Density (Specific Gravity) of Liquids by Bingham Pycnometer. Annual Book of Standards.* ASTM International, West Conshohocken, Pennsylvania.

ASTM D1266. 2021. *Standard Test Method for Sulfur in Petroleum Products (Lamp Method). Annual Book of Standards*. ASTM International, West Conshohocken, Pennsylvania.

ASTM D1298. 2021. *Standard Test Method for Density, Relative Density, or API Gravity of Crude Petroleum and Liquid Petroleum Products by Hydrometer Method. Annual Book of Standards*. ASTM International, West Conshohocken, Pennsylvania.

ASTM D1318. 2021. *Standard Test Method for Sodium in Residual Fuel Oil (Flame Photometric Method). Annual Book of Standards*. ASTM International, West Conshohocken, Pennsylvania.

ASTM D1480. 2021. *Standard Test Method for Density and Relative Density (Specific Gravity) of Viscous Materials by Bingham Pycnometer. Annual Book of Standards*. ASTM International, West Conshohocken, Pennsylvania.

ASTM D1481. 2021. *Standard Test Method for Density and Relative Density (Specific Gravity) of Viscous Materials by Lipkin Bicapillary Pycnometer. Annual Book of Standards*. ASTM International, West Conshohocken, Pennsylvania.

ASTM D1552. 2021. *Standard Test Method for Sulfur in Petroleum Products (High-Temperature Method). Annual Book of Standards*. ASTM International, West Conshohocken, Pennsylvania.

ASTM D1757. 2021. *Standard Test Method for Sulfur in Ash from Coal and Coke. Annual Book of Standards*. ASTM International, West Conshohocken, Pennsylvania.

ASTM D2007. 2021. *Standard Test Method for Characteristic Groups in Rubber Extender and Processing Oils and Other Petroleum-Derived Oils by the Clay-Gel Absorption Chromatographic Method. Annual Book of Standards*. ASTM International, West Conshohocken, Pennsylvania.

ASTM D2161. 2021. *Standard Practice for Conversion of Kinematic Viscosity to Saybolt Universal Viscosity or to Saybolt Furol Viscosity. Annual Book of Standards*. ASTM International, West Conshohocken, Pennsylvania.

ASTM D2265. 2021. *Standard Test Method for Dropping Point of Lubricating Grease Over Wide Temperature Range. Annual Book of Standards*. ASTM International, West Conshohocken, Pennsylvania.

ASTM D2269. 2025. *Standard Test Method for Evaluation of White Mineral Oils by Ultraviolet Absorption. Annual Book of Standards*. ASTM International, West Conshohocken, Pennsylvania.

ASTM D2270. 2021. *Standard Practice for Calculating Viscosity Index from Kinematic Viscosity at 40 and 100°C. Annual Book of Standards*. ASTM International, West Conshohocken, Pennsylvania.

ASTM D2500. 2021. *Standard Test Method for Cloud Point of Petroleum Products. Annual Book of Standards*. ASTM International, West Conshohocken, Pennsylvania.

ASTM D2622. 2021. *Standard Test Method for Sulfur in Petroleum Products by Wavelength Dispersive X-ray Fluorescence Spectrometry. Annual Book of Standards*. ASTM International, West Conshohocken, Pennsylvania.

ASTM D2887. 2021. *Standard Test Method for Boiling Range Distribution of Petroleum Fractions by Gas Chromatography. Annual Book of Standards*. ASTM International, West Conshohocken, Pennsylvania.

ASTM D3117. 2021. *Standard Test Method for Wax Appearance Point of Distillate Fuels. Annual Book of Standards*. ASTM International, West Conshohocken, Pennsylvania.

ASTM D3177. 2021. *Standard Test Methods for Total Sulfur in the Analysis Sample of Coal and Coke. Annual Book of Standards*. ASTM International, West Conshohocken, Pennsylvania.

ASTM D3178. 2021. *Standard Test Methods for Carbon and Hydrogen in the Analysis Sample of Coal and Coke. Annual Book of Standards*. ASTM International, West Conshohocken, Pennsylvania.

ASTM D3179. 2021. *Standard Test Methods for Nitrogen in the Analysis Sample of Coal and Coke. Annual Book of Standards*. ASTM International, West Conshohocken, Pennsylvania.

ASTM D3238. 2021. *Standard Test Method for Calculation of Carbon Distribution and Structural Group Analysis of Petroleum Oils by the n-d-M Method. Annual Book of Standards*. ASTM International, West Conshohocken, Pennsylvania.

ASTM D3340. 2021. *Standard Test Method for Lithium and Sodium in Lubricating Greases by Flame Photometer. Annual Book of Standards*. ASTM International, West Conshohocken, Pennsylvania.

ASTM D4045. 2021. *Standard Test Method for Sulfur in Petroleum Products by Hydrogenolysis and Rateometric Colorimetry. Annual Book of Standards*. ASTM International, West Conshohocken, Pennsylvania.

ASTM D4052. 2021. *Standard Test Method for Density, Relative Density, and API Gravity of Liquids by Digital Density Meter. Annual Book of Standards*. ASTM International, West Conshohocken, Pennsylvania.

ASTM D4057. 2021. *Standard Practice for Manual Sampling of Petroleum and Petroleum Products. Annual Book of Standards*. ASTM International, West Conshohocken, Pennsylvania.

ASTM D4124. 2021. *Standard Test Method for Separation of Asphalt into Four Fractions. Annual Book of Standards*. ASTM International, West Conshohocken, Pennsylvania.

ASTM D4294. 2021. *Standard Test Method for Sulfur in Petroleum and Petroleum Products by Energy Dispersive X-ray Fluorescence Spectrometry. Annual Book of Standards.* ASTM International, West Conshohocken, Pennsylvania.

ASTM D4530. 2021. *Standard Test Method for Determination of Carbon Residue (Micro Method). Annual Book of Standards.* ASTM International, West Conshohocken, Pennsylvania.

ASTM D4628. 2021. *Standard Test Method for Analysis of Barium, Calcium, Magnesium, and Zinc in Unused Lubricating Oils by Atomic Absorption Spectrometry. Annual Book of Standards.* ASTM International, West Conshohocken, Pennsylvania.

ASTM D5002. 2021. *Standard Test Method for Density and Relative Density of Crude Oils by Digital Density Analyzer. Annual Book of Standards.* ASTM International, West Conshohocken, Pennsylvania.

ASTM D5291. 2021. *Standard Test Methods for Instrumental Determination of Carbon, Hydrogen, and Nitrogen in Petroleum Products and Lubricants. Annual Book of Standards.* ASTM International, West Conshohocken, Pennsylvania.

ASTM D5307. 2021. *Standard Test Method for Determination of Boiling Range Distribution of Crude Petroleum by Gas Chromatography. Annual Book of Standards.* ASTM International, West Conshohocken, Pennsylvania.

ASTM D5853. 2021. *Standard Test Method for Pour Point of Crude Oils. Annual Book of Standards.* ASTM International, West Conshohocken, Pennsylvania.

ASTM D5949. 2021. *Standard Test Method for Pour Point of Petroleum Products (Automatic Pressure Pulsing Method). Annual Book of Standards.* ASTM International, West Conshohocken, Pennsylvania.

ASTM D5950. 2021. *Standard Test Method for Pour Point of Petroleum Products (Automatic Tilt Method). Annual Book of Standards.* ASTM International, West Conshohocken, Pennsylvania.

ASTM E258. 2021. *Standard Test Method for Total Nitrogen in Organic Materials by Modified Kjeldahl Method. Annual Book of Standards.* ASTM International, West Conshohocken, Pennsylvania.

ASTM E385. 2021. *Standard Test Method for Oxygen Content Using a 14-MeV Neutron Activation and Direct-Counting Technique. Annual Book of Standards.* ASTM International, West Conshohocken, Pennsylvania.

ASTM E777. 2021. *Standard Test Method for Carbon and Hydrogen in the Analysis Sample of Refuse-Derived Fuel. Annual Book of Standards.* ASTM International, West Conshohocken, Pennsylvania.

Baltus, R.E., and Anderson, J.L. 1984. Comparison of GPC Elution and Diffusion Coefficients of Asphaltenes. *Fuel*, 63: 530.

Blondel-Telouk, A., Loiseleur, H., Barreau, A., Béhar, E., and Jose, J. 1995. Determination of the Average Molecular Weight of Petroleum Cuts by Vapor Pressure Depression. *Fluid Phase Equilibria*, 110: 315–339.

Bouquet, M., and Bailleul, A. 1982. Nuclear Magnetic Resonance in the Petroleum Industry. In: *Petroanalysis '81. Advances in Analytical Chemistry in the Petroleum Industry 1975–1982.* G.B. Crump (Editor). John Wiley & Sons, Chichester, United Kingdom.

Carbognani, L. 1997. Fast Monitoring of C_{20}-C_{160} Crude Oil Alkanes by Size-Exclusion Chromatography-Evaporative Light Scattering Detection Performed with Silica Columns. *Journal of Chromatography A*, 788: 63–73.

Carbognani, L., Díaz-Gómez, L., Oldenburg, T.B.P., and Pereira-Almao, P. 2012. Determination of Molecular Masses for Petroleum Distillates by Simulated Distillation. *CT&F - Ciencia, Tecnología y Futuro*, 4(5): 43–55.

Colin, J.M., and Vion, G. 1983. Routine Hydrocarbon Group-Type Analysis in Refinery Laboratories by High-Performance Liquid Chromatography. *Journal of Chromatography*, 280: 152–158.

Cooper, A.R. 1989. *Determination of Molecular Weight.* John Wiley & Sons Inc., Hoboken, New Jersey.

Del Río, J.C., and Philp, R.P. 1999. Field Ionization Mass Spectrometric Study of High Molecular Weight Hydrocarbons in a Crude Oil and a Solid Bitumen. *Organic Geochemistry*, 30: 279–286.

Dobele, G., Urbanovich, I., Volpert, A., Kampars, V., and Samulis, E. 2007. Fast Pyrolysis – Effect of Wood Drying on the Yield and Properties of Bio-oil. *Bioresources*, 2: 699–706.

Dolbear, G.E., Tang, A., and Moorehead, E.L. 1987. Upgrading Studies with California, Mexican, and Middle Eastern Heavy Oils. In: *Metal Complexes in Fossil Fuels.* R.H. Filby and J.F. Branthaver (Editors). Symposium Series No. 344. American Chemical Society, Washington, DC. Page 220.

Ebert, L.B., Scanlon, J.C., and Mills, D.R. 1984. X-Ray Diffraction of N-Paraffins and Stacked Aromatic Molecules: Insights into the Structure of Petroleum Asphaltenes. *Liquid Fuels Technology*, 2(3): 257–286.

Ebert, L.B., Mills, D.R., and Scanlon, J.C. 1987. Preprints. *Division of Fuel Chemistry American Chemical Society*, 32(2): 419.

Ebert, L.B. 1990. Comment on the Study of Asphaltenes by X-Ray Diffraction. *Fuel Science and Technology International*, 8: 563–569.

Felix, G., Bertrand, C., and Van Gastel, F. 1985. Hydroprocessing of Heavy Oils and Residua. *Chromatographia*, 20(3): 155–160.

Gary, J.G., Handwerk, G.E., and Kaiser, M.J. 2007. *Petroleum Refining: Technology and Economics*. 5th Edition. CRC Press, Taylor & Francis Group, Boca Raton, Florida.

Hasan, M., Ali, M.F., and Arab, M. 1989. Structural Characterization of Saudi Arabian Extra Light and Light Crudes by 1-H and 13-C NMR Spectroscopy. *Fuel*, 68: 801–803.

Hsu, C.S., and Robinson, P.R. (Editors). 2017. *Handbook of Petroleum Technology*. Springer, Cham, Switzerland.

Koots, J.A., and Speight, J.G. 1975. The Relation of Petroleum Resins to Asphaltenes. *Fuel*, 54: 179.

Long, R.B., and Speight, J.G. 1998. The Composition of Petroleum. In: *Petroleum Chemistry and Refining*. J.G. Speight (Editor). Taylor & Francis, Washington, DC. Chapter 2.

Long, R.B., and Speight, J.G. 1989. Studies in Petroleum Composition. I: Development of a Compositional Map for Various Feedstocks. *Revue de l'Institut Francais du Pétrole*, 44: 205.

MacAllister, D.J., and DeRuiter, R.A. 1985. Further Development and Application of Simulated Distillation for Enhanced Oil Recovery. Paper No. SPE 14335. 60th Annual Technical Conference. Society of Petroleum Engineers, Las Vegas, September 22–25.

Matsushita, S., Tada, Y., and Ikushige, T. 1981. Rapid Hydrocarbon Group Analysis of Gasoline by High-Performance Liquid Chromatography. *Journal of Chromatography A*, 208: 429–432.

Miller, R.L., Ettre, L.S., and Johansen, N.G. 1983. Quantitative Analysis of Hydrocarbons by Structural Group Type in Gasoline and Distillates. Part II. *Journal of Chromatography*, 259: 393.

Misra, M.K., Ragland, K.W., and Baker, A.J. 1993. Wood Ash Composition as a Function of Furnace Temperature. *Biomass and Bioenergy*, 4(2): 103–116.

Modenbach, A.A., and Nokes, S.E. 2014. Effects of Sodium Hydroxide Pretreatment on Structural Components of Biomass. Biosystems and Agricultural Engineering Faculty Publications. 84. https://uknowledge.uky.edu/bae_facpub/84

Neer, L.A., and Deo, M.D. 1995. Simulated Distillation of Oils with a Wide Carbon Number Distribution. *Journal of Chromatographic Science*, 33: 133–138.

Norris, T.A., and Rawdon, M.G. 1984. Determination of Hydrocarbon Types in Petroleum Liquids by Supercritical Fluid Chromatography with Flame Ionization Detection. *Analytical Chemistry*, 56: 1767–1769.

Pandey, A., Negi, S., Binod, P., and Larroche, C. (Editors) 2015. *Pretreatment of Biomass: Processes and Technologies*. Elsevier BV, Amsterdam, Netherlands.

Parkash, S. 2003. *Refining Processes Handbook*. Gulf Professional Publishing, Elsevier, Amsterdam, Netherlands.

Prins, M.J., Ptasinski, K.J., and Janssen, F.J.J.G. 2006a. More Efficient Biomass Gasification via Torrefaction. *Energy*, 31(15): 3458–3470.

Prins, M.J., Ptasinski, K.J., and Janssen, F.J.J.G. 2006b. Torrefaction of Wood: Part 1. Weight Loss Kinetics. *Journal of Analytical and Applied Pyrolysis*, 77(1): 28–34.

Prins, M.J., Ptasinski, K.J., and Janssen, F.J.J.G. 2006c. Torrefaction of Wood: Part 2. Analysis of Products. *Journal of Analytical and Applied Pyrolysis*, 77(1): 35–40.

Rawdon, M. 1984. Modified Flame Ionization Detector for Supercritical Fluid Chromatography. *Analytical Chemistry*, 56: 831–832.

Reynolds, J.G., and Biggs, W.R. 1988. Analysis of Residuum Demetallation by Size Exclusion Chromatography with Element Specific Detection. *Fuel Science and Technology International*, 6: 329.

Romanowski, L.J., and Thomas, K.P. 1985. Steamflooding of Preheated Tar Sand. Report No. DOE/FE/60177-2326. United States Department of Energy, Washington, DC.

Schwartz, H.E., and Brownlee, R.G. 1986. Use of Reversed-Phase Chromatography in Carbohydrate Analysis. *Journal of Chromatography*, 353: 77.

Schwartz, H.E., Brownlee, R.G., Boduszynski, M.M., and Su, F. 1987. Simulated Distillation of High-Boiling Petroleum Fractions by Capillary Supercritical Chromatography and Vacuum Thermal Gravimetric Analysis. *Analytical Chemistry*, 59: 1393–1401.

Shafizadeh, A., McAteer, G., and Sigmon, J. 2003. High-Acid Crudes. Proceedings of the Crude Oil Quality Group Meeting, New Orleans, Louisiana, January 30.

Sheridan, M. 2006. California Crude Oil Production and Imports. Staff Paper, Report No. CERC-600-2006-006. Fossil Fuels Office, Fuels and Transportation Division, California Energy Commission. Sacramento, California, April.

Snape, C.E., Ladner, W.R., and Bartle, K.D. 1979. Survey of Carbon-13 Chemical Shifts – Application to Coal-Derived Materials. *Analytical Chemistry*, 51: 2189–2198.

Speight, J.G., Wernick, D.L., Gould, K.A., Overfield, R.E., Rao, B.M.L., and Savage, D.W. 1985. Molecular Weights and Association of Asphaltenes: A Critical Review. *Revue de l'Institut Francais du Petrole*, 40: 27.

Speight, J.G. 1987. Initial Reactions in the Coking of Residua. Preprints, *American Chemical Society, Division of Fuel Chemistry*, 32(2): 413.

Speight, J.G. 1994. Chemical and Physical Studies of Petroleum Asphaltenes. In *Asphaltenes and Asphalts, I. Developments in Petroleum Science*, 40. T.F. Yen and G.V. Chilingarian (Editors). Elsevier, Amsterdam, Netherlands. Chapter 2.

Speight, J.G. 2000. *The Desulfurization of Heavy Oils and Residua*. 2nd Edition. Marcel Dekker Inc., New York.

Speight, J.G. 2001. *Handbook of Petroleum Analysis*. John Wiley & Sons Inc., Hoboken, New Jersey.

Speight, J.G. 2005. Upgrading and Refining of Natural Bitumen and Heavy Oil. In: *Coal, Oil Shale, Natural Bitumen, Heavy Oil and Peat. Encyclopedia of Life Support Systems (EOLSS). Developed under the Auspices of the UNESCO*. EOLSS Publishers, Oxford, UK, http://www.eolss.net.

Speight, J.G. 2009. *Enhanced Recovery Methods for Heavy Oil and Tar Sands*. Gulf Publishing Company, Houston, Texas.

Speight, J.G. (Editor). 2011. *The Biofuels Handbook*. The Royal Society of Chemistry, London, United Kingdom.

Speight, J.G. 2013a. *The Chemistry and Technology of Coal*. 3rd Edition. CRC-Taylor & Francis Group, Boca Raton, Florida.

Speight, J.G. 2013b. *Heavy Oil Production Processes*. Gulf Professional Publishing, Elsevier, Oxford, United Kingdom.

Speight, J.G. 2013c. *Oil Sand Production Processes*. Gulf Professional Publishing, Elsevier, Oxford, United Kingdom.

Speight, J.G. 2013d. *Heavy and Extra Heavy Oil Upgrading Technologies*. Gulf Professional Publishing, Elsevier, Oxford, United Kingdom.

Speight, J.G. 2014. *The Chemistry and Technology of Petroleum*. 5th Edition. CRC Press, Taylor & Francis Group, Boca Raton, Florida.

Speight, J.G. 2015. *Handbook of Petroleum Product Analysis*. 2nd Edition. John Wiley & Sons Inc., Hoboken, New Jersey.

Speight, J.G. 2016. *Introduction to Enhanced Recovery Methods for Heavy Oil and Tar Sands*. 2nd Edition. Gulf Professional Publishing, Elsevier, Oxford, United Kingdom.

Speight, J.G. 2017. *Handbook of Petroleum Refining*. CRC Press, Taylor & Francis Group, Boca Raton, Florida.

Speight, J.G. 2021. *Refinery Feedstocks*. CRC Press, Taylor & Francis Group, Boca Raton, Florida.

Speight, J.G. 2022. *Biomass Processes and Chemicals*. Elsevier, Amsterdam, The Netherlands.

Suatoni, J.C., and Garber, H.R. 1980. HPLC Preparative Group-Type Separation of Olefins from Synfuels. *Journal of Chromatographic Science*, 18(8): 375–378.

Trajano, H.L., and Wyman, C.E. 2013. Fundamentals of Biomass Pretreatment at Low pH. In: *Aqueous Pretreatment of Plant Biomass for Biological and Chemical Conversion to Fuels and Chemicals*. C.E. Wyman (Editor). John Wiley & Sons, Inc., Hoboken, New Jersey. Chapter 6. Page 103–128.

Wallace, D. (Editor). 1988. *A Review of Analytical Methods for Bitumens and Heavy Oils*. Alberta Oil Sands Technology and Research Authority, Edmonton, Alberta, Canada.

Wallace, D., and Carrigy, M.A. 1988. New Analytical Results on Oil Sands from Deposits throughout the World. Proceedings of the 3rd UNITAR/UNDP International Conference on Heavy Crude and Tar Sands. R.F. Meyer (Editor). Alberta Oil Sands Technology and Research Authority, Edmonton, Alberta, Canada.

Weinberg, V.L., Yen, T.F., Gerstein, B.C., and Murphy, P.D. 1981. Characterization of Pyrolyzed Asphaltenes by Diffuse Reflectance-Fourier Transform Infrared and Dipolar Dephasing-Solid State [13]C Nuclear Magnetic Resonance Spectroscopy. Preprints. *Division of Petroleum Chemistry American Chemical Society*, 26(4): 816–824.

Yen, T.F. 1984. Characterization of Heavy Oil. In: *The Future of Heavy Crude and Tar Sands*. R.F. Meyer, J.C. Wynn, and J.C. Olson (Editors). McGraw-Hill, New York.

4 Thermal Cracking Processes

4.1 INTRODUCTION

Distillation has remained a major refinery process and a process to which almost every crude oil that enters the refinery is subjected. However, not all crude oils yield the same distillation products. In fact, the composition and properties of the feedstock dictate the processes that may be required for refining and balancing product yield with demand is a necessary part of refinery operations (Parkash, 2003; Gary et al., 2007; Speight, 2014; Hsu and Robinson, 2017; Speight, 2017). Basic processes for this are still the so-called cracking processes in which relatively high-boiling constituents carbons are cracked, that is, thermally decomposed into lower-molecular-weight, smaller, lower-boiling molecules, although reforming alkylation, polymerization, and hydrogen-refining processes have wide applications in making premium-quality products (Parkash, 2003; Gary et al., 2007; Speight, 2014; Hsu and Robinson, 2017; Speight, 2017).

After 1910 and the conclusion of World War I, the demand for automotive (and other) fuels began to outstrip the market requirements for kerosene and refiners, needing to stay abreast of the market pull, were pressed to develop new technologies to increase gasoline yields. There being finite amounts of straight-run distillate fuels in crude oil, refiners had, of necessity, the urgency to develop processes to produce additional amounts of these fuels. The conversion of coal and oil shale to liquid through the agency of cracking had been known for centuries, and the production of various spirits from crude oil through thermal methods had been known since at least the inception of Greek fire in earlier centuries.

The discovery that higher-molecular-weight (higher-boiling) materials could be decomposed to lower-molecular-weight (lower-boiling) products was used to increase the production of kerosene and was called cracking distillation. In the process, a batch of crude oil was heated until most of the kerosene was distilled from it, and the overhead material became dark in color. At this point, the still fires were lowered, the rate of distillation decreased, and the feedstock was held in the hot zone, during which time some of the large hydrocarbons were decomposed and rearranged into lower-molecular-weight products. After a suitable time, the still fires were increased and distillation continued in the normal way. The overhead product, however, was low boiling (low viscosity) suitable for kerosene instead of the high-boiling (high-viscosity) oil that would otherwise have been produced. Thus, it was not surprising that such technologies were adapted for the fledgling crude oil industry.

The earliest processes, which involved thermal cracking, consisted of heating heavier oils (for which there was a low market requirement) in pressurized reactors and thereby cracking, or splitting, their large molecules into the smaller ones that form the lighter, more valuable lower-boiling product such as naphtha and kerosene. Naphtha manufactured by thermal cracking processes performed better in automobile engines than gasoline derived from straight distillation of crude oil. The development of more powerful aircraft engines in the late 1930s gave rise to a need to increase the combustion characteristics of gasoline to improve engine performance. Thus during World War II and the late 1940s, improved refining processes involving the use of catalysts led to further improvements in the quality of transportation fuels and further increased their supply. These improved processes, including catalytic cracking of residua and other viscous feedstocks, alkylation, isomerization, and polymerization, enabled the crude oil industry to meet the demands of high-performance combat aircraft and, after the war, to supply increasing quantities of transportation fuels.

Catalytic reforming of naphtha has replaced the earlier thermal reforming process and became the leading process for upgrading fuel qualities to meet the needs of higher-compression engines.

DOI: 10.1201/9781003184904-4

Hydrocracking, a catalytic cracking process conducted in the presence of hydrogen, was developed to be a versatile manufacturing process for increasing the yields of either gasoline or jet fuels.

In the early stages of thermal cracking process development, processes were generally classified as either liquid phase, high pressure (350–1,500 psi), low temperature (400°C–510°C, 750°F–950°F) or vapor phase, low pressure (less than 200 psi): high temperature (540°C–650°C, 1,000°F–1,200°F). In reality, the processes were mixed phase with no process really being entirely liquid or vapor phase but the classification (like many classifications of crude oil and related areas were still used as a matter of convenience) (Parkash, 2003; Gary et al., 2007; Speight, 2014; Hsu and Robinson, 2017; Speight, 2017).

The early processes were classified as liquid-phase processes and had the following advantages over vapor-phase processes: (1) large yields of gasoline of moderate octane number, (2) low gas yields, (3) ability to use a wide variety of charge stocks, (4) long cycle time due to low coke formation, and (5) flexibility and ease of control. However, the vapor-phase processes had the advantages of operation at lower pressures and the production of a higher-octane gasoline due to the increased production of olefins and light aromatics. However, there were many disadvantages that curtailed the development of vapor-phase processes: (1) temperatures were required which the steel alloys available at the time could not tolerate, (2) there were high gas yields and resulting losses since the gases were normally not recovered, and (3) there was a high production of olefinic compounds that created naphtha with poor stability (increased tendency to form undesirable gum) (Mushrush and Speight, 1995; Speight, 2014) that required subsequent treating of the gasoline to stabilize it against gum formation. The vapor-phase processes were not considered suitable for the production of large quantities of gasoline but did find application in petrochemical manufacture due to the high concentration of olefins produced.

It is generally recognized that the most important part of any refinery, after the distillation units, is the gasoline (and liquid fuels) manufacturing facilities; other facilities are added to manufacture additional products as indicated by technical feasibility and economic gain. More equipment is used in the manufacture of gasoline, the equipment is more elaborate, and the processes more complex than for any other product. Among the processes that have been used for liquid fuel production are thermal cracking, catalytic cracking, thermal reforming, catalytic reforming, polymerization, alkylation, coking, and distillation of fractions directly from crude oil (Figure 4.1). Each of these processes may be carried out in a number of ways, each of which differ in details of operation or essential equipment, or both (Parkash, 2003; Gary et al., 2007; Speight, 2014; Hsu and Robinson, 2017; Speight, 2017).

Thermal processes are essentially processes that decompose, rearrange, or combine hydrocarbon molecules by the application of heat. The major variables involved are feedstock type, time, temperature, and pressure and, as such, are usually considered in promoting cracking (thermal decomposition) of the heavier molecules to lighter products and in minimizing coke formation. Thus, one of the earliest processes used in the crude oil industry is the non-catalytic conversion of higher-boiling crude oil stocks into lower-boiling products, known as thermal cracking.

The thermal decomposition (cracking) of high molecular weight hydrocarbons to lower molecular weight and normally more valuable hydrocarbons have been practiced since the early days of the crude oil refining industry. The processes are designed to increase the yield of lower-boiling products obtainable from crude oil either directly (by means of the production of naphtha components from higher-boiling feedstocks) or indirectly (by production of olefins and the like, which are precursors of the gasoline components). These processes may also be characterized by the physical state (liquid and/or vapor phase) in which the decomposition occurs. The state depends on the nature of the feedstock as well as conditions of pressure and temperature (Parkash, 2003; Gary et al., 2007; Speight, 2014; Hsu and Robinson, 2017; Speight, 2017).

Conventional thermal cracking is the thermal decomposition, under pressure, of high molecular weight constituents (higher molecular weight and higher boiling than gasoline constituents) to form lower-molecular-weight (and lower-boiling) species. Thus, the thermal cracking process is designed

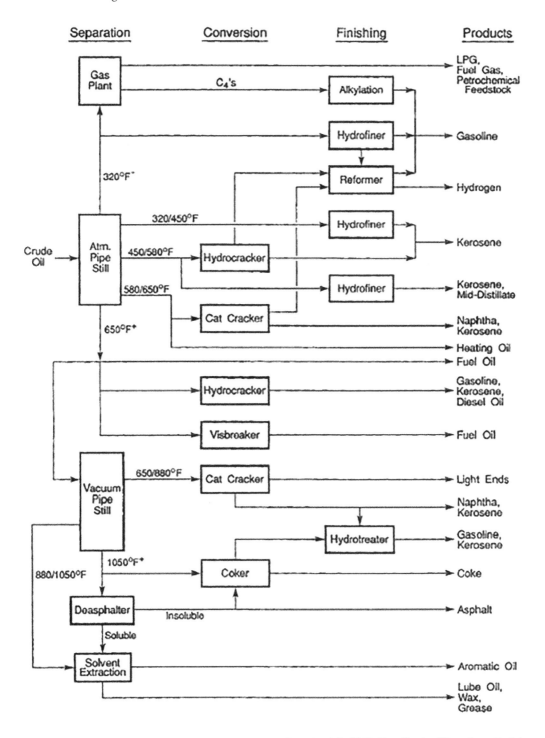

FIGURE 4.1 Schematic representation of a refinery. (Speight, J.G. 2017. *Handbook of Petroleum Refining.* CRC Press, Taylor & Francis Group, Boca Raton, Florida. Figure 8.1, page 293.)

to produce gasoline from higher-boiling charge stocks, and any unconverted or mildly cracked charge components (compounds which have been partially decomposed but are still higher boiling than gasoline) are usually recycled to extinction to maximize gasoline production. A moderate quantity of light hydrocarbon gases is also formed. As thermal cracking proceeds, reactive unsaturated molecules are formed that continue to react and can ultimately create higher-molecular-weight species that are relatively hydrogen deficient and readily form coke. Thus they cannot be recycled without excessive coke formation and are therefore removed from the system as cycle fuel oil.

When crude oil fractions are heated to temperatures over 350°C (660°F), the rates of the thermal decomposition proceed at significant rates (Speight, 2000, 2014). Thermal decomposition does not require the addition of catalyst; therefore, this approach is the oldest technology available for residue conversion. The severity of the thermal cracking process (such as temperature and residence time of the feedstocks in the hot zone) determines the conversion of the feedstock and the characteristics of the product(s). Thermal treatment of residues ranges from mild treatment for reduction of viscosity to ultrapyrolysis (high-temperature cracking at very short residence time) for complete conversion to olefins and light ends. The higher the temperature, the shorter the time required to achieve a given conversion but, in many cases, with a change in the chemistry of the reaction. The severity of the process conditions is the combination of reaction time and temperature to achieve a given conversion.

Thermal reactions, however, can give rise to a variety of different reactions so that selectivity for a given product changes with temperature and pressure. The mild- and high-severity processes are frequently used for processing of residua, while conditions similar to ultrapyrolysis (high temperature and very short residence time) are only used commercially for cracking ethane, propane, butane, and light distillate feeds to produce ethylene and higher olefins. Sufficiently high temperatures convert oils entirely to gases and coke; cracking conditions are controlled to produce as much as possible of the desired product, which is usually gasoline but may be cracked gases for petrochemicals or a lower viscosity oil for use as a fuel oil. The feedstock, or cracking stock, may be almost any fraction obtained from crude oil, but the greatest amount of cracking is carried out on gas oils, a term that refers to the portion of crude oil boiling between the fuel oils (kerosene and/or stove oil) and the residuum. Residua are also cracked, but the processes are somewhat different from those used for gas oils.

Thus, thermal conversion processes are designed to increase the yield of lower-boiling products obtainable from crude oil either directly (by means of the production of gasoline components from higher-boiling feedstocks) or indirectly (by production of olefins and the like, which are precursors of the gasoline components). These processes may also be characterized by the physical state (liquid and/or vapor phase) in which the decomposition occurs. The state depends on the nature of the feedstock as well as conditions of pressure and temperature.

From the chemical viewpoint, the products of cracking are very different from those obtained directly from crude oil – the product is non-indigenous to crude oil because it is created from crude oil by application of an external force (heat). When a 12-carbon atom hydrocarbon typical of straight-run gas oil is cracked or broken into two parts, one may be a six-carbon paraffin hydrocarbon and the other a six-carbon olefin hydrocarbon:

$$CH_3(CH_2)_{10}CH_3 \rightarrow CH_3(CH_2)_4CH_3 + CH_2 = CH(CH_2)_3CH_3$$

The paraffin may be the same as is found in straight-run (distilled) gasoline, but the olefin is new. Furthermore, the paraffin has an octane number approaching 0, but the olefin has an octane number approaching 100. Hence naphtha formed by cracking (cracked gasoline) has a higher octane number than straight-run gasoline. In addition to a large variety of olefins, cracking produces high-octane aromatic and branched-chain hydrocarbons in higher proportions than are found in straight-run gasoline. Diolefins are produced but in relatively small amounts; they are undesirable in gasoline because they readily combine to form gum. The overall complexity of such a reaction is illustrated by the following equations in which the products are subject to the position of bond scission within the starting molecule:

$$CH_3(CH_2)_{10}CH_3 \rightarrow CH_3(CH_2)_8CH_3 + CH_2{=}CH_2$$

$$CH_3(CH_2)_{10}CH_3 \rightarrow CH_3(CH_2)_7CH_3 + CH_2{=}CHCH_3$$

$$CH_3(CH_2)_{10}CH_3 \rightarrow CH_3(CH_2)_6CH_3 + CH_2{=}CHCH_2CH_3$$

$$CH_3(CH_2)_{10}CH_3 \rightarrow CH_3(CH_2)_5CH_3 + CH_2{=}CH(CH_2)_2CH_3$$

$$CH_3(CH_2)_{10}CH_3 \rightarrow CH_3(CH_2)_4CH_3 + CH_2{=}CH(CH_2)_3CH_3$$

$$CH_3(CH_2)10CH_3 \rightarrow CH_3(CH_2)3CH_3 + CH_2{=}CH(CH_2)4CH_3$$

$$CH_3(CH_2)10CH_3 \rightarrow CH3(CH_2)2CH_3 + CH_2{=}CH(CH_2)5CH_3$$

$$CH_3(CH_2)10CH_3 \rightarrow CH_3CH_2CH_3 + CH_2{=}CH(CH_2)6CH_3$$

$$CH_3(CH_2)10CH_3 \rightarrow CH_3CH_3 + CH_2{=}CH(CH_2)7CH_3$$

$$CH_3(CH_2)10CH_3 \rightarrow CH_4 + CH_2{=}CH(CH_2)8CH_3$$

Furthermore, the primary products (unless the reaction conditions are monitored carefully) will react further to yield secondary, tertiary, and even quaternary products.

The hydrocarbons with the least thermal stability are the paraffins, and the olefins produced by the cracking of paraffins are also reactive. Cycloparaffin derivatives (naphthene derivatives) are less easily cracked, their stability depending mainly on any side chains present, but ring splitting may occur, and dehydrogenation can lead to the formation of unsaturated naphthenes and aromatics. Aromatics are the most stable (refractory, i.e., less prone to thermal cracking) hydrocarbons, the stability depending on the length and stability of side chains. Very severe thermal cracking of high molecular weight constituents can result in the production of excessive amounts of coke.

The higher-boiling oils produced by cracking are low-boiling (low-viscosity) and high-boiling (high-viscosity) gas oils as well as a residual oil, which in the case of thermal cracking is usually (erroneously) called tar and in the case of catalytic cracking is called cracked fractionator bottoms. The residual oil may be used as heavy fuel oil, and gas oils from catalytic cracking are suitable as domestic and industrial fuel oils or as diesel fuels if blended with straight-run gas oils. Gas oils from thermal cracking must be mixed with straight-run (distilled) gas oils before they become suitable for domestic fuel oils and diesel fuels.

The gas oils produced by cracking are an important source of gasoline, and in a once-through cracking operation, all of the cracked material is separated into products and may be used as such. However, cracked gas oils are more resistant to cracking (more refractory) than straight-run gas oils but can still be cracked to produce more gasoline. This is done in a recycling operation in which the cracked gas oil is combined with fresh feed for another trip through the cracking unit. The operation may be repeated until the cracked gas oil is almost completely decomposed (cracking to extinction) by recycling (recycling to extinction) the higher-boiling product, but it is more usual to withdraw part of the cracked gas oil from the system according to the need for fuel oils. The extent to which recycling is carried out affects the amount or yield of cracked gasoline resulting from the process.

The gases formed by cracking are particularly important because of their chemical properties and their quantity. Only relatively small amounts of paraffinic gases are obtained from crude oil, and these are chemically inactive. Cracking produces both paraffinic gases (e.g., propane, C_3H_8) and olefinic gases (e.g., propene, C_3H_6); the latter are used in the refinery as the feed for polymerization plants where high-octane polymer gasoline is made. In some refineries, the gases are used to make alkylate, a high-octane component for aviation gasoline and for motor gasoline. In particular, the cracked gases are the starting point for many petrochemicals (Speight, 2014).

The importance of solvents in coking has been recognized for many years (Langer et al., 1961, 1962), but their effects have often been ascribed to hydrogen-donor reactions rather than phase behavior. The separation of the phases depends on the solvent characteristics of the liquid. Addition of aromatic solvents will suppress phase separation while paraffins will enhance separation. Microscopic examination of coke particles often shows evidence for the presence of a mesophase; spherical domains that exhibit the anisotropic optical characteristics of liquid crystal. This phenomenon is consistent with the formation of a second liquid phase; the mesophase liquid is denser than the rest of the hydrocarbon, has a higher surface tension, and likely wets metal surfaces better than the rest of the liquid phase. The mesophase characteristic of coke diminishes as the liquid phase becomes more compatible with the aromatic material (Speight, 1990, 2000).

Thermal cracking of higher-boiling materials to produce motor gasoline is now becoming an obsolete process, since the antiknock requirement of modern automobile engines has outstripped the ability of the thermal cracking process to supply an economical source of high-quality fuel. New units are rarely installed, but a few refineries still operate thermal cracking units built in previous years.

In summary, the cracking of crude oil constituents can be visualized as a series of simple thermal conversions. The reactions involve the formation of transient highly reactive species that may react further in several ways to produce the observed product slate (Germain, 1969; Speight, 2000, 2014). Thus, even though chemistry and physics can be used to explain feedstock reactivity, the main objective of feedstock evaluation is to allow a degree of predictability of feedstock behavior in thermal processes (Chapter 3) (Speight, 2014, 2015b, 2017). And in such instances, chemical principles must be combined with engineering principles to understand feedstock processability and predictability of feedstock behavior. In the simplest sense, process planning can be built on an understanding of the following three parameter groups (Speight, 2015a): (1) feedstock properties, (2) process parameters, and (3) equipment parameters.

Feedstock properties such as carbon residue (potential coke formation), sulfur content (hydrogen needs for desulfurization), metallic constituents (catalyst rejuvenation), nitrogen content (catalyst rejuvenation), naphthenic or paraffinic character through use of a characterization factor or similar indicator (potential for cracking in different ways to give different products), and, to a lesser extent, asphaltene content (coke formation) since this last parameter is related to several of the previous parameters. Process parameters such as time-temperature-pressure relationships (distillate and coke yields), feedstock recycle ratio (distillate and coke yields plus overall conversion), and coke formation (lack of liquid production when liquids are the preferred products). Equipment parameters such as batch operation, semi-continuous operation, or continuous operation (residence time and contact with the catalyst, if any), coke removal, and unit capacity that also dictates residence time. However, it is not the purpose of this text to present the detail of these three categories, but they should be borne in mind when considering and deciding upon the potential utility of any process presented throughout this and subsequent chapters.

Although catalytic cracking has generally replaced thermal cracking, non-catalytic cracking processes using high temperature to achieve decomposition are still in operation. In several cases, thermal cracking processes to produce specific desired products or to dispose of specific undesirable charge streams are being operated or installed. The purpose of this chapter is to provide information that will assist the practicing engineer/crude oil refiner to: (1) determine if a particular thermal cracking process would be suitable for a specific application and could fit into the overall operation, (2) develop a basic design for a thermal cracking process, and (3) operate an existing or proposed process.

4.2 THERMAL CRACKING

As the demand for gasoline increased with the onset of automobile sales, the issue of how to produce more gasoline from less crude oil was solved in 1913 by the incorporation of cracking units into refinery operations in which fractions higher boiling than gasoline were converted into gasoline by thermal decomposition.

Thermal cracking processes (i.e., cracking without the presence or effects of a catalyst) are noncatalytic processes that decompose, rearrange, or combine the feedstock constituents by use of temperatures in excess of over 350°C (660°F), the thermal or free radical reactions start to give way to cracking of the mixture at significant rates. Thermal conversion does not require the addition of a catalyst; therefore, this approach is the oldest technology available for residue conversion. The severity of thermal processing determines the degree of feedstock conversion and the product characteristics.

For example, the thermal cracking of propane in tubular reactors is used to produce olefin derivatives for use as feedstocks for the production of petrochemicals. On the other hand, the thermal cracking of viscous feedstocks (i.e., non-volatile feedstocks) is used to produce distillates and coke. As expected, the distillates contain considerable proportions of olefin derivatives which are then sent to other parts of the refinery (typically for hydrotreatment) to produce valuable products. In fact, without the thermal cracking (or catalytic cracking) of the viscous feedstocks, refineries would not always be able to meet the necessary product slate as demanded by the market.

There are three types of thermal cracking processes, which are (1) mild thermal cracking, of which the prominent example is the visbreaking process in which mild heating is applied to crack the viscous feedstock residue sufficiently enough to lower its viscosity and also to produce limited amounts of distillate products; (2) the delayed coking process in which moderate thermal cracking converts the viscous feedstock into distillate products and coke; and (3) fluid coking in which the feedstock is cracked in a fluidized-bed reactor to produce distillate products and coke; and (4) flexicoking, which is a modified version of the fluid coking process in which steam is used to gasify most of the coke (Parkash, 2003; Gary et al., 2007; Speight, 2014; Hsu and Robinson, 2017; Speight, 2017).

Bio-oils can also be produced from biomass by thermal cracking and the primary constituents obtained from such processes are alkane derivatives, alkene derivatives, alkadiene derivatives, aromatic derivatives, and oxygenated derivatives.

However, thermal cracking processes can give rise to a variety of different reactions so that selectivity for a given product changes with temperature and with pressure, which can lead to fouling of the reactor system (Chapter 6). Thus, relatively mild-severity and high-severity processes are frequently used for processing viscous feedstocks. More extreme conditions are only used commercially for cracking ethane, propane, butane, and low-boiling naphtha feedstocks (i.e. light distillate feedstocks) to produce ethylene ($CH_2=CH_2$) and higher olefins.

The major variables involved in thermal cracking processes are (1) feedstock type, (2) feedstock properties, (3) residence time in the hot zone, (4) the temperature of the hot zone, and (5) the pressure, typically on the order of 100–1,000 psi, which need to be given careful consideration and, as such, are usually considered in promoting cracking (thermal decomposition) of the heavier molecules to lighter products and in minimizing coke formation. The severity of thermal processing, especially the visbreaking process, determines the conversion and the product characteristics.

Thermal cracking is one of the first conversion processes used in the crude oil industry and has been used since 1913 when different fuels and high molecular weight hydrocarbons were heated under pressure in large drums until reaching their thermal fracture into lower molecular size products with a lower boiling point. The various processes offer options for the conversion of viscous feedstocks, while they do require a high operating temperature, they require a low operating pressure without requiring expensive catalysts. Currently, the most widely operated feedstock conversion processes are (1) the visbreaking process and (2) the delayed coking which have the potential to continue well into the 21st century (Parkash, 2003; Gary et al., 2007; Speight, 2014; Hsu and Robinson, 2017; Speight, 2017).

The majority of regular thermal cracking processes use temperatures of 455°C–540°C (850°F–1,005°F) and pressures of 100–1,000 psi. However, more generally, the thermal cracking of various feedstocks (including the viscous feedstocks) ranges from mild treatment for reduction of viscosity (i.e., visbreaking) to ultrapyrolysis (high-temperature cracking at very short residence time) (Hulet et al., 2005; Speight, 2012). The higher temperature requires a shorter time to achieve a given conversion, but, in many cases, there can be a change in the chemistry of the reaction so merely raising the temperature does not necessarily accomplish the same overall goals in terms of product slate and product yields

For example, temperatures in excess of 500°C (930°F) (and low pressures that are typically less than 100 psi) tend to produce lower-molecular-weight hydrocarbon derivatives than those produced at temperatures below 500°C (930°F) and at higher pressure (within the range 400–1,000 psi). The reaction time is also important relative to the viscous feedstocks. For example, light feeds (lower-density feedstocks such as gas oil fractions) and recycle oils require longer reaction times than the readily cracked viscous feedstocks.

Mild cracking conditions (defined here as a low conversion per cycle) favor a high yield of naphtha constituents with low gas production and decreased coke production, but the naphtha quality is not high (such as, for example, a relatively low-octane rating), whereas more severe conditions yield increased gas production and increased coke production as well as a reduced yield of naphtha (but of higher quality, i.e., a higher octane number). With limited conversion per cycle, the more viscous feedstocks must be recycled, but the recycled feedstock constituents can become increasingly refractory upon repeated exposure to the high temperature of the cracking reactor, and if such recycled feedstocks are not required as a blend stock for fuel oil production stock, they may be subjected to a coking operation to increase the overall yield of naphtha or refined by means of a hydrotreating process or a hydrocracking process.

Although new thermal cracking units are now under development for viscous feedstocks, processes that can be regarded as having evolved from the original concept of thermal cracking are visbreaking and various coking processes (Parkash, 2003; Gary et al., 2007; Speight, 2011, 2014; Hsu and Robinson, 2017; Speight, 2017), and the importance of solvents to mitigate coke formation has been recognized for many years (e.g. Langer et al., 1961), the effects have often been ascribed to hydrogen-donor reactions rather than phase behavior. The separation of the phases depends on the solvent characteristics of the liquid. Addition of aromatic solvents will suppress phase separation, while paraffin-based solvents will enhance separation.

In summary, there is a need to improve conversion of the viscous feedstocks, and part of the future growth may even be at or near to the recovery sites where heavy crude oil, extra heavy crude oil, and tar sand bitumen are brought to the surface.

The purpose of this chapter is to present these processes in the light of their use in modern refineries and the information that should be borne in mind when considering and deciding upon the potential utility of any process presented throughout this and subsequent chapters.

4.3 HISTORY

The origins of the cracking process are unknown. There are records that illustrate the use of naphtha in Greek fire almost 2,000 years ago (Speight, 2014) but whether the naphtha was produced naturally by distillation or by cracking distillation is not clear. Cracking was used commercially in the production of oil from coal and from oil shale before the beginning of the modern crude oil industry. From this, the discovery that the higher-boiling materials from crude oil could be decomposed to lower-molecular-weight products was used to increase the production of kerosene and was called cracking distillation (Kobe and McKetta, 1958).

The precise origins of the modern version of cracking distillation, as applied in the modern crude oil industry, are also unknown. However, it is essential to recognize that the production of volatile product by the destructive distillation of wood and coal was known for many years, if not decades or centuries, before the birth of the modern crude oil industry. Indeed, the production of spirits of fire (i.e., naphtha, the flammable constituent of Greek fire) was known from early times. The occurrence of bitumen at Hit (Mesopotamia) that was used as mastic by the Assyrians was further developed for use in warfare through the production of naphtha by destructive distillation.

At the beginning of the 20th century, the yields of gasoline and kerosene fractions were usually markedly increased by means of cracking distillation, but the technique was not entirely suitable for gasoline production. As the need for gasoline arose, the necessity of prolonging the cracking process became apparent and led to a process known as pressure cracking which is a batch operation in which feedstock was heated to approximately 425°C (800°F) in stills (shell stills), especially reinforced to operate at pressures as high as 95 psi for 24 hours. Distillation was then started, and during the next 48–72 hours, a low-boiling distillate was obtained, which was treated with sulfuric acid to remove unstable gum-forming components (olefins and diolefins) and then redistilled to produce a naphtha (cracked gasoline, boiling range <205°C, <400°F) and residual fuel oil (Stephens and Spencer, 1956).

The Burton cracking process for the large-scale production of naphtha (cracked gasoline) was first used in 1912. The process employed batch distillation in horizontal shell stills and operated at approximately 400°C (ca. 750°F) and 75–95 psi and was the first successful method of converting higher-boiling feedstocks to gasoline. However, batch heating gas oil was considered inefficient, and during the years (1914–1922), a number of successful continuous cracking processes were developed. In these processes, gas oil was continuously pumped through a unit that heated the gas oil to the required temperature, held it for a time under pressure, and then the cracked product was discharged into a distillation unit for separation into gases, gasoline, gas oil, and cracked residuum (often called tar).

The tube and tank cracking process is typical of the early continuous cracking processes. Gas oil, preheated by exchange with the hot products of cracking, was pumped into the cracking coil (up to several hundred feet long) that lined the inner walls of a furnace where oil or gas burners raised the temperature of the gas oil to 425°C (800°F). The hot gas oil passed from the cracking coil into a reaction chamber (soaker) where the gas oil was held under these temperature and pressure conditions until the cracking reactions were completed. The cracking reactions formed coke, which over the course of several days filled the soaker. The gas oil stream was then switched to a second soaker, and drilling operations similar to those used in drilling an oil well cleaned out the first soaker. The cracked material (other than coke) left the on-stream soaker to enter an evaporator (tar separator) maintained under a much lower pressure than the soaker, where, because of the lower pressure, all the cracked material except the tar became vaporized. The vapor left the top of the separator, where it was distilled into separate fractions: gases, gasoline, and gas oil. The tar that was deposited in the separator was pumped out for use as asphalt or as a heavy fuel oil.

Shortly thereafter, in 1921, a more advanced thermal cracking process which operated at 750°F–860°F (400°C–460°C) was developed (Dubbs process). In the process, a reduced crude (such as an atmospheric residuum or a toped crude oil) was the feedstock and the process also employed the concept of recycling in which the gas oil was combined with fresh (viscous) feedstock for further cracking. In a typical application of conventional thermal cracking process, the feedstock (reduced crude, i.e., residuum or flashed crude oil) is preheated by direct exchange with the cracked products in the fractionating columns. Cracked gasoline and middle distillate fractions were removed from the upper section of the column. Low-boiling and high-boiling distillate fractions were removed from the lower section and are pumped to separate heaters. Higher temperatures were used to crack the more refractory light distillate fraction. The streams from the heaters were combined and sent to a soaking chamber, where additional time is provided to complete the cracking reactions. The cracked products were then separated in a low-pressure flash chamber where a heavy fuel oil is removed as the non-volatile fraction. The remaining cracked products were sent to the fractionating columns.

As refining technology evolved throughout the 20th century, the preferred feedstocks for cracking processes became gas oil and/or the residuum from a distillation unit. In addition, the residual oil produced as the end product of distillation processes, and even some of the higher-boiling crude oil constituents often contain substantial amounts of asphaltic materials, which preclude use of the residuum as fuel oils or lubricating stocks (Parkash, 2003; Gary et al., 2007; Speight, 2014; Hsu and Robinson, 2017; Speight, 2017).

However, subjecting these residua directly to thermal processes has become economically advantageous since, on the one hand, the end result is the production of lower-boiling products, but, on the other hand, the asphaltene constituents and the resin constituents that are concentrated in residua are precursors to high yields of thermal coke (i.e., coke formed in non-catalytic processes) (Parkash, 2003; Gary et al., 2007; Speight, 2014; Hsu and Robinson, 2017; Speight, 2017).

Low pressures (<100 psi) and temperatures in excess of 500°C (930°F) tend to produce lower-molecular-weight hydrocarbons than those produced at higher pressures (400–1,000 psi) and at temperatures below 500°C (930°F). The reaction time is also important; light feeds (gas oils) and recycle oils require longer reaction times than the readily cracked viscous residues. Recycle of the light oil (middle distillate or fuel oil) fraction also affects the product slate of the thermal cracker. Mild cracking conditions (defined here as a low conversion per cycle) favor a high yield of gasoline components with low gas and coke production, but the gasoline quality is not high, whereas more severe conditions give increased gas and coke production and reduced gasoline yield (but of higher quality). With limited conversion per cycle, the heavier residues must be recycled. However, the recycled oils become increasingly refractory upon repeated cracking, and if they are not required as a fuel oil stock, they may be subjected to a coking operation to increase gasoline yield or refined by means of a hydrogen process.

Thus, the purpose of this chapter is to present descriptions of the thermal processes that are available for the conversion of various feedstocks (especially the viscous feedstocks) and to place the thermal cracking processes in the perspective of the other processes in the refinery.

Over the next three-to-five decades, thermal methods of separation may assume a higher prominence in crude oil refining. The actual process may appear more as a visbreaker in which the operating parameter takes the unit to a point beyond the typical operating point so that pre-coke and coke are separated from the body of the oil. The amount of material separated by such a process will be feedstock and temperature dependent and solvent may or may not be required. The process may be compared to a thermal deasphalter with integrated asphaltene coking. There are a wide range of possible refining application scenarios, including feedstock pre-treatment prior to use in downstream catalytic processes that are sensitive to contaminants that occur to varying extents in the heavier feedstocks.

The main limitation of thermal (non-catalytic) processing is that the products tend to be unstable and very amenable to fouling through sediment or gum production. Thermal cracking at low pressure produces olefins, particularly in the naphtha fraction, which olefins give a very unstable product insofar as the naphtha, which tends to undergo further reactions (often called polymerization reactions) to form gum and tar (polymerization fouling). The high-boiling constituents can form solids or sediments which also depend upon feedstock composition.

4.4 DISTILLATE CRACKING

The application of thermal cracking to a variety of refinery distillates is an integral and necessary part of refinery operations. For example, the production of ethylene (an important feedstock for the petrochemical industry) is manufactured in greater amounts and is typically produced using a steam cracking process. This is a thermal process where distillate constituents are cracked (decomposed) into lower-molecular-weight products that are then used to manufacture more useful (and valuable) chemicals. In the petrochemical industry, two of the main feedstocks for steam crackers are naphtha and ethane.

Steam cracking is a process in which a feedstock such as ethane, CH_3CH_3, propane, $CH_3CH_2CH_3$, butane $CH_3CH_2CH_2CH_3$, or naphtha is thermally cracked through the use of steam in a steam cracking furnace to produce lower-molecular-weight lighter hydrocarbon derivatives.

In the process, a gaseous or liquid hydrocarbon feedstock is diluted with steam and briefly heated in a furnace in the absence of oxygen. Typically, the reaction temperature is very high, at around 850°C (1,560°F), and the residence time is on the order of milliseconds. The products produced depend on (1) the composition of the feedstock, (2) the feedstock-steam ratio, (3) the cracking temperature, and (4) the residence time of the feedstock in the furnace. Hydrocarbon feedstocks such as ethane or low-boiling naphtha yield low-boiling olefin derivatives which include ethylene, propylene, and butadiene ($CH_2=CHCH=CH_2$). Higher-boiling hydrocarbon feedstocks (such as full-range naphtha and high-boiling naphtha as well as other refinery products) may yield some of these same products as well as aromatic hydrocarbon derivatives that are suitable as gasoline blend stock or fuel oil.

However, the actual cracking reaction must be carefully tailored depending on the composition of the feedstock as well as the desired end product. The temperature in the cracking furnace, the residence time of the feedstock, and the quenching process used all affect the quality of the produced hydrocarbon stream. Once the feedstock has been cracked, the product stream must be purified and separated, using various processes depending on the feedstock. These furnaces must occasionally be de-coked, which requires injecting flammable gas into the furnace in order to burn residue (coke) off of the tubes carrying the hydrocarbon feedstock.

A higher cracking temperature (also referred to as higher severity) favors the production of ethylene and benzene, whereas a lower temperature (also referred to as lower severity) produces higher amounts of propylene, C4-hydrocarbon derivatives, and liquid products. The process also results in the slow deposition of coke on the reactor walls which, over time, diminishes the efficiency of the reactor, and reaction conditions must be designed to minimize coke deposition. Decoking requires the furnace to be isolated from the process, and then a flow of steam or a steam/air mixture is passed through the furnace which coils converts the coke to carbon monoxide and carbon dioxide. Once this reaction is complete, the furnace can be returned to service.

4.5 VISBREAKING

Visbreaking (viscosity reduction, viscosity breaking), a mild form of thermal cracking, was developed in the late 1930s to produce more desirable and valuable products (Parkash, 2003; Gary et al., 2007; Joshi et al., 2008; Stell et al., 2009a, 2009b; Carrillo and Corredor, 2013; Speight, 2014; Hsu and Robinson, 2017; Speight, 2017). The process can be regarded as having evolved from the original concept of thermal cracking are visbreaking and the various coking processes (Table 4.1).

The process is a relatively mild, liquid-phase thermal cracking process used to convert high-viscosity feedstocks to lower viscosity product that are suitable for use in heavy fuel oil. This ultimately results in less production of fuel oil since less cutter stock (low viscosity diluent) is required for blending to meet fuel oil viscosity specifications. The cutter stock no longer required in fuel oil may then be used in more valuable products. A secondary benefit from the visbreaking operation is the production of gas oil and gasoline streams that usually have higher product values than the visbreaker charge. Visbreaking produces a small quantity of light hydrocarbon gases and a larger amount of gasoline and remains a process of promise for the viscous feedstocks (Stark and Falkler, 2008; Stark et al., 2008; Speight, 2014, 2017).

The process can also be used as the first step in upgrading viscous feedstocks (Schucker, 2003; Speight, 2014, 2017). In such a process, the viscous feedstock is first thermally cracked using visbreaking technology or hydrovisbreaking technology (i.e., visbreaking with hydrogen present) technology to produce a product that is lower in molecular weight and boiling point than the feed. The product is then deasphalted using an alkane solvent at a solvent to feed ratio of less than 2 wherein separation of solvent and deasphalted oil from the asphaltenes is achieved through the use of a two-stage membrane separation system in which the second stage is a centrifugal membrane.

TABLE 4.1

Comparison of the Various Thermal Cracking Processes

Thermal Cracking

Purpose: thermal decomposition of feedstock and conversion to products

High conversion

Process configuration: various feedstock dependent

Visbreaking

Purpose: viscosity reduction of feedstock

Low conversion (ca. 10% w/w) to products boiling less than 220°C (430°F)

Temperature: 470°C–495°C (880°F–920°F)

Pressure: 50–200 psi

Thermal reactions quenched before going to completion

Heated coil or soaker drum

Delayed Coking

Purpose: conversion of feedstock to distillates

Complete conversion of the feedstock

Moderate – short residence time in hot zone (480°C–515°C; 900°F–960°F)

Pressure: on the order of 90 psi

Thermal reactions allowed to proceed to completion out in the hot zone

Soak drums (450°C–480°C; 845°F–900°F) used in pairs

One drum on-stream; one-off stream for decoking

Coke yield: 20%–40% w/w (feedstock dependent)

Fluid Coking

Purpose: conversion of feedstock to distillates

Complete conversion of the feedstock

Severe – long residence time in hot zone (480°C–565°C; 900°F–1,050°F

Pressure: on the order of 10 psi

Thermal reactions allowed to proceed to completion in hot zone

Coke bed fluidized with steam

Heat dissipated throughout the fluid bed

Higher yields of hydrocarbon gases ($<C_5$) than delayed coking

Less coke yield than delayed coking (for one particular feedstock)

Source: Speight, J.G. 2017. *Handbook of Petroleum Refining.* CRC Press, Taylor & Francis Group, Boca Raton, Florida. Table 8.2, page 300.

Visbreaking, unlike conventional thermal cracking, typically does not employ a recycle stream. Conditions are too mild to crack a gas oil recycle stream, and the unconverted residual stream, if recycled, would cause excessive heater coking. The boiling range of the product residual stream is extended by visbreaking so that light gas oil and heavy gas oil can be fractionated from the product residual stream, if desired. In some present applications, the heavy gas oil stream is recycled and cracked to extinction in a separate higher-temperature heater with the production of product that are lower boiling than the original feedstock (Ballard et al., 1992; Parkash, 2003; Negin and Van Tine, 2004; Gary et al., 2007; Speight, 2014; Hsu and Robinson, 2017; Speight, 2017). Low residence times are required to avoid polymerization and coking reactions, although additives can help to suppress coke deposits on the tubes of the furnace.

The process includes investigations of the following parameters such as (1) the effect of feedstock properties on the process, (2) the effect of feedstock properties on fuel oil stability, (3) chemical pathways and mechanisms, (4) reaction kinetics, (5) coking and fouling, (6) sensitivity of the operating variables, such as temperature, pressure, and residence time, (7) visbreaker design, (8) liquid-phase mixing, and (9) mathematical modeling of the visbreaker, which includes the behavior of the coil visbreaker and the soaker visbreaker.

Visbreaking conditions range from 455°C to 510°C (850°F to 950°F) at a short residence time and from 50 to 300 psi at the heating coil outlet. It is the short residence time that brings to visbreaking the concept of being a mild thermal reaction in contrast to, for example, the delayed coking process where residence times are much longer and the thermal reactions are allowed to proceed to completion. The visbreaking process uses a quench operation to terminate the thermal reactions. Liquid-phase cracking takes place under these low-severity conditions to produce some naphtha, as well as material in the kerosene and gas oil boiling range. The gas oil may be used as additional feed for catalytic cracking units, or as heating oil.

Two visbreaking processes are commercially available which are (1) the coil visbreaking process and (2) the soaker visbreaking process.

4.5.1 COIL VISBREAKING

The coil visbreaking process (Figure 4.2) differs from soaker visbreaking insofar as the coil process achieves conversion by high-temperature cracking within a dedicated soaking coil in the furnace.

With conversion primarily achieved as a result of temperature and residence time, coil visbreaking is described as a high-temperature, short-residence-time route. The main advantage of the coil-type design is the two-zone fired heater that provides better control of the material being heated,

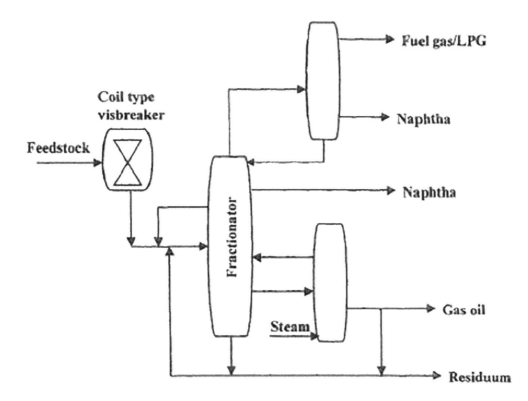

FIGURE 4.2 The coil visbreaking process. (Speight, J.G. 2017. *Handbook of Petroleum Refining*. CRC Press, Taylor & Francis Group, Boca Raton, Florida. Figure 8.4, page 303.)

and with the coil-type design, decoking of the heater tubes is accomplished more easily by the use of steam-air decoking.

The coil visbreaking process achieves conversion by high-temperature cracking within a dedicated soaking coil in the furnace. Since the feedstock conversion is achieved as a result of temperature and residence time, coil visbreaking is often described as a high-temperature, short-residence-time process. The main advantage of the coil-type design is the two-zone fired heater that provides better control of the material being heated, and with the coil-type design, decoking of the heater tubes is accomplished more easily by the use of steam-air decoking.

The coil visbreaker has superseded the soaker visbreaker in some refineries because the soaker visbreaker requires periodically decoking of the soaker drum. Moreover, in some cases, the soaker process does not conveniently adjust to changes in feedstock quality because of the need to adjust the temperature and residence time. Recent development of the coil visbreaker design has provided the coil process with a competitive advantage over the traditional soaker visbreaker process and advances in visbreaker coil heater design now allow for the isolation of one or more passes through the heater for decoking, thereby eliminating the need to shut the entire visbreaker down for furnace decoking.

Furthermore, the higher heater outlet temperature specified for a coil visbreaker is considered to be an important advantage of the process. The higher heater outlet temperature is used to recover significantly higher quantities of visbroken gas oil. Typically, this capability cannot be achieved by a soaker visbreaker without the addition of a vacuum flasher.

4.5.2 Soaker Visbreaking

In the alternative soaker visbreaking process (Figure 4.3), the bulk of the cracking reaction occurs not in the furnace but in a drum located after the furnace (the soaker) in which the heated feedstock is held at an elevated temperature for a predetermined period of time to allow cracking to occur

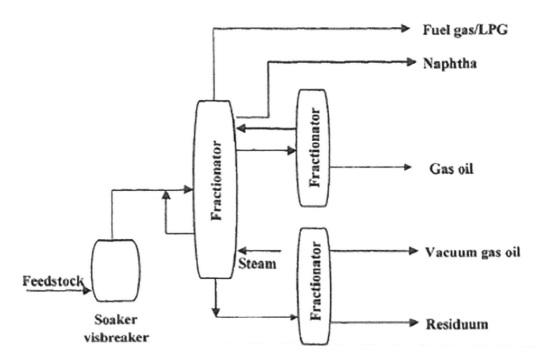

FIGURE 4.3 The soaker visbreaking process. (Speight, J.G. 2017. *Handbook of Petroleum Refining*. CRC Press, Taylor & Francis Group, Boca Raton, Florida. Figure 8.5, page 303.)

before being quenched and then passed to a fractionator. In the soaker visbreaking process, a lower temperature is used than in the higher-temperature coil visbreaking process.

Consequently, the soaker visbreaking process is described as a low-temperature, high-residence-time route. By providing the residence time required to achieve the desired reaction, the soaker drum design allows the heater to operate at a lower outlet temperature.

In the process, the viscous feedstock is passed through a furnace where it is heated to a temperature of 480°C (895°F) under an outlet pressure of approximately 100 psi, and the product is then passed into a flash-distillation chamber. The overhead material from this chamber is then fractionated to produce naphtha and light gas oil. The liquid products from the flash chamber are cooled and then sent to a vacuum fractionator. This yields a heavy gas oil distillate and a residuum of reduced viscosity – a 5%–10% v/v conversion of residuum to naphtha is usually sufficient to afford at least an approximate five-fold reduction in viscosity. Reduction in viscosity is also accompanied by a reduction in the pour point, and an option is to use a lower furnace temperatures and longer residence time which can be achieved by installing a soaking drum between the furnace and the fractionator. The disadvantage of this approach may be the need to remove a coke-like product from the soaking drum.

The higher heater outlet temperature specified for a coil visbreaker is an important advantage of coil visbreaking which is used to recover significantly higher quantities of high-boiling visbroken gas oil. This capability cannot be achieved with a soaker visbreaker without the addition of a vacuum flasher. In terms of product yield, each option can offer significant advantages in particular situations. For example, the cracking reaction forms coke as a byproduct. In coil visbreaking, this lays down in the tubes of the furnace and will eventually lead to fouling or blocking of the tubes. The lower temperatures used in the soaker approach mean that these units use less fuel. In cases where a refinery buys fuel to support process operations, any savings in fuel consumption could be extremely valuable. In such cases, soaker visbreaking may be advantageous, and in fact, many of the existing visbreaker units in refineries appear to be the soaker type of unit and utilize a soaker drum in conjunction with a fired heater to achieve conversion, which reduces the temperature required to achieve conversion while producing a stable residue product, thereby increasing the heater run length and reducing the frequency of unit shut down for heater decoking.

Decoking is accomplished by a high-pressure water jet. First, the top and bottom heads of the coke drum are removed after which a hole is drilled in the coke from the top to the bottom of the vessel and a rotating stem is lowered through the hole, spraying a water jet sideways. The high-pressure jet cuts the coke into lumps, which fall out the bottom of the drum for subsequent loading into trucks or railcars for shipment to customers. Typically, coke drums operate on fixed cycles that depend upon the feedstock and the coking parameters. Cokers produce no liquid residue but yield up to 30% coke by weight. Much of the low-sulfur product is employed to produce electrodes for the electrolytic smelting of aluminum.

The main disadvantage is the decoking operation of the heater and soaker drum, and although decoking requirements of the soaker drum design are not as frequent as those of the coil-type design, the soaker design requires more equipment for coke removal and handling. The customary practice of removing coke from a drum is to cut it out with high-pressure water, thereby producing a significant amount of coke-laden water that needs to be handled, filtered, and then recycled for use again.

The soaker visbreaking process achieves some conversion within the heater, but the majority of the conversion occurs in a reaction vessel or soaker that holds the two-phase effluent at an elevated temperature for a predetermined length of time. Soaker visbreaking is described as a low-temperature, high-residence-time route. Product quality and yields from the coil and soaker drum design are essentially the same at a specified severity being independent of visbreaker configuration. By providing the residence time required to achieve the desired reaction, the soaker drum design allows the heater to operate at a lower outlet temperature (thereby saving fuel), but there are disadvantages.

4.5.3 OTHER VARIANTS

Variations of visbreaking technology include the Tervahl-T and Tervahl H processes (Parkash, 2003; Gary et al., 2007; Speight, 2014; Hsu and Robinson, 2017; Speight, 2017). The Tervahl-T alternative includes only the thermal section to produce a synthetic crude oil with better transportability by having reduced viscosity and greater stability. The Tervahl H alternative adds hydrogen that also increases the extent of the desulfurization and decreases the carbon residua.

The main limitation of the visbreaking process, and for that matter a limitation of all thermal processes, is that the products can be unstable. For example, thermal cracking at low pressure produces olefin derivatives, particularly in the naphtha fraction. These olefins give a very unstable product, which tends to undergo secondary reactions to form gum and intractable residua. Product stability of the visbreaker residue is a main concern in selecting the severity of the visbreaker operating conditions. Severity, or the degree of conversion, can cause phase separation of the fuel oil even after cutter stock blending. Increasing the severity of the operation and the percent conversion will initially lead to a reduction in the visbroken fuel oil viscosity, but the stability of the visbroken product may decrease as the level of severity (and, hence, the conversion) is increased, dependent on the characteristics of the feedstock.

The instability of the visbroken fuel oil is related to the asphaltene constituents and their thermal presence in the residuum. Asphaltene constituents are the high molecular weight non-volatile constituents of a feedstock that can be classified according to the solubility of these constituents in various solvents. The asphaltene constituents can be thermally altered during visbreaking operations. In addition, during visbreaking, some of the high molecular weight constituents, including some of the asphaltene constituents, are converted to lower-boiling and medium-boiling paraffinic components, some of which are removed from the residuum. The asphaltenes and thermally altered, being unchanged, are thus concentrated in the product residuum (that may contain new paraffinic material), and if the extent of the visbreaking reaction is too high, the asphaltene constituents or altered asphaltene constituents will phase tend to precipitate in the product fuel oil, creating an unstable fuel oil.

A common method of measuring the amount of the asphaltene fraction in crude oil is by addition of a low-boiling liquid hydrocarbon such as n-pentane (ASTM D893, ASTM D2007) or n-heptane (ASTM D3279, ASTM D4124) by which treatment the asphaltenes separate as a solid (Speight, 2000, 2014). Since the amount of asphaltenes in the visbreaking unit charge residuum may limit the severity of the visbreaking operations, the n-pentane insoluble content or n-heptane-insoluble content of the feedstock is used as the correlating parameter in various visbreaking correlations. However, these correlations can be visbreaker and feedstock dependent, and application from one unit to another and one feedstock to another may be misleading.

Sulfur in the visbroken residuum can also be an issue since the sulfur content of the visbreaker residuum is often higher (approximately 0.5% w/w or greater) than the sulfur in the feedstock. Therefore, it can be difficult to meet the commercial sulfur specifications of the refinery product residual fuel oil, and blending with low-sulfur cutter stocks may be required.

Visbreaking, like thermal cracking, is a first-order reaction. However, due to the visbreaking severity limits imposed by fuel oil instability, operating conditions do not approach the level where secondary reactions, polymerization and condensation, occur to any significant extent. The first-order reaction rate equation altered to fit the visbreaking reaction is

$$K = (1/t)(\ln 100/X_1)$$

where K = first-order reaction velocity constant, 1/s; t = time at thermal conversion conditions, s; and X_1 = 900°F + visbroken residuum yield, volume %.

The central piece of equipment in any thermal process is, and visbreaking is with no exception, the heater. The heater must be adequate to efficiently supply the heat required to accomplish the desired degree of thermal conversion. A continually increasing temperature gradient designed to give most of the temperature increase in the front part of the heater tubes with only a slow rate of increase near the outlet is preferred. Precision control of time and temperature is usually not critical in the processes covered in this article. Usually, all that is required is to design to some target temperature range and then adjust actual operations to achieve the desired cracking. In the higher-temperature processes (e.g., ethylene manufacture), temperature control does become of prime importance due to equilibrium considerations.

Equipment design to minimize coke formation is of importance. The excessive production of coke adversely affects the thermal cracking process in the following ways: (1) reduces heat-transfer rates, (2) increases pressure drops, (3) creates overheating, (4) reduces run time, and (5) requires the expense of removing the coke from the equipment. In addition, the metallurgy of the equipment, specifically the heater tubes and pumps, in the high temperature, corrosive environments must be adequate to prevent expensive destruction and replacement of equipment. In the early days of thermal cracking, the metallurgy of the heater tubes was not of sufficient quality to permit extended periods of high temperatures. Modern improvements in the quality of steel have extended the durability of the thermal cracking equipment.

The advances in heater design have reached a point where very efficient furnace and heating tube arrangements can be built that give the refiner the desired thermal cracking operation. The practice of the refiner is to set the specifications the heater is expected to meet for the specific application and have a heater manufacturer prepare a suitable design. Proper tube size selection depends upon minimizing pressure drop while obtaining good turbulence for proper heat transfer that is also dependent upon the charge rate that ultimately affects the residence time and, therefore, the extent of the conversion.

The charge stock liquid velocity should be sufficient to provide enough turbulence to ensure a good rate of heat transfer and to minimize coking. A minimum linear cold 15.6°C (60°F) velocity on the order of 5 ft/second for a 100% liquid charge rate is considered to be sufficient. The maximum velocity would be limited to approximately 10 ft/second due to excessive pressure drop. The velocities at the higher cracking temperatures would, of course, be greater due to the partial vaporization of the charge.

Most of the heat supplied to the charge stock is radiant heat. The convection section of the heater is used primarily to supply preheat to the charge prior to the main heating in the radiant section. The heat-transfer rate in the convection section will range from 3,000 to 10,000 Btu/ft^2 of tube outside area per hour with an average rate of 5,000 Btu/ft^2/h. The heating rates in the radiant section will range from 8,000 to 20,000 Btu/ft^2/h depending upon the charge stock, with heavier oil generally requiring the lower heating rate.

The heating tube outlet temperature will depend upon the charge stock being processed and the degree of thermal conversion required. The outlet temperature will vary from a minimum of 425°C (800°F for visbreaking to a maximum of 595°C (1,100°F) for thermal reforming. The combustion chamber temperature will range from 650°C to 870°C (1,200°F to 1,600°F) at a point approximately 1 ft below the radiant tubes. Flue gas temperatures are usually high (425°C–595°C, 800°F–1,100°F) particularly since the viscous feedstock is usually entering the heater at a high temperature. An exception to the charge entering at a high temperature would be when charging gasoline to a thermal reformer. However, since thermal reforming requires high temperatures, flue gas temperatures will also be high.

Since it is desirable to maintain different temperature increase rates throughout the charge heating, i.e., rapid increase at the beginning of the heating coil and a lower rate near the outlet, zone temperature control within the furnace is desired. A three-zone furnace is preferred with the first zone giving the greatest rate of temperature increase and the last zone the least.

Coke formation limits the operation of the heater, and techniques should be employed to minimize coke formation in the heater tubes. Coking occurs on the walls of the tubes, particularly where turbulence is low and temperature is high. Maintaining sufficient turbulence assists in limiting coke formation. Baffles within the tubes are sometimes used, but water injection into the charge stream is the preferred method. Water is usually injected at the inlet although water also may be injected at additional points along the heater tubes. The water, in addition to providing turbulence in the heater tubes as it is vaporized to steam, also provides a means to control temperature. The optimum initial point of water injection into the heater tubes is at the point of incipient cracking where coke would start to form. An advantage to this injection point is the elimination of the additional pressure drop that would have been created by the presence of water between the heater inlet and the point of incipient cracking.

The preferred method to remove coke (de-coke) from the heater tubes is to burn off the coke using a steam-air mixture. The heater tubes, therefore, should be capable of withstanding temperatures up to 760°C (1,400°F) at low pressures for limited time periods. The heater tubes along with the tube supports should be designed to handle the thermal expansion extremes that would be encountered. Mechanical means, such as drills, can also be used to remove coke, but most modern heaters use the steam-air combustion technique. Parallel heaters may be employed so that one can be decoked while permitting cracking to proceed in the other heater(s).

The metallurgy of thermal cracking units is variable although alloy steel tubes of 7%–9% chromium are usually satisfactory to resist sulfur corrosion in thermal cracking heaters. If the hydrogen sulfide content of the cracked products exceeds 0.1% in the cracking zone, a higher alloy steel may be required. Stabilized stainless steel, such as Type 321 or 347, would be suitable in this case. Other alloys, such as the Inconel or Incoloy alloys, could also be used. Seamless tubes with welded return bends are now normally used in heaters. Flanged return bends were used in earlier thermal cracking units to facilitate cleaning. However, use of steam air to burn out the coke essentially eliminates the need for flanged fittings that, in turn, reduces the possibility of dangerous leaks.

A useful tool to aid in the design and operation of thermal cracking units is the soaking volume factor (SVF). This factor combines time, temperature, and pressure of thermal cracking operations into a single numerical value. The SVF is defined as the equivalent coil volume in cubic feet per daily barrel of charge (fresh plus recycle) if the cracking reaction had occurred at 425°C (800°F) and 750 psi.

$$SVF_{750psi/800F} = 1/F \ RK_p dV$$

where $SVF_{750psi/800F}$ is the SVF at base reaction conditions of 750 psi gauge pressure and 800°F, cubic feet of coil volume per total charge throughput in barrels per day, F is the charge (fresh plus recycle) throughput rate, barrels per day, R is the ratio of reaction velocity constant at temperature Y and reaction velocity constant at 800°F, K_T/K_{800}, K_p is the pressure correction factor for pressures other than 750 psi gauge, and dV is the incremental coil volume, cubic feet.

When an additional soaking drum is used, the SYF for the soaking drum should be added to the coil SVF. The SVF for the drum may be determined from:

$$SVF_D = DV/F(K_{TD})(K_p)$$

SVF_D = the SVF of the drum; DV = volume of drum, ft³; F = charge (fresh plus recycle) throughput rate, bbl/day; K_{TD} is the reaction velocity constant for the mean drum temperature; and K_p is the pressure correction factor for the mean drum pressure.

The SVF will range from 0.03 for visbreaking of viscous residual feedstocks to approximately 1.2 for light gas oil cracking. The SVF is a numerical expression of cracking rate and thus can be correlated with product yield and quality SVF may also be translated into cracking coils and still volumes of known dimensions under design conditions of temperature and pressure.

A cracking unit seldom operates very long at design conditions. Charge stock quality changes, desired product yields, and qualities change, or additional capacity is required. These changes require an SVF that is different from the design SVF. The SVF may be varied by (1) varying pressure at constant temperature and feed rate; (2) varying temperature at constant feed rate, the pressure gradient varying with the effect upon cracking rate and fluid density in the cracking coil; and (3) varying the soaking volume at constant temperature and pressure by varying heater feed rate and/or varying the number of tubes in the section above 425°C (800°F).

With the advent of higher firing rate and better efficiency heaters, the use of external soaking drums to provide additional reaction times is of less importance in thermal cracking operations. In modern units, the coil in the heater is usually sufficient to provide the temperature-time relationships required. A possible exception would be the case where it is desirable to crack a considerable amount of viscous residuum. The temperature required probably could not be successfully obtained in a heater coil without excessive coking. A reaction chamber (soaking drum) is employed where the hotter, cleaner light gas oil is used to supply heat to the heavier dirty oil stream in a soaking drum. A low-temperature light gas oil stream is also frequently used to wet the walls of the soaking drum to minimize coking. Parallel soakers could be used to allow one to be decoked, while the other is used for cracking operations.

The pumps used in thermal cracking operations must be capable of operation for extended periods handling a high temperature (above 230°C, 450°F, and up to 345°C, 650°F) corrosive liquid. In addition, since coke particles are formed in thermal cracking, the pumps must be able to withstand the potential erosion of the metal parts by the coke particles. In the early days of thermal cracking, reciprocating pumps were commonly used. In later units, centrifugal pumps have been used. A preferred centrifugal pump would be of the coke-crushing type or may have open impellors with case wear plates substituted for the front rings. The metal should be 12% chromium steel alloy or a higher alloy if serious corrosion is potential.

Heat exchangers should be constructed to provide easy cleaning since the high temperatures and coke particles can create extensive fouling of the exchangers. The downstream processing equipment (flash drums, separators, fractionating towers) are typically of standard design, and no special design specifications are required other than minimizing potential coke buildup. This can be accomplished by designing the equipment so there would be no significant holdup or dormant spots in the process equipment where coke could accumulate.

In thermal cracking operations, there is a considerable amount of excess heat that cannot be economically utilized within the cracking unit itself. When a thermal cracking unit is being considered, it is desirable to construct the unit in conjunction with some other unit, such as a crude still, which could utilize the excess heat to preheat the crude oil charge. Alternatively, the excess heat could be used in steam generation facilities.

Visbreaking may be the most under-estimated and/or under-valued process in a refinery. The process may find rejuvenated use not only for viscous feedstocks (including tar sand bitumen) but also for biomass-derived feedstocks. These visbreaking process possesses sufficient hardware flexibility to accommodate feedstock blending (crude oil feedstocks and bio-feedstocks) and of the unit as well as a high measure of reliability and predictive operations/maintenance, thereby minimizing unplanned shutdowns.

The severity of visbreaker operation is generally limited by the stability requirement of the product as well as the extent of fouling and coke laydown in the visbreaker heater (Speight, 2015a). The former requirement means that the stability of the residue must be sufficient to ensure that the finished fuel resulting from blending with diluents (that are less aromatic than the residue) is stable and that asphaltene flocculation does not occur. Where the residue is converted to an emulsion, blend stability is improved and severity/conversion can be increased, subject to acceptable levels of heater fouling and coke deposition (Miles, 2009). Operational modifications, such as increasing steam injection or recycling high-boiling distillates from the visbreaker fractionator, may help mitigate coking tendency and enhance yield while some relatively low-cost options to increase heater capacity might be implemented in certain instances.

In terms of processing bio-feedstocks, many bio-feedstocks have a high oxygen content and high minerals content which could (even when blended) disqualify the use of the biomaterial as a feedstock to a hydroprocessing unit. Refiners are very wary of high-oxygen and high-mineral feedstocks because of the increased hydrogen requirements (hydrogen is an expensive refinery commodity) to remove the oxygen from the hydrocarbon products with the appearance of the additional hydrogen as water. However, blending a bio-feedstock with a resid as feedstock to a visbreaking unit to produce additional fuel products is a concept that could pay dividends and provide refineries with a source of fuels to supplement crude oil feedstocks. In the visbreaker, the feedstock is converted to overhead (volatile products) and coke (if the unit is operated beyond the typical operating point or coke-forming threshold) (Speight, 2014). The majority of the nitrogen, sulfur, and minerals appear in the coke. Oxygen often appears in the volatile product as water and carbon dioxide, unfortunately removing valuable hydrogen from the internal hydrogen management system.

Alternatively, another option is the preparation of a modified feedstock that is more acceptable to a modern refinery than the original (unchanged) feedstock (Speight and Moschopedis, 1979). In particular, any process that reduces the mineral matter in the bio-feedstock and reduces the oxygen content in the bio-feed would be a benefit.

This can be accomplished by one or two preliminary treatment steps (such as the visbreaking process) in which the feedstock is de-mineralized and the oxygen constituents are removed as overhead (volatile) material giving the potential for the production of a fraction rich in oxygen functions that may be of some use to the chemical industry. Such a process might have to be established at a bio-feedstock production site unless the refinery has the means by which to accommodate the feedstock in an already existing unit.

In a manner similar to the visbreaking process where the bio-feedstock is blended with a residuum, the bio-feedstock alone would be heated in a visbreaker-type reactor (at a lower temperature than the conventional visbreaking temperature) to the point where hydrocarbons (or alcohols) are evolved and coke starts to form. As the coke forms, the mineral matter is deposited with the coke, and the oxygen constituents are deoxygenized leaving a (predominantly) hydrocarbon product, which is a liquid that will ensure easy separation from the coke and mineral matter.

In summary, the visbreaking process has much potential and, in fact, remains an important, relatively inexpensive bottom-of-the-barrel upgrading process in many areas of the world. Most of the existing visbreakers are the soaker type, which utilizes a soaker drum in conjunction with a fired heater to achieve conversion and reduces the temperature required to achieve conversion while producing a stable residue product, thereby increasing the heater run length and reducing the frequency of unit shut down for heater decoking.

Finally, in the context of this book as it relates to the visbreaking process, an alternative option for the initial refining of the viscous feedstocks is to use a lower furnace temperature and a longer residence time of the feedstock in the reactor which can be achieved by installing a soaking drum between the furnace and the fractionator. The disadvantage of this approach may be the need for periodic removal of coke from the soaking drum.

4.6 COKING

A coking process, as the term is used in the crude oil industry, is a process for converting non-distillable fractions (residua) of crude oil and viscous feedstocks (i.e., heavy crude oil, extra heavy crude oil, tar sand bitumen, residua, or blends thereof) to lower-boiling products and coke. The coking processes generally utilize longer reaction times than thermal cracking processes. To accomplish this, drums or chambers (reaction vessels) are employed, but it is necessary to use two or more such vessels so that coke removal can be accomplished in those vessels not on-stream without interrupting the semi-continuous nature of the process. The coke can be used as fuel within the refinery confines, but processing for specialty uses, such as electrode manufacture, production of chemicals, and metallurgical coke, is also possible. For these latter uses, the coke may require treatment to remove sulfur and metal impurities – calcined crude oil coke can be used for making anodes for aluminum manufacture, and a variety of carbon or graphite products such as brushes for electrical equipment.

Coking is often used in preference to catalytic cracking because of the presence of metals and nitrogen components that poison catalysts. There are actually several coking processes, such as delayed coking, fluid coking, and flexicoking, as well as several other variations (Moschopedis et al., 1998; Parkash, 2003; Gary et al., 2007; Speight, 2014; Hsu and Robinson, 2017; Speight, 2017). The products are gases, naphtha, light and heavy gas oil, and coke. The gas oil may be the major product of a coking operation and serves primarily as a feedstock for catalytic cracking units. The coke obtained is usually used as fuel, but processing marketing for specialty uses, such as electrode manufacture, production of chemicals, and metallurgical coke, is also possible and increases the value of the coke. For these uses, the coke may require treatment to remove sulfur and metal impurities. Furthermore, the increasing attention paid to reducing atmospheric pollution has also served to direct some attention to coking, since the process not only concentrates such pollutants as feedstock sulfur in the coke but usually yields products that can be conveniently subjected to desulfurization processes.

Coking processes generally utilize longer reaction times than thermal cracking processes, and unlike the visbreaking process, the reactions are allowed to proceed to completion (Table 4.2).

To accomplish this, drums or chambers (reaction vessels) are employed, but it is necessary to use two or more such vessels so that coke removal can be accomplished in those vessels not on-stream without interrupting the semi-continuous nature of the process. At the end of a cocking cycle, hydraulic cutters are used to remove the coke from the drum. Also, coking processes have the virtue of eliminating the residue fraction of the feedstock, at the cost of forming coke – a solid carbonaceous product.

The yield of coke in a given coking process tends to be proportional to the carbon residue of the feed (measured as the Conradson carbon residue) (Speight, 2001, 2014, 2015a). The formation of large quantities of coke is a severe drawback unless the coke can be put to use. Calcined crude oil coke can be used for making anodes for aluminum manufacture, and a variety of carbon or graphite products such as brushes for electrical equipment. These applications, however, require a coke that is low in mineral matter and sulfur.

If the feedstock produces a high-sulfur, high-ash, high vanadium coke, one option for use of the coke is combustion of the coke to produce process steam (and large quantities of sulfur dioxide

---

TABLE 4.2

Reaction Sequences for the Asphaltene Constituents[a] of a Feedstock

Primary Products	Secondary Products	Tertiary Products
Gas	Gas[b]	
Liquid	Gas	
	Liquid	Gas[b]
		Liquid
Solids		
Carbene[c]	Gas	
	Liquid	Gas
		Liquid
Carboid[c]	Gas	Gas
	Coke	

[a] The resin constituents also follow similar reaction paths but with lower yields of coke than is obtained from the asphaltene constituents.

[b] The formation of gases as secondary products and tertiary product may include the additional formation of olefin derivatives.

[c] Please see Chapter 2 for the separation scheme that presents the names of these fractions (especially Figure 2.6).

unless the coke is first gasified or the combustion gases are scrubbed). Another option is stockpiling. For some feedstocks (particularly the viscous feedstocks), the combination of poor coke properties for anode use limits sulfur dioxide emissions, and loss of liquid product volume have tended to relegate coking processes to a strictly secondary role in any new upgrading facilities.

Two types of crude oil coking processes are presently operating: (1) delayed coking, which uses multiple coking chambers to permit continuous feed processing wherein one drum is making coke and one drum is being decoked; and (2) fluid coking, which is a fully continuous process where product coke can be withdrawn as a fluidized solid.

Crude oil residua obtained from the vacuum distillation tower as a non-volatile (bottoms) fraction, heavy crude oil, extra heavy crude oil, tar sand bitumen, or residua are the typical feedstocks to coking units. Atmospheric tower bottoms (long residua) may be charged to coking units, but it is generally not attractive to thermally degrade the gas oil fraction contained in the longer residua. Other feedstocks to coking units are deasphalter bottoms (often referred to as deasphalter pitch) and tar sand bitumen, and cracked residua (thermal tars). The products are gases, naphtha, fuel oil, gas oil, and coke. The gas oil may serve primarily as a feedstock for catalytic cracking units while the coke is typically used an onsite fuel. Processing the coke for specialty uses, such as electrode manufacture, production of chemicals, and metallurgical coke, is also possible and increases the value of the coke.

The data illustrate how the yield of coke from delayed and fluid coking varies with Conradson carbon residue of the feedstock (Parkash, 2003; Gary et al., 2007; Speight, 2014; Hsu and Robinson, 2017; Speight, 2017).

4.6.1 Delayed Coking

Delayed coking is the oldest, most widely used process and has changed very little since the process was first brought on-stream approximately 80 years ago. It is a semi-continuous (semi-batch) process in which the heated charge is transferred to large coking (or soaking) drums that provide the long residence time needed to allow the cracking reactions to proceed to completion (McKinney, 1992; Feintuch and Negin, 2004). The process is widely used for treating residua and is particularly attractive when the green coke produced can be sold for anode or graphitic carbon manufacture or when there is no market for fuel oils. The process uses long reaction times in the liquid phase to convert the residue fraction of the feed to gases, distillates, and coke. The condensation reactions that give rise to the highly aromatic coke product also tend to retain sulfur, nitrogen, and metals, so that the coke is enriched in these elements relative to the feed.

In the process (Figure 4.4), the heated feedstock is charged to the fractionator and subsequently mixed with an amount of recycle material (usually approximately 10% v/v, but as much as 25% v/v, of the total feedstock) from the coker fractionator through a preheater and then to one of a pair of coke drums; the heater outlet temperature varies from 480°C to 515°C (895°F to 960°F) to produce the various products.

The cracked products leave the drum as overheads to the fractionator, and coke deposits form on the inner surface of the drum. The majority of the sulfur originally in the feedstock remains in the coke (Parkash, 2003; Gary et al., 2007; Speight, 2014; Hsu and Robinson, 2017; Speight, 2017). A pair of coke drums is used so that while one drum is on stream, the other is being cleaned allowing continuous processing, and the drum operation cycle is typically 48 hours. The temperature in the coke drum ranges from 415°C to 450°C (780°F to 840°F) at pressures from 15 to 90 psi.

In terms of the process parameters, the temperature is used to control the severity of coking. In delayed coking, the temperature controls the quality of the coke produced. High temperature will remove more volatile materials, and the yield of coke decreases as temperature increases. If the furnace temperature is high, this might lead to coke formation in the furnace. A low inlet furnace temperature will lead to incomplete coking. Short cycle time will increase capacity but will give lower amounts of liquid products and will shorten the drum lifetime. Increasing pressure will

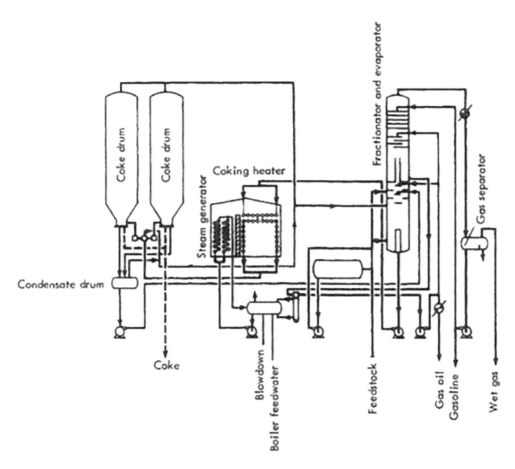

FIGURE 4.4 The delayed coking process. (Speight, J.G. 2017. *Handbook of Petroleum Refining*. CRC Press, Taylor & Francis Group, Boca Raton, Florida. Figure 8.15, page 317.)

increase coke formation and slightly increase gas yield. Recycle ratio is used to control the endpoint of the coker gas oil. It has the same effect as pressure. Feedstock variables are the characterization factor and the Conradson carbon which affect yield production. Sulfur and metal content are usually retained in the coke produced. Engineering variables also affect the process performance which include (1) the mode of operation, (2) the capacity of the unit, and (3) the coke removal and handling equipment.

Fractionators separate the overhead products from the coke drum into fuel gas (low-molecular-weight gases up to and including ethane), propane and propylene ($CH_3CH_2CH_3$–$CH_3CH=CH_2$), butane-butene ($CH_3CH_2CH_2CH_3$–$CH_3CH_2CH=CH_2$), naphtha, light gas oil, and heavy gas oil. Yields and product quality vary widely due to the broad range of feedstock used for coking units (Parkash, 2003; Gary et al., 2007; Speight, 2014; Hsu and Robinson, 2017; Speight, 2017).

Coker naphtha typically has a boiling range up to 220°C (430°F), is olefinic, and must be upgraded by hydrogen processing for removal of olefins and sulfur. They are then used conventionally for reforming to gasoline or as chemical feedstocks. Middle distillates, boiling in the range of 220°C–360°C (430°F–680°F), are also hydrogen treated for improved storage stability, sulfur removal, and nitrogen reduction. They can then be used for either diesel or burner fuels or thermally processed to lower-boiling naphtha. The gas oil boiling up to approximately 510°C (950°F) end point may be charged to a fluid catalytic cracking unit immediately or after hydrogen upgrading when low sulfur is a requirement.

As noted above, the coke drums are on a 48-hour cycle. The coke drum is usually on-stream for approximately 24 hours before becoming filled with porous coke after which time the coke is removed by the following procedure: (1) the coke deposit is cooled with water; (2) one of the heads of the coking drum is removed to permit the drilling of a hole through the center of the deposit; and (3) a hydraulic cutting device, which uses multiple high-pressure water jets, is inserted into the hole and the wet coke is removed from the drum. Typically, 24 hours is required to complete the cleaning operation and to prepare the coke drum for subsequent use on-stream (Parkash, 2003; Gary et al., 2007; Speight, 2014; Hsu and Robinson, 2017; Speight, 2017).

A well-designed delayed coker will have an operating efficiency of better than 95%, although delayed coking units are generally scheduled for shutdown for cleaning and repairs on a 12–18 month schedule, depending on the most economical cycle for the refinery. In terms of process efficiency, the feedstock heater and the coke drums are the most critical parts of the delayed coking process. The function of the heater or furnace is to preheat the charge quickly, to avoid preliminary decomposition, to the required temperature. Since coking is endothermic, the furnace outlet temperature must be approximately 55°C (100°F) higher than the coke drum temperature to provide the necessary process heat. The heater run length is a function of coke laydown in heater tubes, and careful design is necessary to avoid premature shutdown with cycle lengths preferably at least 1 year. When the charge stock is derived from crude distillation, double desalting is desirable since salt deposits will shorten heater cycles.

The heater for a delayed coking unit does not require as broad an operating range as a thermal cracking or visbreaking heater where both contact time and temperature can be varied to achieve the desired level of conversion. The coker heater must reach a fixed outlet temperature for the required coke drum temperatures. Thus the coker heater requires a short residence time, high radiant heat flux, and good control of heat distribution.

The function of the coke drum is to provide the residence time required for the coking reactions to proceed to completion and to accumulate the coke. In sizing coke drums, a superficial vapor velocity in the range of 0.3–0.5 ft/second is used, and coke drums with heights of 97 ft (30 m) have been constructed and approach a practical limit for hydraulic coke cutting. Drum diameters up to 26 ft (8 m) have been commonly used, and larger drums are feasible for efficient processing. Various types of level detectors are used to permit drum filling to within 7–8 ft (2–2.5 m) of the upper tangent line of the drum monitor coke height in the drum during on-stream service.

Hydraulic cutters are used to remove coke from the drum, and the first step is to bore a vertical pilot hole through the coke after which cutting heads with horizontally directed nozzles then undercut the coke and drop it out of the bottom of the drum. Hydraulic pressures in the range of 3,000–3,600 psi are used in the 26-ft diameter coking drums.

In regard to the process parameters and products yields, an increase in the coking temperature (1) decreases coke production, (2) increases liquid yield, and (3) increases gas oil end point. On the other hand, increasing pressure and/or recycle ratio (1) increases gas yield, (2) increases coke yield, (3) decreases liquid yield, and (4) decreases gas oil end point. As an example, increasing the pressure from the currently designed 15–35 psi causes the higher-boiling products to remain in the hot zone longer causing further decomposition and an increase in yield of the naphtha fraction, a decrease in the yield of the middle distillate-gas oil faction, and an increase in the yield of coke.

In the past, many delayed coking units were designed to provide complete conversion of a viscous feedstock (such as an atmospheric residuum) to naphtha, kerosene, and other low-boiling products. However, some units have been designed to minimize coke and produce heavy coker gas oil (HCGO) that is catalytically upgraded. In fact, the product slate for a delayed coker can be varied to meet a variety of objectives through the selection of the operating parameters of the process. Furthermore, delayed coking has an increasingly important role to play in the integration of modem crude oil refineries because of the inherent flexibility of the process to handle even a variety of high-viscosity feedstocks (Parkash, 2003; Gary et al., 2007; Speight, 2014; Hsu and Robinson, 2017; Speight, 2017).

The process provides essentially complete rejection of metals and precursors to term coke (precursors to Conradson carbon) while providing partial or complete conversion to naphtha and diesel. In the past, many cokers were designed to provide complete conversion of atmospheric residua to naphtha and kerosene – modern cokers may still operate in this mode. However, most recent cokers have been designed to minimize coke and produce heavy coker gas oil (HCGO) that is upgraded to naphtha and kerosene in a catalytic cracking unit. The economics of delayed coking are driven by the differential between transportation fuels and high-sulfur residual fuel oil. The yield slate for a delayed coker can be varied to meet a refiner's objectives through the selection of operating parameters. Coke yield and the conversion of heavy coker gas oil are reduced, as the operating pressure and recycle are reduced and to a lesser extent as temperature is increased.

The low-pressure coking process is designed for a once-through, low-pressure operation and is, in fact, similar to delayed coking except that recycling is not usually practiced and the coke chamber operating conditions are 435°C (815°F), 25 psi. Excessive coking is inhibited by the addition of water to the feedstock in order to quench and restrict further reactions of the reactive intermediates.

High-temperature coking is a semi-continuous process designed to convert asphaltic residua to gas oil and coke as the primary products. In the process, the feedstock is heated (at 370°C, 700°F, 30 psi) prior to entry into the coking unit where temperatures may be as high as 980°C–1,095°C (1,800°F–2,000°F). The volatile products exiting the coking unit are fractionated, and after the cycle is complete, the coke is collected for sulfur removal before storage.

Delayed coking has been, and is likely to remain, the most commonly used thermal cracking processes for the foreseeable future for upgrading viscous feedstocks. To this end, in recent years, many process developments have taken place which include (1) development of automated coke drum unheeding devices, allowing the operator to carry out the decoking procedure safely from a remote location, (2) understanding of process parameters affecting yields, coker product qualities and coke qualities (e.g., shot coke), and (3) design and operation of major equipment items, in particular coke drums (allowing shorter coking cycles) and the delayed coker heater (online spalling/decoking and minimization of coking in furnace tubes).

4.6.2 FLUID COKING

Throughout the history of the refining industry, with only short-term exceptions, there has been a considerable economic driving force for upgrading residua. This has led to the development of processes to reduce residua yields such as thermal cracking, visbreaking, delayed coking, vacuum distillation, and deasphalting. The process is also valuable for conversion of heavy crude oil, extra heavy crude oil, and tar sand bitumen to distillate products (Parkash, 2003; Gary et al., 2007; Speight, 2014; Hsu and Robinson, 2017; Speight, 2017). In the process, the hot feedstock is sprayed on to a fluidized bed of hot, fine coke particles, which permits the coking reactions to be conducted at higher temperatures and shorter contact times than can be employed in delayed coking. These conditions result in decreased yields of coke; greater quantities of more valuable liquid product are recovered in the fluid coking process.

As a brief history, in the late 1940s and early 1950s, there was a large incentive to develop a continuous process to convert vacuum residua into lighter, more valuable products. During this period, fluid coking using the principle of fluidized solids was developed, and contact coking, using the principle of a moving solid bed, was also developed and the first commercial fluid coker went on-stream in late 1954. During the late 1960s, environmental considerations indicated that, in many areas, it would no longer be possible to utilize high-sulfur coke as a boiler fuel. This and other environmental considerations resulted in the development of flexicoking to convert the coke product from a fluid coker into clean fuel. The first commercial Flexicoking unit went on-stream in 1976.

Fluid coking (Figure 4.5) is a continuous process that uses the fluidized solids technique to convert residua, including vacuum residua and cracked residua, to more valuable products (Roundtree, 1997).

This coking process allows improvement in the yield of distillates by reducing the residence time of the cracked vapors and also allows simplified handling of the coke product. Heat for the process is supplied by partial combustion of the coke with the remaining coke being drawn as product. The new coke is deposited in a thin fresh layer (ca. 0.005 mm, 5 μm) on the outside surface of the circulating coke particle, giving an onion skin effect.

The equipment for the fluid coking process is similar to that used in fluid catalytic cracking and follows comparable design concepts except that the fluidized coke solids replace more expensive catalyst (Parkash, 2003; Gary et al., 2007; Speight, 2014; Hsu and Robinson, 2017; Speight, 2017). Small particles of coke made in the process circulate in a fluidized state between the vessels and are the heat-transfer medium, and thus, the process requires no high-temperature preheat furnace.

Fluid coking uses two vessels: a reactor and a burner; coke particles are circulated between these to transfer heat (generated by burning a portion of the coke) to the reactor (Figure 4.5) (Blaser, 1992; Parkash, 2003; Gary et al., 2007; Speight, 2014; Hsu and Robinson, 2017; Speight, 2017). The reactor holds a bed of fluidized coke particles, and steam is introduced at the bottom of the reactor to fluidize the bed. The feed coming from the bottom of a vacuum tower at, for example, 260°C–370°C (500°F–700°F) is injected directly into the reactor. The temperature in the coking vessel ranges from 480°C to 565°C (900°F to 1,050°F), with short residence times of the order of 15–30 seconds, and the pressure is substantially atmospheric so the incoming feed is partly vaporized and partly deposited on the fluidized coke particles. The material on the particle surface then cracks and vaporizes, leaving a residue that dries to form coke. The vapor products pass through cyclones that remove most of the entrained coke.

Vapor products leave the bed and pass through cyclones that are necessary for removal of the entrained coke. The cyclones discharge the vapor into the bottom of a scrubber, and any coke dust remaining after passage through the cyclones is scrubbed out with a pump-around stream, and the products are cooled to condense the viscous tar. The resulting slurry is recycled in the reactor. The scrubber overhead vapors are sent to a fractionator where they are separated into wet gas, naphtha, and various gas oil fractions. The wet gas is compressed and further fractionated into desired components.

In the reactor, the coke particles flow down through the vessel into the stripping zone. The stripped coke then flows down a standpipe and through a slide valve that controls the reactor-bed level. A riser carries the cold coke to the burner. Air is introduced to the burner to burn part of the coke to provide reactor heat. The hot coke from the burner flows down a standpipe through a slide valve that controls coke flow and thus the reactor-bed temperature. A riser carries the hot coke to the top of the reactor bed. Combustion products from the burner bed pass through two stages of cyclones to recover coke fines and return them to the burner bed.

Coke is withdrawn from the burner to keep the solids inventory constant. To aid in keeping the coke from becoming too coarse, large particles are selectively removed as product in a quench elutriator drum, and coke fines are returned to the burner. The product coke is quenched with water in the quench elutriator drum and pneumatically transported to storage. A simple jet attrition system in the reactor provides additional seed coke to maintain a constant particle size within the system.

Due to the higher thermal cracking severity used in fluid coking compared to delayed coking, the products are somewhat more olefinic than the products from delayed coking. In general, products are handled for upgrading in a comparable manner from both coking processes.

Coke, being a product of the process, must be withdrawn from the system to keep the solid inventory from increasing. The net coke produced is removed from the burner bed through a quench elutriator drum, where water is added for cooling and cooled coke is withdrawn and sent to storage. During the course of the coking reaction, the particles tend to grow in size. The size of the coke particles remaining in the system is controlled by a grinding system within the reactor.

The coke product from the fluidized process is a laminated sphere with an average particle size of 0.17–0.22 mm (170–220 μm), readily handled by fluid transport techniques. It is much harder and denser than delayed coke, and in general, it is not as desirable for manufacturing formed products.

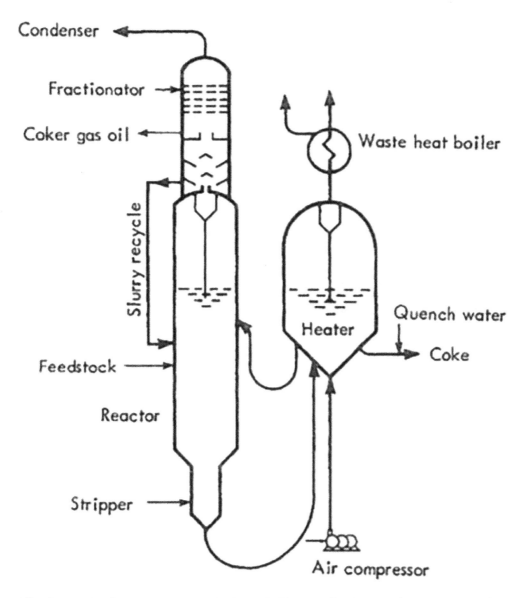

FIGURE 4.5 The fluid coking process. (Speight, J.G. 2017. *Handbook of Petroleum Refining.* CRC Press, Taylor & Francis Group, Boca Raton, Florida. Figure 8.16, page 322.)

The yields of products are determined by the feed properties, the temperature of the fluid bed, and the residence time in the bed. The lower limit on operating temperature is set by the behavior of the fluidized coke particles. If the conversion to coke and light ends is too slow, the coke particles agglomerate in the reactor, a condition known as bogging. The use of a fluidized bed reduces the residence time of the vapor-phase products in comparison to delayed coking, which in turn reduces cracking reactions. The yield of coke is thereby reduced, and the yield of gas oil and olefins increased. An increase of 5°C (9°F) in the operating temperature of the fluid-bed reactor typically increases gas yield by 1% w/w and naphtha by approximately 1% w/w.

The disadvantage of burning the coke to generate process heat is that sulfur from the coke is liberated as sulfur dioxide (SO_2). The gas stream from the coke burner also contains carbon monoxide

(CO), carbon dioxide (CO_2), and nitrogen (N_2.). An alternate approach is to use a coke gasifier to convert the carbonaceous solids to a mixture of carbon monoxide (CO), carbon dioxide (CO_2), and hydrogen (H_2).

The liquid products from the coker can, following cleanup via commercially available gas oil hydrodesulfurization technology, provide large quantities of low-sulfur fuel (<0.2% by weight sulfur) (Parkash, 2003; Gary et al., 2007; Speight, 2014; Hsu and Robinson, 2017; Speight, 2017). The incentive for fluid coking or flexicoking increases relative to alternate processing, such as direct hydroprocessing, as feedstock quality (Conradson carbon, metals, sulfur, nitrogen, etc.) decreases. Changes in yields and product quality result from a change from a low cut point, high reactor temperature operation, to a high cut point operation with a lower reactor temperature (Parkash, 2003; Gary et al., 2007; Speight, 2014; Hsu and Robinson, 2017; Speight, 2017).

Fluid coke is used in electrodes for aluminum manufacture, in silicon carbide manufacture, in ore sintering operations, and as fuel. The coke from a feedstock containing a large amount of contaminants may not be suitable for these uses, either from a product contamination or environmental standpoint. The flexicoking process overcomes this problem by converting part of the gross coke to a gas that can be burned in process furnaces and boilers. The coke fines from a flexicoking unit contain most of the metals in the feedstock and may be suitable for metals recovery.

The fluid coking processes can be used to produce a high yield of low-sulfur fuel oil as well as to completely eliminate residual fuel and asphalt from the refinery product slate (Parkash, 2003; Gary et al., 2007; Speight, 2014; Hsu and Robinson, 2017; Speight, 2017). The different distributions are obtained by varying the fluid coker/flexicoker operating conditions and changing the downstream processing of the coker reactor products. In fact, there are many process variations that can be used to adapt the process to particular refining situations. Once-through or partial recycle coking can be used where there is a small market for heavy fuel oil, or where a quantity of high-sulfur material can be blended into the fuel oil pool.

In reference to the process parameters, the reactor temperature is normally set at 510°C–540°C (950°F–1,000°F). Low temperature favors high liquid yields and reduces the unsaturation of the gas but increases the reactor holdup requirements. The burner temperature is normally 55°C–110°C (100°F–200°F) above the reactor temperature. Regulating the amount of coke sent to the reactor from the burner controls the reactor temperature. Burner temperature is controlled by the air rate to the burner.

Low pressure provides maximum gas oil recycle cut point, minimizes steam requirements, and reduces air blower horsepower. Reactor pressure normally adjusts to the gas compressor suction pressure but is higher due to the pressure drop through the piping, the condenser, the fractionation tower, and the reactor cyclone. The unit pressure balance required for coke circulation and is normally controlled at a fixed differential pressure relative to the reactor sets the burner pressure. Reactor coke level is controlled by the cold coke slide valve on the transfer line from the reactor to the burner, and burner coke level is controlled by the coke withdrawal rate.

In all coking processes, product yields are a function of feed properties, the severity of the operation, and recycle cut point (Parkash, 2003; Gary et al., 2007; Speight, 2014; Hsu and Robinson, 2017; Speight, 2017). Severity is a function of time and temperature since low severity and high gas oil cut point favor high liquid yields, whereas high severity and low gas oil cut point increase coke and gas yields. Data from these sources indicate that the gross coke yield is directly related to feedstock Conradson carbon residue. Coke quality and gas quality are also important (Parkash, 2003; Gary et al., 2007; Speight, 2014; Hsu and Robinson, 2017; Speight, 2017).

In most cases, high liquid yield and minimum coke and gas yields are required, and in theory, two cracking rates should be considered. The first is the rate at which the liquid cracks and vaporizes after initially laying down on the coke particles that determine the reactor holdup. The second is that vapor-phase cracking determines the distribution of the products between gas, naphtha, and gas oil. The vapor residence time can be determined from the reactor volume and the volume flow of hydrocarbon vapor and steam, and can be divided into time in the fluid bed and time in the disperse phase. The former is a function of the coke holdup or weight space velocity (W/H/W) that is normally expressed as reciprocal hours. For maximum liquid yield, the secondary cracking time should be kept at a minimum, and thus, it is normally desirable to design the unit for the maximum operable W/H/W.

The maximum rate at which feed can be injected into a fluid coker is limited by a condition known as bogging. The conditions required to avoid a bogging are: (1) the feedstock must be uniformly distributed over the entire surface of the heat-transfer medium; (2) the layer of feed material on the particles should not be too great; the thickness of the sticky plastic layer depends on the specific flow rate of feedstock, its coking factor, and the recirculation rate of the heat-transfer medium; (3) the bed temperature and the initial temperature of the heat-transfer medium should be sufficiently high that the first stage of the process is completed in a short time; and (4) the heat-transfer medium should not consist of particles that are too fine. The heat reserve of the granules should be sufficient to cover the entire energy requirements in connection with heating the feedstock, supplying the energy for the endothermic cracking reaction, and evaporating the decomposition products. If the feed injection rate exceeds the vaporization rate for an extended period of time, the thickness of the tacky oil film on the particles will increase until the particles rapidly agglomerate, causing the bed to lose fluidity. When fluidization is lost, the heat-transfer rate is greatly reduced, further aggravating the condition. Coke circulation cannot be maintained due to the loss of reactor fluidization.

For comparative purposes because of the similarity of the processes, there are some notable differences between the operation of a fluid coker and a fluid catalytic cracking unit (FCCU), and some of these differences tend to make the fluid coker easier to operate. The fluid coker heat balance is very easy to maintain, as there is always an excess of carbon to burn whereas a fluid catalytic cracking unit has a sensitive interaction between heat balance and intensity balance, and therefore between carbon burned and carbon produced, which complicates control, especially during operating changes, startup, and shutdown.

In addition, recovery from upsets caused by loss of utilities such as steam and air is normally easier and faster with a fluid coker than with a fluid catalytic cracking unit. The fluid coker normally operates well at low feed rates, and turndown to low rate is normally limited by the ability of the tower to maintain fractionation of the products. The fluid coker proper can operate at any feed rate that will provide enough coke to heat balance.

However, the fluid coker has some inherent features that can create problems if proper precautions are not followed. The high-viscosity residuum can solidify if the lines are not properly heated traced and insulated. Low reactor temperature results in reactor bogging. If the particle size of the circulating coke is not properly controlled, the size can grow to the point that coke circulation problems are encountered. The feed nozzles must be maintained and occasionally cleaned to prevent poor feed distribution followed by excessive agglomerate formation. Control of the reactor-bed level is critical since an excessively high-bed level will flood the reactor cyclone and allow coke to be carried to the scrubber where it will plug the heavy oil lines.

Along similar lines to the fluid coking process, the rapid thermal processing process (RTP process, now the HTL process or heavy-to-light upgrading technology) was developed by Ivanhoe Energy Inc. in the 1980s. The process uses a circulating transport bed of hot sand to rapidly induce thermal cracking of the viscous feedstock in the absence of air to produce a light synthetic crude oil (Veith et al., 2007; Koshka et al., 2008; Silverman et al., 2011).

4.6.3 Flexicoking

Flexicoking is a direct descendent of fluid coking (Figure 4.6) and uses the same configuration as the fluid coker but includes a gasification section in which excess coke can be gasified to produce refinery fuel gas (Roundtree, 1997; Marano, 2003).

The flexicoking process was designed during the late 1960s and the 1970s as a means by which excess coke could be reduced in view of the gradual incursion of heavier feedstocks in refinery

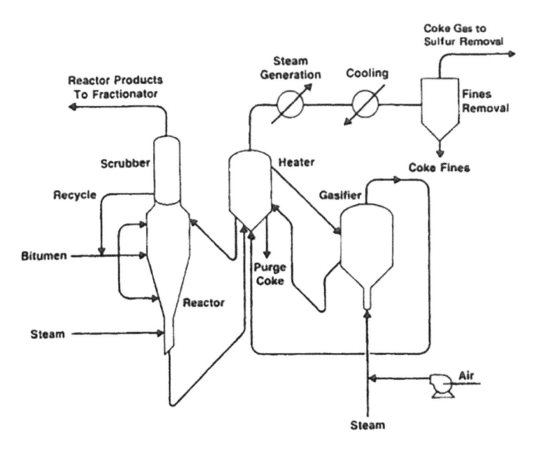

FIGURE 4.6 The flexicoking process. (Speight, J.G. 2017. *Handbook of Petroleum Refining.* CRC Press, Taylor & Francis Group, Boca Raton, Florida. Figure 8.18, page 329.)

operations. Such feedstocks are notorious for producing high yields of coke (>15% by weight) in thermal and catalytic operations.

The process was designed during the late 1960s and the 1970s as a means by which excess coke could be reduced in view of the gradual incursion of the viscous feedstocks in refinery operations. Such feedstocks are notorious for producing high yields of coke (>15% by weight) in thermal and catalytic operations. The flexicoking process unit uses the same equipment configuration as the fluid coker but includes a gasification section in which excess coke can be gasified to produce refinery fuel gas or petrochemical gases. In the process, excess coke-make is reduced and offers an attractive option to combat the incursion of the heavier feedstocks into refinery operations. In fact, gasification units may become a staple of refinery operations in the future, especially as refineries may well accept biomass as part of the feedstock slate.

In the process, the viscous feedstock enters the scrubber for direct contact heat exchange with the overhead product vapors from the reactor. The higher-boiling products (>525°C/>975°F) present in the overhead condense in the scrubber and return to the reactor as a recycle stream with fresh feedstock. Lower-boiling overhead constituents in the scrubber go to a conventional fractionator and also to light-ends recovery. The feedstock is thermally cracked in the reactor fluidized bed to a range of gas and liquid products, and coke. The coke inventory is maintained by circulating the bed coke from the reactor to the heater via the cold coke transfer line. In the heater, the coke is heated by the gasifier products and circulated back to the reactor via the hot coke transfer line to supply the heat that sustains the thermal cracking process.

Excess coke is converted to a low-heating value gas in a fluid-bed gasifier with steam and air. The air is supplied to the gasifier to maintain temperatures of 830°C–1,000°C (1,525°F–1,830°F) but is insufficient to burn all of the coke. The gasifier products, consisting of a gas and coke mixture, return to the heater to heat up the coke. The gas exits the heater overhead and may be used for (1) steam generation of steam, (2) dry/wet particulate removal, and (3) fuel in refinery boilers and furnaces. The majority (on the order of 95% w/w) of the coke generated in the reactor is converted in the process. Only a small amount of product coke is collected as fines from the flexi-gas and purged from the heater to extract feed metals.

A typical gas product, after removal of hydrogen sulfide, contains carbon monoxide (CO, 18%), carbon dioxide (CO_2, 10%), hydrogen (H_2, 15%), nitrogen (N_2, 51%), water (H_2O, 5%), and methane (CH_4, 1%). The heater is located between the reactor and the gasifier, and it serves to transfer heat between the two vessels. The heater temperature is controlled by the rate of coke circulation between the heater and the gasifier. Adjusting the air rate to the gasifier controls the unit inventory of coke, and the gasifier temperature is controlled by steam injection to the gasifier.

The yields of liquid products from flexicoking are the same as from fluid coking because the coking reactor is unaltered. The units are designed to gasify 60%–97% w/w of the coke from the reactor. The flexicoking process produces a clean fuel gas with a heating value on the order of 90 Btu/ft^3. The coke gasification can be controlled to burn approximately 95% w/w of the coke to maximize production of coke gas or at a reduced level to produce both gas and a coke that has been desulfurized to remove approximately 65% w/w of the sulfur.

Fluid coking and flexicoking are versatile processes that are applicable to a wide range of viscous feedstocks (such as vacuum residua, visbreaker residuum, extra heavy crude oil, and tar sand bitumen) to provide a variety of products.

4.7   THERMAL CRACKING CHEMISTRY

With the dramatic increases in the number of gasoline-powered vehicles, distillation processes (Speight, 2014, 2017) were not able to completely fill the increased demand for gasoline.

As a historical note, in 1913, the thermal cracking process was developed and is the phenomenon by which higher-boiling (higher-molecular-weight) constituents in crude oil are converted into lower-boiling (lower-molecular-weight) products application of elevated temperatures (usually on the order of >350°C, >660°F). Thermal cracking is the oldest and, in principle, the simplest conversion process employed in a refinery. The process parameters (i.e., the temperature, the pressure, and the residence time) are feedstock dependent as well as the product requirements.

Thermal cracking is a phenomenon by which higher-boiling constituents (higher-molecular-weight constituents) in crude oil are converted into lower-boiling (lower-molecular-weight) products. However, certain products may interact with one another to yield products having higher molecular weights than the constituents of the original feedstock. Some of the products are expelled from the system as, say, gases, gasoline-range materials, kerosene-range materials, and the various intermediates that produce other products such as coke. Materials that have boiling ranges higher than gasoline and kerosene may (depending upon the refining options) be referred to as recycle stock, which is recycled in the cracking equipment until conversion is complete.

Thus, thermal cracking processes allow the production of lower-molecular-weight products such as the constituents of liquefied petroleum gas (LPG) and naphtha/gasoline constituents from higher-molecular-weight fraction such as gas oils and residua. The simplest thermal cracking process – the visbreaking process – is used to upgrade fractions such as distillation residua and other aforementioned viscous feedstocks to produce feedstocks for other refinery processes or products (such as fuel oil) that meet specifications for sales (Parkash, 2003; Gary et al., 2007; Speight, 2014; Hsu and Robinson, 2017; Speight, 2017).

4.7.1 GENERAL CHEMISTRY

As a brief introduction to the subject, in the process of thermal cracking catalytic cracking or hydrocracking, the high molecular weight constituents of the feedstock decomposed to produce (1) gases, such as methane (CH_4) ethane (CH_3CH_3), ethylene ($CH_2=CH_2$), propane ($CH_3CH_2CH_3$), propylene ($CH_3CH=CH_2$), butane ($CH_3CH_2CH_2CH_3$), and the butylene isomers ($CH_3CH_2CH=CH_2$, $CH_3CH=CHCH_3$), all of which can be used as feedstocks for petrochemicals production, (2) lower-molecular-weight more volatile liquids of which naphtha (a precursor to gasoline) is an example, (3) middle boiling-range liquids of which kerosene is an example and which is used in diesel fuel, (4) low-boiling and high-boiling gas oils which are used to produce the various grades of fuel oil, and (5) coke, which is a solid carbonaceous product that can be used as a refinery fuel or in a gasification process to produce synthesis gas (Parkash, 2003; Gary et al., 2007; Speight, 2014; Hsu and Robinson, 2017).

Furthermore, the relative reactivity of crude oil constituents can be assessed on the basis of bond energies, but the thermal stability of an organic molecule is dependent upon the bond strength of the weakest bond. And even though the use of bond energy data is a method for predicting the reactivity or the stability of specific bonds under designed conditions, the reactivity of a particular bond is also subject to its environment. Thus, it is not only the reactivity of the constituents of crude oil that are important in processing behavior, but it is also the stereochemistry of the constituents as they relate to one another that is also of some importance (Speight, 2014). It must be appreciated that the stereochemistry of organic compounds is often a major factor in determining reactivity and properties.

More specifically, thermal cracking is a free radical mechanism in which an atom or group of atoms possessing an unpaired electron (the free radical) is very reactive (often difficult to control), and it is the mode of reaction of free radicals that determines the product distribution during thermal cracking (i.e., non-catalytic thermal decomposition). In addition, a significant feature of hydrocarbon free radicals is the resistance to isomerization during the existence of the radical. For example, thermal cracking does not produce any degree of branching in the products (by migration of an alkyl group) other than that already present in the feedstock. Nevertheless, the classical chemistry of free radical formation and behavior involves the following chemical reactions – it can only be presumed that the formation of free radicals during thermal (non-catalytic) cracking follows similar paths:

1. Initiation reaction, where a single molecule breaks apart into two free radicals. Only a small fraction of the feedstock constituents may actually undergo initiation, which involves breaking the bond between two carbon atoms, rather than the thermodynamically stronger bond between a carbon atom and a hydrogen atom.

$$CH_3CH_3 \rightarrow 2CH_3\bullet$$

2. Hydrogen abstraction reaction in which the free radical abstracts a hydrogen atom from another molecule:

$$CH_3\bullet + CH_3CH_3 \rightarrow CH_4 + CH_3CH_2\bullet$$

3. Radical decomposition reaction in which a free radical decomposes into an alkene:

$$CH_3CH_2\bullet \rightarrow CH_2=CH_2 + H\bullet$$

4. Radical addition reaction in which a radical reacts with an alkene to form a single, larger free radical:

$$CH_3CH_2\bullet + CH_2=CH_2 \rightarrow CH_3CH_2CH_2CH_2\bullet$$

5. Termination reaction in which two free radicals react with each other to produce the products – two common forms of termination reactions are recombination reactions (in which two radicals combine to form one molecule) and disproportionation reactions (in which one free radical transfers a hydrogen atom to the other to produce an alkene and an alkane):

$$CH_3 \bullet + CH_3CH_2 \bullet \rightarrow CH_3CH_2CH_3$$

$$CH_3CH_2 \bullet + CH_3CH_2 \bullet \rightarrow CH_2=CH_2 + CH_3CH_3$$

The smaller free radicals, hydrogen, methyl, and ethyl are more stable than the larger radicals. They will tend to capture a hydrogen atom from another hydrocarbon, thereby forming a saturated hydrocarbon and a new radical. In addition, in many thermal cracking processes many different chemical reactions occur simultaneously. Thus an accurate explanation of the mechanism of the thermal cracking reactions is difficult. The primary reactions are the decomposition of higher-molecular-weight species into lower-molecular-weight products.

As the molecular weight of the hydrocarbon feedstock increases, the reactions become much more complex leading to a wider variety of products. For example, using a more complex hydrocarbon (dodecane, $C_{12}H_{26}$) as the example, two general types of reaction occur during cracking:

1. The decomposition of high molecular weight constituents into lower-molecular-weight constituents (primary reactions):

$$CH_3(CH_2)10CH_3 \rightarrow CH_3(CH_2)8CH_3 + CH_2=CH_2$$

$$CH_3(CH_2)10CH_3 \rightarrow CH_3(CH_2)7CH_3 + CH_2=CHCH_3$$

$$CH_3(CH_2)10CH_3 \rightarrow CH_3(CH_2)6CH_3 + CH_2=CHCH_2CH_3$$

$$CH_3(CH_2)10CH_3 \rightarrow CH_3(CH_2)5CH_3 + CH_2=CH(CH_2)2CH_3$$

$$CH_3(CH_2)10CH_3 \rightarrow CH_3(CH_2)4CH_3 + CH_2=CH(CH_2)3CH_3$$

$$CH_3(CH_2)10CH_3 \rightarrow CH_3(CH_2)3CH_3 + CH_2=CH(CH_2)4CH_3$$

$$CH_3(CH_2)10CH_3 \rightarrow CH_3(CH_2)2CH_3 + CH_2=CH(CH_2)5CH_3$$

$$CH_3(CH_2)10CH_3 \rightarrow CH_3CH_2CH_3 + CH_2=CH(CH_2)6CH_3$$

$$CH_3(CH_2)10CH_3 \rightarrow CH_3CH_3 + CH_2=CH(CH_2)7CH_3$$

$$CH_3(CH_2)10CH_3 \rightarrow CH_4 + CH_2=CH(CH_2)8CH_3$$

2. Reactions by which some of the primary products interact to form higher-molecular-weight materials (secondary reactions):

$$CH_2=CH_2 + CH_2=CH_2 \rightarrow CH_3CH_2CH=CH_2$$

$$RCH=CH_2 + R_1CH=CH_2 \rightarrow \text{cracked residuum} + \text{coke} + \text{other products}$$

Thus, from the chemistry of the thermal decomposing of pure compounds (and assuming little interference from other molecular species in the reaction mixture), it is difficult but not impossible to predict the product types that arise from the thermal cracking of various feedstocks. However, during thermal cracking, all of the reactions illustrated above can and do occur simultaneously and to some extent are uncontrollable. However, one of the significant features of hydrocarbon free radicals is their resistance to isomerization, for example, migration of an alkyl group, and as a result, thermal cracking does not produce any degree of branching in the products other than that already present in the feedstock.

Data obtained from the thermal decomposition of pure compounds indicate certain decomposition characteristics that permit predictions to be made of the product types that arise from the thermal cracking of various feedstocks. For example, normal paraffins are believed to form, initially, higher-molecular-weight material, which subsequently decomposes as the reaction progresses. Other paraffinic materials and (terminal) olefins are produced. An increase in pressure inhibits the formation of low-molecular-weight gaseous products and therefore promotes the formation of higher-molecular-weight materials.

Furthermore, for saturated hydrocarbons, the connecting link between gas-phase pyrolysis and liquid-phase thermal degradation is the concentration of alkyl radicals. In the gas phase, alkyl radicals are present in low concentration and undergo unimolecular radical decomposition reactions to form α-olefins and smaller alkyl radicals. In the liquid phase, alkyl radicals are in much higher concentration and prefer hydrogen abstraction reactions to radical decomposition reactions. It is this preference for hydrogen abstraction reactions that gives liquid-phase thermal degradation a broad product distribution.

Branched paraffins react somewhat differently to the normal paraffins during cracking processes and produce substantially higher yields of olefins having one fewer carbon atom than the parent hydrocarbon. Cycloparaffins (naphthenes) react differently to their non-cyclic counterparts and are somewhat more stable. For example, cyclohexane produces hydrogen, ethylene, butadiene, and benzene: Alkyl-substituted cycloparaffins decompose by means of scission of the alkyl chain to produce an olefin and a methyl or ethyl cyclohexane.

The aromatic ring is considered fairly stable at moderate cracking temperatures (350°C–500°C, 660°F–930°F). Alkylated aromatics, like the alkylated naphthenes, are more prone to dealkylation that to ring destruction. However, ring destruction of the benzene derivatives occurs above 500°C (930°F), but condensed aromatics may undergo ring destruction at somewhat lower temperatures (450°C, 840°F).

Generally, the relative ease of cracking of the various types of hydrocarbons of the same molecular weight is given in the following descending order: (1) paraffins, (2) olefins, (3) naphthenes, and (4) aromatics. To remove any potential confusion, paraffins are the least stable and aromatics are the most stable.

Within any type of hydrocarbon, the higher-molecular-weight hydrocarbons tend to crack easier than the lighter ones. Paraffins are by far the easiest hydrocarbons to crack with the rupture most likely to occur between the first and second carbon bonds in the lighter paraffins. However, as the molecular weight of the paraffin molecule increases, rupture tends to occur nearer the middle of the molecule. The main secondary reactions that occur in thermal cracking are polymerization and condensation

Two extremes of thermal cracking in terms of product range are represented by high-temperature processes, namely (1) steam cracking or (2) pyrolysis. Steam cracking is a process in which feedstock is decomposed into lower-molecular-weight (often-unsaturated) products saturated hydrocarbons. In the process, a gaseous or liquid hydrocarbon feed such as ethane or naphtha is diluted with steam and briefly heated in a furnace (at approximately 850°C, 1,560°F) in the absence of oxygen at a short residence time (often on the order of milliseconds). After the cracking temperature has been reached, the products are rapidly quenched in a heat exchanger. The products produced in the reaction depend on (1) the properties and composition of the feedstock, (2) the feedstock/steam ratio, (3) the cracking temperature, and (4) the residence time.

Typically, pyrolysis processes require temperatures on the order of 750°C–900°C (1,380°F–1,650°F) to produce high yields of low-molecular-weight products, such as ethylene, for petrochemical use. Delayed coking, which uses temperature on the order of 500°C (930°F), is used to produce distillates from non-volatile residua as well as coke for fuel and other uses – such as the production of electrodes for the steel and aluminum industries.

4.7.2 ASPHALTENE CHEMISTRY

An important aspect of any thermal cracking process in which viscous feedstocks (i.e., heavy crude oils, extra heavy crude oils, tar sand bitumen, and residua) are employed is the thermal chemistry of the asphaltene constituents and the resin constituents which can give rise to instability and/or incompatibility of the reaction mix leading to the occurrence of deposits (Table 4.3) (Speight, 2014).

The phenomenon of instability is often referred to as incompatibility, and more commonly known as sludge formation, sediment formation, or deposit formation which can lead to fouling of equipment (Chapter 6). In crude oil and its products, instability often manifests itself in various ways (Tables 5.1 and 5.2) (Mushrush and Speight, 1995; Speight, 2014). Hence, there are different ways of defining each of these terms, but the terms are often used interchangeably. For the most part, especially in the current context, instability and incompatibility of viscous feedstocks are attributable to the presence of asphaltene constituents in the feedstocks which can undergo thermal decomposition to yield (intermediate or final) products that are incompatible with the other constituents of the reaction mix. The most obvious example of incompatibility (*non-miscible*) is the inability of hydrocarbon fuels and water to mix. In the present context, incompatibility usually refers to the presence of various polar functions (i.e., heteroatom function groups containing nitrogen, oxygen, sulfur, and even various combinations of the heteroatoms) in the feedstock or product mix.

TABLE 4.3
Examples of the Events That Occur during Thermal Decomposition

Temperature	Event
<100°C (<212°F)	Volatiles evolved, including some water.
	Heat-sensitive substances may partially decompose.
100°C (212°F)	Water remaining absorbed in the material is evolved.
	Water trapped in crystal structure of hydrates may be evolved at higher temperatures.
	Some solid substances, such as fats, sugars, and waxes may tend to melt.
100°C–500°C (212°F–930°F)	Organic molecules decompose.
	Most sugar derivatives start decomposing at 160°C–180°C (320°F–355°F).
	Cellulose decomposes at approximately 350°C (660°F).
	Lignin starts decomposing at approximately 350°C (660°F); continues releasing volatile products up to 500°C (930°F).
	The decomposition products usually include water, carbon monoxide, and/or carbon dioxide as well as organic compounds.
	The non-volatile residues typically become richer in carbon and form large, disordered constituents, with colors ranging between brown and black.
200°C–300°C (390°F–570°F)	When oxygen has not been excluded, the carbonaceous residue may start to burn in a highly exothermic reaction releasing carbon dioxide and/or monoxide.
	At this stage, some of the nitrogen still remaining in the residue may be oxidized to nitrogen oxides (such as NO_2 and N_2O).
	Sulfur and other elements (such as nitrogen) may be oxidized and volatilized.

Source: Speight, J.G. 2017. *Handbook of Petroleum Refining*. CRC Press, Taylor & Francis Group, Boca Raton, Florida. Table 8.2, page 300.

The asphaltene constituents are defined as the heptane-insoluble asphaltene constituents and resin constituents that, because of the content of polynuclear aromatic compounds and polar functionalities, provide hurdles to conversion. Some workers prefer to use n-pentane as the solvent for the separation of the asphaltene fraction, but it must be remembered that the heptane asphaltene and the pentane asphaltene have different amounts of the various constituents and must be recognized that the behavior of each fraction will exhibit differences (Speight, 2014, 2015b).

The high thermal stability of polynuclear aromatic systems prevents thermal decomposition to lower-boiling-point products and usually results in the production of substantial yields of thermal coke. Furthermore, the high concentrations of heteroatom compounds (nitrogen, oxygen, sulfur) and metals (vanadium and nickel) in heavy oils and residua have an adverse effect on catalysts. Therefore, process choice often favors thermal process, but catalytic processes can be used as long as catalyst replacement and catalyst regeneration is practiced.

Asphaltene constituents and, to a lesser extent, resin constituents can cause major problems in refineries through unanticipated coke formation and/or through excessive coke formation. Recognition of this is a step in the direction of mitigating the problem and improvement in the conversion of viscous feedstocks may be sought through the use of specific chemical additives. However, to improve the conversion of viscous feedstocks, it is necessary to understand the chemistry of conversion.

The thermal decomposition of the more complex asphaltene constituents has received some attention (Magaril and Aksenova, 1967, 1968; Magaril and Ramazaeva, 1969; Magaril, and Aksenova, 1970a, 1970b; Magaril et al., 1970, 1971; Magaril and Aksenova, 1972; Schucker and Keweshan, 1980; Speight, 2014). The thermal reaction is believed to be first order, although there is the potential that it is, in fact, a multi-order reaction process, but because of the multiplicity of the reactions that occur, it appears as a pseudo first-order process. However, it is definite that there is an induction period before coke begins to form that seems to be triggered by phase separation of reacted asphaltene product (Magaril and Aksenova, 1967, 1968; Magaril and Ramazaeva, 1969; Magaril, and Aksenova, 1970a, 1970b; Magaril et al., 1970, 1971; Magaril and Aksenova, 1972; Wiehe, 1992, 1993; Speight, 2014; Wiehe, 1994). The organic nitrogen originally in the asphaltene constituents invariably undergoes thermal reaction to concentrate in the non-volatile coke . In scheme, the chemistry of asphaltene coking has been suggested to involve the thermolysis of thermally labile bonds to form reactive species that react with each other (condensation) to form coke. However, not all the original aromatic carbon in the asphaltene constituents forms coke. Volatile aromatic species are eliminated during thermal decomposition, and it must be assumed that some of the original aliphatic carbon plays a role in coke formation.

It is more likely that the initial reactions of asphaltene constituents involves thermolysis of pendant alkyl chains to form lower-molecular-weight, higher-polar species that are often referred to as carbenes and carboids (Speight, 2014, 2015b) which then react to form coke. The reactions involve unimolecular thermolysis of aromatic-alkyl systems of the asphaltene constituents to produce volatile species (paraffins and olefins) and non-volatile species (aromatics) (Speight, 1987, 1994; Wiehe, 1994).

It is also interesting to note that although the aromaticity of the resin and asphaltene constituents is approximately equivalent to the yield of thermal coke, not all the original aromatic carbon in the asphaltene constituents forms coke. Volatile aromatic species are eliminated during thermal decomposition, and it must be assumed that some of the original aliphatic carbon plays a role in coke formation. Its precise nature has yet to be determined, but the process can be represented as involving a multi-reaction process involving series and parallel reactions (Speight, 2014).

As examples of thermal cracking, in the delayed coking process, the feedstock is heated to high temperatures (480°C–500°C; 895°F–930°F) in a furnace and then reaction is allowed to continue in a cylindrical, insulated drum. The volatile products pass overhead into a fractionator and coke accumulates in the drum. Any high-boiling liquid product from the fractionator is recycled to the coker furnace. When the drum fills up with coke, the reacting feedstock is directed to a second drum. The coke is removed from the first drum by hydraulic drilling and cutting after which the drum is ready for the next 16–24 hour reaction cycle. During this process, the asphaltene and resin constituents in the feedstock are converted to coke in accordance with their respective carbon residue values (ca. 50% w/w for asphaltene constituents and ca. 35% w/w for resin constituents) (Speight, 2014).

Nitrogen species also appear to contribute to the pattern of the thermolysis insofar as the carbon-carbon bonds adjacent to ring nitrogen undergo thermolysis quite readily (Fitzer et al., 1971; Speight, 1998). Thus, the initial reactions of asphaltene decomposition involve thermolysis of aromatic-alkyl bonds that are enhanced by the presence of heterocyclic nitrogen (Speight, 1987). Thus, the molecular species within the asphaltene fraction, which contain nitrogen and other heteroatoms (and have lower volatility than the pure hydrocarbons), are the prime movers in the production of coke. Such species, containing various polynuclear aromatic systems, can be denuded of the attendant hydrocarbon moieties and are undoubtedly insoluble (Bjorseth, 1983; Dias, 1987, 1988) in the surrounding hydrocarbon medium. The next step is gradual carbonization of these heteroatom-rich entities to form coke.

Thus, coke formation is a complex thermal process involving both chemical reactions and thermodynamic behavior. The challenges facing process chemistry and physics are determining (1) the means by which crude oil constituents thermally decompose, (2) the nature of the products of thermal decomposition, (3) the subsequent decomposition of the primary thermal products, (4) the interaction of the products with each other, (5) the interaction of the products with the original constituents, and (6) the influence of the products on the composition of the liquids.

The goal is to mitigate coke formation by elimination or modification of the prime chemical reactions in the formation of incompatible products during the processing of feedstocks containing asphaltene constituents, particularly those reactions in which the insoluble lower-molecular-weight products (arbitrarily called carbenes and carboids) are formed (Chapter 3) (Wiehe, 1992, 1993; Speight, 2014, 2015b). Thus, the challenges facing process chemistry and physics are determining (1) the means by which crude oil constituents thermally decompose, (2) the nature of the products of thermal decomposition, (3) the subsequent decomposition of the primary thermal products, (4) the interaction of the products with each other, (5) the interaction of the products with the original constituents, and (6) the influence of the products on the composition of the liquids.

4.7.3 Biomass Chemistry

The utilization of biomass to produce valuable products by thermal processes is an important aspect of biomass technology. Biomass pyrolysis gives usually rise to three phases: (1) gases, (2) condensable liquids, and (3) char/coke. However, there are various types of related kinetic pathways ranging from very simple paths to more complex paths, and all usually include several elementary processes occurring in series or in competition. As anticipated, the kinetic paths are different for cellulose, lignin, and hemicelluloses (biomass main basic components) and also for usual biomasses according to their origin, composition, and inorganic contents.

The main biomass constituents – hemicellulose, cellulose, and lignin – can be selectively devolatilized into value-added chemicals. This thermal breakdown is guided by the order of thermochemical stability of the biomass constituents that ranges from hemicellulose (as the least stable constituent) to the more stable – lignin exhibits an intermediate thermal degradation behavior. Thus, wood constituents are decomposed in the order of hemicellulose-cellulose-lignin, with a restricted decomposition of the lignin at relatively low temperatures. With prolonged heating, condensation of the lignin takes place, whereby thermally largely stable macromolecules develop. Whereas both hemicellulose and cellulose exhibit a relatively high devolatilization rate over a relatively narrow temperature range, thermal degradation of lignin is a slow-rate process that commences at a lower temperature when compared to cellulose.

Since the thermal stabilities of the main biomass constituents partially overlap and the thermal treatment is not specific, a careful selection of temperatures, heating rates, and gas and solid residence times is required to make a discrete degasification possible when applying a stepwise increase in temperature. Depending on these process conditions and parameters such as composition of the biomass and the presence of catalytically active materials, the product mixture is expected to contain degradation products from hemicellulose, cellulose, or lignin.

4.8 PROCESS PARAMETERS

Thermal cracking of viscous feedstocks invariably results in the formation of coke in the process which, though undesirable, is an important part of the fuel balance of the refinery.

Most thermal cracking units are operated to maximize conversion to gases and naphtha, which is particularly true when building gasoline inventory for peak season demand or reducing clarified oil yield due to low market demand. Maximum conversion of a specific feedstock is usually limited by both (1) the unit design constraints, such as regenerator temperature and (2) the processing objectives. Typically, each unit that is operated for maximum conversion at constant fresh feed quality has an optimum conversion point beyond which a further increase in conversion reduces the yield of naphtha and increases the yield of liquefied petroleum gas, and the optimum conversion point is referred to as the over-cracking point.

4.8.1 THE REACTOR

The thermal cracking reactor (also known as the pyrolysis reactor) is, put simply, a vessel that can contain the feedstock and allow the product to flow out of the reactor for further processing. Typically, the reactor consists of heating the feedstock in an inert atmosphere, promoting thermal bond scission in the feedstock to produce a variety of low-molecular-weight (volatile) hydrocarbon derivatives, i.e., a product mix fraction that includes paraffin derivatives, olefin derivatives, naphthene derivatives, and aromatic derivatives as well as the non-volatile coke (Parkash, 2003; Gary et al., 2007; Speight, 2014; Hsu and Robinson, 2017; Speight, 2017).

Depending on the feedstock characteristics (such as the heteroatom-containing constituents, the degree of aromaticity, and the bond dissociation energy), different cracking mechanisms take place.

4.8.2 TEMPERATURE

There are primary variables available to the operation of thermal cracking units for maximum unit conversion for a given feedstock quality that can be divided into two groups: (1) process variables and (2) variables such as pressure, reaction time, and reactor temperature. Higher conversion and coke yield are thermodynamically favored by higher pressure. However, pressure is usually varied over a very narrow range due to limited air blower horsepower. Conversion is not significantly affected by unit pressure since a substantial increase in pressure is required to significantly increase conversion.

An increase in reaction time available for cracking also increases conversion. Fresh feed rate, riser steam rate, recycle rate, and pressure are the primary operating variables which affect reaction time for a given unit configuration. Conversion varies inversely with these stream rates due to limited reactor size available for cracking. Conversion has been increased by a decrease in rate in injection of fresh feedstock. Under these circumstances, over-cracking of gasoline to liquefied petroleum gas and to dry gas may occur due to the increase in reactor residence time. One approach to offset any potential gasoline over-cracking is to add additional riser steam to lower hydrocarbon partial pressure for more selective cracking. Alternatively, an operator may choose to lower reactor pressure or increase the recycle rate to decrease residence time.

Increased reactor temperature increases feedstock conversion, primarily through a higher rate of reaction for the endothermic cracking reaction and also through increased cat/oil ratio. A 5.6°C (10°F) increase in reactor temperature can increase conversion by 1%–2% absolute, but, again, this is feedstock dependent. Higher reactor temperature also increases the amount of olefins in gasoline and in the gases. This is due to the higher rate of primary cracking reactions relative to secondary hydrogen transfer reactions.

4.8.3 Coke Formation

The formation of coke deposits has been observed in virtually every unit in operation, and the deposits can be very thick with thicknesses up to 4 ft have been reported. Coke has been observed to form where condensation of hydrocarbon vapors occurs. The reactor walls and plenum offer a colder surface where hydrocarbons can condense. Higher-boiling constituents in the feedstock may be very close to their dew point, and they will readily condense and form coke nucleation sites on even slightly cooler surfaces.

Unvaporized feed droplets readily collect to form coke precursors on any available surface since the high-boiling feedstock constituents do not vaporize at the mixing zone of the riser. Thus, it is not surprising that residuum processing makes this problem even worse. Low residence time cracking also contributes to coke deposits since there is less time for heat to transfer to feed droplets and vaporize them. This is an observation in line with the increase in coking when short contact time riser crackers (q.v.) were replacing the longer residence time fluid-bed reactors.

Higher-boiling feedstocks that have high aromaticity result in higher yields of coke. Furthermore, polynuclear aromatics and aromatics containing heteroatoms (i.e., nitrogen, oxygen, and sulfur) are more facile coke makers than simpler aromatics (Hsu and Robinson, 2006; Speight, 2008). However, feed quality alone is not a foolproof method of predicting where coking will occur. However, it is known that feedstock hydrotreaters rarely have coking problems. The hydrotreating step mitigates the effect of the coke formers, and coke formation is diminished.

Thermal cracking results from extended residence times of hydrocarbon vapors in the reactor disengaging area and leads to high dry gas yields via non-selective free radical cracking mechanisms. On the other hand, dilute phase catalytic cracking results from extended contact between catalyst and hydrocarbon vapors downstream of the riser. While much of this undesirable cracking was eliminated in the transition from bed to riser cracking, there is still a substantial amount of non-selective cracking occurring in the dilute phase due to the significant catalyst holdup.

Once coke is formed, it is a matter of where it will appear. Coke deposits are most often found in the reactor (or disengager), transfer line, and slurry circuit and cause major problems in some units such as increased pressure drops, when a layer of coke reduces the flow through a pipe, or plugging, when chunks of coke spall off and block the flow completely. Deposited coke is commonly observed in the reactor as a black deposit on the surface of the cyclone barrels, reactor dome, and walls. Coke is also often deposited on the cyclone barrels 180° away from the inlet. Coking within the cyclones can be potentially very troublesome since any coke spalls going down into the dip leg could restrict product flow. Coke formation also occurs at nozzles which can increase the nozzle pressure drop. It is possible for steam or instrument nozzles to be plugged completely, a serious problem in the case of unit instrumentation.

Coking in the transfer line between the reactor and main fractionator is also common, especially at the elbow where it enters the fractionator. Transfer line coking causes pressure drop and spalling and can lead to reduced throughput. Furthermore, any coke in the transfer line which spalls off can pass through the fractionator into the circulating slurry system where it is likely to plug up exchangers, resulting in lower slurry circulation rates and reduced heat removal. Pressure balance is obviously affected if the reactor has to be run at higher pressures to compensate for transfer line coking. On units where circulation is limited by low slide valve differentials, coke laydown may then indirectly reduce circulation. The risk of a flow reversal is also increased.

Shutdowns and startups can aggravate problems due to coking. The thermal cycling leads to differential expansion and contraction between the coke and the metal wall that will often cause the coke to spall in large pieces. Another hazard during shutdowns is the possibility of an internal fire when the unit is opened up to the atmosphere. Proper shutdown procedures which ensure that the internals have sufficiently cooled before air enters the reactor will eliminate this problem. In fact, the only defense against having coke plugging problems during start up is to thoroughly clean the unit during the turnaround and remove all the coke. If strainers on the line(s), they will have to be cleaned frequently.

The two basic principles to minimize coking are to avoid dead spots and prevent heat losses. An example of minimizing dead spots is using purge steam to sweep out stagnant areas in the disengager system. The steam prevents collection of high-boiling condensable products in the cooler regions. Steam also provides a reduced partial pressure or steam distillation effect on the high-boiling constituents and cause enhanced vaporization at lower temperatures. Steam for purging should preferably be superheated since medium-pressure low-velocity steam in small pipes with high heat losses is likely to be very wet at the point of injection and will cause more problems. Cold spots are often caused by heat loss through the walls in which case increased thermal resistance might help reduce coking. The transfer line, being a common source of coke deposits, should be as heavily insulated as possible, provided that stress-related problems have been taken into consideration.

4.9   OPTIONS FOR VISCOUS FEEDSTOCKS

The decarbonizing thermal process is designed to minimize coke and gasoline yields but, at the same time, to produce maximum yields of gas oil. The process is essentially the same as the delayed coking process, but lower temperatures and pressures are employed. For example, pressures range from 10 to 25 psi, heater outlet temperatures may be 485°C (905°F), and coke drum temperatures may be of the order of 415°C (780°F). Decarbonizing in this sense of the term should not be confused with propane decarbonizing, which is essentially a solvent deasphalting process (Parkash, 2003; Gary et al., 2007; Speight, 2014; Hsu and Robinson, 2017; Speight, 2017).

The low-pressure coking process is designed for a once-through, low-pressure operation and is similar to the delayed coking process except that recycling is not usually practiced, and the coke chamber operating conditions are 435°C (815°F) and a pressure on the order of 25 psi. Excessive coking is inhibited by the addition of water to the feedstock in order to quench and restrict the reactions of the reactive intermediates.

On the other hand, the high-temperature coking process is a semi-continuous thermal conversion process designed for high-melting viscous feedstocks (such as tar sand bitumen and various residua) that yields coke and gas oil as the primary products. The coke may be treated to remove sulfur to produce a low-sulfur coke (≤5%), even though the feedstock contained as much as 5% w/w sulfur. In the process, the feedstock is transported to the pitch accumulator, then to the heater (370°C, 700°F, 30 psi), and finally to the coke oven, where temperatures may be as high as 980°C–1,095°C (1,800°F–2,000°F). Volatile materials are fractionated, and after the cycle is complete, coke is collected for sulfur removal and quenching before storage.

Mixed-phase cracking (also called liquid-phase cracking) is a continuous thermal decomposition process for the conversion of viscous feedstocks to products boiling in the gasoline range. The process generally employs rapid heating of the feedstock (kerosene, gas oil, reduced crude, or even whole crude), after which it is passed to a reaction chamber and then to a separator where the vapors are cooled. Overhead products from the flash chamber are fractionated to gasoline components and recycle stock, and flash chamber bottoms are withdrawn as a heavy fuel oil. Coke formation, which may be considerable at the process temperatures (400°C–480°C, 750°F–900°F), is minimized by use of pressures in excess of 350 psi.

Vapor-phase cracking is a high-temperature (545°C–595°C, 1,000°F–1,100°F), low-pressure (<50 psi) thermal conversion process that favors dehydrogenation of feedstock (gaseous hydrocarbons to gas oils) components to olefins and aromatics. Coke is often deposited in heater tubes, causing shutdowns. Relatively large reactors are required for these units.

4.9.1 Aquaconversion Process

The aquaconversion process is a hydrovisbreaking technology that uses catalyst-activated transfer of hydrogen from water added to the feedstock. Reactions that lead to coke formation are suppressed, and there is no separation of asphaltene-type material (Marzin et al., 1998; Parkash, 2003; Gary et al., 2007; Speight, 2014; Hsu and Robinson, 2017; Speight, 2017). The important aspect of the aquaconversion technology is that it does not produce any solid byproduct such as coke nor requires any hydrogen source or high-pressure equipment. In addition, the aquaconversion process can be implanted in the production area, and thus the need for external diluent and its transport over large distances is eliminated.

4.9.2 Asphalt Coking Technology Process

The asphalt coking technology process (often referred to as the ASCOT process) is a residual oil upgrading process that integrates the delayed coking process and the deep solvent deasphalting process (low energy deasphalting process, the LEDA process) (Bonilla, 1985; Bonilla and Elliot, 1987). Removing the deasphalted oil fraction prior to application of the delayed coking process has two benefits: (1) in the coking process this fraction is thermally cracked to extinction, degrading this material as an FCC feedstock, and (2) thermally cracking this material to extinction results in conversion of a significant portion to coke.

In the process, the vacuum residuum is brought to the desired extraction temperature and then sent to the extractor where the solvent (straight-run naphtha, coker naphtha) flows upward and extracts soluble material from the down-flowing feedstock. The solvent-deasphalted phase leaves the top of the extractor and flows to the solvent recovery system where the solvent is separated from the deasphalted oil and recycled to the extractor. The deasphalted oil is sent to the delayed coker where it is combined with the heavy coker gas oil (from the coker fractionator) and then sent to the heavy coker gas oil stripper where low-boiling hydrocarbons are stripped off and returned to the fractionator. The stripped deasphalted oil/heavy coker gas oil mixture is removed from the bottom of the stripper and used to provide heat to the naphtha stabilizer-reboiler before being sent to battery limits as a cracking stock. The raffinate phase containing the asphalt and some solvent flows at a controlled rate from the bottom of the extractor and is charged directly to the coking section.

The solvent contained in the asphalt and in the deasphalted oil is condensed in the fractionator overhead condensers, where it can be recovered and used as lean oil for a propane/butane recovery system thereby eliminating the need to recirculate lean oil from the naphtha stabilizer. The solvent introduced in the coker heater and coke drums results in a significant reduction in the partial pressure of asphalt feed, compared win a regular delayed coking unit. The low asphalt partial pressure results in low coke and high liquid yields in the coking reaction.

With the ASCOT process, there is a significant reduction in byproduct fuel as compared to either solvent deasphalting or delayed coking, and the process can be tailored to process a specific quantity or process to a specific quality of cracking stock (Parkash, 2003; Gary et al., 2007; Speight, 2014; Hsu and Robinson, 2017; Speight, 2017).

4.9.3 Cherry-P Process

The Cherry-P process (comprehensive heavy ends reforming refinery process) is a process for the conversion of heavy crude oil or residuum into distillate and a cracked residuum (Ueda, 1976, 1978). In this process, the principal aim is to upgrade heavy crude oil residues at conditions between those of conventional visbreaking and delayed coking. Although coal is added to the feedstock, it is not intended to be a co-processing feedstock, but the coal is intended to act as a scavenger to prevent the buildup of coke on the reactor wall. The use of scavengers in the process is projected to increase (Stark and Falkler, 2008; Stark et al., 2008; Speight, 2014).

In the process, the feedstock is mixed with coal powder in a slurry mixing vessel (without a catalyst or hydrogen), heated in the furnace, and fed to the reactor where the feedstock undergoes thermal cracking reactions for several hours at a temperature higher than 400°C–450°C (750°F–840°F) and under pressure (70–290 psi) with a residence time on the order of 1–5 hours. Gas and distillate from the reactor are sent to a fractionator, and the cracked residuum residue is extracted out of the system after distilling low-boiling fractions by the flash drum and vacuum flasher to adjust its softening point. Distillable product yields of 44% by weight on total feed are reported (Parkash, 2003; Gary et al., 2007; Speight, 2014; Hsu and Robinson, 2017; Speight, 2017). Since this yield is obtained when using anthracite, the proportion that is derived from the coal is likely to be very low and unlikely to cause compatibility reactions in downstream reactors due to the presence of phenols and other polar species (Speight, 1990, 2014).

4.9.4 CONTINUOUS COKING PROCESS

A new coking process that can accept viscous feedstock and continuously discharge vapor and dry crude oil coke particles has been developed (Sullivan, 2011). The process promotes a rapid recovery of volatiles from the resid enabling recovery of more volatiles. It also causes the carbonization reactions to proceed more rapidly, and it produces uniform composition and uniform size of coke particles that have a low volatiles content.

The new process uses a kneading and mixing action to continuously expose new resid surface to the vapor space and causes a more complete removal of volatiles from the produced crude oil coke. Not only are more valuable volatiles recovered, but the volatiles are also likely to be richer in middle distillates. As a result of kneading/mixing action by the reactor/devolatilizer, new surfaces of the residuum mass are continuously exposed to the gas phase, enhancing the rapid mass transfer of volatiles into the gas phase. The volatiles are then rapidly cooled to retard degradation. With the rapid reduction of volatiles content in the resid mass, the carbonization reaction rates are accelerated, enabling continuous and rapid production of solid crude oil coke particles. The short contact time of the volatile products with the hot residuum minimizes the thermal degradation of the volatile products.

Concurrently with the carbonization reactions and the formation of coke, some cracking of side chains of the larger molecules likely occurs. These smaller, low-boiling molecules produced from cracking reactions join the population of the indigenous volatiles. Some volatiles may be generated even after the solid coke is formed. In the delayed coking process, many of these late-forming volatiles remain trapped in the coke. The process promotes the release of these late-forming cracked volatiles, allowing them to escape into the gas phase by breaking the solid coke into small particles.

4.9.5 DECARBONIZING PROCESS

The thermal decarbonizing process (not to be confused with the propane decarbonizing process, which is a deasphalting process) is designed to minimize coke and gasoline yields but, at the same time, to produce maximum yields of gas oil. Decarbonizing in this sense of the term should not be confused with propane decarbonizing, which is essentially a solvent deasphalting process (Parkash, 2003; Gary et al., 2007; Speight, 2014; Hsu and Robinson, 2017; Speight, 2017). Thermal decarbonizing is, in many respects, similar to the delayed coking process, but lower temperatures and pressures are employed. For example, heater outlet temperatures may be 485°C (905°F) and coke drum temperatures may be of the order of 415°C (780°F) while pressures range from 10 to 25 psi.

4.9.6 DEEP THERMAL CONVERSION PROCESS

The shell deep thermal conversion (DTC) process offers a bridge between visbreaking and coking and provides maximum distillate yields by applying deep thermal conversion to vacuum residua followed by vacuum flashing of the products.

In the process, the heated viscous feedstock residuum is charged to the heater and from there to the soaker where conversion occurs. The products are then led to an atmospheric fractionator to produce gases, naphtha, kerosene, and gas oil. The fractionator residuum is sent to a vacuum flasher that recovers additional gas oil and distillate. The next steps for the coke are dependent on its potential use and it may be isolated as liquid coke (pitch, cracked residuum) or solid coke (Parkash, 2003; Gary et al., 2007; Speight, 2014; Hsu and Robinson, 2017; Speight, 2017).

4.9.7 ET-II Process

The ET-II process is a thermal cracking process for the production of distillates and cracked residuum for use as metallurgical coke and is designed to accommodate feedstocks such as heavy oils, atmospheric residua, and vacuum residua (Kuwahara, 1987). The distillate (often referred to as cracked oil) is suitable as a feedstock to hydrocracker and fluid catalytic cracking.

In the process, the feedstock is heated up to 350°C (660°F) by passage through the preheater and fed into the bottom of the fractionator, where it is mixed with recycle oil, and the high-boiling fraction of the cracked oil. The ratio of recycle oil to feedstock is within the range 0.1%–0.3% by weight. The feedstock mixed with recycle oil is then pumped out and fed into the cracking heater, where the temperature is raised to approximately 490°C–495°C (915°F–925°F), and the outflow is fed to the stirred-tank reactor where it is subjected to further thermal cracking. Both cracking and condensation reactions take place in the reactor.

The cracked oil and gas products, together with steam from the top of the reactor, are introduced into the fractionator where the oil is separated into two fractions, cracked light oil and vacuum gas oil, and pitch (Parkash, 2003; Gary et al., 2007; Speight, 2014; Hsu and Robinson, 2017; Speight, 2017).

4.9.8 Eureka Process

The Eureka process is a thermal cracking process to produce a cracked oil and aromatic residuum from viscous feedstocks (Aiba et al., 1981; Parkash, 2003; Gary et al., 2007; Ohba et al., 2008; AlHumaidan et al., 2013a, 2013b; Speight, 2014; Hsu and Robinson, 2017; Speight, 2017). The cracking reactions occur under lower cracked oil partial pressure by introducing steam into the reactor. The unconverted cracked residuum (pitch) in the reactor behaves as a homogeneous system that provides stable and trouble-free operating conditions. The cracked oil is further hydrotreated, cracked, and/or hydrocracked to produce marketable fuels, and the cracked residuum is utilized as a boiler fuel or as a gasification (partial oxidation) feedstock for hydrogen production or synthesis gas production (Parkash, 2003; Gary et al., 2007; Speight, 2014; Hsu and Robinson, 2017; Speight, 2017).

In this process (Figure 4.7), the viscous feedstock is fed to the preheater and then enters the bottom of the fractionator, where it is mixed with the recycle oil. The mixture is then fed to the reactor system that consists of a pair of reactors operating alternately.

In the reactor, thermal cracking reaction occurs in the presence of superheated steam which is injected to strip the cracked products out of the reactor and supply a part of heat required for cracking reaction. At the end of the reaction, the bottom product is quenched. The oil and gas products (and steam) pass from the top of the reactor to the lower section of the fractionator, where a small amount of entrained material is removed by a wash operation. The upper section is an ordinary fractionator, where the heavier fraction of cracked oil is drawn as a side stream. The process bottoms (pitch) can be used as boiler fuel, as partial oxidation feedstock for producing hydrogen and carbon monoxide, and as binder pitch for manufacturing metallurgical coke (Parkash, 2003; Gary et al., 2007; Speight, 2014; Hsu and Robinson, 2017; Speight, 2017).

The process reactions proceed at lower cracked oil partial pressure by injecting steam into the reactor, keeping crude oil pitch in a homogeneous liquid state, and unlike a conventional delayed

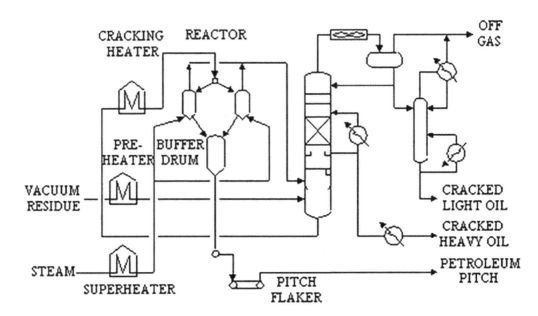

FIGURE 4.7 The eureka process. (Speight, J.G. 2017. *Handbook of Petroleum Refining*. CRC Press, Taylor & Francis Group, Boca Raton, Florida. Figure 8.20, page 335.)

coker, a higher cracked oil yield can be obtained. A wide range of viscous feedstock can be used and included atmospheric residua, vacuum residua, various cracked residua, asphalt-type products from solvent deasphalting, as well as tar sand bitumen. After hydrotreating, the cracked oil is used as feedstock for a fluid catalytic cracker or hydrocracker.

The original Eureka process uses two batch reactors, while the newer ET-II and the HSC process both employ continuous reactors.

4.9.9 Fluid Thermal Cracking Process

The process (fluid thermal cracking process (FTC process) is a heavy oil and residuum upgrading process in which the feedstock is thermally cracked to produce distillate and coke (Miyauchi et al., 1981, 1987; Parkash, 2003; Gary et al., 2007; Speight, 2014; Hsu and Robinson, 2017; Speight, 2017).

The feedstock, mixed with recycle stock from the fractionator, is injected into the cracker, is immediately absorbed into the pores of the particles by capillary force, and is subjected to thermal cracking. In consequence, the surface of the non-catalytic particles is kept dry, and good fluidity is maintained allowing a good yield of, and selectivity for, middle distillate products. Hydrogen-containing gas from the fractionator is used for the fluidization in the cracker. Excessive coke caused by the metals accumulated on the particle is suppressed under the presence of hydrogen. The particles with deposited coke from the cracker are sent to the gasifier, where the coke is gasified and converted into carbon monoxide (CO), hydrogen (H_2), carbon dioxide (CO_2), and hydrogen sulfide (H_2S) with steam and air. Regenerated hot particles are returned to the cracker.

4.9.10 High-Conversion Soaker Cracking Process

The high-conversion soaker cracking process (HSC process) is a cracking process designed for moderate conversion, higher than visbreaking but lower than coking (Watari et al., 1987; Washimi, 1989; Parkash, 2003; Gary et al., 2007; Speight, 2014; Hsu and Robinson, 2017; Speight, 2017). The process is an advanced continuous thermal cracking technology with a proprietary soaking drum, featuring a wide range conversion of levels between visbreaking and coking while producing pumpable liquid residue at process temperature. A broad range of viscous feedstocks such as heavy crude, long and short residue with high contents of sulfur and heavy metals and even visbroken residue can be charged to the HSC process. The cracked distillates from the HSC process are mostly light and high-boiling (high-viscosity) gas oil fractions with fewer unsaturated compounds than coker distillates. The heavy gas oil fraction serves as the feedstock to the fluid catalytic cracking unit, and the cracked residue can be used as the fuel for boiler at the power station. The process uses no hydrogen, no catalyst, and no high-pressure equipment. The process economics is benefited by low investment cost and low utilities consumptions due to its simple process scheme as visbreaking process. The process features less gas-make and a higher yield of distillate compared to other thermal cracking processes. The process can be used to convert a wide range of feedstocks with high-sulfur and -metal content, including heavy crude oil, extra heavy crude oil, tar sand bitumen, distillation residua, and visbroken residua. As a note of interest, the HSC process employs continuous reactors whereas the original Eureka process often is designed to employ two batch reactors.

In the process, the preheated feedstock enters the bottom of the fractionator, where it is mixed with the recycle oil after which the mixture is pumped up to the charge heater and fed to the soaking drum (at approximately atmospheric pressure with steam injection at the top and bottom of the drum) where there is sufficient residence time is provided (depending upon the feedstock characteristics) to complete the thermal cracking. In the soaking drum, the feedstock and some product flow downward passing through a number of perforated plates while steam with cracked gas and distillate vapors flow through the perforated plates counter-currently.

The volatile products from the soaking drum enter the fractionator where the distillates are fractionated into desired product streams, including a high-boiling (and somewhat viscous) gas oil fraction. The cracked gas product is compressed and used as refinery fuel gas after sweetening, and after hydrotreating, the cracked oil product can be used as a feedstock for a fluid catalytic cracking unit or for a hydrocracker unit. The residuum is suitable for use as boiler fuel, road asphalt, binder for the coking industry, and as a feedstock for partial oxidation.

4.9.11 MIXED-PHASE CRACKING PROCESS

The mixed-phase cracking process (also called liquid-phase cracking) is a continuous thermal decomposition process for the conversion of viscous feedstocks to products boiling in the gasoline range. The process generally employs rapid heating of the feedstock (kerosene, gas oil, reduced crude, or even whole crude), after which it is passed to a reaction chamber and then to a separator where the vapors are cooled. Overhead products from the flash chamber are fractionated to gasoline components and recycle stock, and flash chamber bottoms are withdrawn as a heavy fuel oil. Coke formation, which may be considerable at the process temperatures (400°C–480°C, 750°F–900°F), is minimized by the use of pressures in excess of 350 psi.

4.9.12 SELECTIVE THERMAL CRACKING PROCESS

The selective thermal cracking process (STC process) is a thermal conversion process that utilizes different conditions depending on the nature of the feedstock. For example, a heavy oil may be cracked at 495°C–515°C (920°F–960°F) and 300–500 psi; a lower-boiling gas oil may be cracked at 510°C–530°C (950°F–990°F) and 500–700 psi (Figure 4.8). Each feedstock has its own particular characteristics that dictate the optimum conditions of temperature and pressure for maximum yields of the products (Parkash, 2003; Gary et al., 2007; Speight, 2014; Hsu and Robinson, 2017; Speight, 2017).

These factors are utilized in the selective combination of cracking units in which the more refractory feedstocks are cracked for longer periods of time or at higher temperatures than the less stable feedstocks, which are cracked at lower temperatures.

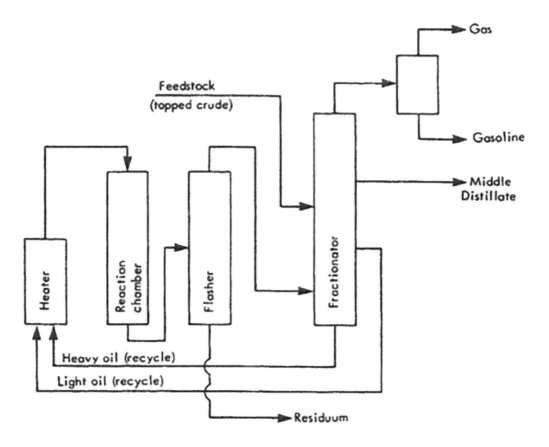

FIGURE 4.8 The selective thermal cracking process. (Speight, J.G. 2017. *Handbook of Petroleum Refining*. CRC Press, Taylor & Francis Group, Boca Raton, Florida. Figure 8.19, page 331.)

The process eliminates the accumulation of stable low-boiling material in the recycle stock and also minimizes coke formation from high-temperature cracking of the higher-boiling material. The end result is the production of fairly high yields of gasoline, middle distillates, and olefin gases.

The thermal cracking of naphtha involves the upgrading of low-octane fractions of catalytic naphtha to higher-quality material. The process is designed, in fact, to upgrade the heavier portions of naphtha, which contain virgin feedstock, and to remove naphthenes, as well as paraffins. Some heavy aromatics are produced by condensation reactions, and substantial quantities of olefins occur in the product streams.

The limitations of processing the more complex difficult-to-convert heavy oil, residua, and tar bitumen depend to a large extent on the amount of non-volatile higher-molecular-weight constituents, which also contain the majority of the heteroatoms (i.e., nitrogen, oxygen, sulfur, and metals such as nickel and vanadium) (Chapter 2). The chemistry of the thermal reactions of some of these constituents dictates that certain reactions, once initiated, cannot be reversed and proceed to completion and coke is the eventual product (Parkash, 2003; Gary et al., 2007; Speight, 2014; Hsu and Robinson, 2017; Speight, 2017).

Upgrading residua, which are similar in character to some heavy oils and tar sand bitumen, began with the introduction of desulfurization processes that were designed to reduce the sulfur content of residua as well as some heavy crude oils and products therefrom. In the early days, the goal was desulfurization, but, in later years, the processes were adapted to a 10%–30% partial conversion operation, as intended to achieve desulfurization and obtain low-boiling fractions simultaneously, by increasing severity in operating conditions.

Refinery evolution has seen the introduction of a variety of viscous feedstock cracking processes (some use catalysts and are, of necessity, included here). These processes are different from one another in cracking method, cracked product patterns, and product properties, and will be employed in refineries according to their respective features.

Each feedstock has its own particular characteristics that dictate the optimum conditions of temperature and pressure for maximum yields of the products. These factors are utilized in selective combination of cracking units in which the more refractory feedstocks are cracked for longer periods of time or at higher temperatures than the less stable feedstocks, which are cracked at lower temperatures.

The process eliminates the accumulation of stable low-boiling material in the recycle stock and also minimizes coke formation from high-temperature cracking of the higher-boiling material. The end result is the production of fairly high yields of gasoline, middle distillates, and olefin gases.

4.9.13 Tervahl-T Process

The Tervahl-T process offers options that allow the process to accommodate differences in the feedstock as well as the desired sale of products. In the process (LePage et al., 1987; Peries et al., 1988; Parkash, 2003; Gary et al., 2007; Speight, 2014; Hsu and Robinson, 2017; Speight, 2017), the feedstock is heated to the desired temperature using the coil heater and heat recovered in the stabilization section and held for a specified residence time in the soaking drum. The soaking drum effluent is quenched and sent to a conventional stabilizer or fractionator where the products are separated into the desired streams (Parkash, 2003; Gary et al., 2007; Speight, 2014; Hsu and Robinson, 2017; Speight, 2017). The gas produced from the process is used for fuel.

In the related Tervahl H process (a hydrogenation process but covered here for convenient comparison with the Tervahl-T process), the feedstock and hydrogen-rich stream are heated using heat recovery techniques and fired heater and held in the soak drum as in the Tervahl-T process. The gas and oil from the soaking drum effluent are mixed with recycled hydrogen and separated in the hot separator where the gas is cooled passed through a separator and recycled to the heater and soaking drum effluent. The liquids from the hot and cold separator are sent to the stabilizer section where purge gas and synthetic crude are separated. The gas is used as fuel and the synthetic crude oil can now be stored for future use or transported to a refinery unit.

4.9.14 Other Options

The decarbonizing thermal process is designed to minimize coke and gasoline yields but, at the same time, to produce maximum yields of gas oil. The process is essentially the same as the delayed coking process, but lower temperatures and pressures are employed. For example, pressures range from 10 to 25 psi, heater outlet temperatures may be 485°C (905°F), and coke drum temperatures may be of the order of 415°C (780°F). Decarbonizing in this sense of the term should not be confused with propane decarbonizing, which is essentially a solvent deasphalting process (Parkash, 2003; Gary et al., 2007; Speight, 2014; Hsu and Robinson, 2017; Speight, 2017).

Low-pressure coking is a process designed for a once-through, low-pressure operation. The process is similar to delayed coking except that recycling is not usually practiced and the coke chamber operating conditions are 435°C (815°F), 25 psi. Excessive coking is inhibited by the addition of water to the feedstock in order to quench and restrict the reactions of the reactive intermediates.

High-temperature coking is a semi-continuous thermal conversion process designed for high-melting asphaltic residua that yields coke and gas oil as the primary products. The coke may be treated to remove sulfur to produce a low-sulfur coke (≤5%), even though the feedstock contained as much as 5% w/w sulfur. In the process, the feedstock is transported to the pitch accumulator, then to the heater (370°C, 700°F, 30 psi), and finally to the coke oven, where temperatures may be as high as 980°C–1,095°C (1,800°F–2,000°F). Volatile materials are fractionated, and after the cycle is complete, coke is collected for sulfur removal and quenching before storage.

Mixed-phase cracking (also called liquid-phase cracking) is a continuous thermal decomposition process for the conversion of viscous feedstocks to products boiling in the gasoline range. The process generally employs rapid heating of the feedstock (kerosene, gas oil, reduced crude, or even whole crude), after which it is passed to a reaction chamber and then to a separator where the vapors are cooled. Overhead products from the flash chamber are fractionated to gasoline components and recycle stock, and flash chamber bottoms are withdrawn as a heavy fuel oil. Coke formation, which may be considerable at the process temperatures (400°C–480°C, 750°F–900°F), is minimized by the use of pressures in excess of 350 psi.

Vapor-phase cracking is a high-temperature (545°C–595°C, 1,000°F–1,100°F), low-pressure (<50 psi) thermal conversion process that favors dehydrogenation of feedstock (gaseous hydrocarbons to gas oils) components to olefins and aromatics. Coke is often deposited in heater tubes, causing shutdowns. Relatively large reactors are required for these units.

Selective thermal cracking is a thermal conversion process that utilizes different conditions depending on the nature of the feedstock. For example, a heavy oil may be cracked at 494°C–515°C (920°F–960°F) and 300–500 psi; a lighter gas oil may be cracked at 510°C–530°C (950°F–990°F) and 500–700 psi. Each feedstock has its own particular characteristics that dictate the optimum conditions of temperature and pressure for maximum yields of the products. These factors are utilized in selective combination of cracking units in which the more refractory feedstocks are cracked for longer periods of time or at higher temperatures than the less stable feedstocks, which are cracked at lower temperatures.

The process eliminates the accumulation of stable low-boiling material in the recycle stock and also minimizes coke formation from high-temperature cracking of the higher-boiling material. The end result is the production of fairly high yields of gasoline, middle distillates, and olefin gases.

The thermal cracking of naphtha involves the upgrading of low-octane fractions of catalytic naphtha to higher-quality material. The process is designed, in fact, to upgrade the heavier portions of naphtha, which contain virgin feedstock, and to remove naphthenes, as well as paraffins. Some heavy aromatics are produced by condensation reactions, and substantial quantities of olefins occur in the product streams.

The OrCrude process was developed specifically for the upgrading of Canadian tar sand bitumen. The process used exiting technology to convert tar sand bitumen to products.

In the process, (1) the separated bitumen is distilled – atmospheric and vacuum, (2) the residuum is deasphalted – deasphaltened to remove the asphaltene fraction, (3) the deasphalted oil is sent to a thermal cracking unit, and (4) the cracked products are recycled into the feedstock to the atmospheric distillation, so as to recover the distillates and separate the asphaltene component produced by the thermal treatment.

The process produces distillates and a heavy residue (asphaltene fraction) which is used as feedstock for the gasification plant that produces synthesis gas (also referred to as syngas) to generate energy and the steam needed for extraction of the bitumen from the deposit, as well as the hydrogen needed for further upgrading of the products.

REFERENCES

Aiba, T., Kaji, H., Suzuki, T., and Wakamatsu, T. 1981. The Eureka Process. *Chemical Engineering Progress*, February: 37.

AlHumaidan, F., Hauser, A., Al-Rabiah, H., Lababidi, H., and Bouresli, R. 2013a. Studies on Thermal Cracking Behavior of Vacuum Residues in Eureka Process. *Fuel*, 109: 635–646.

AlHumaidan, F., Haitham, M.S., Lababidi, H., and Al-Rabiah, H. 2013b. Thermal Cracking Kinetics of Kuwaiti Vacuum Residues in Eureka Process. *Fuel*, 109: 923–931.

ASTM D893. 2015. *Standard Test Method for Insolubles in Used Lubricating Oils. Annual Book of Standards*. ASTM International, West Conshohocken, Pennsylvania.

ASTM D2007. 2015. *Standard Test Method for Characteristic Groups in Rubber Extender and Processing Oils and Other Petroleum-Derived Oils by the Clay-Gel Absorption Chromatographic Method. Annual Book of Standards*. ASTM International, West Conshohocken, Pennsylvania.

ASTM D3279. 2015. *Standard Test Method for n-Heptane Insolubles. Annual Book of Standards*. ASTM International, West Conshohocken, Pennsylvania.

ASTM D4124. 2015. *Standard Test Method for Separation of Asphalt into Four Fractions. Annual Book of Standards*. ASTM International, West Conshohocken, Pennsylvania.

Ballard, W.P., Cottington, G.I., and Cooper, T.A. 1992. Cracking, Thermal. In: *Petroleum Processing Handbook*. J.J. McKetta (Editor). Marcel Dekker Inc., New York. Page 309.

Bjorseth, A. 1983. *Handbook of Polycyclic Aromatic Hydrocarbons*. Marcel Dekker Inc., New York.

Blaser, D.E. 1992. Coking, Petroleum (Fluid). In: *Petroleum Processing Handbook*. J.J. McKetta (Editor). Marcel Dekker Inc., New York. Page 255.

Bonilla, J. 1985. Energy Progress. *December*, 5(4): 239–244.

Bonilla, J., and Elliott, J.D. 1987. Asphalt Coking Method. United States Patent 4,686,027. August 11.

Carrillo, J.A., and Corredor, L.M. 2013. Heavy Crude Oil Upgrading: Jazmin Crude. *Advances in Chemical Engineering and Science*, 3: 46–55.

Dias, J.R. 1987. *Handbook of Polycyclic Hydrocarbons. Part A. Benzenoid Hydrocarbons*. Elsevier, New York.

Dias, J.R. 1988. *Handbook of Polycyclic Hydrocarbons. Part B. Polycyclic Isomers and Heteroatom Analogs of Benzenoid Hydrocarbons*. Elsevier, New York.

Feintuch, H.M., and Negin, K.M. 2004. FW Delayed Coking Process. In: *Handbook of Petroleum Refining Processes*. 2nd Edition. R.A. Meyers (Editor). McGraw-Hill, New York. Chapter 12.33.

Fitzer, E., Mueller, K., and Schaefer, W. 1971. The Chemistry of the Pyrolytic Conversion of Organic Compounds to Carbon. *Chemistry & Physics of Carbon*, 7: 237–383.

Gary, J.G., Handwerk, G.E., and Kaiser, M.J. 2007. *Petroleum Refining: Technology and Economics*. 5th Edition. CRC Press, Taylor & Francis Group, Boca Raton, Florida.

Germain, J.E. 1969. *Catalytic Conversion of Hydrocarbons*. Academic Press Inc., New York.

Hsu, C.S., and Robinson, P.R. (Editors). 2017. *Handbook of Petroleum Technology*. Springer, Cham, Switzerland.

Hulet, C., Briens, C., Berruti, F., and Chan, E.W. 2005. A Review of Short Residence Time Cracking Processes. *International Journal of Chemical Reactor Engineering*, 3(1): R1. https://www.degruyter.com/document/doi/10.2202/1542-6580.1139/html

Joshi, J.B., Pandit, A.B., Kataria, K.L., Kulkarni, R.P., Sawarkar, A.N., Tandon, D., Ram, Y., and Kumar, M.M. 2008. *Industrial & Engineering Chemistry Research*, 47: 8960–8988.

Kobe, K.A., and McKetta, J.J. 1958. *Advances in Petroleum Chemistry and Refining*. Interscience, New York.

Koshka, E., Kuhach, J., and Veith, E. 2008. Improving Athabasca Bitumen Development Economics through Integration with HTL Upgrading. Proceedings of the World Heavy Oil Congress, Edmonton, Alberta, Canada. Alberta Department of Energy, Edmonton, Alberta, Canada. March 2008.

Kuwahara, I. 1987. The ET-II Process. *Koagaku Kogaku*, 51: 1.

Langer, A.W., Stewart, J., Thompson, C.E., White, H.T., and Hill, R.M. 1961. Thermal Hydrogenation of Crude Residua. *Industrial and Engineering Chemistry*, 53: 27–30.

Langer, A.W., Stewart, J., Thompson, C.E., White, H.T., and Hill, R. M. 1962. Hydrogen Donor Diluent Visbreaking of Residua. *Industrial and Engineering Chemistry Process Design and Development*, 1: 309–312.

LePage, J.F., Morel, F., Trassard, A.M., and Bousquet, J. 1987. Preprints Div. *Fuel Chemistry*, 32: 470.

Magaril, R.Z., and Akensova, E.I. 1967. Mechanism of Coke Formation during the Cracking of Petroleum Tars. *Izvestiia vysshikh uchebnykh zavedeniĭ. Neft' i gaz*, 10(11): 134–136.

Magaril, R.Z., and Akensova, E.I. 1968. Study of the Mechanism of Coke Formation in the Cracking of Petroleum Resins. *International Chemical Engineering*, 8(4): 727–729.

Magaril, R.Z., and Aksenova, E.I. 1970a. Mechanism of Coke Formation in the Thermal Decompositon of Asphaltenes. *Khimiya i Tekhnologiya Topliv i Masel*, 15(7): 22–24.

Magaril, R.Z., and Aksenova, E.I. 1970b. Kinetics and Mechanism of Coking Asphaltenes. *Khim. Izvestiia vysshikh uchebnykh zavedeniĭ. Neft' i gaz*, 13(5): 47–53.

Magaril, R.Z., and Aksenova, E.I. 1972. Coking Kinetics and Mechanism of Asphaltenes. *Khim. Kim Tekhnol., Tr. Tyumen Ind. Inst*. Page 169–172.

Magaril, R.Z., and Ramazaeva, L.F. 1969. Study of Carbon Formation in the Thermal Decomposition of Asphaltenes in Solution. *Izvestiia vysshikh uchebnykh zavedeniĭ. Neft' i gaz*, 12(1): 61–64.

Magaril, R.Z., Ramazaeva, L.F., and Aksenova, E.I. 1970. Kinetics of Coke Formation in the Thermal Processing of Petroleum. *Khimiya i Tekhnologiya Topliv i Masel*, 15(3): 15–16.

Magaril, R.Z., Ramazaeva, L.F., and Aksenova, E.I. 1971. Kinetics of the Formation of Coke in the Thermal Processing of Crude Oil. *International Chemical Engineering*, 11(2): 250–251.

Marano, J.J. 2003. Refinery Technology Profiles: Gasification and Supporting Technologies. Report Prepared for the United States Department of Energy, National Energy Technology Laboratory. United States Energy Information Administration, Washington, DC, June.

Marzin, R., Pereira, P., McGrath, M.J., Feintuch, H.M., and Thompson, G. 1998. A New Option for Residue Conversion and Heavy Oil Upgrading. *Oil & Gas Journal*, 97(44): 79.

McKinney, J.D. 1992. Coking, Petroleum (Delayed and Fluid). In: *Petroleum Processing Handbook*. J.J. McKetta (Editor). Marcel Dekker Inc., New York. Page 245.

Miles, J. 2009. Maximizing Distillate Yields and Refinery Economics – An Alternative Solution to Conventional Fuel Oil Production or Residue Conversion. Proceedings of the Session A. 14th Annual Meeting – European Refining Technology Conference, November 11.

Miyauchi, T., Furusaki, S., and Morooka, Y. 1981. Upgrading Resid. In *Advances in Chemical Engineering*. Academic Press Inc., New York. Chapter 11.

Miyauchi, T., Tsutsui, T., and Nozaki, Y. 1987. A New Fluid Thermal Cracking Process for Upgrading Resid. Paper 65B. Proceedings of the Spring National Meeting. American Institute of Chemical Engineers, Houston, March 29.

Moschopedis, S.E., Ozum, B., and Speight, J.G. 1998. Upgrading Heavy Oils. *Reviews in Process Chemistry and Engineering*, 1(3): 201–259.

Mushrush, G.W., and Speight, J.G. 1995. *Petroleum Products: Instability and Incompatibility*. Taylor & Francis Publishers, Philadelphia, Pennsylvania.

Negin, K.M., and Van Tine, F.M. 2004. FW/UOP Visbreaking Process. In: *Handbook of Petroleum Refining Processes*. 2nd Edition. R.A. Meyers (Editor). McGraw-Hill, New York. Chapter 12.3.

Ohba, T., Shibutani, I., Watari, R., Inomata, J., and Nagata, H. 2008. The Advanced EUREKA Process: Environment Friendly Thermal Cracking Process. Paper No. WPC-19-2856. Proceedings of the 19th World Petroleum Congress, Madrid, Spain, June 29–July 3.

Parkash, S. 2003. *Refining Processes Handbook*. Gulf Professional Publishing, Elsevier, Amsterdam, Netherlands.

Peries, J.P., Quignard, A., Farjon, C., and Laborde, M. 1988. Thermal and Catalytic ASVAHL Processes under Hydrogen Pressure for Converting Heavy Crudes and Conventional Residues. *Revue Institut Français Du Pétrole*, 43(6): 847–853.

Roundtree, E.M. 1997. Fluid Coking. In: *Handbook of Petroleum Refining Processes*. 2nd Edition. R.A. Meyers (Editor). McGraw-Hill, New York. Chapter 12.1.

Schucker, R.C., and Keweshan, C.F. 1980. Reactivity of Cold Lake Asphaltenes. Preprints. *Division of Fuel Chemistry American Chemical Society*, 25: 155.

Schucker, R.C. 2003. Heavy Oil Upgrading Process. United States Patent 6,524,469. February 25.

Silverman, M.A., Pavel, S.K., and Hillerman, M.D. 2011. HTL Heavy Oil Upgrading: A Key Solution for Heavy Oil Upstream and Midstream Operations. Paper No. WHOC11-419. Proceedings. World Heavy Oil Congress, Edmonton, Alberta, Canada.

Speight, J.G., and Moschopedis, S.E. 1979. The Production of Low-Sulfur Liquids and Coke from Athabasca Bitumen. *Fuel Processing Technology*, 2: 295.

Speight, J.G. 1990. Tar Sands. In: *Fuel Science and Technology Handbook*. J.G. Speight (Editor), Marcel Dekker Inc., New York. Chapters 12–16.

Speight, J.G. 1998. Thermal Chemistry of Petroleum Constituents. In: *Petroleum Chemistry and Refining*. J.G. Speight (Editor). Taylor & Francis, Washington, DC. Chapter 5.

Speight, J.G. 2000. *The Desulfurization of Heavy Oils and Residua*. 2nd Edition. Marcel Dekker Inc., New York.

Speight, J.G. 2008. *Synthetic Fuels Handbook: Properties, Processes and Performance*. McGraw-Hill, New York.

Speight, J.G. 2011. *The Refinery of the Future*. Gulf Professional Publishing, Elsevier, Oxford, United Kingdom.

Speight, J.G. 2012. Visbreaking: A Technology of the Past and the Future. *Scientia Iranica*, 19(3): 569–573.

Speight, J.G. 2014. *The Chemistry and Technology of Petroleum*. 5th Edition. CRC Press, Taylor & Francis Group, Boca Raton, Florida.

Speight, J.G. 2015a. *Fouling in Refineries*. Gulf Professional Publishing, Elsevier, Oxford, United Kingdom.

Speight, J.G. 2015b. *Handbook of Petroleum Product Analysis*. 2nd Edition. John Wiley & Sons Inc., Hoboken, New Jersey.

Speight, J.G. 2017. *Handbook of Petroleum Refining*. CRC Press, Taylor & Francis Group, Boca Raton, Florida.

Stark, J.L., and Falkler, T. 2008. Method for Improving Liquid Yield during Thermal Cracking of Hydrocarbons. United States Patent 7,425,259. September 16, 2008.

Stark, J.L., Falkler, T., Weers, J.J., and Zetlmeisl, M.J. 2008. Method for Improving Liquid Yield during Thermal Cracking of Hydrocarbons. United States Patent 7,416,654. August 26.

Stell, R.C., Balinsky, G.J., McCoy, J.N., and Keusenkothen, P.F. 2009a. Process and Apparatus for Cracking Hydrocarbon Feedstock Containing Resid. United States Patent 7,588,737. September 15.

Stell, R.C., Dinicolantonio, A.R., Frye, J.M., Spicer, D.B., McCoy, J.N., and Strack, R.D. 2009b. Process for Steam Cracking Heavy Hydrocarbon Feedstocks. United States Patent 7,578,929. August 25.

Stephens, M.M., and Spencer, O.F. 1956. *Petroleum Refining Processes*. Penn State University Press, University Park, Pennsylvania.

Sullivan, D.W. 2011. New Continuous Coking Process. Proceedings of the 14th Topical Symposium on Refinery Processing. AIChE Spring Meeting and Global Congress on Process Safety. Chicago, Illinois. March 13–17.

Ueda, K. 1976. *Journal of the Japan Petroleum Institute*, 19(5): 417.

Ueda, H. 1978. *Journal of the Fuel Society of Japan*, 57: 963.

Veith, E.J. 2007. Performance of Heavy-to-Light-Crude-Oil Upgrading Process. Proceedings of the SPE International Oil Conference and Exhibition, Veracruz, Mexico. June 27–30. Society of Petroleum Engineers, Richardson, Texas.

Washimi, K. 1989. *Hydrocarbon Processing*, 68(9): 69.

Watari, R., Shoji, Y., Ishikawa, T., Hirotani, H., and Takeuchi, T. 1987. Annual Meeting. National Petroleum Refiners Association. San Antonio, Texas. Paper AM-87-43.

Wiehe, I.A. 1992. A Solvent-Resid Phase Diagram for Tracking Resid Conversion. *Industrial & Engineering Chemistry Research*, 31: 530–536.

Wiehe, I.A. 1993. A Phase-Separation Kinetic Model for Coke Formation. *Industrial & Engineering Chemistry Research*, 32: 2447–2454.

Wiehe, I.A. 1994. The Pendant-Core Building Block Model of Petroleum Residua. *Energy & Fuels*, 8: 536–544.

5 Catalytic Cracking Processes

5.1 INTRODUCTION

The main feature of catalytic cracking is, as the name implies, the use of a catalyst to accomplish the conversion and produce the necessary a non-random slate of products. The mechanism of catalytic cracking is also different insofar as the mechanism of the thermal process involves free radical intermediates whereas the mechanism of the catalytic process involves ionic intermediates (Table 5.1). However, there has been the claim that, in some cases, the feedstock constituents decompose by thermolysis before the species may come into contact with the catalyst.

Catalytic cracking is widely used to convert high-boiling feedstocks into more valuable products such as naphtha (a blend stock for gasoline manufacture and used for the solvents and petrochemical products) as well as the production of other low-boiling products. As the demand for gasoline increased, catalytic cracking replaced thermal cracking with the evolution of catalytic cracking. Fluid catalytic cracking (FCC) refers to the behavior of the catalyst during this process insofar as the fine, powdered catalyst (typically zeolites, which have a particle size on the order of 70 µm) takes on the properties of a fluid when it is mixed with the vaporized feed (Sadeghbeigi, 1995, 2011). Fluidized catalyst particles circulate continuously between the reaction zone and the regeneration zone.

In terms of process parameters, catalytic cracking is typically performed at temperatures ranging from 485°C to 540°C (900°F to 1,000°F) and pressures up to 100 psi. Feedstocks for the process have typically been gas oil fractions, but the focus is shifting to blends of gas oil with viscous feedstocks (Shidhaye et al., 2015). In some cases, a viscous (high boiling) feedstock has been blended with the minimum amount of gas oil (added as a flux) as the feedstock to catalytic cracking units. In the process, the feedstock enters the unit at temperatures on the order of 485°C–540°C (900°F–1,000°F), and the circulating catalyst provides heat from the regeneration zone to the oil feed. Carbon (coke) is burned off the catalyst in the regenerator, raising the catalyst temperature to 620°C–735°C (1,150°F–1,350°F, before the catalyst returns to the reactor.

The preferred feedstock to a FCC unit has been and continues to be the portion of the crude oil that has an initial boiling point (at atmospheric pressure) of approximately 275°C (525°F) up to the initial boiling point of the atmospheric residuum (345°C, 655°F) (Table 5.2). On occasion, the vacuum gas oil (boiling range: 345°C–510°C or 345°C–565°C; 655°F–950°F or 655°F–1,050°F) may also be used as feedstock to the FCC unit.

However, the changing slate of refinery feedstocks has caused this to change. Currently, the feedstocks for catalytic cracking can be any one (or blends) of the following: (1) straight-run gas oil, (2) vacuum gas oil, (3) atmospheric residuum, and (4) vacuum residuum, with special emphasis on the heavier feedstocks (Lifschultz, 2005; Ross et al., 2005; Fahim et al., 2010; Speight, 2014). If blends of the above feedstocks are employed, compatibility of the constituents of the blends (i.e., no phase separation) must be assured or excessive coke (and metals) will be laid down on to the catalyst.

In addition, there are several pretreatment options for the feedstocks that offer process benefits, and these are: (1) deasphalting to prevent excessive coking on catalyst surfaces; (2) demetallization, i.e., removal of nickel, vanadium, and iron to prevent catalyst deactivation; (3) use of a short residence time as a means of preparing the feedstock; (4) hydrotreating or mild hydrocracking to prevent excessive coking in the FCC unit; (5) blending with an aromatic gas oil-type to prevent phase separation; and (6) staged partial conversion (Birch and Ulivieri, 2000; Speight, 2000; Patel et al., 2002; Parkash, 2003; Patel et al., 2004; Speight, 2004; Gary et al., 2007; Dziabala et al., 2011;

DOI: 10.1201/9781003184904-5

TABLE 5.1
Comparison of Thermal Cracking and Catalytic Cracking

Process	Description
Thermal cracking	No catalyst
	Free radical reaction mechanisms
	Moderate yields of naphtha and other distillates
	Gas yields feedstock dependent
	Low-to-moderate product selectivity
	Alkanes produced but yields are feedstock dependent
	Naphtha has a low-octane number
	Some chain-branching in the alkane products
	Low-to-moderate yield of C_4 olefins
	Low-to-moderate yields of aromatics
Catalytic cracking	Uses a catalyst
	More flexible in terms of product slate
	Ionic reaction mechanisms
	Good integration of cracking and regeneration
	High yields of high-octane naphtha and other distillates
	Low-gas yields
	High product selectivity
	Chain-branching and high yield of C_4 olefins
	High yields of aromatics

TABLE 5.2
Preferred Composition of the Feedstock for a FCC Unit

Test	Data Range
Gravity, API	19.5–23.0
Density @ 15°C, kg/L	0.9153–0.9366
Distillation (D 1160) °C (°F)	
Initial boiling point	275 (525)
End point	345 (655)
Flash point	116–143 (240–290)
Pour point	17–38 (60–100)
Viscosity @ 50°C (122°F), cSt	20–50
Sulfur, % w/w	1.1–1.4
Carbon residue, % w/w	0.1–0.5
Aniline point	73–79 (163–174)
Asphaltene content, % w/w	<2
Nitrogen, ppm	1,200–1,700
Basic nitrogen, ppm	400–600
Vanadium, ppm	<0.05
Nickel, ppm	<0.10

TABLE 5.3

Feedstock and Product Data for the Fluid Catalytic Process with and without Feedstock Hydrotreating

Feedstock (>370°C, >700°F)	Without Feedstock Hydrotreating	With Feedstock Hydrotreating
API	15.1	20.1
Sulfur, % w/w	3.3	0.5
Nitrogen, % w/w	0.2	0.1
Carbon residue, % w/w	8.9	4.9
Nickel+vanadium, ppm	51.0	7.0
Products		
Naphtha (C5-221°C, C5-430°F), % v/v	50.6	58.0
Light cycle oil (221–360°C, 430–680°F), % v/v	21.4	18.2
Residuum (>360°C, >680°F), % w/w	9.7	7.2
Coke, % w/w	10.3	7.0

Speight, 2014; Hsu and Robinson, 2017; Speight, 2017). Hydrotreating the feedstock to the fluid catalytic cracker improves the yield and quality of naphtha (Table 5.3) and reduces the sulfur oxide (SO_x) emissions from the catalytic cracker unit (Sayles and Bailor, 2005).

On a global basis, the effect of declining crude quality (Speight, 2011a, 2014) may be looked upon as influencing FCC feedstock quality and amount. However, this will be a secondary factor compared with the changes required in the refined products slate (Benazzi and Cameron, 2005). In addition to the heavier viscous crude oils – as a blend or as a hydrotreated feedstock – the production of synthetic crude oil from tar sand bitumen will increase dramatically in the next decade (Patel, 2007; Speight, 2008, 2009, 2011a). For example, the synthetic crude oil from Canadian tar sand sources is projected to increase to 3.0 million bpd by 2015. With Canadian reserves in excess of 170 billion barrels (170×10^9 barrels) of viable oil, economic forecasts predict that tar sand deposits will continue to be a significant crude source (and, hence, feedstock to the catalytic cracking unit) for the foreseeable future (Schiller, 2011). With the increasing focus to reduce sulfur content in fuels, the role of desulfurization in the refinery becomes more and more important. Currently, the process of choice is the hydrotreater in which the unfinished fuel is hydrotreated to remove sulfur from the fuel. Hydrotreating of the feedstock to the catalytic cracking unit can increase conversion by 8%–12% v/v and with most feedstocks (Salazar-Sotelo et al., 2004).

Finally, the use of bio-feedstocks (such as namely animal fats, vegetable oils, cellulosic materials, and lignin) in the FCC unit will be used to increase the yield of light cycle oil and will also provide high-quality products in terms of cetane number (Speight, 2008). Practical implementation in a refinery will be accompanied by blending with vacuum gas oil or resid (Speight, 2011a). From a strategic point of view, refiners should not try to compete with biofuels producers, but rather try to use renewable feedstocks in traditional crude oil refining processes and make products that are compatible with conventional hydrocarbon fuels (Speight, 2008, 2011a).

Furthermore, FCC technology represents one of the most expanded processes producing the precursors to liquid fuels (naphtha and kerosene) and automobile fuels from gas oil distillates and from high-boiling (high-viscosity) feedstocks. A key factor is the use of active, stable, and selective (tailor-made) catalysts to convert specific feedstocks (especially the high-boiling high-viscosity feedstocks) into desired products. Thus, the refinery process (Figure 5.1) can be applied to a variety of feedstocks ranging from gas oil to the more viscous feedstocks.

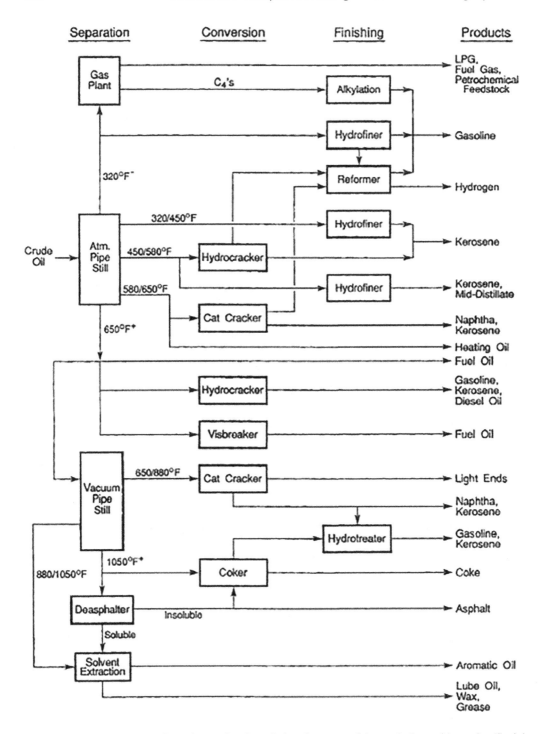

FIGURE 5.1 Generalized refinery layout showing relative placement of the catalytic cracking units. (Speight, J.G. 2017. *Handbook of Petroleum Refining*. CRC Press, Taylor & Francis Group, Boca Raton, Florida. Figure 8.1, page 293.)

It is one of several practical applications used in a refinery that employ a catalyst to improve process efficiency (Parkash, 2003; Gary et al., 2007; Speight, 2014; Hsu and Robinson, 2017; Speight, 2017). The original incentive to develop cracking processes arose from the need to increase gasoline supplies, and since cracking could virtually double the volume of naphtha from a barrel of crude oil, the purpose of cracking was wholly justified.

In the 1930s, thermal cracking units produced approximately half the total naphtha manufactured, the octane number of which was approximately 70 compared to approximately 60 for straight-run naphtha. These were usually blended together with light ends and sometimes with polymer gasoline and reformatted to form a gasoline base stock with an octane number of approximately 65. The addition of tetraethyllead (ethyl fluid) increased the octane number to approximately 70 for regular-grade gasoline and 80 for premium-grade gasoline. The thermal reforming and polymerization processes that were developed during the 1930s could be expected to further increase the octane number of gasoline to some extent, but something new was needed to break the octane barrier that threatened to stop the development of more powerful automobile engines. In 1936, a new cracking process – the catalytic cracking process – was developed to produce higher-octane naphtha and, hence, higher-octane gasoline; this process was catalytic cracking. Since that time, the use of catalysts in the crude oil industry has spread to other processes (Bradley et al., 1989).

Catalytic cracking is basically the same as thermal cracking, but it differs by the use of a catalyst that is not (in theory) consumed in the process (Table 5.4). The catalyst directs the course of the cracking reactions to produce more of the desired products that can be used for the production of better-quality gasoline and other liquid fuels (Avidan and Krambeck, 1990).

The naphtha produced by catalytic cracking has a higher octane number and consists largely of iso-paraffin derivatives and aromatic derivatives. The iso-paraffin derivatives and the aromatic

---

TABLE 5.4
Summary of Catalytic Cracking Processing Conditions

Process Parameters

Solid acid catalyst
Silica-alumina (SiO_2-Al_2O_3), zeolite
Temperature: 480°C–535°C (900°F–1,000°F)
Pressure: 10–20 psi

Process Reactors

Fixed bed
Moving bed
Fluid bed

Feedstocks

Naphtha, middle distillates, gas oil, and residua
Prepared to remove metals and asphaltene (and resin) fraction

Products

C_2–C_4 gases (with some methane – feedstock dependent)
Paraffin derivatives and iso-paraffin derivatives
Some aromatic derivatives
Coke

Source: Speight, J.G., 2017. *Handbook of Petroleum Refining.* CRC Press, Taylor & Francis Group, Boca Raton Florida. Table 9.5, page 345.

---

hydrocarbon derivatives have high-octane numbers and greater chemical stability than mono-olefin derivatives and di-olefin derivatives. The olefin derivatives and di-olefin derivatives are present in much greater quantities in thermally cracked naphtha. Furthermore, olefin derivatives (e.g., $RCH=CH_2$ where R=H or an alkyl group) and smaller quantities of methane (CH_4) and ethane (CH_3CH_3) are produced by catalytic cracking and are suitable for petrochemical use (Parkash, 2003; Gary et al., 2007; Speight, 2014; Hsu and Robinson, 2017; Speight, 2017). Sulfur compounds are changed in such a way that the sulfur content of naphtha produced by catalytic cracking is lower than the sulfur content of naphtha produced by thermal cracking. Catalytic cracking produces less residuum and more useful gas oil constituents than thermal cracking. Finally, the process has considerable flexibility, permitting the manufacture of both automobile gasoline and aviation gasoline and a variation in the gas oil production to meet changes in the fuel oil market.

The reactions that occur during catalytic cracking are complex (Germain, 1969), but there is a measure of predictability now that catalyst activity is better understood. The major catalytic cracking reaction exhibited by paraffin derivatives is carbon-carbon bond scission into a lighter paraffin and olefin. Bond rupture occurs at certain definite locations on the paraffin molecule, rather than randomly as in thermal cracking. For example, paraffin derivatives tend to crack toward the center of the molecule, the long chains cracking in several places simultaneously. The n-paraffin derivatives usually crack at carbon-carbon bonds or still nearer the center of the molecule. On the other hand, iso-paraffin derivatives tend to rupture between carbon atoms that are, respectively, next (alpha) to a tertiary carbon.

As in thermal cracking (Chapter 4), high molecular weight constituents usually crack more readily than small molecules, unless there has been some recycle and the constituents of the recycle stream have become more refractory and are less liable to decompose. Paraffin derivatives having more than six carbon atoms may also undergo rearrangement of their carbon skeletons before cracking, and a minor amount of dehydrocyclization also occurs, yielding aromatic derivatives and hydrogen.

Olefin derivatives are the most reactive class of hydrocarbon derivatives in catalytic cracking and tend to crack from 1,000 to 10,000 times faster than in thermal processes. Severe cracking conditions destroy olefin derivatives almost completely, except for those in the low-boiling naphtha and gaseous hydrocarbon range, and as in the catalytic cracking of paraffin derivatives, iso-olefin derivatives crack more readily than n-olefin derivatives. The olefin derivatives tend to undergo rapid isomerization and yield mixtures with an equilibrium distribution of double-bond positions. In addition, the chain-branching isomerization of olefin derivatives is fairly rapid and often reaches equilibrium. These branched-chain olefin derivatives can then undergo hydrogen transfer reactions with naphthene derivatives and other hydrocarbon derivatives. Other olefin reactions include polymerization and condensation to yield aromatic molecules, which in turn may be the precursors of coke formation.

In catalytic cracking, the cycloparaffin (naphthene) species crack more readily than paraffin derivatives but not as readily as olefin derivatives. Naphthene cracking occurs by ring rupture and by rupture of alkyl chains to yield olefin derivatives and paraffin derivatives, but the formation of methane and the C_2 hydrocarbon derivatives (ethane, CH_3CH_3, ethylene, $CH_2=CH_2$, and acetylene, $CH\equiv CH$) is relatively minor.

Aromatic hydrocarbon derivatives exhibit wide variations in their susceptibility to catalytic cracking. The benzene ring is relatively inert, and condensed-ring compounds, such as naphthalene, anthracene, and phenanthrene, crack very slowly. When these aromatic derivatives crack a substantial part of their conversion is reflected in the amount of coke deposited on the catalyst. Alkylbenzenes with attached groups of C_2 or larger primarily form benzene and the corresponding olefin derivatives, and heat sensitivity increases as the size of the alkyl group increases.

In terms of the bed types, the process that employs a fixed bed (also known as a static bed) of the catalyst in several reactors allows a continuous flow of feedstock to be maintained. The catalyst, which may be an activated natural or synthetic material, is employed in bead, pellet, or microspherical form and can be used as a fixed-bed, moving-bed, or fluid-bed configurations. Thus, in this process, the cycle of operations consists of (1) flow of feedstock through the catalyst bed, (2) discontinuance of feedstock flow and removal of coke from the catalyst by burning, and (3) insertion of the reactor on-stream. The moving-bed process employs a reactor in which cracking takes place as well as a kiln in which the spent catalyst is regenerated, and catalyst movement between the vessels is provided by various means.

There is also the fluid-bed process which differs from the fixed-bed and the moving-bed processes insofar as the powdered catalyst is circulated essentially as a fluid with the feedstock (Sadeghbeigi, 1995, 2011; Hudec, 2014). The several FCC processes in use differ primarily in mechanical design, and there are configurations involving (1) a side-by-side reactor-regenerator configuration or (2) a configuration where the reactor is either above or below the regenerator are the main mechanical variations. From an operational perspective, all of the FCC processes contact the feedstock (and any recycle streams) with the finely divided catalyst in the reactor (Parkash, 2003; Gary et al., 2007; Speight, 2014; Hsu and Robinson, 2017; Speight, 2017).

Feedstocks may range from naphtha fractions (included in normal heavier feedstocks for upgrading) to an atmospheric residuum (reduced crude). Feed preparation (to remove metallic constituents and high-molecular weight non-volatile materials) is usually carried out through the application of any one of several other processes: coking, propane deasphalting, furfural extraction, vacuum distillation, viscosity breaking, thermal cracking, and hydrodesulfurization (Speight, 2000).

The major process variables are temperature, pressure, catalyst-oil ratio (ratio of the weight of catalyst entering the reactor per hour to the weight of oil charged per hour), and space velocity (weight or volume of the oil charged per hour per weight or volume of catalyst in the reaction zone). Wide flexibility in product distribution and quality is possible through control of these variables along with the extent of internal recycling is necessary. Increased conversion can be obtained by applying higher temperature or higher pressure. Alternatively, lower space velocity and higher catalyst-oil ratio will also contribute to an increased conversion.

When cracking is conducted in a single stage, the more reactive hydrocarbon derivatives may be cracked, with a high conversion to gas and coke, in the reaction time necessary for reasonable conversion of the more refractory hydrocarbon derivatives. However, in a two-stage process, gas and naphtha from a short-reaction-time, high-temperature cracking operation are separated before the main cracking reactions take place in a second-stage reactor. For the short time of the first stage, a flow line or vertical riser may act as the reactor, and some conversion is effected with minimal coke formation. Cracked gases are separated and fractionated; the catalyst and residue, together with recycle oil from a second-stage fractionator, pass to the main reactor for further cracking. The products of this second-stage reaction are gas, naphtha and gas oil streams, and recycle oil.

Most fluid catalytic cracking units (FCCs or FCCUs) are operated to maximize conversion to naphtha and LPG (Speight, 2011a, 2014). In the current context, the catalyst, which may be an activated natural or a synthetic material, is employed in bead, pellet, or microspherical form in any one (or all) of the several available or fluidized-bed (fluid-bed) configurations, which differ primarily in mechanical design (Sadeghbeigi, 1995; Hemler, 1997; Hunt, 1997; Johnson and Niccum, 1997; Ladwig, 1997; Hemler and Smith, 2004; Sadeghbeigi, 2011; Speight, 2011a, 2014, 2017). In addition, as the worldwide consumption of fuels has increased, product demand pattern has continued to shift toward distillate fuels such as gasoline, diesel, and kerosene-jet fuel with varying demand for various categories of low-viscosity and high-viscosity fuels (Ross et al., 2005). On the other hand, the octane number of the naphtha is also enhanced by over-cracking the middle boiling point fraction with low-octane number. This technique is more effective in the case of that octane number enhancement in FCC naphtha and if an increase in propylene yield has a priority over naphtha production (Buchanan et al., 1996; Imhof et al., 2005).

The last 70 years have seen substantial advances in the development of catalytic processes. This has involved not only rapid advances in the chemistry and physics of the catalysts themselves but also major engineering advances in reactor design, for example, the evolution of the design of the catalyst beds from fixed beds to moving beds to fluidized beds. Catalyst chemistry and physics and bed design have allowed major improvements in process efficiency and product yields (Sadeghbeigi, 1995, 2011). Most important, in terms of catalyst use, the most important concerns of the crude oil refining industry in the near future are: (1) meeting the growing market of cleaner fuels; (2) gradual substitution of scarce light low-sulfur refinery feedstocks by the heavier high-sulfur feedstocks; (3) the decreasing demand for high viscosity fuel oil; and (4) the need to update processing operations (Parkash, 2004; Swaty, 2005; Gary et al., 2007; Gembicki et al., 2007; Bridjanian and Khadem Samimi, 2011; Letzsch, 2011; Parkash, 2003; Gary et al., 2007; Speight, 2014; Gary et al., 2007; Speight, 2011a, 2014; Hsu and Robinson, 2017; Speight, 2017).

As the trend toward processing viscous feedstocks increases (Patel, 2007; Speight, 2011a), evolving environmental mandates require lower levels of sulfur in the final fuel product and a reduction in emissions of sulfur dioxide (EPA, 2010). Reducing the sulfur concentration requires not only more efficient process options and specialized catalysts, but especially options also required to process heavy crude oil, extra heavy crude oil, tar sand bitumen, and residua (Gembicki et al., 2007; Patel, 2007; Runyan, 2007; Speight, 2014, 2017).

Furthermore, stricter environmental regulations are on the horizon. That venerable, almost revered, Bunker fuel oil was, in the past, released to markets that served as an outlet for a large percentage of the organic sulfur in the refinery feedstock. The International Convention on the Reduction of Pollution from Ships (MARPOL, 2005) has mandated a staged reduction in the allowable sulfur content of maritime fuels, which will drop to less than 0.5% w/w by 2020. In short, the continued emphasis for refinery operations in the foreseeable future 2020 will focus on: (1) the production of clean fuels, (2) upgrading viscous feedstocks to produce transportation fuels, and (3) conversion of bio-feedstocks to bio-oil and thence to a variety of products.

5.2 PROCESS TYPES

Catalytic cracking is another innovation that truly belongs to the 20th century (Table 5.5). It is the modern method for converting high-boiling crude oil fractions, such as gas oil, into gasoline and other low-boiling fractions.

TABLE 5.5
Historical Timeline for Catalytic Cracking Processes

Process (Year Introduced)

McAfee (1915)
Batch reactor catalytic cracking to produce light distillates
Catalyst: a Lewis acid, electron acceptor

Houdry (1936)
Continuous feedstock **flow** with multiple **fixed-bed** reactors
Cracking/catalyst regeneration cycles
Catalyst: clays, natural alumina/silica particles

Thermafor (1942)
Continuous feedstock **flow** with **moving-bed** catalyst
Catalyst: synthetic alumina/silica particles
Higher thermal efficiency by process integration

FCC (1942)
Continuous feedstock **flow** with **fluidized-bed**
catalyst
Catalyst: synthetic alumina/silica and zeolites (1965)

Several processes currently employed in catalytic cracking differ mainly in the method of catalyst handling, although there is an overlap with regard to catalyst type and the nature of the products. The catalyst, which may be an activated natural or synthetic material, is employed in bead, pellet, or microspherical form and can be used as a *fixed bed*, *moving-bed*, or *fluid-bed* configurations.

Feedstocks may range from naphtha to atmospheric residuum (*reduced crude*). Feed preparation (to remove *metallic constituents* and *high-molecular weight non-volatile materials*) is usually carried out through any one of the following ways: (1) coking, (2) propane deasphalting, (3) furfural extraction, (4) vacuum distillation, (5) visbreaking, (6) thermal cracking, and (7) hydrodesulfurization (Parkash, 2003; Gary et al., 2007; Speight, 2014; Hsu and Robinson, 2017; Speight, 2017).

The major process variables are temperature, pressure, catalyst-feedstock ratio (ratio of the weight of catalyst entering the reactor per hour to the weight of feedstock charged per hour), and space velocity (weight or volume of the feedstock charged per hour per weight or volume of catalyst in the reaction zone). Wide flexibility in product distribution and quality is possible through control of these variables along with the extent of internal recycling is necessary. Increased conversion can be obtained by applying higher temperature or higher pressure. Alternatively, lower space velocity and higher catalyst-feedstock ratio will also contribute to an increased conversion.

When cracking is conducted in a single stage, the more reactive hydrocarbon derivatives may be cracked, with a high conversion to gas and coke, in the reaction time necessary for reasonable conversion of the more refractory hydrocarbon derivatives. However, in a two-stage process, gas and gasoline from a short-reaction-time, high-temperature cracking operation are separated before the main cracking reactions take place in a second-stage reactor.

5.2.1 Fixed-Bed Processes

The fixed-bed process was the first to be used commercially and uses a static bed of catalyst in several reactors, which allows a continuous flow of feedstock to be maintained. Thus the cycle of operations consists of (1) flow of feedstock through the catalyst bed, (2) discontinuance of feedstock flow and removal of coke from the catalyst by burning, and (3) insertion of the reactor on-stream.

Although fixed-bed catalytic cracking units are not used to the same extent as the other processes (in fact, in some refineries, the fixed-bed processes have been phased out of existence), they represented an outstanding chemical engineering commercial development by incorporating a fully automatic instrumentation system which provided a short-time reactor/purge/regeneration cycle, a novel molten salt heat transfer system, and a flue gas expander for recovering power to drive the regeneration air compressor. Historically, the Houdry fixed-bed process (Figure 5.2), which went on-stream in June 1936, was the first of the modern catalytic cracking processes. Only the McAfee batch process that employed a metal halide catalyst but which has long since lost any commercial significance preceded it.

In a fixed-bed process, the catalyst in the form of small lumps or pellets is made up of layers or beds in several (four or more) catalyst-containing drums called converters. Feedstock vaporized at approximately 450°C (840°F) and less than 7–15 psi pressure is passed through one of the converters where the cracking reactions take place. After a short time, the deposition of coke on the catalyst renders the catalyst ineffective, and using a synchronized valve system, the feed stream was passed into a converter while the catalyst in the first converter was regenerated by carefully burning the coke deposits with air. After approximately 10 minutes, the catalyst is ready to go on-stream again.

The requirement of complete vaporization necessarily limited feeds to those with a low-boiling range and higher-boiling feedstock constituents are retained in a separator before the feed is passed into the bottom of the upflow fixed-bed reactors. The catalyst consisted of a pelletized natural silica alumina catalyst and was held in reactors or cases approximately 11 ft (3.4 m) in diameter and 38 ft (11.6 m) length for a 15,000-bbl/day unit. Cracked products are passed through the preheat exchanger, condensed, and fractionated in a conventional manner. The reactors operated at approximately 30 psi and 480°C (900°F).

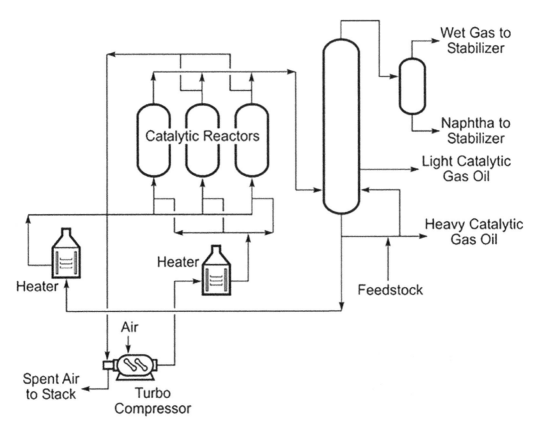

FIGURE 5.2 The Houdry fixed-bed catalytic cracking process. (Speight, J.G. 2017. *Handbook of Petroleum Refining*. CRC Press, Taylor & Francis Group, Boca Raton, Florida. Figure 9.3, page 349.)

The heat of reaction and some of the required feed circulating a molten salt through vertical tubes distributed through the reactor beds. The reaction cycle of an individual reactor was approximately 10 minutes, after which the feed was automatically switched to a new reactor that had been regenerated. The reactor was purged with steam for approximately 5 minutes and then isolated by an automatic cycle timer. Regeneration air was introduced under close control, and carbon was burned off at a rate at which the recirculating salt stream could control the bed temperature. This stream comprised a mixture of potassium nitrate (KNO_3) and sodium nitrate ($NaNO_3$), which melts at 140°C (285°F), and was cooled in the reactors through which feed was being processed. The regeneration cycle lasted approximately 10 minutes. The regenerated bed was then purged of oxygen and automatically cut back into cracking service. There were three to six reactors in a unit. Naphtha yields diminished over the life of the catalysts (18 months) from 52% by volume to 42% by volume, based on fresh feedstock.

Equilibrium was never reached in this cyclic process. The gas oil conversion, i.e., the amount of feed converted to lighter components, was high at the start of a reaction cycle and progressively diminished as the carbon deposit accumulated on the catalyst until regeneration was required. Multiple parallel reactors were used to approach a steady-state process. However, the resulting process flows were still far from steady state. The reaction bed temperature varied widely during reaction and regeneration periods, and the temperature differential within the bed during each cycle was considerable.

Fixed-bed catalytic cracking units have now generally been replaced by moving-bed or fluid-bed processes.

5.2.2   Moving-Bed Processes

The fixed-bed process had obvious capacity and mechanical limitations that needed improvement and such improvement and, thus, were replaced by a moving-bed process in which the hot salt systems were eliminated. The catalyst was lifted to the top of the reactor system and flowed by gravity down through the process vessels. The plants were generally limited in size to units processing up to approximately 30,000; these units have been essentially replaced by larger fluid solids units.

The *moving-bed process* uses a reaction vessel in which cracking takes place, and a kiln in which the spent catalyst is regenerated, catalyst movement between the vessels is provided by various means.

In the moving-bed processes, the catalyst is a pelletized form (approximately 0.125 in. (3 mm) diameter beads that flow by gravity from the top of the unit through a seal zone to the reactor that operates at approximately 10 psi 455°C–495°C (850°F–925°F). The catalyst then flows down through another seal and countercurrent through a stripping zone to the regenerator or kiln that operates at a pressure that is close to atmospheric. In early moving-bed units, built around 1943, bucket elevators were used to lift the catalyst to the top of the structure. In later units, built after 1949, a pneumatic lift was used. This pneumatic lift permitted higher catalyst circulation rates, which in turn permitted injection of all liquid feedstocks, as well as feedstocks that had a higher boiling range. A primary air stream was used to convey the catalyst, and a secondary air stream was injected through an annulus into which the catalyst could flow. Varying the secondary air rate varied the circulation rate.

The lift pipe is tapered to a larger diameter at the top and minimize erosion and catalyst attrition at the top. This taper is also designed so that total collapse of circulation will not occur instantaneously when a specific concentration or velocity of solids, below which particles tend to drop out of the flowing gas stream, is experienced. The taper can be designed so that this potential separation of solids is preceded by a pressure instability that can alert the operators to take corrective action.

The airlift thermofor catalytic cracking process (the Socony Airlift TCC process) (Figure 5.3) is a moving-bed, reactor-over-generator continuous process for conversion of high boiling gas oil into lighter high-quality naphtha and middle distillate fuel oil.

Feedstock preparation may consist of flashing in a tar separator to obtain vapor feed, and the tar separator bottoms may be sent to a vacuum tower from which the liquid feed is produced.

The gas-oil vapor-liquid flows downward through the reactor concurrently with the regenerated synthetic bead catalyst. The catalyst is purged by steam at the base of the reactor and gravitates into the kiln, or regeneration is accomplished by the use of air injected into the kiln. Approximately 70% of the carbon on the catalyst is burned in the upper kiln burning zone and the remainder in the bottom-burning zone. Regenerated, cooled catalyst enters the lift pot, where low-pressure air transports it to the surge hopper above the reactor for reuse.

The airlift thermofor catalytic cracking process (the Socony Airlift TCC Process) is a moving-bed, reactor-over-generator continuous process for conversion of high-boiling gas oil into lighter high-quality naphtha and middle distillates. Feed preparation may consist of flashing in a separator to obtain vapor feed, and the separator bottoms may be sent to a vacuum tower from which the liquid feed is produced. In the process, the gas-oil vapor-liquid flows downward through the reactor concurrently with the regenerated synthetic bead catalyst. The catalyst is purged by steam at the base of the reactor and gravitates into the kiln, or regeneration is accomplished by the use of air injected into the kiln. Approximately 70% of the carbon on the catalyst is burned in the upper kiln burning zone and the remainder in the bottom-burning zone. Regenerated, cooled catalyst enters the lift pot, where low-pressure air transports it to the surge hopper above the reactor for reuse.

The Houdriflow catalytic cracking process (Figure 5.4) is a continuous, moving-bed process employing an integrated single vessel for the reactor and regenerator kiln. The charge stock, sweet or sour, can be any fraction of the crude boiling between naphtha and soft asphalt.

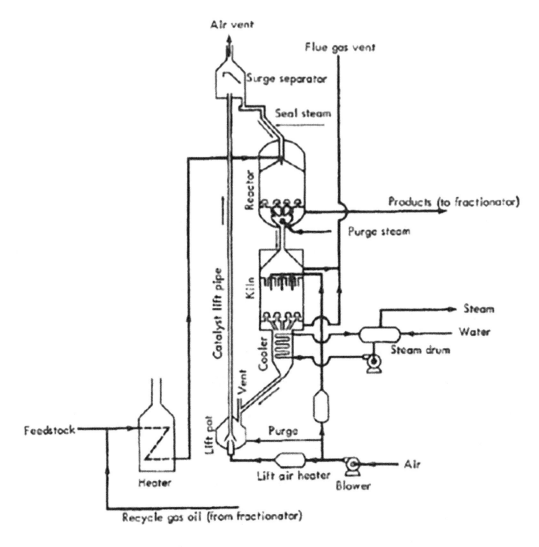

FIGURE 5.3 The airlift thermofor catalytic cracking process. (Speight, J.G. 2017. *Handbook of Petroleum Refining*. CRC Press, Taylor & Francis Group, Boca Raton, Florida. Figure 9.5, page 352.)

　　Also, the reactor feed and catalyst pass concurrently through the reactor zone to a disengager section, in which vapors are separated and directed to a conventional fractionation system. The catalyst is transported from the bottom of the unit to the top in a gas lift employing compressed flue gas and steam. The spent catalyst, which has been steam purged of residual oil, flows to the kiln for regeneration, after which steam and flue gas are used to transport the catalyst to the reactor.

　　The Houdriflow catalytic cracking process is a continuous, moving-bed process employing an integrated single vessel for the reactor and regenerator kiln. The sweet or sour feedstock can be any fraction of the crude boiling between naphtha and atmospheric residua. The catalyst is transported from the bottom of the unit to the top in a gas lift employing compressed flue gas and steam. The feedstock and catalyst pass concurrently through the reactor zone to a disengager section, in which vapors are separated and directed to a conventional fractionation system. The spent catalyst, which has been steam purged of residual oil, flows to the kiln for regeneration, after which steam and flue gas are used to transport the catalyst to the reactor.

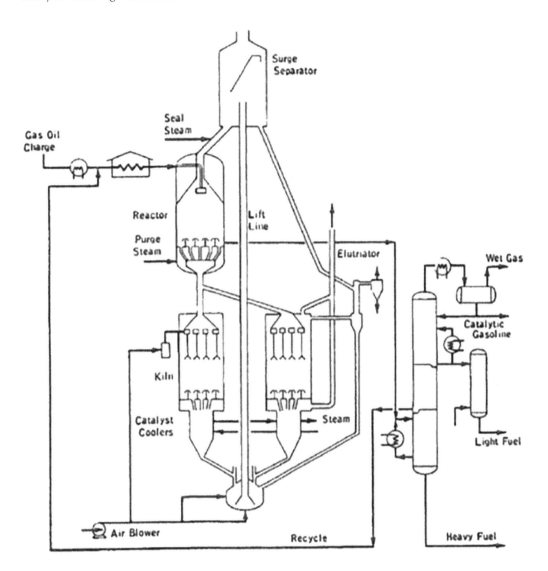

FIGURE 5.4 The Houdriflow moving-bed catalytic cracking process. (Speight, J.G. 2017. *Handbook of Petroleum Refining.* CRC Press, Taylor & Francis Group, Boca Raton, Florida. Figure 9.5, page 352.)

On the other hand, the Houdresid catalytic cracking process (Figure 5.5) is a process that uses a variation of the continuously moving catalyst bed designed to obtain high yields of high-octane naphtha and light distillate from reduced crude charge.

Viscous feedstocks can be employed as the feedstock, and the catalyst is synthetic or natural (Alvarenga Baptista et al., 2010a, b). Although the equipment employed is similar in many respects to that used in Houdriflow units, novel process features modify or eliminate the adverse effects and catalyst and product selectivity usually resulting when heavy metals (such as iron, nickel, copper, and vanadium) are present in the fuel. The Houdresid catalytic reactor and catalyst-regenerating kiln are contained in a single vessel. Fresh feed plus recycled gas oil are charged to the top of the unit in a partially vaporized state and mixed with steam.

The Suspensoid catalytic cracking process (Figure 5.6) was developed from the thermal cracking process carried out in tube and tank units. The process was developed from the thermal cracking process carried out in tube and tank units.

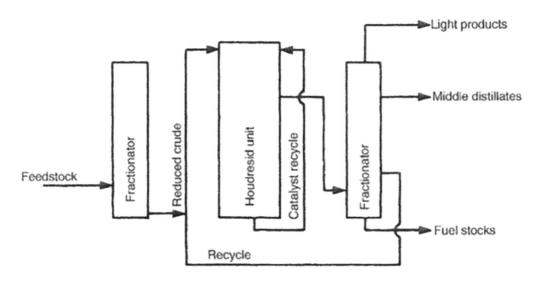

FIGURE 5.5 The Houdresid catalytic cracking process. (Speight, J.G. 2017. *Handbook of Petroleum Refining*. CRC Press, Taylor & Francis Group, Boca Raton, Florida. Figure 9.7, page 353.)

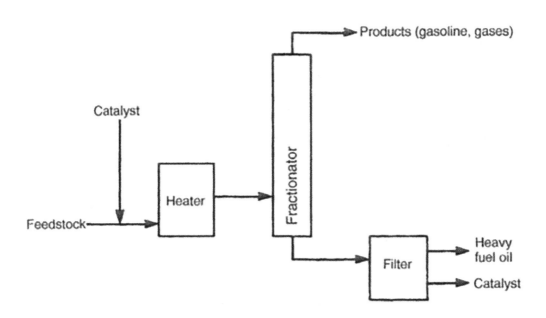

FIGURE 5.6 The suspensoid catalytic cracking process. (Speight, J.G. 2017. *Handbook of Petroleum Refining*. CRC Press, Taylor & Francis Group, Boca Raton, Florida. Figure 9.8, page 354.)

In the process, small amounts of powdered catalyst or a mixture with the feedstock are pumped through a cracking coil furnace at a temperature on the order of 550°C–610°C (1,025°F–1,130°F), with pressures of 200–500 psi. After leaving the furnace, the cracked material enters a tar separator where the catalyst and tar are left behind. The cracked vapors enter a bubble tower where they are separated into two parts: gas oil and pressure distillate. The latter is separated into naphtha and gases. The spent catalyst is filtered from the tar, which is used (either solely or as a blend stock) for industrial fuel oil. The main effect of the catalyst is to allow a higher cracking temperature and to assist mechanically in keeping coke from accumulating on the walls of the tubes. The normal catalyst employed is spent clay obtained from the contact filtration of lubricating oils (2–10 lbs per barrel of feed).

The process is actually a compromise between catalytic and thermal cracking. The main effect of the catalyst is to allow a higher cracking temperature and to assist mechanically in keeping coke from accumulating on the walls of the tubes. The normal catalyst employed is spent clay (2–10 lbs per barrel of feed) that obtained from the contact filtration of lubricating oils

5.2.3 Fluid-Bed Processes

The application of fluidized solids techniques to catalytic cracking resulted in a major process breakthrough. It was possible to transfer all of the regeneration heat to the reaction zone. Much larger units could be built and higher-boiling feedstocks could be processed. Furthermore, there are processes that allow catalytic cracking of viscous feedstocks and residua (Parkash, 2003; Gary et al., 2007; Speight, 2014; Hsu and Robinson, 2006; Speight, 2017).

The first FCCUs were Model I upflow units in which the catalyst flowed up through the reaction and regeneration zones in a riser type of flow regime. Originally, the Model I unit was designed to feed a reduced crude to a vaporizer furnace where all of the gas oil was vaporized and fed, as vapor, to the reactor. The nonvolatile residuum (bottoms) bypassed the cracking section. The original Model I upflow design (1941) was superseded by the Model II downflow design (1944) followed by the Model III (1947) balanced-pressure design with the later introduction of the Model IV low-elevation design (Figure 5.7).

Of the catalytic cracking process concepts, the FCC process is the most widely used process and is characterized by the use of a finely powdered catalyst that is moved through the processing unit (Figure 5.8).

The catalyst particles are of such a size that when aerated with air or hydrocarbon vapor, the catalyst behaves like a liquid and can be moved through pipes. Thus, vaporized feedstock and fluidized catalyst flow together into a reaction chamber where the catalyst, still dispersed in the hydrocarbon vapors, forms beds in the reaction chamber and the cracking reactions take place. The cracked vapors pass through cyclones located in the top of the reaction chamber, and the catalyst powder is thrown out of the vapors by centrifugal force. The cracked vapors then enter the bubble towers where fractionation into gaseous products, naphtha, low-boiling cracked gas oil and high-boiling cracked gas oil occurs.

Since the catalyst in the reactor becomes contaminated with coke, the catalyst is continuously withdrawn from the bottom of the reactor and lifted by means of a stream of air into a regenerator where the coke is removed by controlled burning. The regenerated catalyst then flows to the fresh feed line, where the heat in the catalyst is sufficient to vaporize the fresh feed before it reaches the reactor, where the temperature is approximately 510°C (950°F).

The orthoflow fluid-bed catalytic cracking process (Figure 5.9) uses the unitary vessel design, which provides a straight-line flow of catalyst and thereby minimizes the erosion encountered in pipe bends.

Commercial orthoflow designs are of three types: models A and C, with the regenerator beneath the reactor, and model B, with the regenerator above the reactor. In all cases, the catalyst-stripping section is located between the reactor and the regenerator. All designs employ the heat-balanced principle incorporating fresh feed-recycle feed cracking.

The Universal Oil Products (UOP) fluid-bed catalytic cracking process is adaptable to the needs of both large and small refineries. The major distinguishing features of the process are: (1) elimination of the air riser with its attendant large expansion joints, (2) elimination of considerable structural steel supports, and (3) reduction in regenerator and in air-line size through use of 15–18 psi pressure operation. The UOP process is also designed to produce low molecular weight olefin

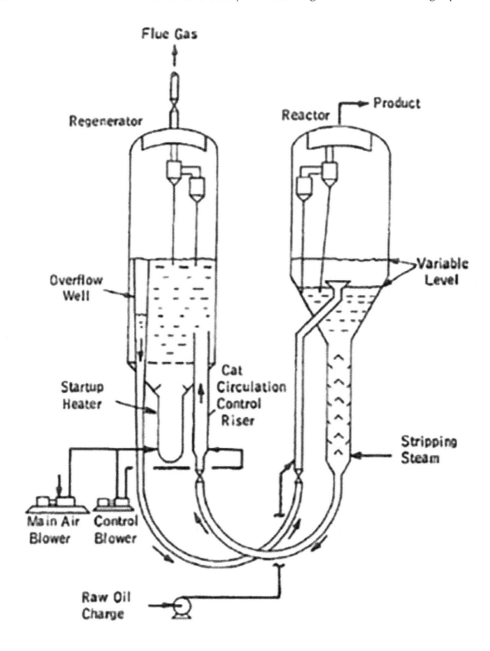

FIGURE 5.7   The model IV catalytic cracking unit. (Speight, J.G. 2017. *Handbook of Petroleum Refining.* CRC Press, Taylor & Francis Group, Boca Raton, Florida. Figure 9.11, page 356.)

derivatives (for alkylation, polymerization, etherification, or petrochemicals), liquefied petroleum gas, high-octane naphtha, distillates, and fuel oils.

In the process, a side-by-side reactor/regenerator configuration and a patented pre-acceleration zone to condition the regenerated catalyst before feed injection – the riser terminates in a vortex separation system. A high efficiency stripper then separates the remaining hydrocarbon derivatives from the catalyst, which is then reactivated in a combustor-style regenerator. The reactor zone

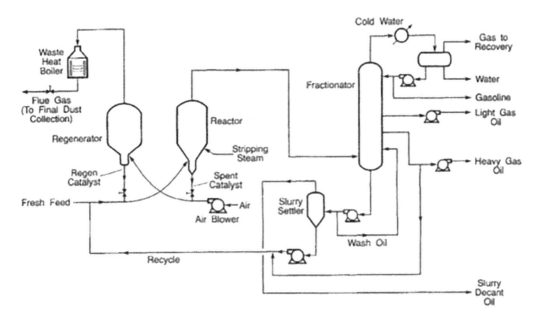

FIGURE 5.8 The fluid-bed catalytic cracking process. (Speight, J.G. 2017. *Handbook of Petroleum Refining.* CRC Press, Taylor & Francis Group, Boca Raton, Florida. Figure 9.12, page 356.)

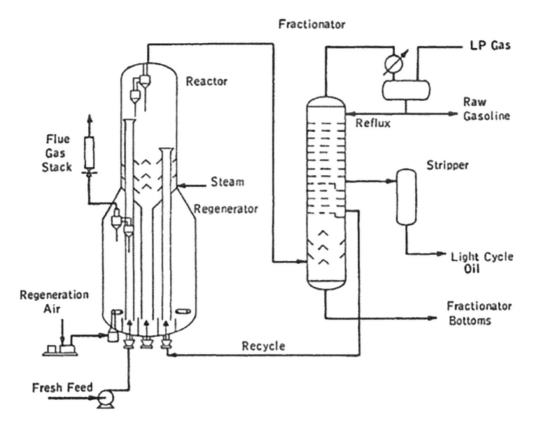

FIGURE 5.9 The orthoflow catalytic cracking unit. (Speight, J.G. 2017. *Handbook of Petroleum Refining.* CRC Press, Taylor & Francis Group, Boca Raton, Florida. Figure 9.13, page 357.)

features a short-contact-time riser, state-of the art riser termination device for quick separation of catalyst and vapor, with high hydrocarbon containment technology) and a portion of the stripped (carbonized) catalyst from the reactor is blended with the hot-regenerated catalyst in a proprietary mixing chamber for delivery to the riser. Additionally, the recycling of cooler partially spent catalyst back to the base of the riser lowers the reactor inlet temperature which results in a reduction of undesirable thermally produced products, including dry gas. The ability to vary the carbonized/ regenerated catalyst ratio provides considerable flexibility to handle changes in feedstock quality and enables a real-time switch between gasoline, olefin derivatives, or distillate operating modes. For viscous feedstocks, a two-stage regenerator is used – in the first stage, the bulk of the carbon is burned from the catalyst, forming a mixture of carbon monoxide and carbon dioxide. In the second stage, where the remaining coke is burned from the catalyst resulting in only low levels of carbon on the regenerated catalyst. A catalyst cooler is located between the stages.

The shell two-stage fluid-bed catalytic cracking process (Figure 5.10) was devised to permit greater flexibility in shifting product distribution when dictated by demand.

Thus, feedstock is first contacted with cracking catalyst in a riser reactor, that is, a pipe in which fluidized catalyst and vaporized oil flow concurrently upward, and the total contact time in this first stage is of the order of seconds. High temperatures 470°C–565°C (875°F–1,050°F) are employed to reduce undesirable coke deposits on catalyst without destruction of naphtha by secondary cracking. Other operating conditions in the first stage are a pressure of 16 psi and a catalyst-oil ratio of 3:1–50:1, and volume conversion ranges between 20% and 70% have been recorded.

All or part of the unconverted or partially converted gas oil product from the first stage is then cracked further in the second-stage fluid-bed reactor. Operating conditions are 480°C–540°C (900°F–1,000°F) and 16 psi with a catalyst-oil ratio of 2:1–12:1. Conversion in the second stage varies between 15% and 70%, with an overall conversion range of 50%–80%.

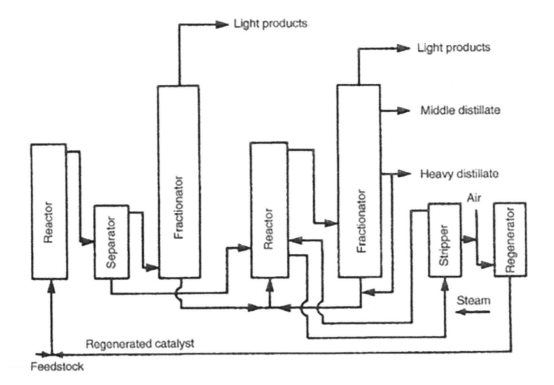

FIGURE 5.10   A two-stage FCC process. (Speight, J.G. 2017. *Handbook of Petroleum Refining*. CRC Press, Taylor & Francis Group, Boca Raton, Florida. Figure 9.15, page 359.)

The residuum cracking unit M.W. Kellogg Company/Phillips Petroleum Company offers conversions of viscous feedstocks up to 85% w/w. The unit is similar to the orthoflow C unit (Figure 5.9), but there are some differences that enhance performance on residua. The catalyst flows from the regenerator through a plug valve that controls the flow to hold the reactor temperature. Steam is injected upstream of the feed point to accelerate the catalyst and disperse it so as to avoid high rates of coke formation at the feed point. The feedstock, atomized with steam, is then injected into this stream through a multiple nozzle arrangement. The flow rates are adjusted to control the contact time in the riser since the effects of metals poisoning on yields are claimed to be largely a function of the time that the catalyst and oil are in contact. Passing the mix through a rough cut cyclone stops the reaction.

The Gulf residuum process consists of cracking a residuum that has been previously hydrotreated to low sulfur and metals levels. In this case, high conversions are obtained, but coke yield and hydrogen yield are kept at conventional levels by keeping metals on catalyst low.

Other processes include the deep catalytic cracking (DCC) process that is a fluidized catalytic process for selectively cracking a variety of feedstocks to low molecular weight olefin derivatives (Chapin and Letzsch, 1994). A traditional reactor/regenerator unit design is employed with a catalyst having physical properties much like those of a catalyst for a FCCU. The unit may be operated in two operational modes: maximum propylene (Type I) or maximum iso-olefin derivatives (Type II), and each mode utilizes a unique catalyst as well as reaction conditions. The Type I unit uses both riser and bed cracking at relatively severe reactor conditions, while the Type II unit uses only riser cracking like a modern FCC unit at milder conditions. The overall flow scheme of the process is similar to that of a conventional FCC unit but changes in the areas of catalyst development, process variable selection and severity, and gas plant design enable the production of higher yields of olefin derivatives than the conventional foud catalytic cracking processes. The products are light olefin derivatives, high-octane naphtha, light cycle oil, dry gas, and coke. Propylene yields over 24% w/w are achievable with paraffin feedstocks.

There is also the flexicracking process which is designed for conversion of gas oils, residua, and deasphalted oils to distillates (Draemel, 1992).

5.3 COKE FORMATION AND ADDITIVES

As in the thermal cracking process, coke is also a product of the catalytic cracking process as a side-product of catalytic cracking due to hydrogen transfer reactions. In this case, the coke is deposited on the catalytic sites which causes deactivation of the catalyst by the direct coverage of active sites.

The current industrial methods for deterring coke formation is, to some extent, an adjustment of the process parameters. On the other hand, removing coke from deactivated catalysts typically employs oxidation or combustion to convert the coke to carbon dioxide and/or carbon monoxide and to avoid sintering of the active sites (metals) at high temperatures, the oxidation conditions have to be mild.

5.3.1 COKE FORMATION

The formation of coke deposits has been observed in virtually every unit in operation, and the deposits can be very thick with thicknesses up to 4 ft have been reported (McPherson, 1984). Coke has been observed to form where condensation of hydrocarbon vapors occurs. The reactor walls and plenum offer a colder surface where hydrocarbon derivatives can condense. Higher boiling constituents in the feedstock may be very close to their dew point, and they will readily condense and form coke nucleation sites on even slightly cooler surfaces.

Non-vaporized feedstock droplets readily collect to form coke precursors on any available surface since the high-boiling feedstock constituents do not vaporize at the mixing zone of the riser. Thus, it is not surprising that residuum processing makes this problem even worse. Low residence time cracking also contributes to coke deposits since there is less time for heat to transfer to feed droplets and vaporize them. This is an observation in line with the increase in coking when short contact time riser crackers (q.v.) were replacing the longer residence time fluid bed reactors.

Higher-boiling feedstocks that have high aromaticity result in higher yields of coke. Furthermore, polynuclear aromatic derivatives and aromatic derivatives containing heteroatoms (i.e., nitrogen, oxygen, and sulfur) are more facile coke makers than simpler aromatic derivatives (Hsu and Robinson, 2006; Speight, 2016). Feedstock quality alone is not a foolproof method of predicting where coking will occur. However, it is known that feedstock hydrotreaters rarely have coking problems. The hydrotreating step mitigates the effect of the coke formers, and coke formation is diminished.

The recognition that significant post-riser cracking occurs in commercial catalytic cracking units resulting in substantial production of dry gas and other low-valued products (Avidan and Krambeck, 1990). There are two mechanisms by which this post-riser cracking occurs, thermal and dilute phase catalytic cracking.

Thermal cracking results from extended residence times of hydrocarbon vapors in the reactor disengaging area and leads to high dry gas yields via non-selective free radical cracking mechanisms. On the other hand, dilute phase catalytic cracking results from extended contact between catalyst and hydrocarbon vapors downstream of the riser. While much of this undesirable cracking was eliminated in the transition from bed to riser cracking, there is still a substantial amount of non-selective cracking occurring in the dilute phase due to the significant catalyst holdup.

Many catalytic cracking units are equipped with advanced riser termination systems to minimize post-riser cracking (Long et al., 1993). However, due to the complexity and diversity of catalytic cracking units, there are many variations of these systems, and many such closed cyclones and many designs are specific to the unit configuration, but all serve the same fundamental purpose of reducing the undesirable post-riser reactions. Furthermore, there are many options for taking advantage of reduced post-riser cracking to improve yields. A combination of higher reactor temperature, higher cat/oil ratio, higher feed rate, and/or poorer quality feed is typically employed. Catalyst modification is also appropriate, and typical catalyst objectives such as low coke and dry gas selectivity are reduced in importance due to the process changes, while other features such as activity stability and bottoms cracking selectivity become more important for the new unit constraints.

Certain catalyst types seem to increase coke deposit formation. For example, these catalysts (some rare-earth zeolites) that tend to form aromatic derivatives from naphthene derivatives as a result of secondary hydrogen transfer reactions and the catalysts contribute to coke formation indirectly because the products that they produce have a greater tendency to be coke precursors. In addition, high zeolite content, low surface areas cracking catalysts are less efficient at cracking the viscous feedstocks than many amorphous catalysts because the non-zeolite catalysts contained a matrix which was better able to crack the more viscous feedstocks as well as conversion of the coke precursors. The active matrix of some modern catalysts serves the same function.

Once coke is formed, it is a matter of where it will appear. The most frequent appearance of the coke is in the reactor (or disengager), the transfer line, and the slurry which can circuit cause major problems in some units such as (1) increased pressure drop that is caused when a layer of coke reduces the flow through a pipe, or (2) plugging, when pieces of coke will break into smaller pieces – often referred to as spalling – and block the flow completely. Deposited coke is commonly observed in the reactor as a black deposit on the surface of the cyclone barrels, reactor dome, and walls. Coke is also often deposited on the cyclone barrels 180° away from the inlet. Coking within the cyclones can be potentially very troublesome since any coke spalls going down into the dipleg could restrict catalyst flow or jam the flapper valve. Either situation reduces cyclone efficiency or can increase catalyst losses from the reactor. Coke formation also occurs at nozzles which can increase the nozzle pressure drop. It is possible for steam or instrument nozzles to be plugged completely, a serious problem in the case of unit instrumentation.

If coking occurs in the transfer line between the reactor and main fractionator thereby causing a pressure drop, there will be a reduction in the throughput. Furthermore, any coke in the transfer line which spalls off can pass through the fractionator into the circulating slurry system where it is likely to plug up exchangers, resulting in lower slurry circulation rates and reduced heat removal. In units where circulation is limited by low slide valve differentials, coke laydown may then indirectly reduce catalyst circulation. The risk of a flow reversal is also increased. Also, in units that contain reactor grids, coking increases grid pressure drop, which can directly affect the catalyst circulation rate.

Shutdowns and startups can aggravate problems due to coking. The thermal cycling leads to differential expansion and contraction between the coke and the metal wall that will often cause the coke to spall in large pieces. Another hazard during shutdowns is the possibility of an internal fire when the unit is opened up to the atmosphere. Proper shutdown procedures which ensure that the internals have sufficiently cooled before air enters the reactor will eliminate this problem. In fact, the only defense against having coke plugging problems during start up is to thoroughly clean the unit during the turnaround and remove all the coke. If strainers on the line(s), they will have to be cleaned frequently.

The two basic principles to minimize coking are to avoid dead spots and prevent heat losses. An example of minimizing dead spots is using purge steam to sweep out stagnant areas in the disengager system. The steam prevents the collection of high-boiling condensable products in the cooler regions. Steam also provides a reduced partial pressure or steam distillation effect on the high-boiling constituents and causes enhanced vaporization at lower temperatures. Steam for purging should preferably be superheated since medium-pressure low-velocity steam in small pipes with high heat losses is likely to be very wet at the point of injection and will cause more problems. Cold spots are often caused by heat loss through the walls in which case increased thermal resistance might help reduce coking. The transfer line, being a common source of coke deposits, should be as heavily insulated as possible, provided that stress-related problems have been taken into consideration.

In some cases, changing catalyst type or the use of an additive (q.v.) can alleviate coking problems. The catalyst types which appear to result in the least coke formation (not delta coke or catalytic coke) contain low- or zero-earth zeolites with moderate matrix activities. Eliminating heavy recycle streams can lead to reduced coke formation. Since clarified oil is a desirable feedstock to make needle coke in a coker, then it must also be a potential coke maker in the disengager.

One of the trends in recent years has been to improve product yields by means of better feed atomization. The ultimate objective is to produce an oil droplet small enough so that a single particle of catalyst will have sufficient energy to vaporize it. This has the double benefit of improving cracking selectivity and reducing the number of liquid droplets which can collect to form coke nucleation sites.

5.3.2 ADDITIVES

In addition to the use of a variety of catalysts that promote the cracking reactions as well as product yields and product properties, a series of additives has been developed that catalyze or otherwise alter the primary catalyst's activity/selectivity or act as pollution control agents. Additives are most often prepared in microspherical form to be compatible with the primary catalysts and are available separately in compositions that (1) enhance naphtha octane and light olefin formation, (2) selectively crack heavy cycle oil, (3) passivate vanadium and nickel present in many heavy feedstocks, (4) oxidize coke to carbon dioxide, and (5) reduce sulfur dioxide emissions.

Both vanadium and nickel deposit on the cracking catalyst and are extremely deleterious when present in excess of 3,000 ppm on the catalyst. Formulation changes to the catalyst can improve tolerance to vanadium and nickel, but the use of additives that specifically passivate either metal is often preferred.

5.4 PROCESS PARAMETERS

There are primary variables available for the operation of catalytic cracking units for maximum unit conversion for a given feedstock quality can be divided into two groups: (1) process variables and (2) catalyst variables. In addition to the catalyst variables (*q.v.*), there are also process variables that include (1) pressure, (2) reaction time, and (3) reactor temperature. Higher conversion and coke yield are thermodynamically favored by higher *pressure*. However, pressure is usually varied over a very narrow range due to limited air blower horsepower. Conversion is not significantly affected by unit pressure since a substantial increase in pressure is required to significantly increase conversion.

However, these variables (presented below in alphabetical order rather than by importance or preference) are not always available for maximizing conversion since most FCC units operate at an optimum conversion level corresponding to a given feed rate, feed quality, set of processing objectives, and catalyst at one or more unit constraints (e.g., wet gas compressor capacity, fractionation capacity, air blower capacity, reactor temperature, regenerator temperature, catalyst circulation). Once the optimum conversion level is found, there are very few additional degrees of freedom for changing the operating variables.

5.4.1 CATALYST ACTIVITY

Catalyst activity as an independent variable is governed by the capability of the unit to control the carbon content of the spent catalyst and the quantity and quality of fresh catalyst that can be continuously added to the unit. The carbon content of the regenerated catalyst is generally maintained at the lowest practical level to obtain the selectivity benefits of low carbon on the catalyst. Thus catalyst addition is, in effect, the principal determinant of catalyst activity.

The deliberate withdrawal of catalyst over and above the inherent loss rate through regenerator stack losses and decant or clarified oil, if fly, and a corresponding increase in fresh catalyst addition rate is generally not practiced as a means of increasing the activity level of the circulating catalyst. If a higher activity is needed, the addition of a higher activity fresh catalyst to the minimum makeup rate to maintain inventory is usually the more economical route. The general effects of increasing activity are to permit a reduction in severity and thus reduce the extent of secondary cracking reactions. Higher activity typically results in more naphtha and less coke. In other cases, higher activity catalysts are employed to increase the feed rate at essentially constant conversion and constant coke production so that the coke burning or regenerator air compression capacities are fully utilized.

5.4.2 CATALYST/FEEDSTOCK RATIO

The dependent variable catalyst/oil ratio is established by the unit heat balance and coke make that in turn are influenced by almost every independent variable. Since catalyst-oil ratio changes are accompanied by one or more shifts in other variables, the effects of catalyst-oil ratio are generally associated with other effects. A basic relation, however, in all catalyst-oil ratio shifts is the effect on conversion and carbon yield. At constant space velocity and temperature, increasing catalyst-oil ratio increases conversion. In addition to increasing conversion, higher catalyst-oil ratios generally increase coke yield at a constant conversion. This increase in coke is related to the hydrocarbon derivatives entrapped in the pores of the catalyst and carried through the stripper to the regenerator. Thus, this portion of the catalyst-oil ratio effect is highly variable and depends not only on the catalyst-oil ratio change but also on the catalyst porosity and stripper conditions. The following changes, in addition to increased coke yield, accompany an increase in the catalyst-oil ratio in the range of 5–20 at constant conversion, reactor temperature, and catalyst activity: (1) decreased hydrogen yield, (2) decreased methane to butane(s) yields, and (3) little effect on the naphtha yield or octane number.

The catalyst-oil ratio (v/v) ranges from 5:1 to 30:1 for the different processes, although most processes are operated to 10:1. By comparison, the catalyst-oil volume ratio for moving-bed processes may be substantially lower than 10:1. In a traditional FCC unit, increasing the catalyst-to-oil ratio to increase conversion also increases the coke yield and catalyst circulation to the regenerator. A high catalyst-oil ratio is necessary to maintain high reaction temperature by transferring enough heat from regenerator to reactor in commercial units. It is possible, using a specific type of catalyst (containing low-acid density zeolite), to suppress hydrogen transfer and maximize olefin production (Soni et al., 2009; Fujiyama et al., 2010).

The UOP RxCat technology provides the ability to increase both conversion and selectivity by recycling a portion of the carbonized catalyst back to the base of the reactor riser. The carbonized catalyst circulated from the stripper back to the base of the riser is effectively at the same temperature as the reactor. Since the recycle catalyst adds no heat to the system, the recycle is heat-balance neutral. For the first time, the catalyst circulation up the riser can be varied independently from the catalyst circulation rate to the regenerator and is de-coupled from the unit heat balance (Wolschlag and Couch, 2010b; Wolschlag, 2011). When process conditions are changed so that an increase in the catalyst-oil ratio occurs, an increase in conversion is also typically observed (Hemler, 1997; Hemler and Smith, 2004). By increasing the catalyst-oil ratio, the effects of operating at high reaction temperature (thermal cracking) are minimized. High catalyst-oil ratio maintains heat balance, thereby achieving high reaction temperature. It also increases conversion and maximizes light olefin derivatives production (Maadhah et al., 2008).

5.4.3 FEEDSTOCK CONVERSION

All of the independent variables in catalytic cracking have a significant effect on conversion that is truly a dependent variable but can be shown as a function of API gravity as well as a variety of other functions (Maples, 2000). The detailed effects of changing conversion depend on which the conversion is changed, i.e., by temperature, space velocity, catalyst-oil ratio, and catalyst activity. Increasing conversion increases yields of naphtha and all light products up to a conversion level of 60%–80% by volume in most cases. At this high conversion level, secondary reactions become sufficient to cause a decrease in the yields of olefin derivatives and naphtha. However, the point at which this occurs is the feedstock, operating conditions, catalyst activity, and other parameters.

5.4.4 FEEDSTOCK PREHEATING

In a heat-balanced commercial operation, increasing the temperature of the feed to a cracking reactor reduces the heat that must be supplied by combustion of the coked catalyst in the regenerator. Feedstock preheating is usually supplied by heat exchange with hot product streams, a fired preheater, or both. When feed rate, recycle rate, and reactor temperature are held constant as feed preheat is increased, the following changes in operation result: (1) the catalyst-oil ratio (catalyst circulation rate) is decreased to hold the reactor temperature constant; (2) conversion and all conversion-related yields, including coke, decline due to the decrease in catalyst-oil ratio and severity; (3) the regenerator temperature will usually increase. Although the total heat released in the regenerator and the air required by the regenerator are reduced by the lower coke yield, the lower catalyst circulation usually overrides this effect and results in an increase in regenerator temperature; and (4) as a result of the lower catalyst circulation rate, residence time in the stripper and overall stripper efficiency is increased liquid recovery is increased and a corresponding decrease in coke usually results. Advantage is usually taken of these feed preheat effects, including the reduced air requirement, by increasing the total feed rate until coke production again requires all of the available air.

5.4.5 FEEDSTOCK PRESSURE

Catalytic cracker pressures are generally set slightly above atmospheric by balancing the yield and quality debits of high pressure plus increased regeneration air compression costs against improved era king and regeneration kinetics, the lower cost of smaller vessels, plus, in some cases, power recovery from the regenerator stack gases (Parkash, 2004; Gary et al., 2007; Dziabala et al., 2011; Speight, 2014; Hsu and Robinson, 2017; Speight, 2017).

Pressure levels in commercial units are generally in the range 15–35 psi (103 to gauge). Lowering the partial pressure of the reacting gases with steam will improve yields somewhat, but the major beneficial effect of feed injection steam is that it atomizes the feed to small droplets that will vaporize and react quickly. If feed is not atomized, it will soak into the catalyst and possibly crack to a higher coke make.

Both pressure and partial pressure of the feedstock, or steam/feedstock ratio, are generally established in the design of a commercial unit and thus are usually not available as independent variables over any significant range. However, in some catalytic cracking units, the injector steam can be varied over a narrow range to balance carbon make with regeneration carbon burn off.

5.4.6 FEEDSTOCK QUALITY

Generally, the ability of any single unit to accommodate wide variations in feedstock is an issue related to the flexibility of the process (Navarro et al., 2015). Initially, catalytic cracking units were designed to process gas oil feedstocks, but many units have been modified successfully, and new units designed, to handle more complex feedstocks and feedstock blends, which may also include one or more of the viscous feedstocks (Chapter 2).

Briefly, gas oil occurs in the refinery in as two fractions which are (1) atmospheric gas oil, often referred to as AGO, and (2) vacuum gas oil, often referred to as VGO (Parkash, 2003; Gary et al., 2007; Speight, 2014; Hsu and Robinson, 2017; Speight, 2017, 2021). Typically, the atmospheric gas oil has a boiling range higher than the kerosene boiling range and is on the order of 300°C (570°F) to 360°C (680°F), which is the temperature just before the distillation is terminated. On the other hand, the vacuum gas oil is the first fraction that is obtained in the vacuum distillation process and has a boiling range on the order of 360°C to (950°F), which is the temperature at which the vacuum distillation process is terminated to produce the vacuum residuum. In some refineries, the vacuum distillation process may be terminated at 565°C (1,050°F). The gas oil fractions vary in composition depending upon the source and character of the crude oil. Generally, the gas oil fractions may be used as a source of fuel oil or fuel oil blending stock or as a source of lubricating oil.

Vacuum gas oil (ibp: 315°C–345°C, 600°F–650°F; fbp: 510°C–565°C, 950°F–1,050°F), as produced by vacuum flashing or vacuum distillation, is the usual feedstock with final boiling point being limited by the carbon forming constituents (measured by the Conradson carbon residue) or metals content since both properties have adverse effects on cracking characteristics. The vacuum residua (565°C+, 1,050°F+) are occasionally included in cat cracker feed when the units (residuum catalytic cracking units) are capable of handling such materials. In such cases, if the residua are either relatively low in terms of carbon-forming constituents and metals (as, for example, a residuum from a waxy crude) so that the effects of these properties are relatively small. Many units also recycle slurry oil (455°C+, 850°F+) and a heavy cycle oil stream. Gas oils from thermal cracking or coking processes (Chapter 8), gas oils from hydrotreating processes (Chapter 10), and gas oils from deasphalting processes (Chapter 12) are often included in feedstocks for catalytic cracking units (Parkash, 2003; Gary et al., 2007; Speight, 2014; Hsu and Robinson, 2006; Speight, 2017).

The general feedstock quality effects can be indicated by characterization factor, K:

$$K=(MABP)1/3/\text{specific gravity @ } 60°F/60°F$$

In this equation, MABP is the mean average boiling point expressed in degrees Rankin (°R = °F + 460). However, the MABP is a single parameter that can only be an indicator of general trends, and even then the accuracy and meaningful nature of the data may be very questionable. In general, the coke yield increases as the characterization factor (K) decreases or as the feed becomes less paraffinic and the API gravity decreases and as the conversion increases (Maples, 2000). With straight-run gas oils, naphtha yield increases as the characterization factor decreases (i.e. the paraffin character of the oil decreases), but the opposite effect is obtained with cracked stocks or cycle oils.

Either molecular weight, or average boiling point, or feed boiling range is an important feedstock characteristic in determining cat cracking yields and product quality. In general, for straight-run fractions, thermal sensitivity (increased thermal decomposition or cracking) increases as molecular weight increases; coke and naphtha production (at constant processing conditions) also increases with the viscous feedstocks but the yield of coke may also increase).

However, the characterization factor and feed boiling range are generally insufficient to characterize a feedstock for any purpose other than approximate comparisons. A more detailed description of the feedstock is needed to reflect and predict the variations in feedstock composition and cracking behavior (Speight, 1999, 2000).

Irrespective of the source of the high-boiling feedstocks, a number of issues typically arise when these materials are processed in a FCC unit, although the magnitude of the problem can vary substantially:

Viscous feedstocks have high levels of contaminants (Table 5.6) that will affect the process and must be removed.

Examples are the carbon-forming constituents that yield high levels of (Conradson) carbon residue, and the overall coke production (as carbon on the catalyst) is high. Burning this coke requires additional regeneration air that might be a constraint that limits the capacity of the unit. Metals in the heavy feedstocks also deposit (almost quantitatively) on the catalyst where two significant effects are caused. First, the deposited metals can accelerate certain metal-catalyzed dehydrogenation reactions, thereby contributing to light-gas (hydrogen) production and to the formation of additional coke. A second, and more damaging, effect is the situation in which the deposition of the metals causes a decline in catalyst activity because of the limited access to the active catalytic sites. This latter effect is normally controlled by catalyst makeup practices (adding and withdrawing catalyst).

TABLE 5.6
Feedstock Contaminants That Affect Catalytic Cracking Processes[a]

Contaminants	Effect on Catalyst	Resolution
Sulfur	Catalyst fouling Deactivation of active sites	Hydrodesulfurization
Nitrogen	Adsorption of basic nitrogen Destruction of active sites	Hydrodemetallization
Metals	Fouling of active sites Fouling of pores	Demetallization
Particulate matter	Deactivation of active sites Pore plugging	Filter/pretreatment
Coke precursors	Coke deposition Catalyst fouling Deactivation of active sites Pore plugging	Remove asphaltene fraction Remove resin fraction

Source: Speight, J.G. 2017. *Handbook of Petroleum Refining.* CRC Press, Taylor & Francis Group, Boca Raton, Florida. Table 9.9, page 367.

[a] Also applicable to hydrocracking processes.

The amount of sulfur and nitrogen in the products, waste streams, and flue gas generally increases when high-boiling feedstocks are processed because these feed components typically have higher sulfur and nitrogen contents than gas oil. However, in the case of nitrogen, the issue is not only one of higher nitrogen levels in the products but also (because of the feedstock nitrogen is basic in character) catalyst poisoning that reduces the useful activity of the catalyst.

Heat-balance control may be the most immediate and troublesome aspect of processing high-boiling feedstocks. As the contaminant carbon increases, the first response is usually to increase in regenerator temperature. Adjustments in operating parameters can be made to assist in this control, but eventually, a point will be reached for heavier feedstocks when the regenerator temperature is too high for good catalytic performance. At this point, some external heat removal from the regenerator is required and would necessitate a mechanical modification like a catalyst cooler.

For the last two decades, demetallized oil (produced by the extraction of a vacuum-tower bottoms stream using a light paraffinic solvent) has been included as a component of the feedstock in FCC units. Modern solvent-extraction processes, such as the Demex process, provide a higher demetallized oil yield than is possible in the propane-deasphalting process that has been used to prepare fluid catalytic cracker feedstock, as well as demetallized feedstocks for other processes. Consequently, the demetallized oil is more heavily contaminated. In general, demetallized oils are still good cracking stocks, but most feedstocks can be further improved by hydrotreating to reduce contaminant levels and to increase their hydrogen content, thereby becoming a more presentable and process-friendly feedstock.

In many cases, atmospheric residua have been added as a blended component to feedstocks for existing FCC units as a means of converting the highest boiling constituents of crude oil. In fact, in some cases the atmospheric residuum has ranged from a relatively low proportion of the total feed all the way to a situation in which it represents the entire feed to the unit. To improve the handling of these high-boiling feedstocks, several units have been revamped to upgrade them from their original gas-oil designs whereas other units have taken a stepwise approach to residuum processing whereby modifications to the operating conditions and processing techniques are made as more experience is gained in the processing of residua.

5.4.7 PROCESS DESIGN

Process design improvement will continue to focus on (1) modification of existing units, (2) commercialized technology changes, and (3) new directions in processing technology to allow processing of a variety of feedstocks (Chen, 2006; Freel and Graham, 2011).

5.4.7.1 Commercial Technology Changes

Dramatic yield improvements have been demonstrated commercially using the latest FCC process technology advances. Riser termination designs, in particular, have received much attention, for new units as well as for revamped units. Significant post-riser cracking occurs in commercial FCC units resulting in substantial production of dry gas and other lower-value products (Avidan and Krambeck, 1990).

Due to the complexity and diversity of existing FCC units as well as new unit design differences, there are many variations of these systems such as (1) closed cyclones, (2) close-coupled cyclones, (3) direct connected cyclones, (4) coupled cyclones, (5) high containment systems, and (6) short contact time systems. There are differences in the specific designs, and some may be more appropriate for specific unit configurations than others, but all serve the same fundamental purpose of reducing the undesirable post-riser reactions.

Proper catalyst selection is essential to realizing the maximum potential benefit from these hardware improvements. Both fresh catalyst properties and catalyst management policy are important variables. These include the use of (1) high activity catalyst for maximum conversion or throughput, (2) high matrix catalyst for reducing slurry yield, and (3) a metals tolerant catalyst for processing resid feed. In all cases, improved yields are achieved when incorporating catalyst change effects, which may not have been possible without the advanced hardware, due to unit constraints.

The downer is a gas-solid co-current down-flow reactor which has the potential to overcome the drawback of a conventional up-flow reactor (or a riser) caused by back-mixing of catalyst. In the downer, gas and solid catalysts move downward together with the assist of the gravity; this can avoid the back-mixing of catalyst in the reactor.

The operation of the downer is affected by various key parameters including (1) recycled catalyst flow rate, (2) superficial gas velocity, (3) spent catalyst flow rate, and (4) carbon content on the spent catalyst. The parameters that affect the temperature of the downer regenerator should be carefully selected as they have the most significant effect to the regeneration process. High regeneration temperature could deactivate the catalyst permanently, but low-temperature operation lowers the regeneration performance (Chuachuensuk et al., 2010).

A down-flow reactor system has been adopted for the high-severity fluid catalytic cracking (HS-FCC) process. The downer permits higher catalyst/oil ratios because the lifting of catalyst by vaporized feed is not required. As with most reactor designs involving competing reactions and secondary product degradation, there is a concern over catalyst-feed contacting, back-mixing, and control of the reaction time and temperature. The down-flow reactor would ensure plug flow without back-mixing (Maadhah et al., 2008).

The development of a highly active zeolite catalyst has led to the reaction time has being decreased to a few seconds in the riser reactor. The short contact time (short residence time, <0.5 second) of feed and product hydrocarbon derivatives in the downer should be favorable for minimizing thermal cracking. Undesirable secondary reactions such as polymerization reactions and hydrogen-transfer reactions, which consume olefin derivatives, are suppressed. In order to attain the short residence time, the catalyst and the products have to be mixed and dispersed at the reactor inlet and separated immediately at the reactor outlet. For this purpose, a high-efficiency product separator has been developed capable of suppressing side reactions (oligomerization and hydrogenation of light olefin derivatives) and coke formation accelerated by condensation (Nishida and Fujiyama, 2000). The short contact time reactor affords (1) minimal back-mixing and erosion, (2) efficient catalyst/oil contacting, (3) reduced hydrogen transfer, and (4) high-yield selectivity.

The over-cracking of naphtha to gases is minimized by reducing the contact time between catalyst and hydrocarbon products. Addition of ZSM-5 additive enhances the octane number in FCC naphtha by over-cracking of naphtha fraction (Section 5.2.1). On the other hand, the octane number enhancement is achieved by over-cracking of the middle boiling point fraction with low octane number (Buchanan et al., 1996). Short-contact-time riser cracking is inherently more flexible than a typical FCC unit because the product slate can be easily adjusted to maximize propylene, maximize naphtha, or produce combinations such as propylene plus ethylene or propylene plus naphtha (Jakkula et al., 1997). This process flexibility is a key variable in maximizing profitability in a given market scenario.

The UOP millisecond catalytic cracking process (MSCC process) involves injection of the feedstock perpendicular to a down-flowing steam of catalyst (Schnaith et al., 1998; Harding et al., 2001). The basic MSCC reactor configuration consists of an injection zone, a central dilute phase disengaging zone, the lower dense phase collection zone, and an upper inertial separation zone. The short contact time combined with the low-volume reaction zone reduces secondary cracking reactions and produces more naphtha and less coke compared with conventional FCC. Another benefit is that the low-coke yield allows heavier feedstocks (conventionally high coke-make feedstocks) to be processed.

Improvements in the design of riser separation or termination systems focus on the rapid disengagement of catalyst from the cracked products in a highly contained system. Product vapors are quickly directed to the fractionation system for thermal quench and recovery (Ross et al., 2005). Using a two-riser system, a heavy feedstock can be treated by use of a two riser system (two-stage riser fluid catalytic cracking, TSRFCC). The spent catalyst from the other of the two risers is fed to the inlet of the first riser to produce relatively mild cracking conditions. Improved total naphtha plus distillate yields are achieved, and the novel two riser system facilitates heat balancing of the system (Krambeck and Pereira, 1986; Shan et al., 2003).

5.4.7.2 Modifications for Existing Units

While the feedstock injection system is an advancement in FCC reactor design, a higher regenerator temperature is used to achieve more complete catalyst regeneration. The typical riser top temperature is on the order of 510°C–565°C (950°F–1,050°F), but typical regenerated catalyst temperature is much higher – on the order of 675°C–760°C (1,250–1,400°F). Feed injection reduces thermal cracking reactions by cooling off the lower riser quickly through rapid mixing and vaporization of the feed.

As the FCC feedstock moves to heavy crude oil, feed vaporization is more difficult. However, the newest generation of side-entry FCC feed nozzles generates more uniform feed distribution (and more rapid mixing) as a result of better control of homogeneity of two-phase flow and atomization at the nozzle exit using two-phase choke flow. Some older FCC units still retain the original feed injection system located at the bottom of the riser (bottom-entry nozzles). A new generation of feed injection technology uses a similar side-entry atomization mechanism. For catalyst circulation, the bottom-entry nozzles have the advantage of reducing pressure drop through the riser. This system also enables longer riser residence time if riser height is limited.

The newest generation feed nozzles optimize the temperature profile in the riser and substantially reduces dry gas yield, thereby increasing naphtha yield. These results are in line with the expectation that better feed injection design reduces thermal cracking reactions, which are the primary source for dry gas. Process/hardware technologies to improve light cycle oil yield include proper feed injection systems and risers/reaction zone designs as claimed by Petrobras, Shell, and Sinopec in their latest commercial processes.

Improved riser reaction termination technology sharpens the termination of reactions by the combination of the unique design of primary stripper cyclones and close coupled secondary cyclones as well as designs to reduce coke formation (Hedrick and Palmas, 2011). Furthermore, as a result of the development of highly active zeolite catalysts, the reaction time has been shortened significantly to a few seconds in the modern riser reactor. Since catalytic cracking reactions can only occur after the vaporization of the liquid hydrocarbon feedstock, mixing and feed vaporization must take place in the riser as quickly as possible; otherwise, thermal cracking reactions will dominate (Chen, 2004). An efficient product separator suppresses side reactions (oligomerization and hydrogenation of light olefin derivatives) and coke formation accelerated by condensation.

The high yields of products generated in the highly selective reactor riser environment must be preserved in the rest of the reaction system. Improvements in the design of riser separation or termination systems focus on the rapid disengagement of catalyst from the cracked products in a highly contained system. Product vapors are quickly directed to the fractionation system for thermal quench and recovery (Ross et al., 2005; Fujiyama et al., 2011). The design improves both the separation system to reduce dry gas and the stripping system to reduce coke. In addition, the pressure drop is extremely low in order to limit dip-leg immersion requirements to seal the positive pressure separator and so that the capacity of the unit will not be limited. The dip-leg size and flux are optimized to minimize gas entrainment with the catalyst and even allow for stripping within the dip-legs.

Prospective techniques for FCC control technology fall into four general categories: (1) hydrodesulfurization, (2) catalyst additives, (3) scrubbing, and (4) chemical reaction such as selective noncatalytic reduction (SNCR) and selective catalytic reduction (SCR) (Bouziden et al., 2002; Couch et al., 2004). Also, hydrotreating of the feedstock will decrease feedstock sulfur and nitrogen, thereby decreasing sulfur emissions form the unit as well as nitrogen in the coke on catalyst, and consequently nitrogen oxide (NOx) emissions from the regenerator.

5.4.7.3   New Directions

Hydrotreating of the feedstock to the catalytic cracking unit can increase conversion by 8%–12% v/v, and with most feeds, it will be possible to reduce the sulfur content of the gasoline/naphtha product to levels low enough to meet the future low sulfur gasoline pool specifications. With the increasing focus to reduce sulfur content in fuels, the role of *desulfurization* in the refinery becomes more and more important. Currently, the process of choice is the hydrotreater, in which hydrogen is added to the fuel to remove the sulfur from the fuel. Some hydrogen may be lost to reduce the octane number of the fuel, which is undesirable. Because of the increased attention for fuel desulfurization, various new process-concepts are being developed with various claims of efficiency and effectiveness.

The major developments in desulfurization three main routes are advanced hydrotreating (new catalysts, catalytic distillation) and reactive adsorption (such as the use of metal oxides that will chemically abstract sulfur) (Babich and Moulijn, 2003). Such concepts, already on stream in some refineries, and the number of units performing chemical desulfurization will increase in the future. There are also options to by-pass the atmospheric and vacuum distillation units by feeding crude oil directly into a thermal cracking process, which would provide sufficient flexibility to supply a varying need of products with a net energy savings.

Light cycle oil from catalytic cracking units will be increased by modifying feedstock composition, introducing improved catalysts and additives, and modifying operating conditions (such as recycle ratio, temperature, catalyst/oil ratio). The addition of an active alumina matrix is a common feature to help refiners increase light cycle oil production when cracking heavy feeds. A comprehensive survey of patent literature in the report found the use of an inorganic additive to occur more than once in catalyst formulations, and metal-doped anionic clays and amorphous silico-alumino-phosphates (SAPO) are cited, among other inventions. There is some overlap in catalysts tailored for the production of light cycle oil and catalysts tailored for residuum feedstocks.

Refineries equipped to process viscous feedstocks have, so far, reported high-refining margins because they can take advantage of less-expensive heavy oils. Resid FCC is an important component in the upgrading of such crudes, with unit profitability depending upon the extent to which heavy hydrocarbon derivatives in the feed are cracked into valuable products. The product slate, in turn, depends upon the feed characteristics, the catalyst, the hardware, and the operating conditions. Exemplifying a trend toward heavier feedstocks, the majority of the optional fluid bed catalytic cracking units are expected to process heavy vacuum gas oil and/or or residuum feedstocks.

Complete process technologies for residuum catalytic cracking units are the most comprehensive approach to improve resid processing operations, but present, as might be anticipated, the most expensive. Product recycle and multiple reaction sections will be the prevalent technology trends. Improving feed injectors, riser termination and catalyst separation devices, strippers, and regenerator components are revamp options for existing units. As feedstocks get heavier, the trend toward a higher stripper residence time and, consequently, increased mass transfer between entrained hydrocarbon derivatives and steam will continue. Moreover, the role of the regenerator continues to evolve because of reduction requirements in carbon dioxide emissions. The need for the development of flue gas treating systems points to the need for continued innovation in that area.

Process/hardware technologies to improve light cycle oil yield from the FCC unit will include improved feed injection systems and riser pipe and reaction zone designs.

The implementation of bio-feedstock processing techniques in crude oil refineries can result in a competitive advantage for both refiners and society at large. First, the processes provide refineries with alternative feeds that are renewable and could be lower in cost than crude oil. Second, they can reduce the costs of producing fuels and chemicals from bio-feedstocks by utilizing the existing production and distribution systems for crude oil-based products and avoiding the establishment of parallel systems. Finally, the use of bio-feedstocks provide a production base for fuels and chemicals that is less threatened by changes in government policies toward fossil fuel feeds and renewable energies.

Bio-feedstocks that are able to be processed in the FCC unit can be categorized as biomass-derived oils (both lignocellulosic materials and free carbohydrates) or triglycerides and their free fatty acids. The operating conditions and catalysts used for each type of feed to achieve a desired product slate vary, and each feed comes with inherent advantages and disadvantages. Most of the research work completed to date has been performed on relatively pure bio-feedstocks as opposed to blends of bio-based materials with traditional catalytic cracker feedstocks. Practical implementation in a refinery will, more than likely, be accompanied by blending with vacuum gas oil or residuum. In fact, biomass constituents can be blended with the feedstock and fed to a FCC unit (Speight, 2011a, b). The acidity of the oil (caused by the presence of oxygen functions) acidity can be reduced by means of a moderate thermal treatment at temperatures in the range of 320°C–420°C (610°F–790°F).

Oils derived from bio-feedstocks oils are generally best upgraded by HZSM-5 or ZSM-5, as these zeolite catalysts promote high yields of liquid products and propylene. Unfortunately, these feeds tend to coke easily, and high acid numbers and undesirable byproducts such as water and carbon are, and will continue to be, additional challenges. Waxy feeds obtained from biomass gasification followed by Fischer-Tropsch synthesis to hydrocarbon liquids and waxes (biomass-to-liquids) are especially suited for increasing light cycle oil production in the catalytic cracking unit, due to the high paraffinic character, low-sulfur content, and low-aromatic derivatives yield of the feed. A major disadvantage for biomass-to-liquid products is the intrinsically low-coke yield that can disrupt the heat balance of the catalytic cracking unit. In terms of processability, triglycerides are the best-suited bio-feedstock for the catalytic cracker. These materials generally produce high-quality diesel, high-octane naphtha, and are low in sulfur.

Finally, the use of microwaves to generate heat in a uniform and controlled fashion is well known, and microwave technology with catalysts is another technology which has been applied recently in upgrading of heavy oil (Mutyala et al., 2010; Lam et al., 2012). For example, catalytic hydroconversion of residue from coal liquefaction by using microwave irradiation with a Ni catalyst (Wang et al., 2008) and microwave-assisted desulfurization of heavy and sour crude oil using iron powder as catalyst (Leadbeater and Khan, 2008). Also upgrading of Athabasca bitumen with microwave-assisted catalytic pyrolysis was carried out in one study. Silicon carbide as used for pyrolysis and nickel and molybdenum nanoparticles as catalysts used to enhance the pyrolysis performance. The results of the work suggest that microwave heating with nanoparticles catalyst can be a useful tool for the upgrading of heavy crudes such as bitumen because of rapid heating and energy efficient (Jeon et al., 2012).

Continued research and technology developments for FCC will focus on (1) widening the boiling range of the feed that can be processed in the unit, (2) maximizing diesel yield and light olefin yield, and (3) providing operational flexibility to allow the unit to take advantage of favorable market opportunities.

A new hydrocarbon conversion process, HS-FCC has been developed to maximize propylene production in oil refineries (Fujiyama et al., 2005; Maadhah et al., 2008). The yield of propylene was maximized using a combination of three factors: (1) catalyst properties, (2) reaction conditions, and (3) reactor design. In addition, a high reaction temperature accelerates catalytic cracking rather than hydrogen transfer, and as a result, the olefin-paraffin ratio of the product is higher at high reaction temperatures.

The special features of this process include (1) rapid feed vaporization, (2) down-flow reactor, (3) high severity, (4) short contact time, and (5) high catalyst/oil ratio. Since the FCC process involves successive reactions, the desired products such as olefin derivatives and naphtha are considered intermediate products. A suppression of back-mixing by using the downer reactor is the key to achieving maximum yield of these intermediates. Compared to conventional FCC processes, the HS-FCC has modifications in the reactor/regenerator and stripper sections (Fujiyama, 1999; Ino and Ikeda, 1999; Fujiyama et al., 2000; Nishida and Fujiyama, 2000; Ino et al., 2003).

Recent enhancements made to resid FCC units permit feeding significant amounts of heavy crudes, while simultaneously improving yields and service factors. Traditional technology has been modified in key areas including: (1) catalyst design to accommodate higher metals feed and to minimize the amount of coke formed on the catalyst, (2) feed injection, (3) riser pipe design and catalyst/oil product separation to minimize over-cracking, (4) regenerator design improvements to handle high coke output and avoid damage to catalyst structure, and (5) overall reactor/regenerator design concepts.

These developments have allowed FCC units to substantially increase residue processing capabilities and substantial portions of refinery residua are processes (as blends with gas oils) in fluidized-bed units, thereby increasing naphtha and diesel production. This will not only continue but also increase in the future.

5.4.8 REACTOR TEMPERATURE

Increased *reactor temperature* increases feedstock conversion, primarily through a higher rate of reaction for the endothermic cracking reaction and also through increased catalyst/oil ratio. A 10°F (5.6°C) increase in reactor temperature can increase conversion by 1%–2% absolute but, again, this is feedstock dependent. Higher reactor temperature also increases the amount of olefin derivatives in gasoline and in the gases. This is due to the higher rate of primary cracking reactions relative to secondary hydrogen transfer reactions which saturate olefin derivatives in the gasoline boiling range and lowers gasoline octane.

The principal effects of increasing reactor temperature at constant conversion are to decrease naphtha yield and coke yield and to increase yields of methane (CH_4), ethane (C_2H_6), propane ($CH_3CH_2CH_3$), and total butane (C_4H_{10}) yield; yields of the pentanes (C_5H_{12}) and higher molecular weight paraffin decrease while olefin yields are increased.

The effect of reactor temperature on a commercial unit is, of course, considerably more complicated as variables other than temperature must be changed to maintain heat balance. For example, in order to increase the temperature of a reactor at constant fresh feed rate, the interrelated changes of recycle rate, space velocity, and feedstock preheat are required to maintain heat balance on the unit by increasing circulation rate and coke yield. The combined effects of higher reactor temperature and higher conversion resulted in the following additional changes: (1) increased yields of butanes and propane, (2) naphtha yield is increased, and (3) the yield of light catalytic cycle oil is decreased. Thus the effects of an increase in reactor temperature on an operating unit reflect not only the effects of temperature per se but also the effects of several concomitant changes such as increased conversion and increased catalyst-oil ratio.

5.4.9 RECYCLE RATE

With most feedstocks and catalyst, naphtha yield increases with increasing conversion up to a point, passes through a maximum, and then decreases. This phenomenon (over cracking) is due to the increased thermal stability (refractory character) of the unconverted feed as conversion increases and the destruction of naphtha through secondary reactions, primarily cracking of olefin derivatives. The onset of secondary reactions and the subsequent leveling off or decrease in naphtha yield can be avoided by recycling a portion of the reactor product, usually a fractionator product with a boiling points on the order of 345°C–455°C (650°F–850°F). Other tests have shown the following effects of increasing recycle rate when space velocity was simultaneously adjusted to maintain conversion constant: (1) the naphtha yield increased significantly; (2) the coke yield decreases appreciably; (3) there is a decrease in the yield of dry gas components, propylene, and propane; (4) the yield of butane decreased while the yield of butylene increased; and (5) the yields of light catalytic cycle oil and clarified oil increased but heavy catalytic cycle oil yield decreased.

With the introduction of high activity zeolite catalysts, it was found that in once-through cracking operations with no recycle the maximum in naphtha yield was located at much higher conversions. In effect, the higher activity catalysts were allowing higher conversions to be obtained at severity levels that significantly reduced the extent of secondary reactions (over cracking). Thus, on many units employing zeolite catalysts, recycle has been eliminated or reduced to less than 15% of the fresh feed rate.

5.4.10 REGENERATOR AIR RATE

The amount of air required for regeneration depends primarily on coke production. Regenerators have been operated with only a slight excess of air leaving the dense phase. With less air, carbon content of the spent catalyst increases and a reduction in coke yield is required of air, burning of carbon monoxide (CO) to carbon dioxide (CO_2) above the dense bed will occur. This after burning must be controlled as extremely high temper the absence of the heat sink provided by the catalyst in the dense phase.

In terms of catalyst regeneration, higher stability catalysts are available and regenerator temperatures can be increased by 38°C–65°C (100°F–150°F) up to the 720°C–745°C (1,325°F–1,370°F) range without significant thermal damage to the catalyst. At these higher temperatures, the oxidation of carbon monoxide to carbon dioxide is greatly accelerated, and the regenerator can be designed to absorb the heat of combustion in the catalyst under controlled conditions. The carbon burning rate is improved at the higher temperatures, and there usually are selectivity benefits associated with the lower carbon on regenerated catalyst; this high temperature technique usually results in carbon on regenerated catalyst levels of 0.05% by weight or less. In addition, the use of catalysts containing promoters for the oxidation of carbon monoxide to carbon dioxide produce a major effect of high temperature regeneration, i.e., low carbon monoxide content regenerator stack gases and the resultant regenerator conditions may result in a lower carbon content regenerated catalyst.

5.4.11 REGENERATOR TEMPERATURE

Catalyst circulation, coke yield, and feedstock preheating are the principal determinants of regenerator temperature that is generally allowed to respond as a dependent variable within limits. Mechanical or structural specifications in the regenerator section generally limit regenerator temperature to a maximum value specific to each unit. However, in some cases, the maximum temperature may be set by catalyst stability. In either case, if regenerator temperature is too high, it can be reduced by decreasing feedstock preheat; catalyst circulation is then increased to hold a constant reactor temperature, and his increased catalyst circulation will carry more heat from the regenerator and lower the regenerator temperature. The sequence of events is actually more complicated as the shift in catalyst/oil ratio, and to a lesser extent, the shift in carbon content of the regenerated catalyst will change coke make and the heat release in the regenerator.

5.4.12 RESIDENCE TIME

An increase in *reaction time* available for cracking also increases conversion. Fresh feed rate, riser steam rate, recycle rate, and pressure are the primary operating variables which affect reaction time for a given unit configuration. Conversion varies inversely with these stream rates due to limited reactor size available for cracking. Conversion has been increased by a decrease in rate in injection of fresh feedstock. Under these circumstances, over-cracking of gasoline to liquefied petroleum gas and to dry gas may occur due to the increase in reactor residence time. One approach to offset any potential gasoline over-cracking is to add additional riser steam to lower hydrocarbon partial pressure for more selective cracking. Alternatively, an operator may choose to lower reactor pressure or increase the recycle rate to decrease residence time. Gasoline over-cracking may be controlled by reducing the availability of catalytic cracking sites by lowering cat/oil ratio.

5.4.13 SPACE VELOCITY

The role of space velocity as an independent variable arises from its relation to catalyst contact time or catalyst residence time. Thus

$$\Theta = 60/(WHSV \times CO)$$

In the equation, Q is catalyst residence time in minutes, WHSV is the weight hourly space velocity on a total weight basis, and C/O is the catalyst/oil weight ratio.

The catalyst-oil ratio is a dependent variable so that catalyst time becomes directly related to the weight hourly space velocity. When catalyst contact time is low, secondary reactions are minimized, thus naphtha yield is improved and light as and coke yields are decreased. In dense bed units the holdup of catalyst in the reactor can be controlled within limits, usually by a slide valve in the spent catalyst standpipe, and feed rate can also be varied within limits. Thus, there is usually some freedom to increase space velocity and reduce catalyst residence time. In a riser type reactor, holdup and feed rate are not independent, and space velocity is not a meaningful term. Nevertheless, with both dense-be and riser-type reactors, contact times are usually minimized to improve selectivity. An important step in this direction was the introduction of high activity zeolite catalyst. These catalysts require short contact times for optimum performance and have generally moved cracking operations in the direction of minimum holdup in dense-bed reactors or replacement of dense-bed reactors with short contact time riser reactors.

Strict comparisons of short contact time riser cracking versus the longer contact time dense-bed mode of operations are generally not available due to differences in cat activity, carbon content of the regenerated catalyst, or factors other than contact time but inherent in the two modes of operation. However, in general, improvements in catalyst activity have resulted in the need for less catalyst in the reaction zone.

Many units are designed with only riser cracking, i.e., no dense-bed catalyst reactor cracking occurs and all cracking is done in the catalyst/oil transfer lines leading into the reactor cyclone vessel. However, in some of these instances the reactor temperature must be increased to the 550°C– 565°C (1,020°F–1,050°F) range in order to increase the intensity of cracking conditions to achieve the desired conversion level. This is because not enough catalyst can be held in the riser zone, since the length of the riser is determined by the configuration and elevation of the major vessels in the unit. Alternatively, super-active catalysts can be used to achieve the desired conversion in the riser.

Significant selectivity disadvantages have not been shown if a dispersed catalyst phase or even a very small dense bed is provided downstream of the transfer line riser cracking zone. In this case, the cracking reaction can be run at a lower temperature, say 510°C (950°F), which will reduce light gas make and increase naphtha yield when compared to the higher temperature (550°C–565°C, 1,020°F–1,050°F) operation.

5.5 CATALYSTS

Commercial synthetic catalysts are amorphous and contain more silica than is called for by the preceding formulae; they are generally composed of 10%–15% alumina (Al_2O_3) and 85%–90% silica (SiO_2). The natural materials, montmorillonite, a non-swelling bentonite, and halloysite, are hydrosilicates of aluminum, with a well-defined crystal structure and approximate composition of Al_2O_3 $4Si_2O.xH_2O$. Some of the newer catalysts contain up to 25% alumina and are reputed to have a longer active life.

However, cracking occurs over many types of catalytic materials, and cracking catalysts can differ markedly in both activity to promote the cracking reaction and in the quality of the products obtained from cracking the feedstocks (Gates et al., 1979; Wojciechowski and Corma, 1986; Stiles and Koch, 1995; Cybulski and Moulijn, 1998; Occelli and O'Connor, 1998; Domokos et al., 2010). Activity can be related directly to the total number of active (acid) sites per unit weight of catalyst and also to the acidic strength of these sites. Differences in activity and acidity regulate the extent of various secondary reactions occurring and thus the product quality differences. The acidic sites are considered to be Lewis- or Brønsted-type acid sites, but there is much controversy as to which type of site predominates.

Briefly, and by way of a historical introduction, the first acid catalyst that was tested for cracking of viscous feedstocks was aluminum chloride ($AlCl_3$), but the problems with catalyst manipulation, corrosion, and waste treatment or disposal put the use of this catalyst at a serious disadvantage. In the 1940s, silica-alumina catalysts were created and showed great improvement over the use of catalytic clay minerals natural clay catalysts. After natural alumino-silicate minerals were found to be adequate to the task, synthetic alumino-silicates were prepared and showed enhanced cracking properties. Both types (natural and synthetic) of alumino-silicates were known for the presence of Lewis acid sites. The early synthetic amorphous alumino-silicate catalysts contained approximately 13% w/w alumina (Al_2O_3), which was boosted to 25% w/w alumina the mid-1950s. In the original fixed-bed process, activated bentonite was used (probably in the form of pellets). For the Thermofor catalytic cracking unit, the catalysts were of spherical shape (diameter: approximately 1–2 mm). In 1948, the first spray-dried catalyst was introduced, and the micro-spherical particles (50–100 μm) were produced with the similar particle-size distribution as the ground catalysts. However, the spherical particles showed both improved fluidization properties as well as significant reduction of attrition losses.

The most significant advance came in 1962 when zeolite catalysts (particularly Zeolite-Y at that time) were incorporated into the silica-alumina structures after which advances in catalysts have produced the greatest overall performance of FCC units over the last 50 years. The presence of the zeolites has resulted in the presence of strong Brønsted acid sites with very easily accessible Lewis acid sites also being present. These new types of catalysts possess the properties required of a successful catalyst: activity, stability, selectivity, correct pore size, resistance to fouling, and low cost. The first commercial zeolite catalysts were introduced in 1964, and zeolite catalysts remain in use in modern refineries. Furthermore, catalyst selection has been an important aspect of bio-oil processing that catalyst selection because the conversion of bio-oil and product distribution are highly dependent upon the type of catalyst used. The two most commonly used and effective zeolite catalysts in FCC processes are H–Y zeolites and HZSM-5 (Al-Sabawi et al., 2012).

From this point on, catalyst development has proceeded at a rapid rate and the development of active and stable catalysts (typically acid catalysts) has paralleled equipment design and development. The ultimate goal is the development of catalysts that are resistant to the obnoxious constituents of heavy feedstocks. Generally, the philosophy of the catalyst preparation for FCC units is to have weak acid centers in macro-porous part of catalyst particles to insure pre-cracking of heavy feedstock constituents to lower molecular weight producers that enter to the mesopores with stronger acidity. Cracking in mesopores leads to even low-molecular weight products that can enter the zeolite micro-pores and crack over strongest zeolite acid centers into the desired products, typically naphtha constituents.

5.5.1 Catalyst Types

The first cracking catalysts were acid leached montmorillonite clays. The acid leach was to remove various metal impurities, principally iron, copper, and nickel that could exert adverse effects on the cracking performance of the catalyst. The catalysts first used in fixed-bed and moving-bed reactor systems in the form of shaped pellets.

The desire to have catalysts that were uniform in composition and catalytic performance led to the development of synthetic catalysts. The first synthetic cracking catalyst consisting of 87% silica (SiO_2) and 13% alumina (Al_2O_3) was used in pellet form and used in fixed-bed units in 1940. Catalysts of this composition were ground and sized for use in FCC units. In 1944, catalysts in the form of beads approximately 2.5–5.0 mm diameter were introduced and comprised approximately 90% silica and 10% alumina and were extremely durable. One version of these catalysts contained a minor amount of chromia (Cr_2O_3) to act as an oxidation promoter.

Neither silica (SiO_2) nor alumina (Al_2O_3) alone is effective in promoting catalytic cracking reactions. In fact, they (and also activated carbon) promote hydrocarbon decompositions of the thermal type. Mixtures of anhydrous silica and alumina ($SiO_2.Al_2O_3$) or anhydrous silica with hydrated alumina ($2SiO_2.2Al_2O_3.6H_2O$) are also essentially non-effective. Also, a prepared from hydrous oxides followed by partial dehydration (calcining) has appreciable cracking activity and the small amount of water remaining is necessary for proper functioning.

The catalysts are porous and highly adsorptive, and their performance is affected markedly by the method of preparation. Two catalysts that are chemically identical but have pores of different size and distribution may have different activity, selectivity, temperature coefficient of reaction rate, and response to poisons. The intrinsic chemistry and catalytic action of a surface may be independent of pore size, but small pores appear to produce different effects because of the manner and time in which hydrocarbon vapors are transported into and out of the interstices.

In addition to synthetic catalysts comprising silica-alumina, other combinations of mixed oxides were found to be catalytically active and were developed during the 1940s. These systems included silica (SiO_2), magnesia (MgO), silica-zirconia (SiO_2-ZrO), silica-alumina-magnesia, silica-alumina-zirconia, and alumina-boria (Al_2O_3-B_2O_3). Of these, only silica-magnesia was used in commercial units but operating difficulties developed with the regeneration of the catalyst which at the time demanded a switch to another catalyst. Further improvements in silica-magnesia catalysts have since been made. High yields of desirable products are obtained with hydrated aluminum silicates. These may be either activated (acid-treated natural clays of the bentonite type) or synthesized silica-alumina or silica-magnesia preparations. Both the natural and synthetic catalysts can be used as pellets or beads, and also in the form of powder; in either case, replacements are necessary because of attrition and gradual loss of efficiency (DeCroocq, 1984; Le Page et al., 1987).

During the period 1940–1962, the cracking catalysts used most widely commercially were the aforementioned acid-leached clays and silica alumina. The latter was made in two versions: (1) low alumina, which is approximately 13% Al_2O_3, and (2) high alumina, which is approximately 25% Al_2O_3. High alumina content catalysts showed a higher equilibrium activity level and surface area.

During the 1958–1960 period, semi-synthetic catalysts of silica-alumina catalyst were used in which approximately 25%–35% kaolin was dispersed throughout the silica-alumina gel. These catalysts could be offered at a lower price and therefore were disposable, but they were marked by a lower catalytic activity and greater stack losses because of increased attrition rates. One virtue of the semi-synthetic catalysts was that a lesser amount of adsorbed, unconverted, high-molecular weight products on the catalyst were carried over to the stripper zone and regenerator. This resulted in a higher yield of more valuable products and also smoother operation of the regenerator as local hot spots were minimized.

The catalysts must be stable to physical impact loading and thermal shocks and must withstand the action of carbon dioxide, air, nitrogen compounds, and steam. They should also be resistant to sulfur compounds; the synthetic catalysts and certain selected clays appear to be better in this regard than average untreated natural catalysts.

Commercial cracking catalysts are insulator catalysts that possess strong protonic (acidic) properties because they function as a catalyst by altering the cracking process mechanisms through an alternative mechanism involving chemisorption by proton donation and desorption, resulting in cracked oil and theoretically restored catalyst.

The catalyst-oil volume ratios range from 5:1 to 30:1 for the different processes, although most processes are operated to 10:1. However, for moving-bed processes the catalyst-oil volume ratios may be substantially lower than 10:1.

Crystalline zeolite catalysts having molecular sieve properties were introduced as selective adsorbents in the 1955–1959 period. In a relatively short time period, all of the cracking catalyst manufacturers were offering their versions of zeolite catalysts to refiners. The intrinsically higher activity of the crystalline zeolites vis-à-vis conventional amorphous silica-alumina catalysts coupled with the much higher yields of naphtha and decreased coke and light ends yields served to revitalize research and development in the mature refinery process of catalytic cracking.

A number of zeolite catalysts have been mentioned as having catalytic cracking properties, such as synthetic faujasite (X- and Y-types), offretite, mordenite, and erionite. Of these, the faujasites have been most widely used commercially. While faujasite is typically produced in the sodium form, a base exchange reaction will replace the sodium with other metal ions that, for cracking catalysts, include magnesium, calcium, rare-earth elements (mixed or individual), and ammonium. In particular, mixed rare-earth elements alone or in combination with ammonium ions have been the most commonly used forms of faujasite in cracking catalyst formulations. Empirically, X-type faujasite has a stoichiometric formula of Na_2O Al_2O_3 $2.5SiO_2$ and Y-type faujasite Na_2O Al_2O_3 $4.8SiO_2$. Slight variations in the silica/alumina (SiO_2/Al_2O_3) ratio exist for each of the types. Rare-earth exchanged Y-type faujasite retains much of its crystallinity after steaming at 825°C (1,520°F) with steam for 12 hours; the rare earth from X-faujasite, while thermally stable in dry air, will lose its crystallinity at these temperatures in the presence of steam.

5.5.2 Catalyst Manufacture

While each manufacturer has developed proprietary procedures for making silica-alumina catalyst, the general procedure consists of (1) the gelling of dilute sodium silicate solution (Na_2O. $3.25SiO_2$. xH_2O) by addition of an acid (H_2SO_4, CO_2) or an acid salt such as aluminum sulfate, (2) aging the hydrogel under controlled conditions, (3) adding the prescribed amount of alumina as aluminum sulfate and/or sodium aluminate, (4) adjusting the pH of the mixture, and (5) filtering the composite mixture. After filtering, the filter cake can either be (1) washed free of extraneous soluble salts by a succession of slurrying and filtration steps and spray dried or (2) spray dried and then washed free of extraneous soluble salts before flash drying the finished catalyst.

There are a number of critical areas in the preparative processes that affect the physical and catalytic properties of the finished catalyst. Principal among them is the concentration and temperature of the initial sodium silicate solution. The amount of acid added to effect gelation, the length of time of aging the gel, the method and conditions of adding the aluminum salt to the gel, and its incorporation therein. Under a given set of conditions, the product catalyst is quite reproducible in both physical properties and catalytic performance.

During the period 1940–1962, a number of improvements were made in silica-alumina catalyst manufacture. These included continuous production lines vs batch-type operation, introduction of spray drying to eliminate grinding and sizing of the catalyst while reducing catalyst losses as fines, improved catalyst stability by controlling pore volume, and improved wash procedures to remove extraneous salts from high alumina content catalysts to improve equilibrium catalyst performance.

Zeolite cracking catalysts are made by dispersing or imbedding the crystals in a matrix. The matrix is generally amorphous silica-alumina gel and may also contain finely divided clay. The zeolite content of the composite catalyst is generally in the range of 5%–16% by weight. If clay (e.g., kaolin) is used in the matrix, it is present in an amount of 25%–45% by weight, the remainder being the silica-alumina hydrogel glue that binds the composite together. The zeolite may be pre-exchanged to the desired metal form and calcined to lock the exchangeable metal ions into position before compositing with the other ingredients. In an alternate scheme, sodium-form zeolite is composited with the other components, washed, and then treated with a dilute salt solution of the desired metal ions before the final drying step.

As stated above, the matrix generally consists of silica-alumina, but several catalysts have been commercialized and which contain (1) silica-magnesia, kaolin and (2) synthetic montmorillonite-mica and/or kaolin as the matrix for faujasite.

5.5.3 CATALYST SELECTIVITY

In the catalytic cracking process, the most abundant products are those having 3, 4, and 5 carbon atoms. On a weight basis, the 4-carbon-atom fraction is the largest. The differences between the catalysts of the mixed oxide type lie in the relative action toward promoting the individual reaction types included in the overall cracking operation. For example, silica-magnesia catalyst under a given set of cracking conditions will give a higher conversion to cracked products than silica-alumina catalyst. However, the products from a silica-magnesia (SiO_2-MgO) catalyst have a higher average molecular weight, hence a lower volatility, lesser amounts of highly branched/acyclic isomers, but more olefin derivatives among the naphtha boiling range products (C_4_220°C, 430°F) than the products from a silica-alumina catalyst. With these changes in composition, the naphtha from cracking with a silica magnesia catalyst is of lower octane number.

These differences between catalysts may also be described as differences in the intensity of the action at the individual active catalytic centers. That is, a catalyst such as silica-alumina would give greater intensity of reaction than silica-magnesia as observed from the nature and yields of the individual cracked products and the automobile gasoline octane number. Titration of these two catalysts shows silica-alumina to have a lower acid titer than silica-magnesia, but the acid strength of the sites is higher.

While each of the individual component parts in these catalysts is essentially non-acidic, when mixed together properly they give rise to a titratable acidity as described above. Many of the secondary reactions occurring in the cracking process may also be promoted with strong mineral acids, such as concentrated sulfuric and phosphoric acids, aluminum halides, hydrogen fluoride, and hydrogen fluoride-boron trifluoride (BF_3) mixtures, which supports the concept of the active catalytic site as being acidic. Also, zeolite catalysts have a much higher active site density (titer) than the amorphous mixed oxides, which may account in large part for their extremely high cracking propensity. In addition, these materials strongly promote complex hydrogen transfer reactions among the primary products so that the recovered cracked products have a much lower olefin and higher paraffin content than are obtained with the amorphous mixed oxide catalysts. This hydrogen transfer propensity of zeolites to saturate primary cracked product olefin derivatives to paraffin derivatives minimizes the reaction of polymerizing the olefin derivatives to form a coke deposit, thus accounting in part for the much lower coke yields with zeolite catalysts than with amorphous catalysts.

Activity of the catalyst varies with faujasite content as does the selectivity of the catalyst to coke and naphtha. As the faujasite content drops below 5% by weight, the catalyst starts to show some of the cracking properties of the matrix, while for zeolite contents of 10% by weight or higher, very little change in selectivity patterns is noted. The various ion exchanged forms of the faujasite can result in slightly different cracking properties, e.g., using high cerium content mixed rare earths improve carbon burning rates in the regenerator, use of H-form of faujasite improves selectivity to propane-pentane fractions, use of a minor amount of copper from faujasite increases light olefin yield and naphtha octanes.

5.5.4 CATALYST DEACTIVATION

A cracking catalyst should maintain its cracking activity with little change in product selectivity as it ages in a unit. A number of factors contribute to degrade the catalyst: (1) the combination of high temperature, steam partial pressure, and time; (2) impurities present in the fresh catalyst; and (3) impurities picked up by the catalyst from the feedstock while in use. Under normal operating conditions, the catalyst experiences temperatures of 480°C–515°C (900°F–960°F) in the reactor and steam stripper zones and temperatures of 620°C–720°C (1,150°F–1,325°F) and higher in the regenerator accompanied by a substantial partial pressure of steam. With mixed oxide amorphous gel catalysts, the plastic nature of the gel is such that the surface area and pore volume decrease rather sharply in the first few days of use and then at a slow inexorable rate thereafter. This plastic flow also results in a loss in the number and strength of the active catalytic sites.

Zeolite catalysts comprising both amorphous gel and crystalline zeolite degrade from instability of the gel, as stated above, and also from loss in crystallinity. The latter also results from the combined effects of time, temperature, and steam partial pressure. When crystallinity is lost, the amorphous residue is relatively low in activity, approximating that of the amorphous gel matrix. The rate of degradation of the amorphous gel component may not be the same as that of the zeolite crystals, e.g., the gel may degrade rapidly and through thermoplastic flow effectively coat the crystals and interfere with the diffusion of hydrocarbon derivatives to the catalytic sites in the zeolite. Catalyst manufacturers try to combine high stability in the matrix with high stability zeolite crystals in making zeolite catalysts.

Residual impurities in freshly manufactured catalysts are principally sodium and sulfate. These result from the use of sodium silicate and aluminum sulfate in making the silica-alumina gel matrix and subsequent washing of the composite catalyst with ammonium sulfate to remove sodium. Generally, the sodium content of the amorphous gel is <01% w/w (as Na_2O), and sulfate <0.5% w/w.

With zeolite catalysts, the residual sodium may be primarily associated with the zeolite, so that sodium levels may range from approximately 0.2%–0.80% for the composite catalyst. Sulfate levels in zeolite catalysts are still <0.5%. An excessive amount of sodium reacts with the silica in the matrix under regenerator operating conditions and serves as a flux to increase the rate of surface area and pore volume loss. Sodium faujasite is not as hydrothermally stable as other metal-exchanged (e.g., mixed rare earths) forms of faujasite. It is most desirable to reduce the sodium content of the faujasite component to <5.0% by weight (as Na_2O) with rare earths or with mixtures of rare earths and ammonium ions.

Finally, catalysts can degrade as a result of impurities picked up from the feed being processed. These impurities are sodium, nickel, vanadium, iron, and copper. Sodium as laid down on the catalyst not only acts to neutralize active acid sites, reducing catalyst activity but also acts as a flux to accelerate matrix degradation. Freshly deposited metals are effective poisons to cracking catalysts because of the loss of active surface area by metal deposition (Otterstedt et al., 1986). Zeolite catalysts are less responsive to metal contaminants than amorphous gel catalysts. Hence equilibrium catalysts can tolerate low levels of these metals so long as they have enough time to become buried. A sudden deposition of fresh metals can cause adverse effects on unit performance. Metals levels on equilibrium catalysts reflect the metals content of the feeds being processed typical ranges are 200–1,200 ppm V, 150–500 ppm Ni, and 5–45 ppm Cu. Sodium levels are in the range of 0.25%–0.8% by weight (as Na_2O).

5.5.5 CATALYST STRIPPING

The catalyst that leaves the reaction zone is fluidized with reactor product vapors that must be removed and recovered with the reactor product. In order to accomplish this removal, the catalyst is passed into a stripping zone where most of the hydrocarbon is displaced with steam.

Stripping is generally done in a countercurrent contact zone where shed baffles or contactors are provided to insure equal vapor flow up through the stripper and efficient contacting. Stripping can be accomplished in a dilute catalyst phase. Generally, a dense phase is used, but with lighter feeds or higher reactor temperature and high conversion operations, a significant portion of the contacting can be done in a dilute phase.

The amount of hydrocarbon carried to the regenerator is dependent upon the amount of stripping steam used per pound of catalyst and the pressure and temperature at which the stripper operates.

Probes at the stripper outlet have been used to measure the composition of the hydrocarbon vapors leaving the stripper. When expressed as percent of coke burned in the regenerator, the strippable hydrocarbon is only 2%–5%. Very poor stripping is shown when the hydrogen content of the regenerator coke is on the order of 10% w/w, or higher. Good stripping is shown by 6%–9% by weight hydrogen levels.

The proper level of stripping is found in many operating units by reducing the stripping steam until there is a noticeable effect or rise in regenerator temperature. Steam is then marginally increased above this rate. In some units, stripping steam is used as a control variable to control the carbon burning rate or differential temperature between the regenerator bed and cyclone inlets.

In summary, the catalytic cracking unit is an extremely dynamic unit, primarily because there are three major process flow streams (the catalyst, hydrocarbon, and regeneration air), all of which interact with each other.

5.5.6 Catalyst Treatment

The latest technique developed by the refining industry to increase naphtha yield and quality is to treat the catalysts from the cracking units to remove metal poisons that accumulate on the catalyst. Nickel, vanadium, iron, and copper compounds contained in catalytic cracking feedstocks are deposited on the catalyst during the cracking operation, thereby adversely affecting both catalyst activity and selectivity. Increased catalyst metal contents affect catalytic cracking yields by increasing coke formation, decreasing naphtha and butane and butylene production, and increasing hydrogen production.

The recent commercial development and adoption of cracking catalyst-treating processes definitely improve the overall catalytic cracking process economics.

5.5.6.1 Demet Process

In this process, the cracking catalyst is subjected to two pretreatment steps (Figure 5.11). The first step effects vanadium removal; the second, nickel removal, to prepare the metals on the catalyst for chemical conversion to compounds (chemical treatment step) that can be readily removed through water washing (catalyst wash step).

The treatment steps include use of a sulfurous compound followed by chlorination with an anhydrous chlorinating agent (e.g., chlorine gas) and washing with an aqueous solution of a chelating

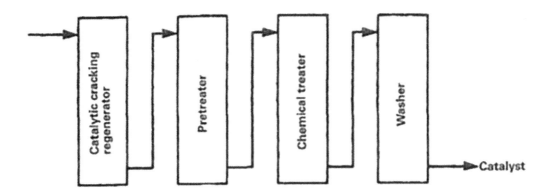

FIGURE 5.11 The demet process. (Speight, J.G. 2017. *Handbook of Petroleum Refining*. CRC Press, Taylor & Francis Group, Boca Raton, Florida. Figure 9.26, page 392.)

agent such as citric acid ($HO_2CCH_2C(OH)(CO_2H)CH_2CO_2H$, 2-hydroxy-1,2,3-propanetricarboxylic acid). The catalyst is then dried and further treated before returning to the cracking unit.

5.5.6.2　Met-X Process

This process consists of cooling, mixing, and ion-exchange separation, filtration, and resin regeneration. Moist catalyst from the filter is dispersed in oil and returned to the cracking reactor in a slurry (Figure 5.12).

On a continuous basis, the catalyst from a cracking unit is cooled and then transported to a stirred reactor and mixed with an ion-exchange resin (introduced as slurry). The catalyst-resin slurry then flows to an elutriator for separation. The catalyst slurry is taken overhead to a filter, and the wet filter cake is slurried with oil and pumped into the catalytic cracker feed system. The resin leaves the bottom of the elutriator and is regenerated before returning to the reactor.

5.5.7　Recent Advances

Recent advances in FCC catalysts have concentrated on modifying zeolite Y for improved coke selectivity, higher cracking activity, and greater stability through manipulation of extra-framework aluminum or through the generation of mesoporous nature of the zeolite crystals. Extra-framework aluminum is introduced either by steaming or via ion exchange. The development of improved FCC catalysts includes modifying a single crystal structure to achieve multiple catalytic objectives (Degnan, 2000).

In order to meet demands for a higher octane product, an increasing number of refiners have switched to using ultra-stable Y (USY) zeolite type catalysts. These catalysts, which feature zeolites

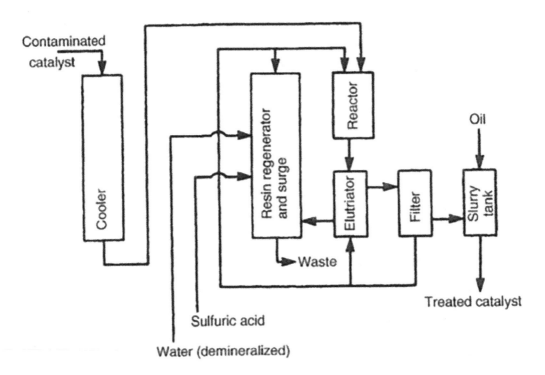

FIGURE 5.12 The met-X process. (Speight, J.G. 2017. *Handbook of Petroleum Refining*. CRC Press, Taylor & Francis Group, Boca Raton, Florida. Figure 9.27, page 392.)

with a reduced unit cell size, are a proven means of increasing the octane number of FCC naphtha (which is used for gasoline production in the blending operation). However, there are tradeoffs associated with using a reduced unit cell size zeolite – one of the more important properties being affected is catalyst activity. In addition, the use of ZSM-5 (a Pentasil zeolite concentrate which typically incorporates up to 25%–40% Pentasil) enhances the octane number in FCC naphtha by over-cracking of naphtha/gasoline fraction. This reaction mainly occurs on the stronger acid site of ZSM-5 than that of the USY catalyst. Comparing to the result with typical FCC catalyst, the yield of LPG is increased by over-cracking and the octane number is enhanced for the mixed catalyst, although the yield of FCC naphtha is decreased. The increase of LPG yield is due to production of propylene in the case of mixed catalyst (Buchanan et al., 1996; Imhof et al., 2005; Li et al., 2007).

5.5.7.1 Matrix, Binder, Zeolite

The matrix in the FCC catalyst is often considered to be that part of the catalyst other than the zeolites – it may or may not have catalytic activity. Very often it does possess sufficient catalytic activity toward some components of the feedstock, and in this case, they are described by the term active matrix. The matrix consists of two main components: firstly a synthetic component like amorphous silica, silica-alumina gel, or silica-magnesia gel, which serve as the binder and also exhibit catalytic properties. The other component is natural or chemically modified clay, such as kaolinite, halloysite, or montmorillonite. The clays provide mechanical stability.

The functions of the fillers and binders incorporated into the FCC catalysts are often similar to those performed by the matrix. Sometimes, additional fillers like kaolin may be provided for physical integrity and as a more efficient fluidizing medium. The binder performs the all-important function of holding the catalyst, the matrix, and the filler glued together. This is especially important when the catalyst contains a higher amount of zeolites. The filler and binder minimize the production of catalyst fines in the reactor-regenerator system and help to control or mitigate catalyst losses.

5.5.7.2 Additives

In addition to what to cracking catalyst described above, a series of additives has been developed that catalyze or otherwise alter the primary catalyst's activity/selectivity or act as pollution control agents. Additives are most often prepared in microspherical form to be compatible with the primary catalysts and are available separately in compositions that (1) enhance naphtha octane and light olefin formation, (2) selectively crack heavy cycle oil, (3) passivate vanadium and nickel present in many heavy feedstocks, (4) oxidize coke to carbon dioxide, and (5) reduce sulfur dioxide emissions. Both vanadium and nickel deposit on the cracking catalyst and are extremely deleterious when present in excess of 3,000 ppm on the catalyst. Formulation changes to the catalyst can improve tolerance to vanadium and nickel, but the use of additives that specifically passivate either metal is often preferred.

The ability of small amounts of ZSM-5 added to the FCC unit to improve naphtha octane number while producing more light olefin derivatives has prompted a substantial amount of process and catalyst research into zeolite-based FCC additives. Significant advances have been made in stabilizing ZSM-5 to harsh FCC regenerator conditions which, in turn, have led to reductions in the level of ZSM-5 needed to achieve desired uplifts and wider use of the less expensive additives (Degnan, 2000).

5.5.7.3 Metal Traps

The performance of FCC catalysts for processing heavy feedstocks is often determined by the tolerance of the catalyst to metal contaminants. Different approaches are used to manage contaminant metals such as riser pipe design (Hedrick and Palmas, 2011). Metals traps have been introduced into catalysts in order to protect the zeolite from poisonous metals that have accumulated in large amounts as a result of lowering the amount of catalyst use, and in order to give the catalyst metal tolerance so that the formation of hydrogen and coke due to the dehydrogenation over poisonous metals would be reduced.

As a result, catalyst manufacturers have developed more stable zeolites and a series of vanadium traps to increase the ability of the zeolite to handle vanadium. These traps are based on barium (Ba), titanium (Ti), rare-earth elements, and other elements. Some are more effective than others, but the basic idea is the same, i.e., to keep the vanadium away from the zeolite by binding to the surface of an inactive particle (Harding et al., 2001).

5.5.7.4 Low Rare Earths

Rare-earth elements inhibit alumina removal from a zeolite, and a higher concentration of acid sites will be found in a rare-earth exchanged catalyst. This improves both the activity and the hydro-thermal stability of the catalyst. On average, these sites are weaker and in closer proximity to each other than those found in a more highly de-aluminated catalyst characterized by lower unit cell size measurements.

The addition of rare earth into the zeolite inhibits the degree of unit cell size shrinkage during equilibration in the regenerator. Steam in the FCC regenerator removes active acidic alumina from the zeolite. Rare earth inhibits the extraction of aluminum from the zeolite's structure (alumina removal) which in turn increases the equilibrium unit cell size for FCC catalysts. Since reducing the equilibrium unit cell size of a FCC catalyst has the effect of improving octane, adding rare earth decreases the octane.

Because rare-earth oxides promote hydrogen transfer, the yield of C_3 and C_4 olefin derivatives in the LPG fraction will be lower. The resulting reduction in the total LPG yield results in a reduction in the wet gas yield, which can have a major effect on plant operations, as compressor capacity is often the limiting factor for FCC unit throughput.

5.5.7.5 Catalysts for Olefin Production

The market for propene derivatives and butene derivatives is cyclical and depends on the local demand for plastics precursors and alkylate feed. There are several key determinants of light olefin selectivity (Harding et al., 2001; Soni et al., 2009). In addition to increasing reactor temperature and using ZSM-5, catalysts can be designed with lower hydrogen transfer (to reduce the conversion of isobutylene to isobutane) and moderate matrix activity (to increase the C4-olefin content of the product).

The addition of ZSM-5 is the single most important method to increase the yields of light olefin derivatives at the expense of naphtha. Catalyst producers have developed methods to increase the concentration of the ZSM-5 in their additives in order to avoid dilution effects at high levels of additive. The breakthrough technologies in this area involve the stabilization of the ZSM-5 to hydrothermal deactivation at higher concentrations in the additive.

Current commercial catalysts range from 10% to 25% ZSM-5, although evolving techniques are expected to allow much higher levels in an additive particle. Several recent efforts have also been made to change the selectivity of ZSM-5 by increasing the Si-Al ratio to increase the ratio of butene to propene. However, this approach reduces the overall activity quite significantly.

5.5.7.6 Catalysts for Jet and Diesel Production

In markets dominated by FCC-based refineries, the need to increase distillate production has taken on a new dimension, posing interesting challenges, while presenting some unique opportunities. Much attention on the possibility of tighter specifications for aromatic derivatives, density, boiling range and cetane number can be expected. New technologies are being developed to address these challenges. For example, a commercialized moderate-pressure hydrocracking (MPHC) suite of technologies can provide an economic solution to desulfurize heavy gas oil (HGO) with options to convert part of the feed into valuable higher-quality distillate products and to maximize production of Jet A1 and diesel products (Degnan, 2000; Hilbert et al., 2008).

The catalysts are large pore, highly de-aluminated Y-zeolites as well as improved base metal combinations that have been tailored to operate well at low hydrogen partial pressures. Processes have been developed around USY-based catalysts for partial conversion of vacuum gas oils, cracked gas oils, deasphalted oil, and FCC light cycle oil. The processes comprise a dual catalyst system consisting of an amorphous hydrotreating catalyst (normally $NiMo/Al_2O_3$) and a metal-containing USY-based hydrocracking catalyst. Single-stage, single-pass conversions are typically in the range of 30%–60% w/w. The process requirements are similar to those used in vacuum gas oil hydrodesulfurization, which has led to several catalytic hydrodesulfurization revamps (Degnan, 2000).

For cracking catalysts, there is usually a trade-off between high activity and high middle distillate selectivity. The objective for new cracking catalyst development is often to increase activity while maintaining selectivity or vice versa. This can be achieved by altering the catalyst's acidic function and/or the hydrogenation function, as is illustrated by the UOP unicracking process (Ackelson, 2004; Abdo et al., 2008).

In addition to control and optimization of process conditions in the unicracking process, catalyst design principles when applied to this problem take advantage of improved molecular-scale definition of feed compositions, appropriate selection of catalytic materials, and enhanced knowledge of the impact of specific process conditions on catalyst performance to deliver the needed level of activity and selectivity to meet cycle length and product yields (Motaghi et al., 2010).

In the early 1990s, Zeolyst introduced Z-603, a zeolite catalyst with high activity and also good middle distillate selectivity. The improved performance of the catalyst versus other zeolite and amorphous silica-alumina catalysts was the result of two major developments: a new de-aluminated Y-zeolite used in conjunction with an amorphous silica-alumina catalysts powder and an improved hydrogenation function. This catalyst has been used successfully at Shell's Godorf (Germany) refinery and at Alliance Refining Company's Rayong (Thailand) refinery. Because of its higher activity and stability, Z-603 has extended the cycle length beyond the capability of an amorphous silica-alumina catalysts catalyst with only a small reduction in middle distillate yield relative to the amorphous silica-alumina catalysts catalyst (Huve et al., 2004).

Continuing development efforts have led to a new range of cracking catalysts (for example, Z-623 and Z-613) with improved middle distillate yields/activities. Z-623 is significantly more active than the Z-603 catalyst with very little middle distillate yield loss. The benefits of the Z-623 catalyst have been demonstrated in a number of commercial cases. At Alliance Refining Company, the catalyst has made it possible to increase the hydrocracker's cycle length from 2 years to more than 3 years. At Shell's Godorf refinery, the catalyst was a key component of the hydrocracker revamp that has produced a 15% v/v increase in throughput of feedstock without any cycle length reduction (Huve et al., 2004).

5.5.7.7 New Directions

With the increasing focus to reduce sulfur content in fuels, the role of desulfurization in the refinery becomes more and more important. Currently, the process of choice is the hydrotreater to remove the sulfur from the fuel. Because of the increased attention for fuel desulfurization, various new process concepts are being developed with various claims of efficiency and effectiveness.

Resid FCC is an important component in the upgrading of heavy oils, with unit profitability depending upon the extent to which heavy hydrocarbon derivatives in the feed are cracked into valuable products. The product slate, in turn, depends upon the feed characteristics, the catalyst, the hardware, and the operating conditions. Exemplifying a trend toward heavier feedstocks, the majority of the fluid-bed catalytic cracking units scheduled to start up before 2015 are expected to process heavy vacuum gas oil and/or or residuum feedstocks.

The major developmental opportunity area in the next decade may well be at the refinery-petrochemical interface. Processes which maximize olefin derivatives and aromatic derivatives while integrating clean fuels production will continue to look attractive to refiners who are seeing their fuels product margins come under increasing competitive pressure.

Furthermore, in order to alleviate growing concerns over energy security and global climate change, the FCC unit will take on two additional roles: (1) acceptance of biomass feedstocks and

(2) reduction in carbon dioxide emissions. With committed efforts by refiners and technology developers, the process will continue to demonstrate adaptability amid changing market requirements through the installation of dual-radius feedstock distributors and spent catalyst distributors (Wolschlag and Couch, 2010a).

Methods to reduce emissions from the FCC unit and other refinery operations lend themselves to two conflicting categories: those that hinder the formation of carbon dioxide and those that aim to produce a pure carbon dioxide stream for capture. Improving unit energy efficiency, reducing coke yields, and shifting vacuum gas oil to the hydrocracking unit belong to the former category, while carbon-capture methods typically comprise the latter (Spoor, 2008). As refiners continue to develop a comprehensive strategy to reduce carbon dioxide emissions, a balance between these two categories must be found. This happy medium will be influenced greatly by the FCC unit heat balance, which determines the amount of coke burned in the regenerator in present unit designs.

5.6 OPTIONS FOR VISCOUS FEEDSTOCKS

FCC is one of the major conversion technologies in the oil refinery industry. FCC currently produces the majority of the gasoline used in the world as well as ethylene and propylene for use in the production of petrochemicals.

In the modern refinery, the older uses still remain, and the prime purpose of converting a higher-boiling feedstock (such as gas oil) to lower-boiling (higher-value) products (such as naphtha and kerosene) is still operative. It is these products that are used to produced gasoline and diesel fuel.

However, the gas oil produced in the vacuum tower, which cannot be distilled in the atmospheric tower (without the potential for decomposition), is heated and sent to the vacuum tower where it is split into gas oil and residuum. The vacuum tower bottoms (the vacuum residuum) can be sent to be processed further in units such as the delayed coking unit, the deasphalting unit, the visbreaking unit, or used as a blend stock for residual fuel oil, or road asphalt.

Thus, the vacuum gas oil was used as feedstock for the catalytic cracking unit, and this feedstock could be used (1) without any further treatment, (2) blended into other FCC feedstocks, (3) partially hydrotreated before introduction to the FCC unit, or (4) fully hydrotreated before introduction to the FCC unit.

Moreover, the FCC unit was (and still is) used to convert low-value, high-boiling feedstocks (such as gas oil) into valuable products such as naphtha and kerosene (or middle distillate). Anu coke that was deposited on the catalyst during the reaction was burned off in the regenerator, which was also a source of heat for the supplying all the heat for the reactor vessel.

However, throughout the past several decades, processes have evolved that are the next generation of the FCC and the residuum catalytic cracking processes. Some of these newer processes use catalysts with different silica/alumina ratios as acid support of metals such as Mo, Co, Ni, and W. In general, the first catalyst used to remove metals from oils was the conventional hydrodesulfurization (HDS) catalyst. Diverse natural minerals are also used as raw material for elaborating catalysts addressed to the upgrading of heavy fractions. Among these minerals are clays; manganese nodules; bauxite activated with vanadium (V), nickel (Ni), chromium (Cr), iron (Fe), and cobalt (Co), as well as the high iron oxide content iron laterites (soils or ores rich in iron and aluminum); clay minerals such as sepiolite minerals (a soft white clay mineral which is a complex magnesium silicate); and mineral nickel and transition metal sulfides supported on silica and alumina. Other kinds of catalysts, such as vanadium sulfide, are generated in situ, possibly in colloidal states.

In the past decades, in spite of the difficulty of handling viscous feedstocks, residuum fluidized catalytic cracking (RFCC) has evolved to become a well-established approach for converting a significant portion of the heavier fractions of the crude barrel into a high-octane gasoline blending component. For example, the residuum fluidized catalytic cracking process, which is an extension of conventional FCC technology for applications involving the conversion of highly contaminated residua (or other viscous feedstocks) can handle a variety of viscous feedstocks to produce gaseous, distillates, and fuel oil-range products.

However, as might be anticipated, the product quality from the residuum fluidized catalytic cracker is directly affected by its feedstock quality. In particular, and unlike hydrotreating, the RFCC redistributes sulfur among the various products but does not remove sulfur from the products unless, of course, one discount the sulfur that is retained by any coke formed on the catalyst. Consequently, tightening product specifications have forced refiners to hydrotreat some, or all, of the products from the resid cat cracking unit. Similarly, in the future, the emissions of sulfur oxides (SO_x) from a resid cat cracker may become more of an obstacle for residue conversion projects. For these reasons, a point can be reached where the economic operability of the unit can be sufficient to justify hydrotreating the feedstock to the cat cracker.

As an integrated conversion block, residue hydrotreating and residuum fluidized catalytic cracking complement each other and can offset many of the inherent deficiencies related to residue conversion.

5.6.1 Asphalt Residual Treating (ART) Process

The asphalt residual treating process (the ART process) is a process for increasing the production of transportation fuels and reduces heavy fuel oil production, without hydrocracking (Bartholic, 1981a, b; Bartholic and Haseltine, 1981; Green and Center, 1985; Bartholic, 1989; Parkash, 2003; Gary et al., 2007; Speight, 2014; Hsu and Robinson, 2006; Speight, 2017). The process is a flexible selective vaporization process that can be used for removing essentially all of the metals and substantial proportion of the carbon residue, nitrogen, and sulfur compounds from the heavy feedstock (Bartholic and Haseltine, 1981). In addition, the process can be considered as an efficient carbon rejection process followed by catalytic hydrogenation.

In the process, the preheated feedstock (which may be whole crude, atmospheric residuum, vacuum residuum, or tar sand bitumen) is injected into a stream of fluidized, hot catalyst (trade name: ArtCat). Complete mixing of the feedstock with the catalyst is achieved in the contactor, which is operated within a pressure-temperature envelope to ensure selective vaporization. The vapor and the contactor effluent are quickly and efficiently separated from each other, and entrained hydrocarbon derivatives are stripped from the contaminant (containing spent solid) in the stripping section. The contactor vapor effluent and vapor from the stripping section are combined and rapidly quenched in a quench drum to minimize product degradation. The cooled products are then transported to a conventional fractionator that is similar to that found in a FCC unit. Spent solid from the stripping section is transported to the combustor bottom zone for carbon burn-off. The process is claimed to require lower hydrogen consumption is claimed in comparison to direct hydrocracking processes (Dennis, 1989; Suchanek and Moore, 1986).

In the combustor, coke is burned from the spent solid that is then separated from combustion gas in the surge vessel. The surge vessel circulates regenerated catalyst streams to the contactor inlet for feed vaporization and to the combustor bottom zone for premixing. The components of the combustion gases include carbon dioxide (CO_2), nitrogen (N_2), oxygen (O_2), sulfur oxides (SO_x), and nitrogen oxides (NO_x) that are released from the catalyst with the combustion of the coke in the combustor. The concentration of sulfur oxides in the combustion gas requires treatment for their removal.

5.6.2 Heavy Oil Treating Process

The HOT process is a catalytic cracking process for upgrading heavy feedstocks such as topped crude oils, vacuum residua, and solvent deasphalted bottoms using a fluidized bed of iron ore particles (arkash, 2003; Gary et al., 2007; Speight, 2014; Hsu and Robinson, 2006; Speight, 2017).

The main section of the process consists of three fluidized reactors, and separate reactions take place in each reactor (cracker, regenerator, and desulfurizer):

$$Fe_3O_4 + \text{asphaltene constituents} \rightarrow \text{coke}/Fe_3O_4 + Oil + Gas \text{ (in the cracker)}$$

$$3FeO + H_2O \rightarrow Fe_3O_4 + H_2 \text{ (in the cracker)}$$

$$\text{Coke}/Fe_3O_4 + O_2 \rightarrow 3FeO + CO + CO_2 \text{ (in the regenerator)}$$

$$FeO + SO_2 + 3CO \rightarrow FeS + 3CO_2 \text{ (in the regenerator)}$$

$$3FeS + 5O_2 \rightarrow Fe_3O_4 + 3SO_2 \text{ (in the desulfurizer)}$$

In the cracker, heavy oil cracking and the steam-iron reaction take place simultaneously under conditions similar to thermal cracking. Any unconverted feedstock is recycled to the cracker from the bottom of the scrubber. The scrubber effluent is separated into hydrogen gas, liquefied petroleum gas (LPG), and liquid products that can be upgraded by conventional technologies to priority products.

In the regenerator, coke deposited on the catalyst is partially burned to form carbon monoxide in order to reduce iron tetroxide and to act as a heat supply. In the desulfurizer, sulfur in the solid catalyst is removed and recovered as molten sulfur in the final recovery stage.

5.6.3 Reduced Crude Oil Conversion Process

In recent years, because of a trend for low-boiling products, most refineries perform the operation by partially blending residua into vacuum gas oil. However, conventional FCC processes have limits in residue processing, so residue FCC processes have lately been employed one after another. Because the residue FCC process enables efficient naphtha production directly from residues, it will play the most important role as a residue cracking process, along with the residue hydroconversion process. Another role of the residuum FCC process is to generate high-quality gasoline blending stock and petrochemical feedstock. Olefin derivatives (propene, the isomeric butene derivatives, and the isomeric pentene derivatives) serve as feed for alkylating processes, for polymer gasoline, as well as for additives for reformulated gasoline.

In the reduced crude oil conversion process (RCC process), the clean regenerated catalyst enters the bottom of the reactor riser where it contacts low-boiling hydrocarbon lift gas that accelerates the catalyst up the riser prior to feed injection (Parkash, 2003; Gary et al., 2007; Speight, 2014; Hsu and Robinson, 2006; Speight, 2017). At the top of the lift gas zone, the feed is injected through a series of nozzles located around the circumference of the reactor riser.

The catalyst/oil disengaging system is designed to separate the catalyst from the reaction products and then rapidly remove the reaction products from the reactor vessel. Spent catalyst from reaction zone is first steam stripped, to remove adsorbed hydrocarbon, and then routed to the regenerator. In the regenerator, all of the carbonaceous deposits are removed from the catalyst by combustion, restoring the catalyst to an active state with a very low carbon content. The catalyst is then returned to the bottom of the reactor riser at a controlled rate to achieve the desired conversion and selectivity to the primary products.

5.6.4 Residue FCC Process

The residue FCC process (HOC process) is a version of the FCC process that has been adapted to conversion of residua that contain high amounts of metal and asphaltene constituents. Depending on quality and product objectives, feedstocks with vanadium-plus-nickel content of 5–30 ppm, and a carbon residue on the order of 5%–10% w/w can be processed without feed pretreatment. The process, when coupled with hydrodesulfurization), provides a particularly effective means for meeting demands for naphtha or diesel fuel blending stock and low-molecular weight olefin derivatives. The asphaltene constituents are converted to coke (which deposits on the catalyst) and distillate from the cracked fragments. The catalyst is regenerated by combusting its carbon deposits, and the heat generated from the combustion of carbon deposits is used to produce high pressure steam.

In the process, a residuum is desulfurized, and the non-volatile fraction from the hydrodesulfurization unit is charged to the residuum FCC unit. The reaction system is an external vertical riser terminating in a closed cyclone system. Dispersion steam in amounts higher than that used for gas oils is used to assist in the vaporization of any volatile constituents of heavy feedstocks. A two-stage stripper is utilized to remove hydrocarbon derivatives from the catalyst. Hot catalyst flows at low velocity in dense phase through the catalyst cooler and returns to the regenerator. Regenerated catalyst flows to the bottom of the riser to meet the feed. The coke deposited on the catalyst is burned off in the regenerator along with the coke formed during the cracking of the gas oil fraction. If the feedstock contains high proportions of metals, control of the metals on the catalyst requires excessive amounts of catalyst withdrawal and fresh catalyst addition. This problem can be addressed by feedstock pretreatment.

The feedstocks for the process are rated on the basis of carbon residue and content of metals: Thus, good-quality feedstocks have less than 5% by weight carbon residue and less than 10 ppm metals. Medium-quality feedstocks have greater than 5% but less than 10% by weight carbon residue and greater than 10 but less than 30 ppm metals. Poor-quality feedstocks have greater than 10 but less than 20% by weight carbon residue and greater than 30 but less than less than 150 ppm metals. Finally, bad-quality feedstocks have greater than 20% by weight carbon residue and greater than 150 ppm metals. One might question the value of this rating of the feedstocks for the HOC process since these feedstock ratings can apply to virtually many FCC processes.

The process is similar to the FCC process in configuration, catalysts, and product handling but differs from gas oil cracking in that the heat release, due to burning the coke produced from the asphaltenes in the charge, is considerably greater. In addition, the much higher content of feedstock metals – particularly nickel and vanadium – requires special consideration in catalyst development and in operation. With the need to convert residual fuels to naphtha and middle distillates, the installation of new heavy oil cracking units, as well as the conversion of FCC units to handle residua, has become a refining necessity.

Special catalysts are required for heavy oil cracking units because of the required specifications for activity and selectivity. Some of the catalyst have a high zeolite content, or they have pore structures which avoid trapping large molecules and causing coke production. Poisons such as sodium and vanadium accelerate the deactivation rate of catalyst, and high amounts of sodium are usually avoided by double desalting of the crude oil. While customized catalysts can improve heavy oil cracking operations, optimization can also take place by closer process control. For example, high cracking activity, if used correctly, can override the adverse dehydrogenation activity of the metals. Thus, low contact time in risers, along with rapid and efficient separation of catalyst and oil vapors, is preferred in reduced crude cracking. Also, heavy catalytic gas oil or slurry oil recycle produces higher yields of coke and higher heat release than it removes through vaporization and, therefore, should be minimized or eliminated. In addition to reducing coke yield, lower regenerator temperatures can be realized by direct heat removal or lower heat generation.

5.6.5 R2R PROCESS

The R2R process is a FCC process for conversion of heavy feedstocks (Heinrich and Mauleon, 1994; Inai, 1994; Parkash, 2003; Gary et al., 2007; Speight, 2014; Hsu and Robinson, 2006; Speight, 2017).

In the process, the feedstock is vaporized upon contacting hot-regenerated catalyst at the base of the riser and lifts the catalyst into the reactor vessel separation chamber where rapid disengagement of the hydrocarbon vapors from the catalyst is accomplished by both a special solids separator and cyclones. The bulk of the cracking reactions takes place at the moment of contact and continues as the catalyst and hydrocarbon derivatives travel up the riser. The reaction products, along with a minute amount of entrained catalyst, then flow to the fractionation column. The stripped spent catalyst, deactivated with coke, flows into the Number 1 regenerator.

Partially regenerated catalyst is pneumatically transferred via an air riser to the Number 2 regenerator, where the remaining carbon is completely burned in a dryer atmosphere. This regenerator is designed to minimize catalyst inventory and residence time at high temperature while optimizing the coke-burning rate. Flue gases pass through external cyclones to a waste heat recovery system. Regenerated catalyst flows into a withdrawal well and after stabilization is charged back to the oil riser.

5.6.6   Shell FCC Process

In the shell FCC process, which is designed to maximize the production of distillates from residua, the preheated feedstock is mixed with the hot regenerated catalyst (Khouw, 1990; Parkash, 2003; Gary et al., 2007; Speight, 2014; Hsu and Robinson, 2006; Speight, 2017).). After reaction in a riser, volatile materials and catalyst are separated after which the spent catalyst is immediately stripped of entrained and adsorbed hydrocarbon derivatives in a very effective multistage stripper. The stripped catalyst gravitates through a short standpipe into a single vessel, simple, reliable, and yet efficient catalyst regenerator. Regenerative flue gas passes via a cyclone/swirl tube combination to a power recovery turbine. From the expander turbine, the heat in the flue gas is further recovered in a waste heat boiler. Depending on the environmental conservation requirements, a de-NO_xing, de-SO_xing, and particulate emission control device can be included in the flue gas train.

There is a claim (Sato et al., 1992) that feedstock pretreatment of bitumen (by hydrogenation) prior to FCC (or for that matter any catalytic cracking process) can result in enhanced yield of naphtha. It is suggested that mild hydrotreating be carried out upstream of a FCC unit to provide an increase in yield and quality of distillate products (Long et al., 1993; Parkash, 2003; Gary et al., 2007; Speight, 2014; Hsu and Robinson, 2006; Speight, 2017). This is in keeping with earlier work (Speight and Moschopedis, 1979) where mild hydrotreating of bitumen was reported to produce low-sulfur liquids that would be amenable to further catalytic processing.

5.6.7   S&W FCC Process

The S&W FCC process is also designed to maximize the production of distillates from residua (Parkash, 2003; Gary et al., 2007; Speight, 2014; Hsu and Robinson, 2006; Speight, 2017). In the process, the heavy feedstock is injected into a stabilized, upward flowing catalyst stream whereupon the feedstock-steam-catalyst mixture travels up the riser and is separated by a high efficiency inertial separator. The product vapor goes overhead to the main fractionator.

The spent catalyst is immediately stripped in a staged, baffled stripper to minimize hydrocarbon carryover to the regenerator system. The first regenerator (650°C–700°C, 1,200–1,290°F) burns 50%–70% of the coke in an incomplete carbon monoxide combustion mode running countercurrently. This relatively mild, partial regeneration step minimizes the significant contribution of hydrothermal catalyst deactivation. The remaining coke is burned in the second regenerator (ca. 775°C, 1,425°F) with an extremely low steam content. Hot clean catalyst enters a withdrawal well that stabilizes its fluid qualities prior to being returned to the reaction system.

5.6.8   Other Options

As the demand for higher octane gasoline has increased, catalytic cracking has replaced thermal cracking as the process of choice but not to the complete exclusion of thermal cracking processes

(Chapter 3) (Parkash, 2003; Gary et al., 2007; Speight, 2014; Hsu and Robinson, 2017; Speight, 2017). In fact, several process innovations have been introduced in the form of varying process options, some using piggyback techniques (where one process works in close conjunction with another process; please see above), there are other options that have not yet been introduced or even invented but may well fit into the refinery of the future.

In the FCC, the major developments are in integration with sulfur removal to produce low-sulfur naphtha without octane loss (Babich and Moulijn, 2003; Rama Rao et al., 2011). This development will build on the development of new catalysts (see above). Furthermore, recent enhancements made to RFCC units permit the use of feedstocks that contain significant amounts of viscous crudes, while simultaneously improving yields and service factors have focused on improved feed injection and dispersion, reduced contact time of products and catalyst, improved separation of products and catalyst, and regenerator heat removal.

Example of a triglyceride.

In terms of processability, feedstocks containing triglyceride derivatives are the best-suited bio-feedstock for catalytic cracking units.

5.7 CHEMISTRY

Catalytic cracking is the thermal decomposition of crude oil constituents in the presence of a catalyst (Pines, 1981; Dwyer and Rawlence, 1993). Thermal cracking has been superseded by catalytic cracking as the process for gasoline manufacture. Indeed, gasoline produced by catalytic cracking is richer in branched paraffin derivatives, cycloparaffin derivatives, and aromatic derivatives, which all serve to increase the quality of the naphtha and, hence, the gasoline. Catalytic cracking also results in production of the maximum amount of butene derivatives and butane derivatives (C_4H_8 and C_4H_{10}) rather than production of ethylene and ethane (C_2H_4 and C_2H_6).

Catalytic cracking processes evolved in the 1930s from research on crude oil and coal liquids. The crude oil work came to fruition with the invention of acid cracking. The work to produce liquid fuels from coal, most notably in Germany, resulted in metal sulfide hydrogenation catalysts. In the 1930, a catalytic cracking catalyst for crude oil that used solid acids as catalysts was developed using acid-treated clay minerals. Clay minerals are a family of crystalline aluminosilicate solids, and the acid treatment develops acidic sites by removing aluminum from the structure. The acid sites also catalyze the formation of coke, and Houdry developed a moving-bed process that continuously removed the cooked beads from the reactor for regeneration by oxidation with air.

Although thermal cracking is a free radical (neutral) process, catalytic cracking is an ionic process involving carbonium ions, which are hydrocarbon ions having a positive charge on a carbon atom. The formation of carbonium ions during catalytic cracking can occur by: (1) addition of a proton from an acid catalyst to an olefin and/or (2) abstraction of a hydride ion (H$^-$) from a hydrocarbon by the acid catalyst or by another carbonium ion. However, carbonium ions are not formed by cleavage of a carbon-carbon bond.

In essence, the use of a catalyst permits alternate routes for cracking reactions, usually by lowering the free energy of activation for the reaction. The acid catalysts first used in catalytic cracking were amorphous solids composed of approximately 87% silica (SiO_2) and 13% alumina (Al_2O_3) and were designated low-alumina catalysts. However, this type of catalyst is now being replaced by crystalline aluminosilicates (zeolites) or molecular sieves.

The first catalysts used for catalytic cracking were acid-treated clay minerals, formed into beads. In fact, clay minerals are still employed as catalyst in some cracking processes (Speight, 2014). Clay minerals are a family of crystalline aluminosilicate solids, and the acid treatment develops acidic sites by removing aluminum from the structure. The acid sites also catalyze the formation of coke, and the development of a moving bed process that continuously removed the cooked beads from the reactor reduced the yield of coke; clay regeneration was achieved by oxidation with air.

Clays are natural compounds of silica and alumina, containing major amounts of the oxides of sodium, potassium, magnesium, calcium, and other alkali and alkaline earth metals. Iron and other transition metals are often found in natural clays, substituted for the aluminum cations. Oxides of virtually every metal are found as impurity deposits in clay minerals.

Clay minerals are layered crystalline materials. They contain large amounts of water within and between the layers (Keller, 1985). Heating the clays above 100°C (212°F) can drive out some or all of this water; at higher temperatures, the clay structures themselves can undergo complex solid-state reactions. Such behavior makes the chemistry of clays a fascinating field of study in its own right. Typical clays include kaolinite, montmorillonite, and illite (Keller, 1985). They are found in most natural soils and in large, relatively pure deposits, from which they are mined for applications ranging from adsorbents to paper making.

Once the carbonium ions are formed, the mode of interaction constitutes an important means by which product formation occurs during catalytic cracking. For example, isomerization either by hydride ion shift or by methyl group shift, both of which occur readily. The trend is for stabilization of the carbonium ion by movement of the charged carbon atom toward the center of the molecule, which accounts for the isomerization of α-olefin derivatives to internal olefin derivatives when carbonium ions are produced. Cyclization can occur by internal addition of a carbonium ion to a double bond which, by continuation of the sequence, can result in aromatization of the cyclic carbonium ion.

Similar to the paraffin derivatives, naphthene derivatives do not appear to isomerize before cracking. However, the naphthene derivatives (from C_9 upward) do produce considerable amounts of aromatic hydrocarbon derivatives (by dehydrogenation) during catalytic cracking. Alkylated benzene derivatives undergo nearly quantitative dealkylation to benzene without apparent ring degradation at temperatures below 500°C (930°F). However, polymethly benzene derivatives undergo disproportionation and isomerization with very little benzene formation.

Catalytic cracking can be represented by simple reaction schemes. However, questions have arisen as to how the cracking of paraffin derivatives is initiated. Several hypotheses for the initiation step in catalytic cracking of paraffin derivatives have been proposed (Cumming and Wojciechowski, 1996). The Lewis site mechanism is the most obvious, as it proposes that a carbenium ion is formed by the abstraction of a hydride ion from a saturated hydrocarbon by a strong Lewis acid site: a tricoordinated aluminum species. On Brønsted sites, a carbenium ion may be readily formed from an olefin by the addition of a proton to the double bond or, more rarely, by the abstraction of a hydride ion from a paraffin by a strong Brønsted proton. This latter process requires the formation of hydrogen as an initial product. This concept was, for various reasons that are of uncertain foundation, often neglected.

It is therefore not surprising that the earliest cracking mechanisms postulated that the initial carbenium ions are formed only by the protonation of olefin derivatives generated either by thermal cracking or present in the feed as an impurity. For a number of reasons, this proposal was not convincing, and in the continuing search for initiating reactions, it was even proposed that electrical fields associated with the cations in the zeolite are responsible for the polarization of reactant paraffin derivatives, thereby activating them for cracking. More recently, however, it has been convincingly shown that a penta-coordinated carbonium ion can be formed on the alkane itself by protonation, if a sufficiently strong Brønsted proton is available (Cumming and Wojciechowski, 1996).

Coke formation is considered, with just cause to a malignant side reaction of normal carbenium ions. However, while chain reactions dominate events occurring on the surface and produce the majority of products, certain less desirable bimolecular events have a finite chance of involving the same carbenium ions in a bimolecular interaction with one another. Of these reactions, most will produce a paraffin and leave carbene/carboid-type species on the surface. This carbene/carboid-type species can produce other products, but the most damaging product will be one which remains on the catalyst surface and cannot be desorbed and results in the formation of coke, or remains in a non-coke form but effectively blocks the active sites of the catalyst.

A general reaction sequence for coke formation from paraffin derivatives involves oligomerization, cyclization, and dehydrogenation of small molecules at active sites within zeolite pores:

Alkanes → alkenes
Alkenes → oligomers
Oligomers → naphthenes
Naphthenes → aromatics
Aromatics → coke

Whether or not these are the true steps to coke formation can only be surmised. The problem with this reaction sequence is that it ignores sequential reactions in favor of consecutive reactions. And it must be accepted that the chemistry leading up to coke formation is a complex process, consisting of many sequential and parallel reactions.

There is a complex and little understood relationship between coke content, catalyst activity, and the chemical nature of the coke. For instance, the atomic hydrogen-carbon ratio of coke depends on how the coke was formed; its exact value will vary from system to system (Cumming and Wojciechowski, 1996). And it seems that catalyst decay is not related in any simple way to the hydrogen-to-carbon atomic ratio of the coke, or to the total coke content of the catalyst, or any simple measure of coke properties. Moreover, despite many and varied attempts, there is currently no consensus as to the detailed chemistry of coke formation. There is, however, much evidence and good reason to believe that catalytic coke is formed from carbenium ions which undergo addition, dehydrogenation and cyclization, and elimination side reactions in addition to the main-line chain propagation processes (Cumming and Wojciechowski, 1996).

The catalytic cracking chemistry of biomass-derived oils involves conventional FCC reactions, such as cleavage of carbon–carbon bonds, hydrogen transfer, and isomerization, as well as deoxygenation reactions, such as dehydration, decarboxylation, and decarbonylation (Al-Sabawi et al., 2012). It is important to note that deoxygenation of these biofeedstocks is essential because oxygen must be removed and replaced by hydrogen to produce liquid fuels. Typical FCC catalysts, which consist of a mixture of an inert matrix, an active matrix, a binder, and a Y-zeolite, are efficient in removing oxygen from biomass-derived molecules. Dehydration reactions occur on the catalyst acid sites during the catalytic cracking of oxygenated compounds. This leads to the formation of water and a dehydrated compound, which may include ketone derivatives, aldehyde derivatives, and alcohol derivatives. Decarboxylation and decarbonylation reactions result in the formation of CO_2 and CO, respectively. The generation of these gases may pose concerns for refiners because they are not typical products of conventional catalytic cracking of crude oil.

Catalyst selection is important in bio-oil upgrading because the conversion of bio-oil and product distribution are highly dependent upon the type of catalyst used. As previously mentioned, the two most commonly used and effective zeolite catalysts in FCC processes are H–Y zeolites and HZSM-5 (Al-Sabawi et al., 2012).

REFERENCES

Abdo, S.F., Thakkar, V., Ackelson, D.B., Wang, L., and Rossi, R.J. 2008. Maximize Diesel with UOP Enhanced Two-Stage Unicracking™ Technology. Proceedings of the 18th Annual Saudi-Japan Symposium. Dhahran, Saudi Arabia.

Ackelson, 2004. UOP Unicracking Process for Hydrocracking. In *Handbook of Petroleum Refining Processes.* R.A. Meyers (Editor). McGraw-Hill, New York. Chapter 7.2.

Al-Sabawi, M., Chen, J., and Ng, S. 2012. Fluid Catalytic Cracking of Biomass-Derived Oils and Their Blends with Petroleum Feedstocks: A Review. *Energy Fuels*, 26: 5355–5372.

Alvarenga Baptista, C.M. de L., Cerqueira, H.S., and Sandes, E.F. 2010a. Process for Fluid Catalytic Cracking of Hydrocarbon Feedstocks with High Levels of Basic Nitrogen. United States Patent 7,736,491 June 15.

Alvarenga Baptista, C.M. de L., Moreira, E.M., and Cerqueira, H.S. 2010b. Process for Fluid Catalytic Cracking of Hydrocarbon Feedstocks with High Levels of Basic Nitrogen. United States Patent 7,744,745 June 29.

Avidan, A.A., and Krambeck, F.J. 1990. FCC Closed Cyclone System Eliminates Post Riser Cracking. Proceedings of the Annual Meeting. National Petrochemical and Refiners Association.

Babich, I.V., and Moulijn, J.A. 2003. Science and Technology of Novel Processes for Deep Desulfurization of Oil Refinery Streams: A Review. *Fuel*, 82: 607–631.

Bartholic, D.B. 1981a. Preparation of FCC Charge from Residual Fractions. United States Patent 4,243,514. January 6.

Bartholic, D.B. 1981b. Upgrading Petroleum and Residual Fractions Thereof. United States Patent 4,263,128. April 21.

Bartholic, D.B. 1989. Process for Upgrading Tar Sand Bitumen. United States Patent 4,804,459. February 14.

Bartholic, D.B., and Haseltine, R.P. 1981. *Oil & Gas Journal*, 79(45): 242.

Benazzi, E., and Cameron, C. 2005. Future Refinements. Fundamentals of the Global Oil and Gas Industry, page 111–113.

Birch, C.H., and Ulivieri, R. 2000. *ULS Gasoline and Diesel Refining Study.* Pervin and Gertz Inc., Houston, Texas.

Bouziden, G., Gentile, K., and Kunz, R.G. 2002. Selective Catalytic Reduction of NOx from Fluid Catalytic Cracking Case Study: BP Whiting Refinery. Paper ENV-03-128. NPRA Meeting. National Environmental and Safety Conference. New Orleans, Louisiana. April 23–24.

Bradley, S.A., Gattuso, M.J., and Bertolacini, R.J. 1989. Characterization and Catalyst Development. Symposium Series No. 411. American Chemical Society, Washington, DC.

Bridjanian, H., and Khadem Samimi, A. 2011. Bottom of the Barrel, an Important Challenge of the Petroleum Refining Industry. *Petroleum & Coal*, 53(1): 13–21. Available online at www.vurup.sk/pc

Buchanan, J.S., Santiesteban, J.G., and Haag, W.O. 1996. Mechanistic Considerations in Acid-Catalyzed Cracking of Olefins. *Journal of Catalysis*, 158: 279–287.

Chapin, L., and Letzsch, W. 1994. Deep Catalytic Cracking Maximize Olefin Production. Paper No. AM-94-43. Proceedings. NPRA Annual Meeting, March 20–22.

Chen, Y-M. 2004. Recent Advances in FCC Technology. Proceedings of the 2004 AIChE Annual Meeting, Austin, Texas, November 7–12.

Chen, Y-M. 2006. Recent Advances in FCC Technology. *Powder Technology*, 163: 2–8.

Chuachuensuk, A., Paengjuntuek, W., Kheawhom, S., and Arpornwichanopa, A. 2010. *Proceedings of the 20th European Symposium on Computer Aided Process Engineering – ESCAPE20.* S. Pierucci and G. Buzzi Ferraris (Editors). Elsevier, Amsterdam, Netherlands.

Couch, K.A., Siebert, K.D., and Van Opdorp, P.J. 2004. Controlling FCC Yield and Emissions. Proceedings of the NPRA Annual Meeting. March.

Cumming, K.A., and Wojciechowski, B.W. 1996. *Catalysis Reviews – Science and Engineering*, 38: 101–157.

Cybulski, A., and Moulijn, J.A. (Editors). 1998. *Structured Catalysts and Reactors.* Marcel Dekker Inc., New York.

DeCroocq, D. 1984. *Catalytic Cracking of Heavy Petroleum Hydrocarbons.* Editions Technip, Paris.

Degnan, T.F. 2000. Applications of Zeolites in Petroleum Refining. *Topics in Catalysis*, 13: 349–356.

Dennis, G.E. 1989. Use of Small Scale ART Processing for Wellhead Viscosity Reduction and Upgrading of Heavy Oils. Proceedings of the 4th UNITAR/UNDP International Conference on Heavy Crude and Tar Sands, 5: 89–102. August 7–12, Edmonton, Alberta, Canada.

Domokos, L., Jongkind, H., Stork, W.H.J., and Van Den Tol-Kershof, J.M.H. 2010. Catalyst Composition, Its Preparation and Use. United States Patent 7,749,937. July 6, 2010.

Draemel, D.C. 1992. Flexicracking IIIR – ER&E's Latest Cat Cracking Design. Proceedings of the JPI Petroleum Refining Conference. Japanese Petroleum Institute, Tokyo, Japan.

Dwyer, J., and Rawlence, D.J. 1993. Fluid Catalytic Cracking: Chemistry. *Catalysis Today*, 18: 487–507.

Dziabala, B., Thakkar, V.P., and Abdo, S.F. 2011. Combination of Mild Hydrotreating and Hydrocracking for Making Low Sulfur Diesel and High Octane Naphtha. United States Patent 8,066,867. November 29.

EPA. 2010. Available and Emerging Technologies for Reducing Greenhouse Gas Emissions from the Petroleum Refining Industry. Sector Policies and Programs Division, Office of Air Quality Planning and Standards, United States Environmental Protection Agency, Research Triangle Park, North Carolina.

Fahim, M.A., Alsahhaf, T.A., and Elkilani, A. 2010. *Fundamentals of Petroleum Refining*. Elsevier BV, Amsterdam, The Netherlands.

Freel, B., and Graham, R.G. 2011. Products Produced from Rapid Thermal Processing of Heavy Hydrocarbon Feedstocks. United States Patent 8,062,503. November 22.

Fujiyama, Y. 1999. Process for Fluid Catalytic Cracking of Oils. United States Patent 5,904,837. May 18.

Fujiyama, Y., Adachi, M., Okuhara, T., and Yamamoto, S. 2000. Process for Fluid Catalytic Cracking of Heavy Fraction Oils. United States Patent 6,045,690. April 4.

Fujiyama, Y., Redhwi, H., Aitani, A., Saeed, R., and Dean, C. 2005. Demonstration Plant for New FCC Technology Yields Increased Propylene. *Oil & Gas Journal*, September 26: 62–67.

Fujiyama, Y., Al-Tayyar, M.H., Dean, C.F., Aitani, A., Redhwi, H.H., Tsutsui, T., and Mizuta, K. 2010. Development of High Severity FCC Process for Maximizing Propylene Production - Catalyst Development and Optimization of Reaction Conditions. *Journal of the Japan Petroleum Institute*, 53(6): 336–341.

Fujiyama, Y., Okuhara, T., and Uchiura, A. 2011. Method of Designing Gas-Solid Separator. United States Patent 8,070,846. December 6.

Gary, J.G., Handwerk, G.E., and Kaiser, M.J. 2007. *Petroleum Refining: Technology and Economics*. 5th Edition. CRC Press, Taylor & Francis Group, Boca Raton, Florida.

Gates, B.C., Katzer, J.R., and Schuit, G.C.A. 1979. *Chemistry of Catalytic Processes*. McGraw-Hill Inc., New York.

Gembicki, V.A., Cowan, T.M., and Brierley, G.R. 2007. Update Processing Operations to Handle Heavy Feedstocks. *Hydrocarbon Processing*, 86(2): 41–53.

Germain, G.E. 1969. *Catalytic Conversion of Hydrocarbons*. Academic Press Inc., New York.

Harding, R.H., Peters, A.W., and Nee, J.R.D. 2001. New Developments in FCC Catalyst Technology. *Applied Catalysis A: General*, 221: 389–396.

Hedrick, B.W., and Palmas, P. 2011. Process for Contacting High Contaminated Feedstocks with Catalyst in an FCC Unit. United States Patent 8,062.506. November 22.

Heinrich, G., and Mauleon, J.-L. 1994. The R2R Process: 21st Century FCC Technology. *Révue Institut Français du Pétrole*, 49(5): 509–520.

Hemler, C.L. 1997. In: *Handbook of Petroleum Refining Processes*. R.A. Meyers (Editor). McGraw-Hill, New York. Chapter 3.3.

Hemler. C.L., and Smith, L.F. 2004. UOP Fluid Catalytic Cracking Process. In: *Handbook of Petroleum Refining Processes*. R.A. Meyers (Editor). McGraw-Hill, New York. Chapter 3.3.

Hilbert, T.L., Chitnis, G.K., Umansky, B.S., Kamienski, P.W., Patel, V., and Subramanian, A., 2008. Consider New Technology to Produce Clean Diesel. *Hydrocarbon Processing*, 87(2): 47–56.

Hsu, C.S., and Robinson, P.R. (Editors). 2017. *Handbook of Petroleum Technology*. Springer, Cham, Switzerland.

Hudec, P. 2014. FCC Catalyst - Key Element in Refinery Technology. Proceedings of the 45th International Petroleum Conference. Bratislava, Slovak Republic. June 13. http://www.vurup.sk/sites/default/files/downloads/46_ft_hudec-fcc.pdf; accessed December 16, 2014.

Hunt, D.A. 1997. In *Handbook of Petroleum Refining Processes*. R.A. Meyers (Editor). McGraw-Hill, New York. Chapter 3.5.

Huve, L.G., Creyghton, E.J., Ouwehand, C., van Veen, J.A.R., and Hanna, A. 2004. New Catalyst Technologies Expand Hydrocrackers' Flexibility and Contributions. Report No. CRI424/0704, Criterion Catalysts and Technologies, Shell Global Solutions International BV, Amsterdam, Netherlands.

Imhof, P., Rautiainen, E.P.H., and Gonzalez, J.A 2005. Maximize Propylene Yields. *Hydrocarbon Processing*, 84(9): 109–114.

Inai, K, 1994. Operation Results of the R2R Process. *Revue Institut Français du Pétrole*, 49(5): 521–527.

Ino, T., and Ikeda, S. 1999. Process for Fluid Catalytic Cracking of Heavy Fraction Oil. Unites States Patent 5,951,850. September 14.

Ino, T., Okuhahra, T., Abul-Hamayel, M., Aitani, A., and Maghrabi, A. 2003. Fluid Catalytic Cracking Process for Heavy Oil. United States Patent No. 6,656,346. December 2.

Jakkula, J., Hiltunen, J., and Niemi, V.M. 1997. Short Contact Time Catalytic Cracking Process: Results from Bench Scale Unit. Paper 29241. Proceedings of the 15th World Petroleum Congress, Beijing, China. October 12–17.

Jeon, S.G., Kwak, N.S., Rho, N.S., Ko, C.H., Na, J-G., Yi, K.B., and Park, S.B. 2012. Catalytic Pyrolysis of Athabasca Bitumen in H2 Atmosphere Using Microwave Irradiation. *Chemical Engineering Research and Design*, 90: 1292–1296.

Johnson, T.E., and Niccum, P.K. 1997. In: *Handbook of Petroleum Refining Processes*. R.A. Meyers (Editor). McGraw-Hill, New York. Chapter 3.2.

Keller, W.D. 1985. Clays. In: *Kirk Othmer Concise Encyclopedia of Chemical Technology*. M. Grayson (Editor). Wiley Interscience, New York. Page 283.

Khouw, F.H.H. 1990. Shell Residue FCC Technology: Challenges and Opportunities in a Changing Environment. JPI Petroleum Refining Conference.

Krambeck, F.J., and Pereira, C.J. 1986. FCC Processing Scheme with Multiple Risers. United States Patent 4,606,810. August 19.

Ladwig, P.K. 1997. In: *Handbook of Petroleum Refining Processes*. R.A. Meyers (Editor). McGraw-Hill, New York. Chapter 3.1.

Lam, S.S., Russell, A.D., Lee, C.L., and Chase, H.A. 2012. Microwave-Heated Pyrolysis of Waste Automotive Engine Oil: Influence of Operation Parameters on the Yield, Composition, and Fuel Properties of Pyrolysis Oil. *Fuel*, 92: 327–339.

Leadbeater, N.E., and Khan, M.R., 2008. Microwave-Promoted Desulfurization of Heavy Oil and a Review of Recent Advances on Process Technologies for Upgrading of Heavy and Sulfur Containing Crude Oil. *Energy Fuel*, 22: 1836–1839.

Le Page, J.F., Cosyns, J., Courty, P., Freund, E., Franck, J.P., Jacquin, Y., Juguin, B., Marcilly, C., Martino, G., Miguel, J., Montarnal, R., Sugier, A., and von Landeghem, H. 1987. *Applied Heterogeneous Catalysis*. Editions Technip, Paris.

Letzsch, W. 2011. Innovation Drives New Catalyst Development. Catalyst. Hydrocarbon Processing, C93–C94.

Li, X., Li, C., Zhang, J., Yang, C., and Shan, H. 2007. Effects of Temperature and Catalyst to Oil Weight Ratio on the Catalytic Conversion of Heavy Oil to Propylene Using ZSM-5 and USY Catalysts. *Journal of Natural Gas Chemistry*, 16(1): 92–99.

Lifschultz. 2005. Oil Refiner's Gathering Storm: Help Is on the Way. *Hydrocarbon Processing*, 84(9): 59–62.

Long, S.L., Johnson, A.R., and Dharia, D. 1993. Advances in Residual Oil FCC. Paper No. AM-93-50. Proceedings of the Annual Meeting. National Petrochemical and Refiners Association.

Maadhah, A.G., Fujiyama, Y., Redhwi, H., Abul-Hamayel, M., Aitani, A., Saeed, M., and Dean, C. 2008. A New Catalytic Cracking Process to Maximize Refinery Propylene. *The Arabian Journal for Science and Engineering*, 33(1B): 17–28.

Maples, R.E. 2000. *Petroleum Refinery Process Economics*. 2nd Edition. PennWell Corporation, Tulsa, Oklahoma.

MARPOL. 2005. International Convention for the Prevention of Pollution from Ships (MARPOL).

Annex I: Regulations for the Prevention of Pollution by Oil (entered into force 2 October 1983).

Annex II: Regulations for the Control of Pollution by Noxious Liquid Substances in Bulk (entered into force 2 October 1983).

Annex III: Prevention of Pollution by Harmful Substances Carried by Sea in Packaged Form (entered into force 1 July 1992).

Annex IV: Prevention of Pollution by Sewage from Ships (entered into force 27 September 2003).

Annex V: Prevention of Pollution by Garbage from Ships (entered into force 31 December 1988).

Annex VI: Prevention of Air Pollution from Ships (entered into force 19 May 2005).

McPherson, L.J. 1984. Causes of FCC Reactor Coke Deposits Identified. *Oil & Gas Journal*, September 10: 139.

Motaghi, M., Shree, K., and Krishnamurthy, S. 2010. Consider New Methods for Bottom of the Barrel Processing – Part 1. *Hydrocarbon Processing*, 89(2): 35–40.

Mutyala, S., Fairbridge, C., Jocelyn Paré, J.R., Bélanger, J.M.R., Ng, S., Hawkins, R. 2010. Microwave Applications to Oil Sands and Petroleum: A Review. *Fuel Processing Technology*, 91: 127–135.

Navarro, U., Ni, M., and Orlicki, DF. 2015. FCC 101: How to Estimate Product Yields Cost-Effectively and Improve Operations. *Hydrocarbon Processing*, 94(2): 41–52.

Nishida, S., and Fujiyama, Y. 2000. Separation Device. United States Patent 6,146,597. November 14.

Occelli, M.L., and O'Connor, P. 1998. *Fluid Cracking Catalysts*. Marcel Dekker Inc., New York.

Otterstedt, J.E., Gevert, S.B., Jaras, S.G., and Menon, P.G. 1986. Processing Heavy Oils. *Applied Catalysis*, 22: 159–179.

Parkash, S. 2003. *Refining Processes Handbook*. Gulf Professional Publishing, Elsevier, Amsterdam, Netherlands.

Patel, R., Zeuthen, P., and Schaldemose, M. 2002. Advanced FCC Feed Pretreatment Technology and Catalysts Improves FCC Profitability. Proceedings of the NPRA Annual Meeting, San Antonio, March 2002. National Petrochemical and Refiners Association, Washington, DC.

Patel, R., Moore, H., and Hamari, B. 2004. FCC Hydrotreater Revamp for Low-Sulfur Gasoline. Proceedings of the NPRA Annual Meeting, San Antonio, Texas. March. National Petrochemical and Refiners Association, Washington, DC.

Patel, S. 2007. Canadian Oil Sands – Opportunities, Technologies, and Challenges. *Hydrocarbon Processing*, 86(2): 65–74.

Pines, H. 1981. *The Chemistry of Catalytic Hydrocarbon Conversions*. Academic Press, New York.

Rama Rao, M., Soni, D., Siele, G.M., and Bhattacharyya, D. 2011. Convert Bottom-of-the-Barrel and Diesel into Light Olefins. *Hydrocarbon Processing*, 90(2): 46–49.

Ross, J., Roux, R., Gauthier, T., and Anderson, L.R. 2005. Fine-Tune FCC operations for Changing Fuels Market. *Hydrocarbon Processing*, 84(9): 65–73.

Runyan, J. 2007. Is Bottomless-Barrel Refining Possible. *Hydrocarbon Processing*, 86(9): 81–92.

Sadeghbeigi, R. 1995. *Fluid Catalytic Cracking: Design, Operation, and Troubleshooting of FCC Facilities*. Gulf Publishing Company, Houston, TX.

Sadeghbeigi, R. 2011. *Fluid Catalytic Cracking Handbook*. Elsevier, Amsterdam, The Netherlands.

Salazar-Sotelo, D., Maya-Yescas, R., Mariaca-Domínguez, E., Rodríguez-Salomón, S., and Aguilera-López, M. 2004. Effect of Hydrotreating FCC Feedstock on Product Distribution. *Catalysis Today*, 98(1–2): 273–280.

Sayles, S., and Bailor, J. 2005. Upgrade FCC Hydrotreating. *Hydrocarbon Processing*, 84(9): 87–90.

Schiller, R. 2011. Effect of Synthetic Crude Feedstocks on FCC Yield. *Refinery Operations*, 2(4): 1–2.

Schnaith, M.W., Sexson, A., Tru, D., Bartholic, D.B., Lee, Y.K., Yoo, I.S., Kang, and H.S. 1998. *Oil & Gas Journal*, 96(25): 53.

Shan, H.H, Zhao, W., He, C.Z., Zhang, J.F., and Yang, C.H. 2003. Maximum FCC Diesel Yield with TSRFCC Technology. Preprints. *Division of Fuel Chemistry, American Chemical Society*, 48(2): 710–711.

Shidhaye, H., Kukade, S., Kumar, P., Rao, P.V.C., and Choudary, N.V. 2015. Improve FCC Margins by Processing More Vacuum Resid in Feed. *Hydrocarbon Processing*, 94(12): 35–38.

Soni, D., Rama Rao, M., Saidulu, G., Bhattacharyya, D., and Satheesh, V.K. 2009. Catalytic Cracking Process Enhances Production of Olefins. *Petroleum Technology Quarterly*, 14(Q4): 95–100.

Speight, J.G., and Moschopedis, S.E. 1979. The Production of Low-Sulfur Liquids and Coke from Athabasca Bitumen. *Fuel Processing Technology*, 2: 295.

Speight J.G. 2000. *The Desulfurization of Heavy Oils and Residua*. 2nd Edition. Marcel Dekker Inc., New York.

Speight, J.G. 2008. *Synthetic Fuels Handbook: Properties, Processes, and Performance*. McGraw-Hill, New York.

Speight, J.G. 2009. *Enhanced Recovery Methods for Heavy Oil and Tar Sands*. Gulf Publishing Company, Houston, Texas.

Speight, J.G. 2011a. *The Refinery of the Future*. Gulf Professional Publishing, Elsevier, Oxford, United Kingdom.

Speight, J.G. (Editor). 2011b. *Biofuels Handbook*. Royal Society of Chemistry, London, United Kingdom.

Speight, J.G. 2014. *The Chemistry and Technology of Petroleum*. 5th Edition. CRC Press, Taylor & Francois Group, Boca Raton, Florida.

Speight, J.G. 2017. *Handbook of Petroleum Refining*. CRC Press, Taylor & Francis Group, Boca Raton, Florida.

Speight, J.G. 2021. *Refinery Feedstocks*. CRC Press, Taylor & Francis Group, Boca Raton, Florida.

Spoor, R.M. 2008. Low-Carbon Refinery: Dream or Reality. *Hydrocarbon Processing*, 87(11): 113–117.

Stiles, A.B., and Koch, T.A. 1995. *Catalyst Manufacture*. Marcel Dekker Inc., New York.

Suchanek, A.J., and Moore, A.S. 1986. Modern Residue Upgrading by ART. Proceedings of the NPRA National Meeting, March 23–25. National Petrochemical and Refiners Association, Washington, DC.

Swaty, T.E. 2005. Global Refinery Industry Trends: The Present and the Future. *Hydrocarbon Processing*, 84(-9): 35–46.

Wang, T.X., Zong, Z.M., Zhang, V.W., Wei, Y.B., Zhao, W., Li, B.M., Wei, X.Y. 2008. Microwave-Assisted Hydroconversion of Demineralized Coal Liquefaction Residues from Shenfu and Shengli Coals. *Fuel*, 87: 498–507.

Wojciechowski, B.W., and Corma, A. 1986. *Catalytic Cracking: Catalysts, Chemistry, and Kinetics*. Marcel Dekker Inc., New York.

Wolschlag, L.M., and Couch, K.A. 2010a. Upgrade FCC Performance. *Hydrocarbon Processing*, 89(9): 57–65.

Wolschlag, L.M., and Couch, K.A. 2010b. UOP FCC Innovations Developed using Sophisticated Engineering Tools. Report No. AM-10-109. UOP LLC, Des Plaines, Illinois.

Wolschlag, L.M. 2011. Innovations Developed using Sophisticated Engineering Tools. Proceedings of the AIChE 2011 Regional Process Technology Conference. American Institute of Chemical Engineers.

6 Fouling during Thermal and Catalytic Processes

6.1 INTRODUCTION

The term "fouling" as it pertains to the crude oil industry refers to a variety of events that can occur during or after refining which include (1) deposit formation and deposition, (2) encrustation, (3) scaling, (4) scale formation, (5) slagging, and (6) sludge formation, all of which can have an adverse effect on recovery operations, refinery operations, as well as transportation of crude oil and crude oil products. More specifically, fouling is the accumulation of unwanted material (usually semi-slid or solid material that forms a separate phase but can also include the formation of a separate liquid phase) within a processing unit or on the solid surfaces of the unit to the detriment of function. The focus of the use of the term in this chapter is the occurrence of fouling during refinery operations.

As an example, when fouling does occur during refinery operations, the major effects include (1) loss of heat transfer as indicated by charge outlet temperature decrease and pressure drop increase, (2) blocked process pipes, (3) under-deposit corrosion and pollution, and (4) localized hot spots in reactors, all of which culminate in production losses and increased maintenance costs. Thus, the separation of solids occurs whenever the solvent characteristics of the liquid phase are no longer adequate to maintain polar and/or high molecular weight material in solution.

6.2 THE CONCEPT OF FOULING

Fouling, as it pertains to petroleum refineries (Parkash, 2003; Gary et al., 2007; Speight, 2014a; Hsu and Robinson, 2017; Speight, 2017), is deposit formation, encrustation, deposition, scaling, scale formation, slagging, and sludge formation, which has an adverse effect on operations. It is the accumulation of unwanted material within a processing unit or on the solid surfaces of the unit to the detriment of function. In addition, the term macrofouling is often used to generally describe the blockage of tubes and pipes, while, on the other hand, microfouling is generally iced to describe scaling on the walls of tubes and pipes. Again, the outcome is a loss of efficiency and output to the refinery.

Fouling during production or transportation or refining can occur in a variety of processes, either inadvertently when the separation is detrimental to the process or by intent (such as in the deasphalting process or in the dewaxing process, in which cases the term fouling is inappropriate). Thus, the separation of solids occurs whenever the solvent characteristics of the liquid phase are no longer adequate to maintain polar and/or high molecular weight constituents in solution.

Examples of such occurrences are (1) separation of asphaltene constituents, which occurs when the paraffin nature of the liquid medium increases, which is analogous to the asphaltene precipitation procedure (Chapter 3); (2) wax separation which occurs when there is a drop in temperature or the aromaticity of the liquid medium increases; and (3) sludge/sediment formation in a reactor which occurs when the solvent characteristics of the liquid medium change so that asphaltic or wax materials separate, coke formation which occurs at high temperatures and commences when the solvent power of the liquid phase is not sufficient to maintain the coke precursors in solution, and sludge/sediment formation in fuel products which occurs because of the interplay of several chemical and physical factors.

In the current context of thermal and catalytic conversion processes, the specific issues that arise from the presence of resin and asphaltene constituents in feedstocks have increased considerably

DOI: 10.1201/9781003184904-6

TABLE 6.1

Instances when Fouling Can Occur in Refineries

Process and Event

Desalting and dewatering

Emulsion formation and deposition of sediment in the alkaline environment.

Asphaltene (and resin) constituents are responsible for the undesired stabilization of emulsions because of the relatively high polarity of these constants and surface activity.

Preheating

Preheating of a heavy feedstock followed by reaction and precipitation of the reacted resin and asphaltene constituents leading to coking and deposition of solids (coke precursors and coke) in the preheater and heat exchanger equipment.

Visbreaking and thermal cracking.

Phase separation and deposition of reacted asphaltene constituents which are more aromatic (because of the loss of aliphatic chains) and less soluble than the unreacted constituents; the same applies to the resin constituents.

Blending

Blending of crude oil with viscous crude oils.

Blending of viscous crude oils.

Blending of viscous crude oils with products, as during recycling of product streams for further treatment.

The change of the polarity (or solubility parameter) of the medium as a result of mixing may cause destabilization of asphaltene (and reacted asphaltene) constituents.

Storage

Phase separation leading to the appearance of sediment and plugging when the resin and asphaltene constituents are exposed to oxygen.

Increased polarity due to the incorporation of oxygen functional groups may cause destabilization of the resin-asphaltene interactions leading aggregation of resin and/or asphaltene constituents followed by sludge/sediment formation.

Can be accelerated if there is bacterial presence in the storage unit (biofouling).

---

because refineries are now accepting more heavy crude oil as well as extra heavy oil and tar sand bitumen to produce the required slate of distillate products. Examples of the issues that arise because of the presence of resin and asphaltene constituents include problems due to flocculation (Speight, 2014a, 2015a, 2015b) (Table 6.1).

In addition, the complexity and variation in the properties of sub-fractions of the resin and asphaltene fraction (Speight, 1994, 2014a) detract from the concept of any form of meaningful average structure for these fractions. Thus, the mechanism of fouling by these constituents is a multi-path approach where each constituent (chemical type) must be considered on the basis of individual chemistry and chemical properties.

6.2.1 TYPES OF FOULING

The majority of the fouling in refineries is caused by means of four general causes: (1) the presence of inorganic contaminants in the crude oil, (2) the formation of separate phases – incompatibility – during blending crude oil feedstocks, (3) coke formation during thermal processes, and (4) the production of insoluble asphaltene (and resin) constituents during or after conversion. Other causes, such as emulsion formation and polymerization through the aerial oxidation of olefin constituents in naphtha and other distillate products, leading to gum formation, also occur. Focus has been mainly on fouling due to the presence of the asphaltene fraction that are caused by the various chemical reactions of the asphaltene constituents (Table 6.2) (Speight, 1994; Speight, 2014a).

Fouling can occur throughout the refinery and can affect almost every unit. Traditionally, the fouling phenomenon is caused by the separation of inorganic and/or organic constituents or the

TABLE 6.2

Reaction Sequences for the Asphaltene Constituents[a] of a Feedstock

Primary Products	Secondary Products	Tertiary Products
Gas	Gas[b]	
Liquid	Gas	
	Liquid	Gas[b]
		Liquid
Solids		
Carbene[c]		Gas
Liquid	Gas	
		Liquid
Carboid[c]	Gas	Gas
		Coke

[a] The resin constituents also follow similar reaction paths but with lower yields of coke than is obtained from the asphaltene constituents.

[b] The formation of gases as secondary products and tertiary products may include the additional formation of olefin derivatives.

[c] Please see Chapter 2 for the separation scheme that presents the names of these fractions (especially Figure 2.6).

reacted products of these constituents in a thermal unit which can manifest itself as early as the tank farm through incompatible crude storage. Further, with exposure to heat, the stability of the crude oil system changes, and any products formed during thermal processes may also appear as a phase-separated sediment which can result in flow restrictions and changes to furnace firing rates (more refinery fuel gas is consumed) in order to maintain process efficiency.

Fouling through the disposition of sediment (either as coke precursors or as coke in the reactor) is real, and the challenge in mitigating the formation of thermal coke is to eliminate or modify the prime chemical reactions in the formation of incompatible products during the processing of feedstocks containing resin and asphaltene constituents, particularly those reactions in which the insoluble lower molecular weight products (carbenes and carboids, which are thermal degradation products of the resin constituents and the asphaltene constituents) are formed (Table 6.3) (Chapter 3).

Typically, the fouling material consists of organic and/or inorganic materials deposited by the feedstock that is deposited by the occurrence of instability or incompatibility of the feedstock (one crude oil) with another during and shortly after a blending operation (Speight, 2014a, 2015b).

Blending is one of the typical operations that a refinery must pursue not only to prepare a product to meet sales specifications but also to blend the different crudes and heavy feedstocks to produce a refinery feedstocks (Parkash 2003; Gary et al., 2007; Speight, 2014a; Hsu and Robinson, 2017). Although simple in principle, the blending operation must be performed with care and diligence with the regular acceptance by refineries of heavy feedstocks as part of the refinery slate. Lack of attention to the properties of the individual feedstocks prior to the blending operations can lead to asphaltene precipitation or phase separation (fouling) due to incompatibility of the different components of the blend. This would result in the occurrence of fouling deposits in heat transfer equipment and reactors as a substantial energy cost to the refinery. Therefore, it is advisable for the refiner to be able to predict the potential for incompatibility by determining not only the appropriate components for the blend but also the ratio of individual crude oils and heavy feedstocks in the blend.

The compatibility of crude oils is generally evaluated by colloidal stability based on bulk composition or asphaltene precipitation. Typically, the test methods are performed to evaluate oil stability at ambient conditions, but applying the data to the potential for fouling under the actual parameters

TABLE 6.3

Feedstock Properties That Contribute to Instability and Incompatibility

Property	Effect
Asphaltene constituents	Interact readily with catalyst
	Thermal alteration leading to polar products
	Polar products lead to phase separation
	Phase separation in paraffinic medium
	Thermal de-alkylation
	Formation of long-chain alkane derivatives (waxes)
	Can also lead to wax deposition
Resin constituents[a]	Interact readily with catalyst
	Thermal alteration leading to polar products
	Polar products lead to phase separation
	Phase separation in paraffinic medium
	Thermal de-alkylation
	Formation of long-chain alkane derivatives (waxes)
	Can also lead to wax deposition
Heteroatom constituents	Thermally labile
	React readily with oxygen leading to formation of highly polar products
	Provide polarity to feedstock
	Provide polarity to thermal products
	Polarity can lead to phase separation
Aromatic constituents	May be incompatible with paraffin medium
	Phase separation of long-chain paraffin derivatives (such as wax deposition)
Non-asphaltene constituents	Thermal alteration can cause changes in polarity
	Generation of polar fragments
	Phase separation of polar species in products

[a] Often have a similar (but lesser) effect than the effects caused by the presence of asphaltene constituents.

used in heat transfer equipment must be done with caution. Fouling is dependent upon not only the conditions of asphaltene separation fluid and the stability of the crude oil/heavy feedstock system (Chapter 4) but also on flow conditions and other parameters. Fouling is concerned with not only asphaltene precipitation. In addition, fouling can also be a consequence of corrosion in a unit when deposits of inorganic solids become evident (Speight, 2014b). With the influx of opportunity crudes, high-acid crudes, heavier crude oils, extra heavy crude oils, and tar sand bitumen into refineries, fouling phenomena are more common and diverse.

In the crude oil industry, the components that may be subject to fouling and the corresponding effects of fouling are: (1) the production zone of crude oil reservoirs and oil wells, which is reflected by a decrease in production with time though the formation of plugs which can lead to the complete cessation of flow; (2) pipes and flow channels, which result in reduced flow, increased pressure drop, increased upstream pressure, slugging in two-phase flow, and flow blockage; (3) heat exchangers surfaces, which result in a reduction in thermal efficiency along with decreased heat flux, increased temperature on the hot side, decreased temperature on the cold side, and under-deposit corrosion; (4) injection/spray nozzles e.g., a nozzle spraying a fuel into a furnace or a reactor, in which the incorrect amount of feedstock is injected; and (5) within a reactor due to uncontrollable chemical and/or physical reaction. In addition, there are macrofouling and microfouling.

Macrofouling is caused by coarse matter from either organic or biological or inorganic origin. Such substances foul the surfaces of heat exchangers and may cause deterioration of the relevant heat transfer coefficient as well as flow blockages. Microfouling is somewhat more complex, and the several distinctive events include the following: (1) particulate fouling, which is the accumulation of particles on a surface; (2) chemical reaction fouling, such as decomposition of organic matter on heating surfaces; (3) solidification fouling, which occurs when components of a flowing fluid with a high-melting point freeze onto a subcooled surface; (4) corrosion fouling, which is caused by corrosion; (5) biofouling, which can often ensure after biocorrosion, is due to the action of bacteria or algae; and (6) composite fouling, whereby fouling involves more than one foulant or fouling mechanism.

Fouling caused by the presence of particulate matter in the feedstock is a common form of fouling and is the process in which particles in the feedstock stream deposit onto heat exchanger surfaces. These particles include particles originally carried by the feedstock before entering the heat exchanger and particles formed in the heat exchanger itself as a result of various reactions, aggregation and flocculation. Particulate fouling increases with particle concentration, and typically particles greater than 1 μ size lead to significant fouling problems.

Fouling of a surface through the formation of deposits does not always develop steadily with time. There may be an induction period when the surface is new or very clean, and the foulant does not accumulate immediately. After the induction period, the fouling rate increases. On the other hand, there is also negative fouling which occurs when relatively small amounts of deposit can improve heat transfer, relative to clean surface, and give an appearance of a negative fouling rate and negative total amount of the foulant. After the initial period of surface roughness control or surface roughness adjustment, the fouling rate may become positive.

In asymptotic fouling, the fouling rate decreases with time until it finally reaches zero, and at this point, the deposit thickness remains constant with time. This often occurs when the deposits are relatively soft or poorly adherent deposits in areas of fast flow or turbulent flow and are usually assigned to the point at which the deposition rate equals the deposit removal rate. However, accelerating fouling is almost the opposite since the fouling rate increases with time, and the rate of deposit buildup accelerates with time until it becomes transport limited. This type of fouling can develop when fouling increases the surface roughness, or when the deposit surface exhibits higher chemical propensity to fouling than the pure underlying metal.

6.2.2 Parameters Affecting Fouling

The fouling process is a dynamic and variable process insofar as several operational and design variables have well-defined effects. These parameters include the fluid flow velocity, the fluid properties, the surface temperature, the surface geometry, the surface material, the surface roughness, and the suspended particles concentration and properties.

Parameters of importance are related to operating conditions and equipment design, such as (1) fluid flow velocity, (2) surface temperature, (3) surface materials, (4) surface roughness, and last but certainly not least (5) fluid properties. All these and other factors that may affect fouling need to be considered and taken into account in order to be able to prevent fouling if possible or to predict the rate of fouling or fouling factor prior to taking the necessary steps for fouling mitigation, control, and removal.

The fouling factor is a means of measuring the performance of a heat exchanger, which in turn is a way of measuring the performance as it deteriorates with time. The fouling may be due to the accumulation of organic material, mineral deposits, rust, or the presence of micro-organism on the heat transfer surfaces. These deposits increase the resistance of heat transfer and cause a decrease in the efficiency of the unit. The resistance is usually represented by a fouling factor, R_f, which measures the thermal resistance introduced by the action of the foulant. The development of fouling depends on number of things, major groups of fouling dependents are: (1) composition of the fluids, (2) operating conditions in the heat exchanger, (3) type and characteristics of the heat exchanger, (4) location of the fouling deposit, and (5) presence of microorganism. However, the occurrence of an induction period before a noticeable amount of mineral deposits has formed so the overall heat transfer coefficient changes noticeably, and thence the rate of fouling increases during the fouling period. In addition to the parameters mentioned above, it is necessary to consider the effect of fouling during the design of heat exchangers so that the units can withstand the effect of fouling up to a certain point without becoming harmful for the intended process.

6.2.2.1 Fluid Flow Velocity

The flow velocity has a strong effect on the fouling rate since it affects both the deposition and removal rates through the hydrodynamic effects at the surface of heat exchangers.

In the refinery, the shell and tube heat exchanger is a commonly used design class of heat exchanger and is aptly suited for high-pressure applications. This type of heat exchanger consists of a large pressure vessel (the shell) with a bundle of tubes inside (Figure 1.1). One fluid runs through the tubes, and another fluid flows through the shell and over the tubes, and heat is transferred between the two fluids. The conventional segment baffle geometry is largely responsible for higher fouling rates. Uneven velocity profiles, back-flows, and flow effects generated on the shell side of a baffled heat exchanger (especially one that is segmented by the presence of baffles) result in higher fouling and shorter run lengths between periodic cleaning and maintenance of tube bundles.

On the other hand, the flow velocity has indirect effects on deposit strength, the mass transfer coefficient, and the adherence of the foulant to the surface. Increasing the flow velocity tends to increase the thermal performance of the exchanger and decrease the fouling rate – a uniform flow of process fluids past the heat transfer surface favors less fouling. Foulants suspended in the process fluids will deposit in low-velocity regions (such as pipe elbows unless the flow is turbulent), particularly where the velocity changes quickly, as in heat exchanger water boxes and on the shell side.

6.2.2.2 Surface Temperature

Generally, the rate of fouling is temperature dependent with different rates of fouling between the feed inlet and outlet sides of the heat exchanger, and fouling will increase with an increase in temperature. This is due to baking-on effect, scaling tendencies, increased corrosion rate, faster reactions, crystal formation and polymerization, and loss in activity by some antifoulants. Lower temperatures produce slower fouling buildup, and usually deposits that are easily removable. However, for some process fluids, low surface temperature promotes crystallization and solidification fouling. As expected, biological fouling is strongly dependent on temperature – there is a temperature below which reproduction and growth rate are arrested, and a temperature above which the organism becomes damaged or killed. If, however, the temperature rises to an even higher level, some heat sensitive cells may die.

6.2.2.3 Surface Material

The selection of surface material is significant to deal with corrosion fouling – carbon steel is corrosive but least expensive while copper exhibits biocidal effects in water and its use is limited in certain applications. Noncorrosive materials such as titanium and nickel will prevent corrosion, but they are expensive and have no biocidal effects. Glass, graphite, and Teflon tubes often resist fouling and/or improve cleaning, but they have low thermal conductivity. Although the construction material is more important to resist fouling, surface treatment by plastics, vitreous enamel, glass, and some polymers will minimize the accumulation of deposits.

6.2.2.4 Surface Roughness

Surface roughness has been noted to have an enhancement on fouling insofar as the rough surface provides sites that enhance laying down the initial deposits of foulant. Rough surfaces encourage particulate deposition and provide a good chance for deposit sticking. After the initiation of fouling, the persistence of the roughness effects will be more a function of the deposit itself. A less rough surface finish has been shown to influence the delay of fouling and ease cleaning. Similarly, non-wetting surfaces delay fouling. However, smooth surfaces may become rough in due course due to scale formation, formation of corrosion products, or erosion.

6.2.2.5 Fluid Properties

In terms of fluid properties, there are also the oft-forgotten chemical and physicochemical aspects of fouling. For example, the structure of the deposit, usually dictated by the chemical species that form the deposit, can lead to different effects, such as localized fouling, under-deposit corrosion of the substrate material, deposit tubercles, and sludge piles. The factors that are most likely to influence deposit structure (and the ensuing effects) include deposit composition and its porosity and permeability, which are all related to feedstock composition. Even minor components of the deposits can sometimes cause severe corrosion of the underlying metal such as the hot corrosion caused by vanadium in the deposits of fired boilers.

However, in addition to feedstock composition, the factors that govern fouling on surface are, in fact, changers are multi-faceted and varied. As already noted, some factors are related to the feedstock properties such as its chemical constituents, API gravity, viscosity, diffusivity, pour point, interfacial properties, and feedstock stability. The propensity of feedstocks to encourage fouling depends on properties such as viscosity and density. Viscosity is an important role for the sublayer thickness where the deposition process is taking place. On the other side, the viscosity and density (usually monitored as API gravity) have a strong effect on the sheer stress, which is the key element in the foulant removal process.

Indeed, the chemical constituents (and their individual or collective behavior) are particularly an important factor that affects the rate and extent of fouling (Chapters 2, 6, and 7). Indeed, the presence in the feed of unsaturated and unstable compounds, inorganic salts, trace elements such as sulfur, nitrogen, and oxygen, as well as the storage conditions (for example, exposure to oxygen during storage) will also affect the nature of the foulant and the rate of fouling.

Moreover, refinery fluids are seldom pure – the intrusion of minute amounts of impurities can initiate or substantially increase fouling, and these impurities can either deposit as a fouling layer or act as catalysts to the fouling processes. For example, chemical reaction fouling may be due to the presence of oxygen and/or trace elements such as nickel, vanadium, and molybdenum. In crystallization fouling, the presence of small particles of impurities may initiate the deposition process by seeding. In addition, impurities such as sand or other suspended particles in the fluid may have a scouring action, which will reduce or remove deposits. Suspended solids promote particulate fouling by sedimentation or settling under gravitation onto the heat transfer surfaces. Since particulate fouling is velocity dependent, prevention is achieved if stagnant areas are avoided. For water, high velocities (above 1 m/s) help prevent particulate fouling. Often it is economical to install an upstream filtration.

In a heat exchanger, the fluid velocity is generally lower on the shell side than on the tube side, less uniform throughout the bundle, and limited by flow-induced vibration. Zero- or low-velocity regions on the shell side serve as ideal locations for the accumulation of foulants. If fouling is expected on the shell side, attention should be paid to the selection of baffle design – segmental baffles have the tendency for poor flow distribution if spacing or baffle cut ratio is not in correct proportions. Too low or too high a ratio results in an unfavorable flow regime that favors fouling.

6.2.3 Asphaltene Deposition

Phase separation leading to solid deposition and fouling from crude oil and from viscous feedstocks is a serious problem as it can cause plugging and malfunction of refinery equipment, and nothing is more sure than the occurrence of such problems during thermal cracking processes and catalytic cracking processes.

The crude oil system (including the various systems of the viscous feedstocks) are delicately balanced system. Disturbance of any one of these systems in the crude oil system can result in irreversible flocculation of asphaltene constituents which can severely reduce the permeability of the reservoir, cause formation damage, and can also plug up the wellbore and tubing. This can occur either during pre-processing operations (such as blending) or during thermal operations.

Thus, the crude oil system and the viscous crude oil systems are systems in which the various constituents are in balance, and any change in the physical and chemical properties of crude oil can be a key factor in the stability of the system. Any factor such as changes in pressure, temperature, or composition that disrupt this adsorption equilibrium and precipitation can cause asphaltene separation and deposition. Precipitation and deposition may occur during primary production and during the displacement of reservoir oil by carbon dioxide, hydrocarbon gas, or water and gas application.

For example, as the degree of conversion in a thermal process increases, the solubility power of the medium toward the heavy and polar molecules decreases due to the formation of saturated products (Speight, 1994). This is reflected in a relative change in the solubility parameters of the dispersed and solvent phases leading to a phase separation.

Association of the asphaltene constituents in any asphaltene-containing feedstock and the appearance of a separate phase (often referred to as flocculation) in paraffinic crude oils or when an asphaltic crude oil (asphaltene-containing crude oil) is often irreversible, even when the conditions are returned to pre-flocculation point. This is the major cause of unrepairable arterial blockage damage to the flow of crude oil fluids. Due to the large size and the adsorption affinity of the asphaltene constituents for solid surfaces, flocculated asphaltene constituents can cause irreversible deposition which is not always easy to remove.

Asphaltene constituents (and resin constituents) in the viscous feedstocks can cause major problems in refineries through the phase separation of insoluble products and through unanticipated coke formation (Speight, 2014a; Yan et al., 2020). The thermal decomposition of these constituents has received some attention with the objective of examining the products and the potential of these products to form coke (Speight, 2014a, 2017). The data contradicted earlier theories that coke formation was predominantly polymerization reaction, and in fact, the initial stages of coke formation involved the separation of lower molecular weight insoluble products from which coke was produced by further reaction of these products. There is an induction period before coke begins to form that seems to be triggered by phase separation of reacted asphaltene. The phase separation is likely triggered by changes in the molecular structure of the resin constituents or changes in the molecular structure of the asphaltene constituents or both, and the products are isolated as insoluble carbene or carboids (Speight, 2014a, 2015a). In addition, the organic nitrogen originally in the resin and asphaltene constituents invariably undergoes thermal reaction to concentrate in the non-volatile coke and is considered likely that carbon-carbon bonds or carbon-hydrogen in a heterocyclic nitrogen ring system may be susceptible to thermal decomposition as an initial event in the thermal decomposition process. When denuded of the attendant hydrocarbon moieties, nitrogen heterocyclic systems and various polynuclear aromatic systems are undoubtedly insoluble (or non-dispersible) in the surrounding hydrocarbon medium. The next step is the gradual carbonization of these entities to form coke (thermal fouling).

One of the effective remediation methods for the organic (fouling) deposits is the use of strong aromatic solvents or naphthenic solvents (such as the naphtha solvents) that could remove the asphaltene deposits. Flocculated asphaltene constituents (aggregates) can form steric colloids in the presence of excessive amounts of resin constituents, but, knowing this, there is still the need to predict when and how phase separation will occur.

A crude oil refinery uses processes that cause physical and chemical changes in constituents of crude oil and involves specialized unit processes that cause changes to the chemical composition of the feedstock and even changes to the chemical structure of the constituents (Chapter 3) (Parkash, 2003; Gary et al., 2007; Speight, 2014a; Hsu and Robinson, 2017; Speight, 2017). Furthermore, understanding refining chemistry not only allows the refiner to understand the means by which these products can be formed from various feedstocks but also offers a chance of process predictability and the means by which fouling may be mitigated.

Even though refining chemistry might be represented by relatively simple equations, the chemistry of conversion process may be quite complex. In fact, the complexity of the individual reactions occurring in an extremely complex mixture, and the interference of the products with those from other components of the mixture or from the original unchanged or partially changed feedstock is unpredictable and also can lead to phase separation and fouling. Similarly, the interference of secondary and tertiary products with the course of a reaction and, hence, with the formation of primary products may also be cause for concern, also leading to fouling.

6.2.4 Wax Deposition

Paraffin wax is a mixture of a range of non-polar high molecular weight alkane derivatives (hence the name paraffin wax) that can crystallize from crude oils or crude oil products or wax-containing solutions primarily due to a decrease in the temperature of the system. The wax is deposited as a solid where there is a change in thermodynamic equilibrium, such that the temperature falls below the cloud point (i.e., the critical point in the rheology of waxy crudes).

More specifically, the wax consists of branched (iso), cyclic, and straight chain (normal) alkane derivatives that have chain lengths in excess of 17 carbon atoms (C_{17+}) and potentially up to C_{100+}. However, despite the fact that crude oils are extremely complex systems containing a multitude of components, it is generally accepted that the crystallizing materials that form the deposits are primarily the n-alkane derivatives. Therefore, in order to obtain a greater insight on the formation of wax deposits to prevent and solve these problems, it is necessary to develop and understand knowledge of the mechanisms involved in the n-paraffins crystallization process.

In terms of the properties, paraffin wax is a white, odorless solid with a typical melting point between approximately 46°C–68°C (115°F–154°F) with a density on the order of approximately 0.9 g/cm³. Wax has low thermal conductivity, has a high heat capacity, and is insoluble in water. While constant deposition of wax can block production lines, it can also act as insulation due to its low thermal conductivity and high heat capacity, resulting in higher arrival temperatures during steady flowing conditions and longer cooling times during equipment shutdown. Paraffin wax is soluble in ether, benzene, and certain esters, while being unaffected by most common chemical reagents. Physically, the hydrocarbon constituents of wax can exist in various states of matter (which are gas, liquid, or solid) depending on their temperature and pressure. When wax freezes and forms crystals, the crystals are commonly referred to as macrocrystalline wax while the crystals formed from naphthenes are known as microcrystalline wax.

Many techniques have been developed to quantify the composition of crude, but by far the most utilized method is by gas chromatography (GC) by which a complete characterization up to carbon number C_{100} can be performed. Detection levels of ppb (parts per billion) can be repeatedly obtained with small sample sizes. In the modern equipment, the inlet, chromatographic column, and detectors can be changed to suit the desired analysis. However, some detectors will not detect nitrogen, carbon dioxide, or anything else that will not burn.

6.2.4.1 Deposition

Wax deposition is a thermodynamic phenomenon that will lead to deposition of solid wax crystals, and fouling by the deposition of paraffin wax has significant effects on crude oil production, transportation, storage, and refining. The buildup of wax deposits decreases flow patterns in the reservoir, decreases the cross-sectional area in pipes and pipelines, restricts operating capacity, and places additional strain on pumping equipment. In addition, wax deposition – the terms deposition and precipitation are used interchangeably in this text – in flowlines and pipelines is an issue impacting the development of deepwater subsea hydrocarbon reservoirs.

Generally, fouling by the deposition of wax can be mild, or it can be severe enough that it is unmanageable, and the sooner the issues is diagnosed, the easier it is to design a preventive or control management plan that will reduce or eliminate technical and economic problems. Technical issues associated with wax deposition include: (1) permeability reduction and formation damage when it occurs in the wellbore and in the vicinity of the wellbore, (2) reduction in the interior diameter and eventual plugging of flow channels and production lines, (3) changes in the crude oil composition and behavior due to phase separation of wax, and (4) additional strain on pumping equipment owing to increased pressure drop along flow channels as wax fouling increases.

Pressure has a significant effect on the temperature at which wax appears in crude oil although it has been proposed that a decrease in the wax appearance temperature is only pronounced at pressures below the bubble point pressure of the crude oil, and that above the bubble point pressure, the wax appearance temperature generally increases with pressure.

The deposition of wax gel has been suggested to follow a process that can be described by the following five steps: (1) gelation of the waxy oil or formation of an incipient gel layer on the cold surface, (2) diffusion of waxy hydrocarbon derivatives with carbon numbers greater than the critical carbon number toward the gel layer from the bulk oil, (3) internal diffusion of the through the trapped oil, (4) precipitation of the waxy hydrocarbon derivatives through the trapped oil, and (5) counter diffusion of de-waxed oil – hydrocarbon derivatives with carbon numbers lower than the critical carbon number – out of the gel deposit layer. The last three steps – steps 3, 4, and 5 – are proposed to be responsible for the increase in solid wax content of the wax gel deposit.

6.2.4.2　Factors Leading to Wax Deposition

Wax can form anywhere in the producing system when the conditions due to temperature and pressure changes become favorable for the precipitation of the paraffin constituents. However, wax deposition typically occurs in wells that produce at lower rates and leads to severe recovery well downtime, and although wax deposition is a commonly encountered problem in production operations, there is no universally effective treatment for the problem. Treatment methods are usually highly case dependent, requiring the proper identification of the mechanisms for wax deposition and the development of a predicting technique that is specific for the target field. Thus, the factors that influence wax deposition and fouling are (1) feedstock composition, (2) temperature, (3) pressure, (4) wax crystallization, and (5) others.

6.2.4.2.1　Feedstock Composition

Refinery feedstocks are variable in composition but is typically composed of fractions that are named according to the method of separation: (1) saturates, (2) aromatics, (3) resin constituents, and (4) asphaltene constituents (Chapters 4 and 5) (Speight, 2001, 2014a; Speight, 2015a). These components are in thermodynamic equilibrium at initial reservoir conditions, and the aromatic constituents serve as solvents for high molecular weight saturates, which are the sources of paraffin waxes in crude oil while the polar components, especially asphaltene constituents, induce wax nucleation. However, the solubility of paraffins in aromatic, naphthenic, and other organic solvents is diminished at low temperatures, and lower-boiling constituents of crude oil may assist in maintaining high molecular weight constituents in solution.

If there is any effects that alter the original composition of the feedstock system, the result can be decreased stability of the system and deposition of the wax constituents. This loss of solubility could lead to the deposition of the wax. Generally, the structural distribution of the paraffin components and the occurrence of other solids such as (1) formation fines, (2) corrosion products, and (3) the presence of asphaltene constituents which could form nucleating sites all contribute to wax precipitation and deposition. Therefore, knowledge of the feedstock composition gives an indication of the feedstock stability and, hence, the propensity of a feedstock for wax deposition.

6.2.4.2.2　Temperature

Wax deposition occurs when the wax constituents contained in the feedstock reach their solubility limit due to one or more changes in the equilibrium conditions (stability or instability) in the feedstock, which causes a decrease in solubility of the wax constituents. The solubility limit for the wax constituents is dependent on the temperature and, as such, can be defined by the temperature, given other specified factors. While some of the factors influence wax precipitation by increasing or decreasing the solubility limit (depending on the temperature), other factors may provide a favorable environment for deposition to occur. Such factors include oil composition plus available solution gas and pressure of the oil which affects the amount of gas in solution. Other factors are flow rate, completion, and pipe or deposition surface roughness. Typically, paraffin solubility increases with increasing temperature and decreases with decreasing temperature. Thus, temperature appears to be the predominant and most critical factor in wax precipitation and deposition due to its direct relationship with the solubility of paraffin hydrocarbon derivatives. Thus, wax deposition occurs when the operating temperature is at or below the wax appearance temperature (cloud point temperature), and wax deposition will not occur until the operating temperature falls to or below the wax appearance temperature. All other factors actually lead to wax deposition when the temperature is already at or below the cloud point.

6.2.4.2.3 Pressure

The wax appearance temperature increases with increase in pressure above the bubble point, at constant composition which indicates that an increase in pressure in the one-phase liquid region (above the bubble point pressure) will favor wax deposition (Brown et al., 1994). The situation is different below the bubble point where there is two-phase existence, and the wax appearance temperature decreases with increase in pressure up to bubble point pressure (Brown et al., 1994) due to dissolution of light ends back into the liquid phase. The wax appearance temperature increases with increase in pressure for some oils (Brown et al., 1994) and increases with increase in pressure for a fixed component liquid mixture (Pan et al., 1997).

6.2.4.2.4 Wax Crystallization

Crystallization is the process of separation of solid phase (which typically appears as wax crystals) from a homogenous solution. Paraffin wax constituents remain in solution as natural components of crude oil until temperature decreases to the point of the wax solubility.

Wax crystal forms through (1) nucleation and (2) growth, in that order but nucleation and growth occur simultaneously in the oil system, with one or the other predominating at a given time. As the solubility limit is approached, the wax molecules form clusters which grow larger and become stable upon reaching a certain critical size. The critical size is dependent upon the prevailing condition, but the clusters re-dissolve when the critical size is not attained and become unstable.

6.2.4.2.5 Other Factors

Although feedstock composition, temperature, and pressure have the most significant effects on wax deposition, other factors that have been identified as contributing to wax deposition include (1) the feedstock flow rate, (2) the gas-feedstock ratio, and (3) pipeline wall roughness.

In terms of flow rate, wax deposition is influenced more by laminar flow than when flow is in the turbulent regime, and increasing the flow rate from laminar to turbulent reduces maximum deposition rate and at the same time lowers the temperature at which maximum deposition rate occurs (Hsu et al., 1994). Low flow rates offer the moving feedstock an increased residence time in the flow channel, which allows more heat loss to the surroundings, leading to a higher chance of the bulk oil temperature falling below the wax appearance temperature and sufficient time for wax separation and deposition. In addition, there is a difference in texture between wax deposited at high flow rates and wax deposited at low flow rates.

The gas/oil ratio influences wax deposition in a manner that depends on the pressure – above the bubble point, where all gases remain in solution, solution gas helps to keep wax in solution. A high gas/oil ratio would result in more expansion and subsequent cooling as pressure of the oil system depletes – a situation that can increase wax deposition.

6.2.4.3 Mechanism

Wax fouling occurs when the temperature of a feedstock (or product) containing wax constituents falls below the cloud point, and the n-paraffin components begin to crystallize into solid wax particles. These particles can adhere to each other and to surfaces such as when the feedstock is in contact with any surface that has a temperature below the wax appearance temperature (WAT), thereby acting as a heat sink.

The predominant mechanisms that have been proposed to describe wax deposition are (1) molecular diffusion and (2) shear dispersion.

The former term (i.e., molecular diffusion) refers to the process by which the radial temperature gradient in the line causes a concentration gradient of dissolved paraffin components in the liquid phase. This concentration gradient causes paraffin to diffuse to the pipe wall, where it is assumed to deposit. The widely recognized transport methods contributing to wax thickness on the pipe wall are molecular diffusion of dissolved wax, particle transport of precipitated wax, and sloughing of previously deposited wax.

The latter term (i.e., shear dispersion) is the relationship between deposition rate and shear rate – shearing of the wax molecules occurs due to the hydrodynamic drag of the flowing fluid which depends on the flow rate and viscosity of the fluid. Higher viscosity and low flow rates result in high wax deposition rates, but in conditions of highly turbulent flow, wax deposition rates decrease with increased flow because wax is mechanically sheared off the deposits on the pipe wall.

An understanding of the phase behavior of the hydrocarbon derivatives in the feedstock will give an indication of the role of phase behavior in wax deposition.

The phase behavior of hydrocarbon derivatives and hydrocarbon mixtures are an important aspect of feedstock crude oil behavior. The production and treatment of the feedstock require knowledge of phase behavior in which pressure-volume-temperature (PVT) properties play an important role throughout the life of a reservoir. Furthermore, accurate PVT properties are required for the design of refineries.

Precipitated wax can be detected in different forms by different techniques such as (1) the form of the quantity of wax precipitated, relative to the oil quantity or composition; (2) the size of the wax crystals; and (3) the number of wax crystals but all of these affect the sensitivity of the measurement techniques. In addition, some techniques detect wax crystals at the microscopic level (nucleation stage of wax crystal formation), while others detect wax crystals at the early stage of growth. However, it has been suggested that none of the available techniques is able to measure the true wax appearance temperature, where the first crystal appears under thermal equilibrium. In this sense, the wax appearance temperature is an indicator of the formation of the solid (wax deposition) phase.

6.3 FOULING DURING THERMAL CRACKING

Briefly, the visbreaking and thermal cracking processes are processes for the conversion of viscous feedstocks into distillate products. Typically, the coke obtained is usually used as fuel, but processing marketing for specialty uses, such as electrode manufacture, production of chemicals, and metallurgical coke, is also possible and increases the value of the coke. For these uses, the coke is not regarded as a foulant but as a saleable product which will require treatment to remove sulfur and metal impurities. Calcined petroleum coke can be used for making anodes for aluminum manufacture and a variety of carbon or graphite products such as brushes for electrical equipment. These applications, however, require a coke that is low in mineral matter and sulfur.

Chemically, the thermal decomposition of the feedstock constituents can be visualized as a series of thermal conversions (Speight, 2014a). The reactions involve the formation of transient free radical species that may react further in several ways to produce the observed product slate. However, because of the multiplicity of possible reactions (as well as secondary and tertiary reactions), the slate of products from thermal cracking is considered difficult to predict.

The available data suggest that thermal conversion (leading to coke formation) is a complex process involving both chemical reactions and thermodynamic behavior (1988; Wiehe, 1993; Speight, 1994; Speight, 2017) (Chapters 3 and 6) and can be summarized as follows: (1) thermal reactions of crude oil constituents result in the formation volatile products; (2) thermal reactions of crude oil constituents also result in the formation of high molecular weight and high polarity aromatic components; and (3) once the concentration of the high molecular weight high polarity material reaches a critical concentration, phase separation occurs giving a denser, aromatic liquid phase.

Reactions that contribute to this process are cracking of side chains from aromatic groups, dehydrogenation of naphthenes to form aromatics, condensation of aliphatic structures to form aromatic, condensation of aromatics to form higher fused-ring aromatics, and dimerization or oligomerization reactions. Loss of side chains always accompanies thermal cracking, while dehydrogenation and condensation reactions are favored by hydrogen-deficient conditions. Formation of oligomers is enhanced by the presence of olefins or diolefins, which themselves are products of cracking. The condensation and oligomerization reactions are also enhanced by the presence of Lewis acids, for example, aluminum chloride ($AlCl_3$).

Operational deposits on the side of a cracking unit will typically be a mixture of organic and inorganic materials consisting of iron sulfate derivatives and iron sulfide derivatives, and the consistency of these deposits can range from a hard coke-like material to a sludge. The coke deposits will usually be found in the heat exchanger bundles in the cycle oil line between the bottom of the fractionator and the reactor. Coke deposits may also be found in the feed stream pre-heat furnace. The foulants on the water side of the heat exchangers will be primarily water-borne scales and corrosion products. The air side of the fin-fan coolers will become fouled with a soot-like coating of ordinary dirt which can usually be removed by spray washing with a hot neutral detergent-based solution containing a suitable surfactant.

The importance of solvents to mitigate coke formation has been recognized for many years, but their effects have often been ascribed to hydrogen-donor reactions rather than phase behavior. The separation of the phases depends on the solvent characteristics of the liquid. The addition of aromatic solvents will suppress phase separation, while paraffins will enhance separation.

Microscopic examination of coke particles often shows evidence for the presence of a mesophase; spherical domains that exhibit the anisotropic optical characteristics of liquid crystal. This phenomenon is consistent with the formation of a second liquid phase; the mesophase liquid is denser than the rest of the hydrocarbon, has a higher surface tension, and likely wets metal surfaces better than the rest of the liquid phase. The mesophase characteristic of coke diminishes as the liquid phase becomes more compatible with the aromatic material.

6.3.1 FOULING DURING VISBREAKING

Loss of side chains always accompanies thermal cracking, while dehydrogenation and condensation reactions are favored by hydrogen deficient conditions (Chapter 3) (Parkash, 2003; Gary et al., 2007; Speight, 2014a; Hsu and Robinson, 2017; Speight, 2017). Formation of oligomers is enhanced by the presence of olefins or diolefins, which themselves are products of cracking. The condensation and oligomerization reactions are also enhanced by the presence of Lewis acids, for example, aluminum chloride ($AlCl_3$).

Generally, free radical reactions which occur in thermal (non-catalytic) processes lack the control that is evident (from product distribution) in catalytic processes (Chapter 3). The efficiency, and economics, of thermal cracking reactions are based mainly on achievement of maximum conversion while controlling the often run-away free radical reactions. The main barrier to process control (and fouling mitigation) is the loss of stability of resin and asphaltene constituents, causing their precipitation to give fouling in the form of coke, which shortens unit run time by deposition at the heater, pre-heat exchangers, and columns. This is particularly true for visbreaker operations where maximum conversion is achieved by controlling the process severity for any processed feed (typically by controlling heater outlet operating temperature), while the use of chemical additives to mitigate fouling in the form of sediment through the deposition of coke particles (Joshi et al., 2008; Speight, 2012).

A recurring issue with the soaker visbreaker is the need to periodically de-coke the soaker drum and the inability of the soaker process to easily adjust to changes in feedstock quality, because of the need to fine-tune two process variables, temperature, and residence time. Recent combinations of visbreaking technology and the addition of new coil visbreaker design features have provided the coil process with a competitive advantage over the traditional soaker visbreaker process. Limitations in heater run length are no longer a problem for the coil visbreaker. Advances in visbreaker coil heater design now allow for the isolation of one or more passes through the heater for decoking, eliminating the need to shut the entire visbreaker down for furnace decoking.

Overall, the main limitation of the visbreaking process, and for that matter, all thermal processes, is that products can be unstable, due to the presence of unsaturated products. For example, thermal cracking at low pressure produces olefins (and di-olefins), particularly in the naphtha fraction. These olefins give a very unstable product, which tends to undergo secondary reactions to form gum and intractable non-volatile tar.

6.3.1.1 Types of Foulants

The types of foulants that are produced during thermal and catalytic processes are very much dependent upon the composition and properties of the feedstock. Thus, the foulants can be either (1) organic material or (2) inorganic material. The organic foulants arise from the decomposition of the constituents of the feedstocks (such as asphaltene constituents and resin constituents) while the inorganic material can arise from inorganic material (such as mineral matter that was not removed during the dewatering and desalting processes – such removal is often difficult when the feedstock is one (or more as a blend) of the viscous feedstocks which can (literally) "hide" the mineral matter because of the high viscosity of the feedstock) and carried into the reactor by the feedstock or the presence of organometallic constituents (such as the porphyrin derivatives) in the feedstock.

Inorganic solids arise from the viscous feedstocks (heavy crude oil, extra heavy crude oil, tar sand bitumen, and distillation residua) as well as corrosion flakes/particles from upstream equipment. Residua must also be included here as a source of inorganic solids because of the tendency for the distillation process to be a concentration process, whereby any solids that have been carried through the dewatering/desalting unit are collected in the residuum from the atmospheric distillation unit (Parkash, 2003; Gary et al., 2007; Speight, 2014a; Hsu and Robinson, 2017; Speight, 2017). In addition, inorganic or organic salts ($RNH_3^+Cl^-$) that form from amine and chloride components are also solids that will influence the operation and outcome of the visbreaker. Other sources of inorganic solids are any type of slop oil that has been included in the feedstock which will enter the process directly and will have immediate consequences.

Also, in the presence of olefin derivatives, the asphaltene constituents have a significant effect on the fouling tendency of thermally processed products, whereas in the absence of the olefins, the impact of the asphaltene constituents on the fouling tendency was limited. Similarly, olefin derivatives have a limited effect on the fouling tendency of visbreaking product in the absence of asphaltene constituents (Xing et al., 2021).

6.3.1.2 Impact of Foulants

As with any process, the foulants can have different effects on the process units and on the ancillary units. The organic solids are primarily products that arise from the thermal decomposition of resin and asphaltene constituents and are: (1) coke particles and/or (2) coke fines and/or (3) incompatible or reacted asphaltene constituents. For the most part, organic solids are typically hydrocarbonaceous solids insofar as they contain carbon and hydrogen as well as metals, sulfur, oxygen, and nitrogen. The volatile products exiting the visbreaker (prior to quenching the hot product mix) can entrain very small tarry particles that unless removed from the volatile stream will foul the fractionator. Different crudes and feedstock blends have a different tar-generating propensity in the visbreaking process – the blends interact differently under different soaker and coil conditions.

The nature of the process virtually dictates that organic solids will be present either because of (1) feedstock composition or (2) because the process operates on the barrier between no coke formation and coke formation. However, inorganic solids may also be present if the heavy feedstocks have come straight into the refinery to the visbreaker, whether or not the feedstocks have been through the desalter.

In such cases, any particles in the feedstock may promote premature coke or sediment and influence the course of the reaction (Chapters 1 and 3). Furthermore, the foulants may originate and occur in the reactor, but (because of the potential for an induction period before phase separation occurs) (Magaril and Aksenova, 1967, 1968; Magaril and Ramazaeva, 1969; Magaril and Aksenova, 1970a, 1970b; Magaril et al., 1970, 1971; Magaril and Akesnova, 1972; Wiehe, 1992; Speight, 1992; Wiehe, 1993; Speight, 1994, 2014a) their effects are felt to a much greater extent in the ancillary equipment. In fact, the movement of the foulants from the reactor is, as is often the case with fast-moving fluids, influenced by the velocity of the volatile materials leaving the soaker (or coil) – there is a high potential for entrainment of solids in the volatile streams. In this respect, the foulants in visbreaking will have some effect on the product distribution but also on any ancillary equipment such as (1) heat exchangers and (2) the coker fractionator.

6.3.1.2.1 Heat Exchangers and Heaters

As with many thermal processes – and visbreakers are no exception – any feedstock preheat units used to increase the temperature of the heavy feedstock will be subject to the presence of potential foulants, particularly inorganic foulants.

Heat exchangers are used to recover heat from product streams to preheat the feedstock which minimizes the fuel that is needed for the furnace to raise the feed to the required process temperature. As a result, since heating of the feedstock reduces the viscosity, this can create accumulations of solids in the low-velocity zones in the pipes – lower viscosity allows solids to settle more readily at a higher rate. Furthermore, high-asphaltene feedstocks will give rise to a similar effect (Chapter 6), and such feedstocks may require frequent removal of the organic or inorganic foulant.

In a crude preheat exchanger system, the hot preheat section is usually the area of greatest concern. The hottest exchangers or exchangers with the highest heat flux typically show the highest fouling rates. This is also the case for the catalytic cracker feedstock/effluent heat exchangers and the slurry heat exchangers.

An additional contributor to fouling is the presence of sodium, which can be present as an inorganic solid (salt, caustic) or an organic (sodium naphthenate from the residuum of high-acid crude oils), which can accelerate fouling in the heater. In addition, iron oxide or any oxygen-containing contaminant will accelerate heater tube fouling. Moreover, iron oxide will react with asphaltene constituents in the delayed coker feed causing rapid fouling in the heater.

6.3.1.2.2 Fractionator

The fractionator is used for the fractionation of the volatile products which may (or typically) contain entrained solids, depending upon the reaction parameters of the visbreaker. If the carryover of fines is the case, the fines will accumulate in low-velocity zones, and such fines may not be filterable – the fractionator bottoms system may act as a strainer (filter) to remove large particles but does not always capture fine particles. As a result, the performance of the fractionator may be severely impacted by the accumulation of fines and sludge in the lower section of the tower, where low-tray velocity can increase the risk of the accumulation of solids (tray fouling). This may result in the need to clean the bottom circulation system filters on a regular basis depending on the unit vapor velocity.

6.3.1.2.3 Visbreaker Bottoms

The bottoms from the visbreaker (visbreaker bottoms, visbreaker tar) have a higher fouling tendency than the unit feed due to the presence of the asphaltene constituents and their potential to precipitate and result in coke and deposits. This loss of stability is related to the thermal cracking of asphaltene constituents and their associated stabilizing resin constituents – the resins and asphaltene constituents de-alkylate to produce lower molecular weight but less soluble, polar aromatic products (ASTM D7157). At the same time, the other dealkylated product (paraffins and olefins that have been sheared from aromatic systems) which remain in the liquid phase approximate the additions of a paraffin solvent as used in the laboratory deasphalting process or in the commercial deasphalting processes (Speight, 2014a, 2014b). Once the concentration of resin product and asphaltene products exceeds a solubility limit that is specific to the feedstock and process parameters (Magaril and Aksenova, 1967, 1968; Magaril and Ramazaeva, 1969; Magaril and Aksenova, 1970a, b; Magaril et al., 1970, 1971; Magaril and Akesnova, 1972; Wiehe, 1992; Speight, 1992; Wiehe, 1993; Speight, 1994, 2014a), separation and fouling occur. In addition, the rate of fouling (sediment formation) increases markedly with temperatures and residence time of the feedstock in the hot zone. Fouling in the heat exchanger by asphaltene or reacted asphaltene deposition (Chapter 6) is another area of concern for visbreaker operations. This is due to the instability of the visbreaker bottoms and the tendency of the reacted resin and asphaltene product to form a separate phase due to the complexity of the various reactions (Chapter 3) (Virzi and Respini, 2014).

The consequence of particle in residua has already been discussed (Chapter 8), but it is worth mentioning again at this point. Under typical visbreaker operations, coke formation is undesirable because the presence of particles in the visbreaker products can have serious consequence for (1) the usability of the gas oil – due to particle entrainment leading to catalyst fouling in a fluid catalytic cracking unit, and (2) particles in the visbreaker tar assuming that the tar might be sent to the asphalt plant – a poor grade asphalt would, more than likely because of though presence of the particles, be priced at a much lower level than a good grade (particle-free) asphalt.

6.3.2 Fouling during Coking

Fouling, as it occurs in coking processes, in not the same type of fouling as occurs in visbreaking processes. Recall, in visbreaking, the goal is to produce liquid product with the process parameters stopping short of the point at this coke formation occurs (Chapter 4). The thermal reactions are not allowed to proceed to completion. On the other hand, the thermal reactions in coking processes are allowed to proceed to completion with the formation of liquid (distillable) products and coke. Thus, during this process, the asphaltene and resin constituents in the feedstock are converted to coke in accordance with their respective carbon residue values (ca. 50% w/w for asphaltene constituents and ca. 35% w/w for resin constituents) (Speight, 2001, 2014a, 2015a). On this basis, the occurrence of fouling might not be expected, but it does occur through the presence and/or generation of fine particulate matter (particulate fouling).

In one form or another, the coking process (initially known as thermal cracking) has been in use for approximately 100 years during that time the process has undergone continuous improvements to make it a primary process to provide the upgrading needs to convert heavy feedstocks to saleable products. Other processing options for heavy feedstocks are equally viable such as fluid catalytic cracking and hydrocracking as well as the non-thermal process solvent deasphalting which is often piggy-backed on to (or just prior to) a thermal (non-catalytic or catalytic) process (Parkash, 2003; Gary et al., 2007; Speight, 2014a; Hsu and Robinson, 2017; Speight, 2017).

In the coking process, the occurrence of solids in, for example, the viscous feedstocks (such as mineral matter and/or any type of particulate matter) can arise from several different sources presenting an opportunity for fouling because of the ability of the solids to reduce the ability of the coker to reach full capacity utilization. Both inorganic and organic solids can significantly impact unit reliability, and in some cases, sudden changes in the refinery or a major upset in the refinery desalter-dewatering unit (Chapter 8) can cause shutdown of the coking unit.

6.3.2.1 Types of Foulants

As for visbreaking operations, foulants vary with the feedstock and may be either (1) organic solids or (2) inorganic solids, and each can have different effects on the process units and on the ancillary units.

For the most part, the organic solids are typically hydrocarbonaceous solids insofar as they contain carbon and hydrogen as well as metals, sulfur, oxygen, and nitrogen. Coke fines are the single largest type of fines in the delayed coker and are a problem in both the primary process operations and the ancillary process operations such as coke cutting and coke handling. When the coke drums are being filled, coke fines can be entrained in the overhead liquids (due to high vapor velocity and high foam front) and thence into the fractionator (fractionator fouling). In addition, fouling can occur due to the formation of sediment because of incompatibility of the cracked products – the causes include separation of reacted asphaltene products, wax formation, or other feedstock constituents that give rise to solid separation as the composition of the feedstock changes. The sedimentation process can be reversible (waxes) or non-reversible (reacted asphaltene separation).

The volatile products exiting the coke drum can also entrain very small pitch particles (in the form of an aerosol spray or mist), which is a function of the vapor velocity, the coke level in the drum, the amount of foam, and the presence of unstabilized coke, at the top of the drum. Different crude blends have different foaming potential in thermal processes – the blends interact differently under different coke drum operations such as temperature, recycle, pressure, cycle time, and drum utilization (coke level in the drum).

Inorganic solids arise from heavy oil, extra heavy oil, and tar sand bitumen as well as catalyst material (from a fluid catalytic cracking unit) or corrosion flakes/particles from upstream equipment. Residua must also be included here as a source of inorganic solids because of the marked tendency for the distillation process to be a concentration process whereby any solids that have been carried through the dewatering/desalting unit are collected in the residuum from the atmospheric distillation unit. In addition, inorganic or organic salts ($RNH_3^+Cl^-$) that form from amine and chloride components are also solids that affect the operation of the coking unit.

Inorganic solids also include catalyst fines for a fluid catalytic cracking unit, corrosion deposits, and other solids in the heavy feedstock (Speight, 2013, 2014b). Other sources of inorganic solids are slops originating from coke drum overhead line quench and grey water used to cool or quench coke in the drum. The solids found in the water are typically coke fines, but the water can also contain a variety of inorganic solids as well. Solids, both organic and inorganic, introduced in the cooling/quench of the coker rarely enter the primary processing operations and are confined to the coke cutting and quench water system and the coke handling system. However, slops injected into the coke drum overhead line as a line quench will enter the process directly and will have immediate consequences.

6.3.2.2 Impact of Foulants

The nature of the delayed coking process virtually dictates that organic fines will be present either because of (1) feedstock composition, especially in the delayed coking process, or (2) coke attrition, especially in the fluid coking and flexicoking processes. However, inorganic solids are not as typical, and it is the interaction of the organic and inorganic solids which cause a majority of the fouling problems.

However, unlike the visbreaking process where the foulant occurs in the reactor, as premature coke or sediment, and influences the course of the reaction in the coking process, the foulants may originate and occur in the reactor, but their effects are felt to a much greater extent in the ancillary equipment. In fact, the movement of the foulants from the reactor (coker) is influenced by the velocity of the volatile materials leaving the coker and the high potential for entrainment of solids in the volatile streams. In this connection, the foulants in delayed and fluid coking while having some effect on the coker itself are similar insofar as the foulants affect (1) heat exchangers and (2) the coker fractionator.

Inorganic foulants will cause the corrosion of process equipment will form ferrous-based corrosion products, such as iron sulfide (FeS) or iron oxide, i.e., ferric oxide (Fe_2O_3) which will deposit in heat exchangers, especially where the feedstock velocity is low. Solid and inorganic contaminants in feedstocks (such as sand and silt) can also deposit in the heat exchanger and cause hydraulic or thermal obstructions. As the feedstock is heated in the preheat train, the viscosity of the feedstock oil is lowered and the deposition of solids increases.

Although most of the salts in the feedstock should be removed via the desalting process, some inorganic salts can remain in the feedstock and cause deposition in the preheat train. To combat this, a caustic solution may be injected into the desalted feedstock for distillation tower overhead corrosion control. However, caution is advised, because this practice is introducing another inorganic material which can enhance the potential for fouling in downstream exchangers and coking in downstream process furnaces. In a fluid catalytic cracking slurry system, catalyst fines can contribute to significant fouling.

Organic fouling results from the precipitation of organic components which become insoluble in the system, such as asphaltene constituents and high-molecular weight hydrocarbon derivatives (i.e., paraffin wax).

The asphaltene constituents can become unstable because of the blending of incompatible crudes and/or the heating of the fluid, and the phase-separated material can adhere to the metal surface and dehydrogenate, forming coke. Coke formation can also result from thermal degradation of feedstock constituents (especially the resin constituents and the asphaltene constituents) due to prolonged period in the hot zone. Organic fouling on the feedstock side of the catalytic feedstock/effluent exchangers and reactors systems is usually caused by chemical reactions that are initiated by the foulant precursors that are present in the feedstock such as unsaturated constituents, carbonyl constituents, nitrogen-containing constituents, and metal-containing constituents.

In a slurry system of a fluid catalytic cracking unit, chemical reactions can occur which lead to the formation of higher molecular weight products and even coke precursors. Furthermore, high-molecular weight polynuclear aromatic derivatives (often referred to as PNAs) can agglomerate and further degrade on the tube surface and contribute to fouling. These high-molecular weight polynuclear aromatic derivatives result from the cleaving of aliphatic side chains from the aromatic system of the naturally occurring asphaltene constituents as well as from recombination reactions of high molecular weight cracked hydrocarbon derivatives within the transfer line and in the quench zone of the main fractionator.

6.3.2.2.1 Heat Exchangers and Heaters

The feed preheat exchangers increase the temperature of the heavy feedstock and, as a result, reduce the viscosity of the stream. In general, lower viscosity allows solids to settle at a higher rate, creating accumulation of solids in the low-velocity zones in the heads and outlet piping. If there are solids in the feedstock, coke formation in furnace tubes tends to be higher than for non-solids containing feedstocks. Although this effect may be assigned to the presence of fines in the feedstock, high-asphaltene feedstocks will give rise to a similar effect (Chapter 6), and such feedstocks may require pigging to remove the coke as steam air decoking and online spalling are insufficient.

Because of the influx of heaver feedstocks, the coker heater is prone to fouling because of the coking propensity of the feedstocks which is a function of heat flux, fluid velocity, and asphaltene content. The presence of organic or inorganic fines increases the degree of fouling (accumulation of solids) within the heater tubes. In the coker heater, the transport velocity (the velocity that is required to move solids) should be above the settling velocity (salting velocity), and in addition, the unvaporized particles of liquid or continuous liquid phase tend to retain the solids due to the adhesion properties of the fluid.

An additional contributor to fouling is the presence of sodium, which can be present as an inorganic solid (salt, caustic) or an organic (sodium naphthenate from a high-acid crude oil), which can accelerate fouling in the delayed coker heater. In addition, iron oxide or any oxygen-free radical contaminant will accelerate heater tube fouling. Similar to the process of using oxygen (air blowing) and increase asphalt viscosity, iron oxide will react with asphaltene constituents in the delayed coker feed causing rapid fouling in the heater.

6.3.2.2.2 Coker Fractionator

The fractionator is used for the fractionation of the coke drum vapors, which may (or typically) contain entrained solids. These vapors coming from the coke drums are oil quenched (which may contain fines), entering the fractionator between the bottom and the wash section. In the wash section, the coke drum vapors contact the hot wash oil stream, and the wash oil quenches the drum vapors and washes out most of the entrained coke fines. The liquid phase leaving the wash zone is the internal recycle, mixing with the fresh feed, and contains solids from various sources. The heavy coker gas oil is typically the heaviest side product drawn from the fractionator. It is steam stripped, filtered, hydrotreated, and used as feed component for a fluid catalytic cracking unit or as a blending component for fuel oil. Depending on the hydrotreater configuration for fluid catalytic cracking unit feedstock, the heavy coker gas oil may be combined with light coker gas oil before use as a feedstock.

Fines that accumulate in low-velocity zones provided the fluid temperature is sufficiently elevated to lower viscosity enough for solids settling. Solids or fines containing streams within the coker are typically not filterable – the fractionator bottoms system may act as a strainer (filter) to remove large coke particles but does not always capture fine particles. Furthermore, system pumps are prone to erosion due to fines and protection of the seals by flush designs is critical if long run lengths are to be attained. In fact, the performance of the fractionator is impacted by the accumulation of fines and sludge in the lower section of the tower, especially in the heavy gas oil area where low-tray velocity or the occurrence of dry trays increases the risk of the accumulation of solids (tray fouling). Although the fractionator bottoms strainer may be designed to collect fine particles, micron size fines may not be removed by the bottoms strainer and will migrate into the heater charge pump and heater tubes. In addition, the lower tower section – the flash zone or wash zone – will have increased coke build-up or fouling.

Coke fines are carried into the bottom of the fractionator along with other fines entering the system in the bottoms feed stream. This results in the need to clean the bottom circulation system filters about every 2–3 weeks depending on the coke drum vapor velocities. If the fractionator filters are typically designed to remove 3/16 in. particles, the filter will be insufficient to remove smaller particles which will pass into the downstream equipment.

6.4 FOULING DURING CATALYTIC CRACKING

Briefly, the catalytic cracking process uses heat, pressure, and a catalyst to convert a variety of feedstocks (especially the viscous feedstocks and not forgetting the lower-boiling feedstocks into feedstocks for petrochemical units) into lower-boiling (low-density) products with product distributions favoring the more valuable naphtha and middle distillates. Typically, in the past, the feedstocks were usually gas oils from atmospheric distillation, vacuum distillation, coking, and deasphalting processes. These feedstocks (i.e., the gas oils) had a boiling range of 340°C–540°C (650°F–1,000°F). In the modern refinery, the range of feedstock for the process has expanded to include viscous feedstocks (atmospheric residua, vacuum residua, heavy crude oil and associated residua, extra heavy crude oil, and tar sand bitumen) (Sadeghbeigi, 2000; Parkash, 2003; Gary et al., 2007; Speight, 2014a; Hsu and Robinson, 2017; Speight, 2017).

The concept of catalytic cracking is basically the same as thermal cracking, but it differs by the use of a catalyst that is not (in theory) consumed in the process, and it is one of several practical applications used in a refinery that employ a catalyst to improve process efficiency and product slate.

The process requires the presence of a catalyst in the reactor to direct the course of the cracking reactions to produce more of the desired products. In theory, the catalyst is not consumed in the process but, in practice, because of fouling by coke and metals lay-down on the catalyst (Speight, 2014a, 2015b, 2017), frequent catalyst renewal is necessary – the timing of the catalyst renewal is feedstock (and fouling) dependent. Several processes (Sadeghbeigi, 2000; Parkash, 2003; Gary et al., 2007; Speight, 2014a; Hsu and Robinson, 2017; Speight, 2017) currently employed in catalytic cracking differ mainly in the method of catalyst handling, although there is overlap with regard to catalyst type and the nature of the products.

However, many catalytic cracking unit exhibit some degree of coking/fouling. Coke can occur on the reactor internal walls, reactor top head section, inside/outside of the reactor cyclones, reactor overhead vapor line, main fractionator bottom, and fouling of the slurry bottoms pump-around circuit. When coking and fouling occur, they become a problem when they impact throughput or cracking severity.

Fouling is especially possible when the resin constituents or asphaltene constituents of the feedstock interact with catalysts, especially acidic support catalysts, through the functional groups, e.g., the basic nitrogen species, just as they interact with adsorbents. There is also possibility for the interaction of the asphaltene with the catalyst through the agency of a single functional group in which the remainder of the asphaltene molecule remains in the liquid phase. The much less desirable option is the interaction in which the asphaltene constituents react with the catalyst at several points of contact, thereby causing immediate incompatibility on the catalyst surface.

In fact, as with any refinery unit, catalytic cracking units are not denied the production of foulants. The influx of heavy feedstock into the unit either as the sole feedstock (unlikely) or as a blend with gas oil (more likely) all serve to introduce foulants to the process. Preheating the feedstock before the process in a form of mild cracking (controlled visbreaking) can cause phase-separation of the most likely foulants before introduction of the feedstock to the catalytic cracker. Alternatively, mild hydrotreating to remove heteroatoms – the heteroatom constituents of heavy feedstock, especially the nitrogen-containing constituents, are the most likely constituents that initially react to produce coke formers (Speight, 1987) – may also cause a reduction in the generation of foulants.

The fluid catalytic cracking unit has an associated vessel known as a slurry settler tank, which collects catalyst fines that have been suspended in slurry oil from the process. The separated slurry oil is decanted from the top of the vessel for re-use, while the catalyst fines precipitate inside of the slurry tank forming a hard, concrete-like solid. The catalyst solids are impregnated with hydrocarbons that make the catalyst fines a hazardous waste. Recent development in extraction chemistry have resulted in a process that is effective in removing low ppm levels of the numerous constituents that fall into the hazardous waste code allow for a non-hazardous disposal of the spent catalyst fines through conventional non-hazardous waste disposal facilities.

6.4.1 TYPES OF FOULANTS

Depending upon the composition and properties of the feedstock, there can be two types of foulants, which are (1) organic foulants and (2) inorganic foulants.

6.4.1.1 Organic Foulants

The most common type of fouling is organic which is subdivided into two general types of fouling: hard-coke fouling and soft-coke fouling.

An example of hard-coke fouling is solid-coke fragments circulating through the slurry exchanger tubes and restricting flow through the exchanger. The coke fragments generally accumulate on the exchanger tube sheets at the inlet to the tubes resulting in an excessive pressure drop and loss of heat-transfer duty. The hard-coke fragments can originate in the reactor overhead line or in the main fractionator and often become dislodged following a unit shutdown. If the coke fragments are small enough to pass through the suction strainers on the slurry pumps, they can eventually foul the first slurry exchanger in the pump-around loop. Alternatively, hard coke can accumulate in the bottom head of the main column, which restricts the suction of the circulating slurry pumps. Smaller coke particles from the main fractionator or which are formed by side reactions in the slurry pump-around circuit at high main fractionator bottoms temperatures can settle on the tube surface and react even further, resulting in a barrier to heat transfer and slurry flow. Poor feed/catalyst contacting can be a significant source of hard coke formation in the vapor line, which is particularly evident in fluid catalytic cracking units that process heavy feedstocks.

Soft-coke fouling occurs when an insulating barrier is commonly deposited inside the exchanger tubes, reducing the exchanger heat-transfer coefficient. This phenomenon is typically attributed to asphaltene fouling and occurs when the resin and asphaltene constituents (or reacted resin and asphaltene constituents) are susceptible to increased thermal reactivity in the slurry oil circuit, become insoluble in the slurry oil, and begin to precipitate onto the tube surface. In addition, the viscosity of the slurry increases at the tube wall due to the locally cooler temperature leading to adherence of the material to the tubes and, thus, heat resulting in exchanger fouling. A slurry with a high paraffin content may be more prone to fouling (wax fouling) due to the inherently higher viscosity.

Preventing resin and asphaltene constituents from phase separating is critical for the prevention of soft-coke fouling. Generally, a higher aromatic content of the slurry tends to keep asphaltenes in solution, and adding a solvent such as light cycle oil can increase the solubility of any troublesome constituents in the tower bottoms. Adjustment of the composition of the bottoms fraction (as well as the temperature) may be necessary on a regular basis to accommodate changes arising from changes in the composition of the unit feedstock. In addition, the slurry produced from paraffin feedstocks is more likely to cause fouling (wax fouling) and requires a lower temperature of the main fractionator bottoms to minimize fouling.

Soft-coke fouling can also be minimized by a change in the feedstock conversion – a decrease in the feedstock conversion could result in higher slurry exchanger fouling – a higher API gravity feedstock may contain more (alkane) saturated constituents, which serve to precipitate asphaltene constituents (Chapters 5 and 6) (Speight, 2001, 2014a, 2015a) and increase soft-coke fouling.

6.4.1.2 Inorganic Foulants

Inorganic fouling can include catalyst particles or precipitated metals and can also include fouling prompted by corrosion products (iron scale). Catalyst particles are often found in tube deposits and can be identified by the presence of alumina (Al_2O_3), silica (SiO_2), and the rare earth elements.

Briefly, the rare earth elements are a group of 15 elements referred to as the lanthanide series in the periodic table of elements (Figure 6.1).

Although they are not true rare earth elements, scandium and yttrium are included in this categorization because they exhibit similar properties to the lanthanide elements and are found in the same ore bodies. In addition, if active or spent catalyst particles occur in the foulant, it may be the result of organic fouling since the catalyst can accumulate on viscous-precipitated asphaltene constituents or other hydrocarbonaceous materials already present in the tubes.

Antifoulant chemicals have been successfully used to prevent slurry exchanger fouling. The antifoulant chemicals include (1) organic dispersants, which prevent the agglomeration and deposition of asphaltene constituents; (2) inorganic dispersants, which prevent the deposition of catalyst fines or other inorganic foulants such as iron compounds; and (3) coke suppressants, which inhibit condensation reactions and lead to hard coke-like deposits in exchangers. However, the source of the fouling should be identified before a specific antifoulant is applied, and any potential downstream effects of using an antifoulant should also be investigated, such as the effect of the antifoulant on catalyst fines settling in the slurry tanks if an inorganic dispersant is used.

Continuous monitoring of the heat exchanger behavior is critical for the early identification of a slurry exchanger fouling problem early. Monitoring slurry properties such as API gravity, ash content, asphaltene content, and viscosity can indicate when the fluid catalytic cracking unit may

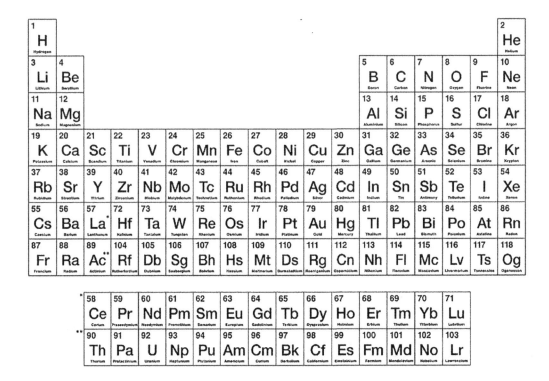

FIGURE 6.1 The periodic table of the elements. *Lanthanide series of elements. **Actinide series of elements.

be more susceptible to fouling. Also, a shift in feedstock properties or unit conversion may also increase fouling.

6.4.2 IMPACT OF FOULING

When fouling does occur in thermal cracking processes or in catalytic cracking processes, the impact is noticeable in (1) heat exchangers and heaters, (2) fractionator, and (3) deactivation of the catalyst.

Thus, the major effects include (1) loss of heat transfer as indicated by charge outlet temperature decrease and pressure drop increase, (2) blocked process pipes, (3) under-deposit corrosion and pollution, and (4) localized hot spots in reactors, all of which culminate in production losses and increased maintenance costs.

6.4.2.1 Heat Exchanger Fouling

Many tubes are needed and the velocity in each of them becomes very low, which both further reduce heat transfer efficiency and make the shell and tube heat exchangers even more prone to fouling, and design options are sought (Andersson, 2007, 2008).

Slurry exchanger fouling is often considered a major issue within the fluid catalytic cracking circle), which causes a reduced heat exchanger efficiency in the slurry/feed preheat exchanger or the steam generators (Hunt et al., 2008). The reduction in the feed preheat temperature that can result from just mild fouling of the FCC slurry/feedstock exchangers can lead to a reduced unit feed rate or conversion (for those units that do not have a fired heater and are air blower limited). Additionally, an excessive pressure drop or inability to cool the slurry to the necessary rundown temperature can require a reduction in the feed rate. When slurry exchangers foul, the feed rate or reactor temperature must often be reduced. Fouling can be generally classified as either organic or inorganic. Several organic or inorganic slurry-fouling mechanisms are possible: (1) organic fouling, which is sub-categorized into hard-coke fouling and soft-coke fouling, and (2) inorganic fouling.

6.4.2.2 Fractionator Fouling

Fractionator fouling is typically the result of particles in fluids. These particles can, for example, be corrosion products from carbon steel equipment in the refinery, sand in seawater, or other particles in poor-quality cooling water and catalyst fines from the reactor chambers in catalytic processes.

However, feedstocks to catalytic cracking units are becoming increasingly heavy, and processing of atmospheric residua is becoming more common – and these residua typically contain traces of chloride salts. This can lead to conditions in which ammonium chloride salts can deposit in the main fractionator and overheads, due to the combination of low temperature and relatively high concentrations of ammonia and hydrogen chloride. The build-up of salt deposits normally occurs in the mid-to-upper sections of the main fractionator column and top pump-around circuits. If not removed, the salts will accumulate and interfere with optimal tower operation, which will ultimately affect the operation of the entire fluid catalytic cracking unit.

Ammonium chloride salt deposition is a particular problem in fractionator towers when sufficient amounts of both ammonia and hydrogen chloride are present. Atmospheric residua and some vacuum gas oils, commonly contain inorganic salts (primarily sodium chloride, NaCl), which dissociate in the riser. The sodium primarily deposits on the catalyst, but the chloride exits the reactor in the form of hydrogen chloride. Since high concentrations of ammonia are usually present in the overhead systems, an increase in hydrogen chloride is usually responsible for driving conditions to the point above the temperature-dissociation constant curve, where deposition will occur. The formation of solid ammonium chloride occurs when the salts separate from the vapor phase:

$$NH_3(g) + HCl(g) \rightarrow NH_4Cl(s)$$

$$g = gas, \ s = solid$$

Ammonium salts are highly water soluble, with negligible solubility in hydrocarbon derivatives. Depending on the level of reactants and the conditions inside the fractionator tower, salts can deposit and accumulate at temperatures above the water dew point, and the accumulated salts can cause severe restrictions in the operation of the fluid catalytic cracking unit. As salt deposits accumulate, the trays in the top section of the tower will be plugged, and as the plugging increases, flooding of the top trays will occur with flooding and a progressive loss of tower efficiency.

In addition, salts such as ammonium chloride are hydroscopic and will absorb water even at temperatures well above the calculated water dew points. Insufficient water washing or poor distribution of water during washing can leave behind damp salts. Localized water condensation can occur due to top pump-around or cold reflux streams returning below the bulk dew point temperature which can lead to under-deposit corrosion of the tower internals as well as corrosion conditions in piping and heat exchanger surfaces. The corrosion by-products can add to the problem of plugged tray and product draws as the corrosion by-products are dislodged by turbulence in the tower. Typically, the shell and trays of the main fractionator columns are constructed of carbon steel, 5 chrome, 410 stainless steel, or 316 stainless steel, which are susceptible to high corrosion rates due to chloride attack (Speight, 2014b).

Methods of preventing salt deposition or removing salt deposits have typically involved (1) temperature adjustment in which the fractionator temperature is increased to a sufficiently high level to ensure the sublimation of all the salts or (2) water washing of the tower. Desalting of the catalytic cracking feedstock is not only an option but a necessity if the continued processing of heavy, contaminated feedstocks is anticipated, especially imported feedstocks that can be contaminated with seawater. Efficient desalting of the crude oil will minimize the salt content of the atmospheric residuum, and there may even be the need to install a desalter immediately prior to the fluid cata-lytic cracking unit. Alternatively, the use of salt dispersant additives acts by chemically binding to the salt deposits, then physically lifting them into the liquid phase, where natural turbulence carries them out of the system. The use of salt dispersants can allow refiners to potentially avoid the prac-tice of water washing of the main fractionator.

As with other processes that require heat through heat exchangers, in the catalytic cracking process, the reactor effluent entering the main fractionator can contain a lot of catalyst fines leading to fouling (Müller-Steinhagen, 1995). The amount depends on the operation of the reactor, but up to 3%–5% w/w of catalyst fines can often be found in the slurry oil from the bottom of the reac-tor, particularly when heavy feedstocks are used. During cooling of the slurry oil, catalyst fines agglomerate and deposit on the heat-exchanger walls, severely reducing heat-transfer performance and increasing the pressure drop. Often, several cleaning periods per year are required for heat exchangers operating in this duty, and during cleaning, a standby heat exchanger is used.

In addition to particle fouling by catalyst fines, chemical fouling is another type of fouling that is primarily found in heat exchangers. When fluids from which salt crystals can form and precipitate, or at high temperatures where there is a risk of coking or of precipitation of asphaltenes, chemical fouling will increase, especially intensify at low-flow velocities and in heat-exchanger dead-zones, such as in the tube turning chamber and/or behind the any baffles. In many shell and tube heat exchangers, the heat-transfer surface area is large.

6.4.2.3   Catalyst Fouling

During the cracking reaction, carbonaceous material is deposited on the catalyst (catalyst fouling) which markedly reduces catalysts activity, and removal of the deposit is very essential to maintain catalyst efficiency – this is usually accomplished by burning the catalyst in the presence of air until catalyst activity is reestablished (the extent of the fouling is reduced). The catalyst, which may be an activated natural or synthetic material, is employed in bead, pellet, or microspherical form and can be used as a fixed bed, moving bed, or fluid bed (Parkash, 2003; Gary et al., 2007; Speight, 2014a; Hsu and Robinson, 2017; Speight, 2017). In terms of the potential for catalyst fouling and reception in catalyst activity and efficiency, each version has advantages and disadvantages which are swayed back and forth by feedstock character.

Several process innovations have been introduced in the form of varying process options, some using piggy-back techniques (where one process works in close conjunction with another process – such as controlled thermal treatment or mild hydrocracking), there are other options that have not yet been introduced or even invented but may well fit into the refinery of the future (Radovanović and Speight, 2011; Speight, 2011a, 2012). Furthermore, recent enhancements made to catalytic cracking units to permit feeding significant amounts of heavy feedstocks to the units, while simul-taneously improving yields and reducing catalyst fouling have focused on improved feed injection and dispersion, reduced contact time of products and catalyst, improved separation of products and catalyst, and regenerator heat removal.

The newer fluid bed catalytic cracking units are expected to process heavy vacuum gas oil and/or feedstocks such as heavy oil as well as extra heavy oil and tar sand bitumen blended with gas oil. As worthy as they are, such use of such blended feedstocks does not mitigate catalyst fouling but, at best, delays it. The feedstocks still contain the precursors that cause catalyst fouling, but at a lower concentration. Furthermore, the technology has been modified in key areas including: (1) catalyst design to accommodate higher metals feed and to minimize catalyst fouling, (2) feed injec-tion, (3) riser pipe design and catalyst/oil product separation to minimize over-cracking, and (4) regenerator design improvements to handle high coke output and avoid damage to catalyst structure (destructive fouling). The use of guard-bed reactors, as used in hydrotreating operations, is another option to reduce catalyst fouling by removal of (1) extraneous particles by filtering and (2) metallic constituents by adsorption and/or by thermal adsorption and subsequent thermal destruction of the filtered/adsorbed materials (Speight, 2000, 2014a).

Over the next two decades, fluid catalytic cracking units will take on two additional roles: (1) acceptance of biomass feedstocks to alleviate concerns over energy supply and security, and (2) reduction in carbon dioxide emissions to alleviate concerns over global climate change (Speight, 2008, 2011a, b). Bio-feedstocks that are able to be processed in the fluid catalytic cracking unit can be categorized as biomass-derived oils (both lignocellulosic materials and free carbohydrates) or triglycerides and their free fatty acids. The operating conditions and catalysts used for each type of feed to achieve a desired product slate vary, and each feed comes with inherent advantages and disadvantages. Caution is needed to ensure that fouling does not increase because of incompatibility issues when biomass constituents are blended with the feedstock and fed to a fluid catalytic cracking unit. The acidity of the oil (caused by the presence of oxygen functions in the biomass) must be reduced by means of a moderate thermal treatment at temperatures in the range of 320°C–420°C (610°F–790°F). Unfortunately, some of the varied constituents of bio-feedstocks tend to coke easily and high acid numbers as well as carbonaceous byproducts – the propensity for catalyst fouling is high. In fact, reactor plugging is a common phenomenon for heavy oil processing due to the deposition of contaminants at the top layer of the catalyst (catalyst fouling), which generates a difference between the reactor entrance and the outlet of the reactor (i.e., called pressure drop). It is also characterized by energy loss and is an important consideration in the design and operation of fixed-bed systems (Ancheyta et al., 2009).

Finally, the catalyst can be formulated with features to minimize fouling by adjustment the catalyst composition to minimize naphthene-aromatics and the paraffin content of the slurry which may improve asphaltene solubility and reduce slurry exchanger fouling. A catalyst with a high tolerance for metal constituents will help avoid fouling and increased catalyst contaminants that result in a decrease in conversion can increase the likelihood of fouling.

6.5   MANAGEMENT AND MITIGATION

It cannot be over-emphasized that management and mitigation of any process unit within the refinery cannot be undertaken without an understanding the operation of the unit. Any changes to feedstock quality, operating conditions, mechanical configuration, or (in the case of catalytic units) the catalyst will affect the operation of the unit and, hence, the management and mitigation of fouling (and the often-ensuing corrosion).

The fouling behavior of the viscous feedstocks (in fact, any feedstock) must be identified and quantified since both fouling by chemical reaction and fouling by the formation of a solid phase have been identified to be responsible for the fouling during the refining of viscous feedstocks, especially on both the shell and tube sides of the shell-and-tube heat exchangers.

Thus, the successful management and mitigation of fouling rely heavily of the ability of the refiner to predict when and where fouling will occur. With the acceptance of the more viscous feedstocks as refinery feedstocks, this is an increasingly challenging task. The approach is to know how the various feedstocks will behave through a series of laboratory studies in order to assess feedstock behavior in thermal processes as well as blend compatibility that is a first approximation of feedstock behavior in the refinery (Speight, 2015b). In many cases, fouling is not a sudden occurrence but may build up relatively slowly and is often managed by exchanger cleaning or improved exchanger design, as well as rapid fouling events. It is imperative for effective refinery operation that rapid events are identified quickly and inferred correctly from plant measurements.

Fouling of the preheat exchangers will increase the heat duty of the furnace up to the point that the maximum furnace capacity is reached, and as that limit is approached, the unit throughput will need to be cut back gradually in order to unload the furnace, and again, significant production losses are incurred. Process stream fouling can lead to increased pressure drop in the exchangers and furnace, and the fouling might be sufficiently severe that the throughput is reduced. In the case of a catalytic unit, the reactor fouling will also cause pressure drop problems resulting in similar consequences. The fouling rate can be greatly influenced by the feedstock type or the feedstock blend, especially if the viscous feedstocks are involved. In fact, such feedstocks (i.e., the viscous feedstocks) can contribute to such severe fouling problems that they may not be considered for processing in certain refineries.

The development of control techniques to be applied to fouling by the asphaltene and wax constituents of feedstocks typically requires a thermodynamically modeling of the asphaltene constituents (or the reacted asphaltene product) or wax precipitation but may not take into consideration any effects such as feedstock of flow rate or other dynamic factors. Dynamic deposition of asphaltene constituents or reacted asphaltene products may describe the formation of the solid phase and the amount of the solid phase deposited on equipment and may take into account effects such as shear rate and flow rate.

When wax fouling occurs, there must be adequate observation of the system to control the production of wax deposits and monitor the wax layer thickness. Mitigation of wax fouling can be either (1) preventive practices and (2) corrective practices.

As might be interpreted, preventive practices are pro-active and involve in-operation plans steps to avoid wax deposits and growth. On the other hand, corrective practices are reactive (not in the chemical sense) and require the periodic removal of any wax deposits. For example, (1) the use of wax inhibitors or dispersants that should be injected into the system above the wax appearance temperature, (2) use of solvents to remove existing wax deposits and dispersants when it is not possible to inject above the wax appearance temperature, (3) use of non-metallic pipe linings and coatings to reduce the frictional drag and thereby reduce the effects of shear dispersion and molecular diffusion, and (4) selection and use of a suitable pig design and periodic pigging of the line. The use of use of such non-metallic lining and coating has an additional benefit insofar as they will also reduce the potential for corrosion of the lines and serve to reduce corrosion fouling (Speight, 2014b). Many refineries use models that involve the assembling of spreadsheets or other approximate models to evaluate refinery (or unit) operations and assess heat transfer efficiency. However, the necessary aspects of the refinery operation that must be taken into account may not give the overall picture but can be used as a guide, for example, for heat exchanger cleaning.

However, when models are developed, it must be remembered that crude oil and crude oil products are complex mixtures of hydrocarbon derivative and impurities (which is emphasized in the viscous crude oils) and is prone to fouling. In fact, many refineries are experiencing new (and previously unknown) fouling problems that are a direct consequence of the increasing amounts of viscous crude oils in the feedstock slate. These problems arise from the changing nature of the slate (a refinery might, for example, experience swings downward of 5°–10° API points on a day-to-day or week-to-week basis) as well as the presence of additives in the feedstocks as well as other issues due to incompatibility between the components of the refinery feedstocks.

In addition, many crude oils from tight formations may have added corrosion inhibitors (usually amine based) and drag reducing agents. These inhibitors can lead to operating issues in the preheat exchangers and furnaces, and in the atmospheric distillation tower where the amine salts are deposited thereby inhibiting heat transfer and fouling in the tower. Increased deposition of sand and particulates is observed in heat exchangers upstream of the desalter. In addition, fouling due to the occurrence of chemical reactions in the typical preheat train operation can dominate the formation of foulants downstream of the desalter. Also, operational irregularities (such as an inefficient desalting operation) can lead to the deposition of inorganic material (such as mineral matter or even prophyrinic material) that has entered the refinery as part of the crude oil (or viscous crude oil). Such problems are emphasized by as the formation of stable emulsions in the desalter which then increase heat exchanger fouling in downstream units. Desalting efficiency can be further challenged by wax precipitation in the cold train exchangers.

The use of dispersants is to limit the particle size of solids in the system. Various dispersants have different efficacies, depending on the components to be dispersed. Dispersant chemistries are available that address deposition problems such as coke particles, asphaltene precipitation, and organic or inorganic deposition. Dispersants will prevent smaller particles from agglomerating to form larger particles which deposit more easily. Similarly, they also prevent the small particles from being attracted to already existing deposits in the system.

Maintaining a high fluid velocity also helps to keep small particles from settling onto the process equipment. Some dispersants can also be surface active providing a surface which hinders a particle's ability to lay down on the metal surface.

Corrosion inhibitors are designed to minimize the contact between the metal surface and the corrosive fluid in order to minimize the formation and deposition of corrosion products in the system.

A metal coordinator (deactivator) will modify the metal ions by complexing, thus reducing the catalytic activity of the metal, so that initiation of polymerization reactions is minimized.

Polymerization inhibitors are designed to react immediately with any radical formed in the system to form a new stable molecule which will no longer contribute to the propagation reactions. These inhibitors will reduce the free radical polymerization of olefins and some of the sulfur compounds and stabilize unstable feedstocks.

Non-free radical polymerization inhibitors reduce the condensation polymerization reactions of carboxylic acids and some of the nitrogen compounds. The most effective and economical treatment program is one that addresses only those mechanisms that cause the fouling problems and at the same time provide sufficient flexibility to handle the typical processing variations.

Thus, following recommendation presented in the above paragraphs relating to the identification and relative importance of each fouling mechanism occurring in a refinery system, a mitigation program can be designed to obtain maximum efficiency in terms of fouling (and corrosion) control.

In summary, the feedstock can be (should be) analyzed to identify specific concern levels for components that reflect a specific fouling mechanism. Besides the physical property data, other data such as analysis of salts, filterable solids, the amount of asphaltene and resin constituents, sulfur, mercaptan derivatives, basic nitrogen derivatives, and the amounts of metals derivatives are necessary.

In addition, there should be an analysis of the deposits that are taken from the unit the sampling should be done to ensure a typical composition of the deposit is obtained. A thermographic analysis will assist in identifying the organic and volatile inorganic components of the deposit as well as constituents of the combustion ash. Extraction by a solvent (such as methylene chloride) is typically applied to identify any entrained hydrocarbon derivative and the degraded feedstock fraction in the sample. The non-extractable portion of the deposit represents the coke and inorganic fraction of the sample. Besides the analysis for metals (such as iron (Fe), sodium (Na), calcium (Ca), magnesium (Mg), and copper (Cu)), elemental analysis for carbon, hydrgoen, nitrogen, and sulfur is also recommended. In theory, on a weight basis, the products in the reactor overhead stream should be equal to the fresh feedstock plus any recycle minus the portion of the feed that was converted to coke. Thus one of the key objectives of conducting the mass balance exercise is to determine the composition of products leaving the reactor thereby indicating the so-called "missing" material that could indicate the beginning of the following path.

Briefly, the key points to be determined as part of the mass balance assessment include (1) the yields and quality of the desired products should be reported and compared with the unit targets, (2) the results of this test run should be compared with the results of previous test runs, and (3) any significant changes in the yields and/or operating parameters. Also, it is advisable that there should be a complete data collection carried out on (at least) a weekly basis – in fact most refineries do collect such data – since changes in refinery units are continuous, regular surveys permit distinction among the effects of feedstock, catalyst, and the operating conditions.

If this type of monitoring is followed, the properties of the feedstock and the analysis of the deposit can give an indication of the major mechanism of foulant production and whether the foulant is inorganic, organic, or a combination of the two. From this information, the management and mitigation of any issues that relates to fouling can be predicted and terminated before unit shutdown is required.

REFERENCES

Ancheyta, J., Trejo, F., and Rana, M.S. 2009. *Asphaltenes: Chemical Transformation during Hydroprocessing of Heavy Oils*. CRC Press, Taylor & Francis Group, Boca Raton, Florida. Chapter 2.

Andersson, E. 2007. Using Compact Plate Heat Exchangers to Optimize Heat Recovery in Refineries. *Petroleum Technology Quarterly*, Q1: 36–40.

Andersson, E. 2008. Minimizing Refinery Costs Using Spiral Heat Exchangers. *Petroleum Technology Quarterly*, Q2: 75–84.

ASTM D7157. 2021. *Standard Test Method for Determination of Intrinsic Stability of Asphaltene-Containing Residues, Heavy Fuel Oils, and Crude Oils (n-Heptane Phase Separation; Optical Detection)*. Annual Book of Standards. ASTM International, West Conshohocken, Pennsylvania.

Gary, J.H., Handwerk, G.E., and Kaiser, M.J. 2007. *Petroleum Refining: Technology and Economics*. 5th Edition. CRC Press, Taylor & Francis Group, Boca Raton, Florida.

Hsu, C.S., and Robinson, P.R. (Editors) 2017. *Practical Advances in Petroleum Processing* Volume 1 and Volume 2. Springer Science, New York.

Hunt, D., Minyard, W., and Koebel, J. 2008. Minimizing FCC Slurry Exchanger. Fouling. Digital Refining. http://www.digitalrefining.com/article/1000132,Minimising_FCC_slurry_exchanger_fouling.html#.VJ7LvV4AA

Joshi, J.B., Pandit, A.B. Kataria, K.L., Kulkarni, R.P. Sawarkar, A.N., Tandon, D., Ram, Y., and Kumar, M.M. 2008. Petroleum Residue Upgradation via Visbreaking: A Review. *Industrial & Engineering Chemistry Research*, 47(23): 8960–8988.

Magaril, R.Z., and Aksenova, E.L. 1967. Mechanism of Coke Formation During the Cracking of Petroleum Tars. *Izv. Vyssh. Zaved., Neft Gaz.*, 10(11): 134–136.

Magaril, R.A., and Aksenova, E.L. 1968. Study of the Mechanism of Coke Formation in the Cracking of Petroleum Resins. *International Chemical Engineering*, 8: 727–729 [first published in *Vysshikh Uchebn. Zavendenii, Neft i Gaz*, 1967, no. 11: 134–136].

Magaril, R.Z., and Ramazaeva, L.F. 1969. Study of Coke Formation in the Thermal Decomposition of Asphaltenes in Solution. *Izv. Vyssh. Ucheb. Zaved., Neft Gaz.*, 12(1): 61–64.

Magaril, R.Z., and Aksenova, E.I. 1970a. Kinetics and Mechanism of Coking Asphaltenes. *Khim. Izv. Vyssh. Ucheb. Zaved., Neft Gaz.*, 13(5): 47–53.

Magaril, R.Z., and Aksenova, E.I. 1970b. Mechanism of Coke Formation in the Thermal Decomposition of Asphaltenes. *Khim. Tekhnol. Topl. Masel.*, 15(7): 22–24.

Magaril, R.Z., Ramazaeva, L.F., and Askenova, E.I. 1970. Kinetics of Coke Formation in the Thermal Processing of Petroleum. *Khim. Tekhnol. Topliv Masel.*, 15(3): 15–16.

Magaril, R.Z., Ramazeava, L.F., and Aksenova, E.I. 1971. Kinetics of Coke Formation in the Thermal Processing of Crude Oil. *International Chemical Engineering*, 11: 250.

Magaril, R.Z., and Aksenova, E.I. 1972. Coking Kinetics and Mechanism of the Thermal Decomposition of Asphaltenes. Khim. Tekhnol. Tr. Tyumen. Ind. Inst. Page 169–172.

Müller-Steinhagen, H. 1995. Fouling of Heat Exchanger Surfaces. *Chemistry & Industry*, 5, 171–174.

Parkash, S. 2003. *Refining Processes Handbook*. Gulf Professional Publishing, Elsevier, Amsterdam, Netherlands.

Radovanović, L., and Speight, J.G. 2011. Visbreaking: A Technology of the Future. Proceedings of the First International Conference – Process Technology and Environmental Protection (PTEP 2011). University of Novi Sad, Technical Faculty Mihajlo Pupin, Zrenjanin, Republic of Serbia. December 7. Page 335–338.

Sadeghbeigi, R. 2000. *Fluid Catalytic Cracking: Design, Operation, and Troubleshooting of FCC Facilities*. Gulf Publishing Company, Houston, Texas.

Speight, J.G. 1987. Initial Reactions in the Coking of Residua. Preprints, *American Chemical Society, Division of Fuel Chemistry*, 32(2): 413.

Speight, J.G. 1992. A Chemical and Physical Explanation of Incompatibility during Refining Operations. Proceedings of the 4th International Conference on the Stability and Handling of Liquid Fuels. US. Department of Energy (DOE/CONF-911102). Page 169.

Speight, J.G. 1994. Chemical and Physical Studies of Petroleum Asphaltenes. In *Asphalts and Asphaltenes*, 1. T.F. Yen and G.V. Chilingarian (Editors). Elsevier, Amsterdam, Netherlands. Chapter 2.

Speight, J.G. 2000. *The Desulfurization of Heavy Oils and Residua*. 2nd Edition. Marcel Dekker Inc., New York.

Speight, J.G. 2001. *Handbook of Petroleum Analysis*. John Wiley & Sons Inc., Hoboken, New Jersey.

Speight, J.G. 2008. *Synthetic Fuels Handbook: Properties, Processes and Performance*. McGraw-Hill, New York.

Speight, J.G. 2011a. *The Refinery of the Future*. Gulf Professional Publishing, Elsevier, Oxford, United Kingdom.

Speight, J.G. (Editor). 2011b. *The Biofuels Handbook*. Royal Society of Chemistry, London, United Kingdom.

Speight, J.G. 2012. Visbreaking: A Technology of the Past and the Future. *Scientia Iranica*, 19(3): 569–573.

Speight, J.G. 2013. *Heavy and Extra Heavy Oil Upgrading Technologies*. Gulf Professional Publishing, Elsevier, Oxford, United Kingdom.

Speight, J.G. 2014a. *The Chemistry and Technology of Petroleum*. 5th Edition. CRC Press, Taylor & Francis Group, Boca Raton, Florida.

Speight, J.G. 2014b. *Oil and Gas Corrosion Prevention*. Gulf Professional Publishing, Elsevier, Oxford, United Kingdom.

Speight, J.G. 2015a. *Handbook of Petroleum Product Analysis*. 2nd Edition. John Wiley & Sons Inc., Hoboken, New Jersey.

Speight, J.G. 2105b. *Fouling in Refineries*. Gulf Professional Publishing Company, Elsevier, Oxford, United Kingdom.

Virzi, M., and Respini, M. 2014. Increasing Conversion and Run Length in a Visbreaker. Digital Refining. http://www.digitalrefining.com/article/1000938,Increasing_conversion_and_run_llength____in_a_visbreaker.html#.VKGOKl4AA; accessed December 8, 2014.

Wiehe, I.A. 1992. A Solvent-Resid Phase Diagram for Tracking Resid Conversion. *Industrial & Engineering Chemistry Research*, 31: 530–536.

Wiehe, I.A. 1993. A Phase-Separation Kinetic Model for Coke Formation. *Industrial & Engineering Chemistry Research*, 32: 2447–2454.

Yan, Y., De Klerk, A., and Prado, G.H.C. 2020. Visbreaking of Vacuum Residue Deasphalted Oil: New Asphaltenes Formation. *Energy Fuels*, 34(5): 5135–5147.

Xing, T., Ali, M., Alem, T., Gieleciak, R., and Chen, J. 2021. Fouling Tendency of Bitumen Visbreaking Products. *Fuel*, 289: 119735. https://www.sciencedirect.com/science/article/abs/pii/S0016236120327319

7 Importance in the Refinery

7.1 INTRODUCTION

Crude oil refining is the separation of crude oil into fractions, and the subsequent treating of these fractions to yield marketable products through the use of a series of unit processes where each unit process carries out a separate function (Parkash, 2003; Gary et al., 2007; Speight, 2014a; Hsu and Robinson, 2017; Speight, 2017). In addition, the corrosion and fouling issues that arise for a particular process (say, for example, the distillation process) will (if allowed to continue unabated) tend to be cumulative throughout the refinery system.

Refinery processes must be selected, and products manufactured to give a balanced operation in which the crude oil feedstock is converted into a variety of products in amounts that are in accord with the demand for each (Parkash, 2003; Gary et al., 2007; Speight, 2014a; Hsu and Robinson, 2017; Speight, 2017). Thus, a refinery is assembled as a group of integrated manufacturing plants which vary in number with the variety of products produced (Chapter 1). Corrosion causes the failure of equipment items as well as dictating the maintenance schedule of the refinery, during which part or the entire refinery must be shut down. Although significant progress in understanding corrosion has been made, it is also clear that the problem continues to exist and will become progressively worse through the introduction of more heavy crude oils, opportunity crudes, and high-acid crudes into refineries (Pruneda et al., 2005; Bhatia and Sharma, 2006; Collins and Barletta, 2012).

The end of the widespread production of liquid fuels and other products from fossil fuel sources within the current refinery infrastructure is considered by some observers to occur during the present century, even as early as during the next five decades but, as with all projections, is very dependent not only upon the remaining reserves but also on petro-politics which is related to the amount of crude oil recovered from the various reservoirs (Speight, 2011a; Speight and Islam, 2016). During this time, fossil fuels (including natural gas and crude oil from tight formations) will be the mainstay of the energy scenarios of many countries.

7.2 THERMAL PROCESSES AND CATALYTIC PROCESSES

The configuration and complexity of a crude oil refinery have evolved from one-pot batch distillation to produce kerosene as the major product (1850s) to the complex modern refinery (that produces a multitude of fuels and petrochemical feedstocks from a wide range of feedstocks) (Parkash, 2003; Gary et al., 2007; Speight, 2014a; Hsu and Robinson, 2017; Speight, 2017, 2019). In spite of this evolution, the importance of thermal processes and catalytic processes that were the mainstay of the original refineries cannot be overstated. In fact, it was the constant demand for hydrocarbon-based products such as liquid fuels that was a major driving force behind the refining industry to produce these products. Another other driving force was the production of a variety of petrochemical products that are the basis of that particular modern industry (Parkash, 2003; Gary et al., 2007; Speight, 2014a; Hsu and Robinson, 2017; Speight, 2017, 2019).

Briefly, a petrochemical is a chemical product developed from crude oil that is useful for the synthesis of products by the modern chemical process industry. The chemical process industry is, in fact, subdivided into other categories that are: (1) the chemicals and allied products industries in which chemicals are manufactured from a variety of feedstocks and may then be put to further use, (2) the rubber and miscellaneous products industries which focus on the manufacture of rubber and plastic materials, and (3) petroleum refining and related industries which, on the basis of prior chapters in this text, is now self-explanatory.

DOI: 10.1201/9781003184904-7

TABLE 7.1
The Various Distillation Fractions of Crude Oil

Product	Lower Carbon Number[a]	Upper Carbon Number[a]	Lower b.p. (°C)[a]	Upper b.p. (°C)[a]	Lower b.p. (°F)[a]	Upper b.p. (°F)[a]
Liquefied petroleum gas	C3	C4	−42	−1	−44	31
Naphtha	C5	C17	36	302	97	575
Middle distillate (kerosene)	C8	C18	126	258	302	575
Atmospheric gas oil	C12	>C20	216	421	345	650
Vacuum gas oil	>C20		345	510	650	930
Residuum	>C20		>510		>930	

[a] The carbon number and boiling point are estimates because of variations in production parameters from refinery to refinery and are inserted for illustrative purposes only.

There is a myriad of products that have evolved through the short life of the petroleum industry, either as bulk fractions or as single hydrocarbon products (Tables 7.1 and 7.2). Moreover, the complexities of product composition have matched the evolution of the products.

In fact, it is the complexity of product composition that has served the industry well and, at the same time, had an adverse effect on product use. However, the same is not true for the viscous feedstocks, hence the need for thermal cracking and catalytic cracking process to reduce the viscous feedstocks to more refinery friendly feedstocks.

The development of refining industry is market-driven, and in order to satisfy the demands of the market demands, technology development will continue to produce the necessary products and feedstocks for the petrochemical industry. These technologies will be based on the tried-and-true technologies of thermal cracking and catalytic cracking in order to prepare the viscous feedstocks for further refining to marketable products. New thermal and catalytic processes and catalytic materials will assist in the improvement of refining technologies (Speight, 2011b).

Thermal cracking was the first commercial conversion process developed in the early 1900s principally to produce more motor gasoline from crude oils and produce high-octane gasoline for aircraft use, initiating an attempt to change the composition of crude oil in crude oil refinery. In the process, a higher boiling feedstock (for which there was a low market demand) was heated in pressurized reactors and thereby cracking (decomposing) the higher molecular weight constituents into lower boiling products such as naphtha (gasoline) and middle distillates.

With the advent of catalytic cracking in the 1930s and 1940s and its capability to produce higher yields of naphtha (often arbitrarily referred to as gasoline since the early days of the industry) with higher octane number, thermal cracking of gas oils has ceased to be an important process for gasoline production in modern refineries. A principal application of thermal cracking of distillate fractions in current refineries is to use naphtha feedstocks for the purpose of producing ethylene (C_2H_4) as a feedstock for the petrochemical industry. However, thermal cracking of residual fractions, particularly vacuum resid, heavy crude oils, extra heavy crude oils, and tar sand bitumen, is still practiced in the form of visbreaking and coking processes in the refineries.

As the refineries progressed and the feedstock changed in composition and properties, the increased demand for gasoline, along with the need to produce high-octane gasoline for increasingly more powerful spark ignition engines, led to the development and maturation of catalytic cracking processes just before and during World War II. Following the development of a fixed-bed process (the Houdry process, 1936) and a moving-bed process (the Thermafor catalytic cracking

TABLE 7.2
Properties of Hydrocarbon Products from Crude Oil

	Molecular Weight	Specific Gravity	Boiling Point (°F)	Ignition Temperature (°F)	Flash Point (°F)	Flammability Limits in Air (% v/v)
Benzene	78.1	0.879	176.2	1,040	12	1.35–6.65
n-Butane	58.1	0.601	31.1	761	−76	1.86–8.41
iso-Butane	58.1		10.9	864	−117	1.80–8.44
n-Butene	56.1	0.595	21.2	829	Gas	1.98–9.65
iso-Butene	56.1		19.6	869	Gas	1.8–9.0
Diesel fuel	170–198	0.875			100–130	
Ethane	30.1	0.572	−127.5	959	Gas	3.0–12.5
Ethylene	28.0		−154.7	914	Gas	2.8–28.6
Fuel oil No. 1		0.875	304–574	410	100–162	0.7–5.0
Fuel oil No. 2		0.920		494	126–204	
Fuel oil No. 4	198.0	0.959		505	142–240	
Fuel oil No. 5		0.960			156–336	
Fuel oil No. 6		0.960			150	
Gasoline	113.0	0.720	100–400	536	−45	1.4–7.6
n-Hexane	86.2	0.659	155.7	437	−7	1.25–7.0
n-Heptane	100.2	0.668	419.0	419	25	1.00–6.00
Kerosene	154.0	0.800	304–574	410	100–162	0.7–5.0
Methane	16.0	0.553	−258.7	900–1,170	Gas	5.0–15.0
Naphthalene	128.2		424.4	959	174	0.90–5.90
Neohexane	86.2	0.649	121.5	797	−54	1.19–7.58
Neopentane	72.1		49.1	841	Gas	1.38–7.11
n-Octane	114.2	0.707	258.3	428	56	0.95–3.2
iso-Octane	114.2	0.702	243.9	837	10	0.79–5.94
n-Pentane	72.1	0.626	97.0	500	−40	1.40–7.80
iso-Pentane	72.1	0.621	82.2	788	−60	1.31–9.16
n-Pentene	70.1	0.641	86.0	569	−	1.65–7.70
Propane	44.1		−43.8	842	Gas	2.1–10.1
Propylene	42.1		−53.9	856	Gas	2.00–11.1
Toluene	92.1	0.867	321.1	992	40	1.27–6.75
Xylene	106.2	0.861	281.1	867	63	1.00–6.00

process, 1941), catalytic cracking process and fluid-bed catalytic cracking process (FCC process, 1942) became the most widely used process worldwide because of the improved thermal efficiency of the process and the high product selectivity achieved, particularly after the introduction of crystalline zeolites as catalysts in the 1960s (Chapter 5).

Thus, the essential driver of the thermal refinery was the shift in demand to gasoline from kerosene because of the introduction of the automobile, the airplane, and electricity. The demand for gasoline rapidly increased when the US declared war on Germany in 1917 and became a party in World War I.

Thermal refinery processes, thermal cracking, thermal reforming, and thermal polymerization enabled the expansion of gasoline supply. With the introduction of tetraethyl lead (TEL) as an octane number boosting additive in 1923, a growing interest was directed to the production of high-performance gasoline which would be defined later as a high-octane number-gasoline after the introduction of a test method to measure the octane number of gasoline as an anti-knock property in 1931. Because of the toxicity of lead, TEL concentration was limited to 3 mL/ gallon of finished gasoline (approximately 800 ppm by volume). The addition of lead to motor gasoline continued until the 1970s in the United States when the mandate for adding catalytic converters to automobiles took effect in accordance with the Clean Air Act to reduce tailpipe emissions, and the unleaded gasoline was introduced. Lead is still added to aviation gasoline used in turboprop aircraft in quantities 0.3–0.56 g/L in a range of avgas grades, and efforts are underway to remove lead from the aviation gasoline as well in the near future. But after the introduction of tetraethyl lead, the contribution of gasoline produced by thermal cracking has steadily increased to reach over 50% of the gasoline pool by 1940. The catalytic refinery at the time of high competition for the production high-performance gasoline and other crude oil-derived evolved during the period leading to and during World War II.

Thus, refining technology has evolved considerably over the last century in response to changing requirements such as (1) demand for gasoline and diesel fuel as well as fuel oil, (2) petrochemicals as building blocks for clothing and consumer goods, (3) and more environmentally friendly processes and products. Moreover, the precise configuration of the refinery of the future is unknown, but it is certain that no two refineries will adapt in exactly the same way. There are small refineries – 1,500–5,000 barrels per day (bpd) – and large refineries that process in excess of 250,000 barrels per day. Some are relatively simple (Speight, 2014a) and produce only fuels while other refineries, such as those with integrated petrochemical processing capabilities, are much more complex. Many refineries are part of large integrated oil companies engaged in all aspects of the crude oil technology – from (1) exploration, (2) production, (3) transportation, (4) refining, and (5) marketing of crude oil products. Historically, this has not always been the case. In the early days of the 20th century, refining processes were developed to extract kerosene for lamps. Any other products were considered to be unusable and were usually discarded. A brief history of crude oil refining is presented in the following paragraphs.

The first crude oil refinery focused on the batch distillation of crude oil in order to maximize the production of kerosene. Technological advancements included, first, the introduction of continuous distillation and then vacuum distillation – developed in 1870 – which greatly facilitated the manufacture of lubricants. The 1890s saw the emergence of the internal combustion engine creates demand for gasoline and diesel fuel as the demand for kerosene (used as lamp oil) declined with the invention and proliferation of the electric light.

Thus, the first refining processes were developed to purify, stabilize, and improve the quality of kerosene and the invention and further development of the internal combustion engine led (at about the time of World War I) to a demand for gasoline for use in increasing quantities as a motor fuel for cars and trucks. This demand on the lower boiling products increased, particularly when the market for aviation fuel developed. Thereafter, refining methods had to be constantly adapted and improved to meet the quality requirements and needs of fuels as well as a variety of other products. This also included the production of improved lubricants that led to the use of processes based on solvent extraction. To make better use of the bottom of the barrel, thermal cracking (in 1913) and visbreaking processes were introduced to crack heavy hydrocarbon molecules to produce more-valuable lighter fractions. The development of thermal cracking was in in response to increased demand for gasoline due to mass production manufacturing of automobiles and the outbreak of World War I. This innovation enabled refineries to produce additional gasoline and distillate fuels by subjecting high boiling crude oil fractions to high pressures and temperatures with the resulting production of lower boiling, lower molecular weight products.

During the 1930s, many advances made to improve gasoline yield and properties as a response to development of higher-compression engines. This involved the development of processes such as (1) catalytic cracking, thermal reforming, and catalytic polymerization to improve octane number; (2) hydroprocesses to remove sulfur; (3) coking processes to produce gasoline blend stocks; (4) solvent extraction processes to the improve viscosity index of lubricating oil; and (5) solvent dewaxing processes to improve pour point the various products. The by-products of these various processes included (1) aromatics, waxes, residual fuel oil, coke, and feedstocks for the manufacture of petrochemicals.

With the onset of World War II and the need for additional supplies of gasoline, the refining industry turned to catalysis for major innovations. Catalytic cracking constituted a step change in the refinery's ability to convert heavy components into highly valued gasoline and distillates. Wartime demand for aviation fuels helped spur development of catalytic alkylation processes (which produced blend stocks for high-octane aviation gasoline) and catalytic isomerization (which produced increased quantities of feedstocks for alkylation units) to create high-octane fuels from lighter hydrocarbons. We redistributed hydrogen content among the refinery's products to improve their properties via catalytic reforming of gasoline, catalytic hydrodesulfurization of distillates, and hydrocracking of midrange streams.

The period in the 1950s to the 1970s saw the development of various reforming processes, which also produced blend stocks that were used to improve gasoline quality and yield. Other processes, such as deasphalting, catalytic reforming, hydrodesulfurization, and hydrocracking, are examples of processes developed during this period. In this time period, refiners making also started further development of the uses for the waste gases from various processes resulting in the expansion of the petrochemical industry. In the latter part of the period, the industry benefitted from a massive infusion of computer-based quantitative methodology that has significantly improved more control over processes and the composition of products. In addition, automation and control enabled the optimization of unit operation and economic performance.

Thus throughout its history, the refining industry has been the subject of the four major forces that affect most industries and which have hastened the development of new crude oil refining processes: (1) the high demand for liquid fuels such as gasoline, diesel, fuel oil, and jet fuel; (2) uncertain feedstock supply, specifically the changing quality of crude oil and geopolitics between different countries and the emergence of alternate feed supplies such as bitumen from tar sand, natural gas, coal, and the ever-plentiful biomass; (3) increasingly stringent recent environmental regulations in relation to sulfur content of liquid fuels; and (4) continued technology development such as new catalysts and processes.

In the process, naphtha cracking converts the naphtha feedstock (produced by various process) (Table 1.4) into ethylene, propylene, benzene, toluene, and xylenes as well as other by-products in a two-step process of cracking and separating. The higher molecular weight products are subsequently processed into fuel oil, light cycle oil and pyrolysis gas by-products. The pyrolysis gas stream can then be fed to the aromatics plants for benzene and toluene production. In addition to recovery of gases in the distillation section of a refinery , gases are also produced in the various thermal processes, thermal cracking processes, and catalytic cracking processes (Figure 1.2), are also available and in processes such as visbreaking and coking and visbreaking (Parkash, 2003; Gary et al., 2007; Speight, 2014; Hsu and Robinson, 2017; Speight, 2017).

7.3   THE NEED FOR THERMAL AND CATALYTIC PROCESSES

By definition, the term "cracking" as used in the crude oil refining industry is thermo-chemcial pathway by which the high molecular constituents are decomposed (cracking, broken up) into lower molecular weight products by means of heat and usually pressure and sometimes catalysts. It is, in fact, the most important process for the commercial production of naphtha and other constituents of fuels. As a consequence, catalytic cracking is the same process only in the presence of a catalyst which can guide to reaction process to other types of low molecular weight products. These processes (thermal cracking and catalytic cracking) processes have been, and continue to be, necessary parts of a crude oil refinery in order to accommodate and maximize the use of the viscous feedstock that many refineries are now accepting for processing.

Thermal cracking processes and catalytic cacking processes are important for two key reasons, which are (1) the processes cracking help balance the availability of products with the demand for those products and (2) when the processes transform higher molecular weight feedstock constituents into lower molecular weight products, the supply of constituents for fuels is increased as is the

TABLE 7.3

Summary of the Importance of Thermal Cracking and Catalytic Craking Processes in the Refinery

Fractional distillation aids in the production of fractions containing high molecular weight (low volatility or non-volatile) constituents.

Viscous feedstocks contain high molecular weight (low volatility or non-volatile) constituents.

Thermal cracking leads to the conversion of high molecular weight constituents to lower molecular weight volatile products.

The volatile products contain alkane derivatives and alkene derivatives.

Alkene derivatives are suitable feedstocks for the manufacture of petrochemical products.

Catalytic cracking produces higher yields of iso-alkane (brand alkane) derivatives that have high octane numbers.

supply of feedstocks for the manufacture of petrochemical products, which also assists balancing demand with supply.

In fact, thermal cracking processes have played a major role in crude oil refining for many years. This is beneficial to the refining process because it allows a larger yield of lighter products to be created from the heavier less desirable products of refining. Moreover, while thermal cracking processes produce shorter straight-chain alkane derivatives and olefin derivatives, there is a lack of the branched iso-alkane derivatives in the products mix. It is for this reason that catalytic cracking is highly favored over thermal cracking in the production of the branched iso-alkane derivatives and, hence, the high-octane naphtha products which are valuable for use as blend stocks for gasoline manufacture (Table 7.3) (Parkash, 2003; Gary et al., 2007; Speight, 2014a; Hsu and Robinson, 2017; Speight, 2017).

Thus, as the need for lower boiling products developed during the 20th century (and continued into the 21st century), the more conventional refinery feedstocks (such as conventional crude oil) that yielded the desired quantities of the lower boiling products became less available, and refineries had to introduce conversion processes to produce greater quantities of lower boiling products from the higher boiling fractions. The means by which a refinery operates in terms of producing the relevant products depends not only on the nature of the refinery feedstock but also on the refinery configuration (i.e., the number of types of the processes that are employed to produce the desired product slate), and the refinery configuration is, therefore, influenced by the specific demands of a market. In fact, refineries need to be constantly adapted and upgraded to remain viable and responsive to ever-changing patterns of crude supply and product market demands.

Furthermore, the need for higher yields of light and middle distillates, i.e., naphtha, kerosene, and diesel fuel, from refinery feedstocks, of necessity led to the incorporation of the necessary process (thermal cracking and catalytic cracking) into refineries. Thus, the evolution of refineries in the four decades between 1910 and 1950 was driven largely by the development of thermal cracking processes and catalytic cracking processes. In addition, the need for the so-called "finishing" processes also became important in this time frame in order to stabilize and purify the products of thermal cracking which enabled the refinery to meet product specifications.

The fractions of the refinery feedstock produced by distillation are rarely ready for sales as high value products (Parkash, 2003; Gary et al., 2007; Speight, 2014a; Hsu and Robinson, 2017; Speight, 2017). The purpose of the thermal and catalytic processes is to convert the feedstocks to product with more desirable products that are focused on the production of intermediate products that can be sued by further refining to produce the valuable products that are ultimately for sale to the consumers.

In processes such as visbreaking processes and the coking processes, a variety of gases are produced. Both catalytic and thermal cracking processes result in the formation of unsaturated hydrocarbon derivatives, particularly ethylene ($CH_2=CH_2$), but also propylene (propene, $CH_3CH=CH_2$), iso-butylene [iso-butene, $(CH_3)_2C=CH_2$], and the n-butenes ($CH_3CH_2CH=CH_2$, and

$CH_3CH=CHCH_3$) in addition to hydrogen (H_2), methane (CH_4),and smaller quantities of ethane (CH_3CH_3), propane ($CH_3CH_2CH_3$), and butanes [$CH_3CH_2CH_2CH_3$, $(CH_3)_3CH$]. Diolefins such as butadiene ($CH_2=CHCH=CH_2$) are also present (Parkash, 2003; Gary et al., 2007; Speight, 2014a; Hsu and Robinson, 2017; Speight, 2017).

7.3.1 Visbreaking, Thermal Cracking, and Coking

Since World War II, the demand for low boiling products (e.g., gasoline, jet, and diesel fuels) has grown, while the requirement for heavy industrial fuel oils has declined. Thus, commencing in 1920, large volumes of residue were being processed in visbreaking units or in thermal cracking units. These simple process units basically consist of a large furnace that heated the feedstock to a temperature on the order of 450°C–500°C (840°F–930°F) at an operating pressure on the order of 150 psi. The residence time in the furnace is carefully limited to prevent much of the reaction from taking place and clogging the furnace tubes. The heated feed is then charged to a reaction chamber, which is kept at a pressure high enough to permit cracking of the large molecules but restrict coke formation. From the reaction chamber the process fluid was cooled to prevent any further cracking and then sent to a distillation column for separation into the various components.

Visbreaking units typically convert approximately 15% w/w (15% v/v) of the feedstock to naphtha and diesel oils and produce a lower-viscosity residual fuel. Thermal cracking units provide more severe processing and often convert as much as 50%–60% w/w (60% v/v) of the incoming feed to naphtha and low-boiling middle distillate fractions.

A potential benefit of the process is the production of a feedstock that is more acceptable to a modern refinery than the original (unchanged) feedstock (Speight and Moschopedis, 1979). In particular, any process that reduces the mineral matter in the bio-feedstock and reduces the oxygen content in the bio-feed would be a benefit. This can be accomplished by one or two preliminary treatment steps (that are modified from the visbreaking process) in which the feedstock is de-mineralized in the initial stages of coke formation in which the newly formed coke forms on the mineral matter and creates a relatively easy separable (solid) phase. Such a process might have to be established at a bio-feedstock production site unless the refinery has the means by which to accommodate the feedstock in an already existing unit.

In particular, any process that reduces the mineral matter in the bio-feedstock and reduces the oxygen content in the bio-feed would be a benefit (Chapter 4). This can be accomplished by one or two preliminary treatment steps (such as the visbreaking process) in which the feedstock is de-mineralized, and the oxygen constituents are removed as overhead (volatile) material giving the potential for then production of a fraction rich in oxygen functions that may be of some use to the chemical industry. Such a process might have to be established at a bio-feedstock production site unless the refinery has the means by which to accommodate the feedstock in an already existing unit.

On the other hand, the coking processes involve severe thermal cracking. In the process, the residue feed is heated to about 475°C–520°C (890°F–970°F) in a furnace with very low residence time and is discharged into the bottom of a large vessel called a coke drum for extensive and controlled cracking. The cracked lower boiling material rises to the top of the drum from which it is sent to the product fractionator for separation into naphtha, middle distillates, and gas oil for further processing in the catalytic cracking unit. The higher boiling product remains in the reactor and, because of the retained heat, cracks ultimately to coke. Once the coke drum is filled with solid coke, it is removed from service and replaced by another coke drum.

Decoking of the reactor (the coking drum) is a necessary occurrence that is accomplished by a high-pressure water jet. First, the top and bottom heads of the coke drum are removed. Next, a hole is drilled in the coke from the top to the bottom of the vessel. Then a rotating stem is lowered through the hole, spraying a water jet sideways which cuts the coke into lumps, which fall out the bottom of the drum for subsequent loading into trucks or railcars for shipment to customers.

Typically, coke drums operate on 24-hour cycles, filling with coke over one 24-hour period followed by cooling, decoking, and reheating over the next 24 hours. The drilling derricks on top of the coke drums are a notable feature of the refinery skyline.

Cokers produce no liquid residue but yield up to 30% w/w coke, and the low-sulfur product is employed, for example, to produce electrodes for the electrolytic smelting of aluminum. Typically, the lower-quality coke is burned as fuel, often as a mixture with coal.

7.3.2 CATALYTIC CRACKING

Catalytic cracking is the modern method for converting high-boiling crude oil fractions, such as gas oil, into gasoline and other low-boiling fractions. Several processes currently employed in catalytic cracking differ mainly in the method of catalyst handling, although there is an overlap with regard to catalyst type and the nature of the products.

The catalyst, which may be an activated natural or synthetic material, is employed in bead, pellet, or microspherical form and can be used as (1) a fixed-bed process, (2) a moving-bed process, or (3) a fluid-bed process.

Feedstocks may range from naphtha to atmospheric residuum (often referred to as reduced crude). Feedstock preparation (to remove metallic constituents and high-molecular weight non-volatile materials) is usually carried out through any one of the following ways: (1) coking, (2) propane deasphalting, (3) furfural extraction, (4) vacuum distillation, (5) visbreaking, (6) thermal cracking, and (7) hydrodesulfurization (Parkash, 2003; Gary et al., 2007; Speight, 2014a; Hsu and Robinson, 2017; Speight, 2017).

Major process variables are temperature, pressure, catalyst-feedstock ratio (ratio of the weight of catalyst entering the reactor per hour to the weight of feedstock charged per hour), and space velocity (weight or volume of the feedstock charged per hour per weight or volume of catalyst in the reaction zone). Wide flexibility in product distribution and quality is possible through control of these variables along with the extent of internal recycling is necessary. Increased conversion can be obtained by applying higher temperature or higher pressure. Alternatively, lower space velocity and higher catalyst-feedstock ratio will also contribute to an increased conversion.

However, throughout the past several decades, processes have evolved that are the next generation of the fluid catalytic cracking and the residuum catalytic cracking processes. Some of these newer processes use catalysts with different silica/alumina ratios as acid support of metals such as molybdenum (Mo), cobalt (Co), nickel (Ni), and tungsten (W). In general, the first catalyst used to remove metals from oils was the conventional hydrodesulfurization (HDS) catalyst.

Among the minerals used as a raw material for catalysts addressed to the upgrading of the higher boiling (viscous) fractions are clay minerals, manganese nodules, bauxite (a naturally occurring, heterogeneous material composed primarily of one or more aluminum hydroxide minerals, plus various mixtures of silica, iron oxide, titania, aluminosilicate, and other impurities in minor or trace amounts) activated with vanadium (V), nickel (Ni), chromium (Cr), iron (Fe), and cobalt (Co), as well as the high iron oxide content iron laterites (minerals rich in iron), clay minerals such as sepiolite minerals, and mineral nickel and transition metal sulfides supported on silica (SiO_2) and/or alumina (Al_2O_3).

Other kinds of catalysts, such as vanadium sulfide, are generated in situ, possibly in the colloidal state.

7.4 REFINERY OF THE FUTURE

The end of the widespread production of liquid fuels and other products within the current refinery infrastructure is considered by some observers to occur during the present century, even as early as the next five decades but, as with all projections, is very dependent upon the remaining reserves and petro-politics (Speight, 2011a; Speight and Islam, 2016). During this time, fossil fuels (including natural gas and crude oil from tight formations) will be the mainstay of the energy scenarios of many counties.

In addition, rightly or wrongly and without much justification but with much speculation, the combustion of fossil fuels is considered as the largest source of anthropogenic emissions of carbon dioxide (CO_2), which is largely blamed for global warming and climate change (Speight, 2020b) although other sources are evident but are often ignored (Speight, 2020). Nevertheless, having been identified as one of the causative agents of climate change, it is necessary to attempt to mitigate the emissions of carbon dioxide from fossil fuel combustion and to offset the depletion of fossil fuels such as the commonly used natural gas, crude oil, and coal. Coal is the current bad boy of the fossil fuel world but still offers many options for energy production, provide the process gases are cleaned rather than vented to the atmosphere (Speight, 2013). On the other hand, oil shale has been of lesser importance having received on-again and off-again popularity as a source of fuels but has never really been recognized as a major source of fuels.

The most popular energy resources (natural gas and the various members of the crude oil family) are currently on a depletion curve with estimates of the longevity of these resources varying up to 50 years (Speight and Islam, 2016). However, seeking alternate sources of energy is of critical importance for long-term security and continued economic growth. Supplementing crude oil consumption with renewable biomass resources is the first step toward this goal. The realignment of the chemical industry from one of crude oil refining to a refinery solely devoted to providing fuels and chemicals from biomass (frequently referred to as a biorefinery), will given time, be feasible and has, in fact, become a national goal of many countries that currently rely on imports of crude oil to sustain their energy needs. However, clearly defined goals are necessary for increasing the use of biomass-derived feedstocks for fuel production and for the production of chemicals, and it is important to keep the goal in perspective. In this context, the increased use of bio-feedstocks and the production of fuels therefrom should be viewed as one of a range of possible measures for achieving self-sufficiency in energy, rather than a panacea (Speight, 2019).

As the refining industry evolves even further and, in many cases, away from natural gas and crude oil as the major feedstocks, a variety of biomass and waste-derived feedstocks will be used as feedstocks for the production of fuels and other valuable low boiling products, and this may be no more evident than the use of bio-feedstocks for gasifier units. Moreover, gasification (1) is a well-established technology, (2) has broad flexibility of feedstocks and operation, and (3) is the most environmentally friendly route for handling these feedstocks for power production.

Alternate biofeedstocks include materials such as wood pellets, and wood chips, waste wood, plastics, municipal solid waste (MSW), refuse-derived fuel (RDF), agricultural and industrial wastes, sewage sludge, switch grass, discarded seed corn, corn stover, and other crop residues will all be used as feedstocks for gasification. In fact, wood is the oldest known biofuel. Burning wood rather than fossil fuels can reduce the carbon dioxide emissions responsible for global climate change. Wood fuel is carbon dioxide (CO_2) neutral. It gives off only as much carbon dioxide when burned as it stores during its lifetime. In addition, wood fuel has very low levels of sulphur, a chemical that contributes to acid rain.

The precise configuration of the refinery of the future is unknown, but, because of feedstock variation and the make-up of feedstock blends, it is certain that no two refineries will adapt in exactly the same way. However, the evolution of the refinery of the future will not be strictly confined to crude oil-based processes but will be based on a variety of feedstocks (Speight 2011b). This will be solved in refinery of the future with the development of deep conversion processing, such as heavy feedstock hydrocracking and the inclusion of processes to accommodate other feedstocks (Speight, 2014a).

Moreover, the future of the petroleum refining industry will be primarily on processes for the production of improved quality products. In addition to *heavy ends deep conversion*, there will also be changes in the feedstock into a refinery. Biomass, liquids from coal, and liquids from oil shale will increase in importance (Speight, 2012, 2013). These feedstocks (1) will be sent to refineries or (2) processed at a remote location and then blended with refinery stocks are options for future development and the nature of the feedstocks. Above all, such feedstock must be compatible with refinery feedstocks and not cause fouling in any form.

Basic refining process for the conversion of residua, heavy oil, and tar sand bitumen to lower-boiling saleable products and the conversion of distillation residues consist of cracking the feedstock constituents to increase the hydrogen content and to decrease the carbon content of the derived products (Speight, 2011b, 2014a). While such processes will continue (at least) for the next 50 years and even throughout the remainder of the 21st century (Speight, 2011b), many refiners are investigating the potential large-scale utilization of biomass as partial feedstocks. For practical reasons, small-capacity refineries might be the first to attempt such uses of biomass in a crude oil refinery complex. The biomass might be used in the form of pre-prepared pellets (obtained from agriculture residues such as forestry residues, corn stock, and straw) as a feedstock blend with heavy oil. The approach could produce benefits such as (1) improvement in the quality of the final market products and of the economics of the entire activity, (2) positive impact on rural development (new jobs and new income for farmers), and (3) decrease in carbon dioxide emissions by substitution of renewable biomass and conversion to hydrocarbon products.

In this respect, the acceptance of viscous feedstocks is expected to have a substantial impact on energy independence and security, as well as on the refining and petrochemical sector. Hydroprocessing (hydrocracking and hydrotreating) and bottom-of-the-barrel strategies are being implemented more extensively to increase the yield of premium products (such as clean fuels) by means of the demand for more hydrogen to produce the premium products. Thus, hydrogen will continue to be an essential part of the refinery of the 21st century.

Hydrogen is typically supplied and balanced in a refinery through a network of hydrogen recovery operations from offgas streams and mostly supplemented by onsite processes for hydrogen production (Parkash, 2003; Gary et al., 2007; Speight, 2014a, 2016; Hsu and Robinson, 2017; Speight, 2017). Moreover, it is more than likely that the future refinery will have a gasification section devoted to the conversion of alternate feedstocks (such as coal, shale oil, and biomass) to Fischer-Tropsch hydrocarbons – perhaps even with rich oil shale added to the gasifier feedstock. Many refineries already have gasification capabilities (for the production of hydrogen), but the trend will increase to the point (over the next two decades) nearly all refineries feel the need to construct a gasification section to handle viscous crude oil-related feedstocks and refinery waste as well as non-crude oil related carbonaceous feedstocks, such as coal, biomass, and non-refinery waste material (Speight, 2014b).

Coal gasification is a tried-and-true technology, and there has been a move in recent years to feedstocks other than coal (Speight, 2012, 2013, 2014b). Among other alternative energy conversion pathways, biomass gasification has great potential because of its flexibility to use a wide range of feedstock and to produce energy and a wide range of fuels and chemicals. Recently, the focus of its application has changed from the production of combined heat and power to the production of liquid transportation fuels. The technical challenges in commercialization of fuels and chemicals production from biomass gasification include increasing the energy efficiency of the system and developing robust and efficient technologies for cleaning the product gas and its conversion to valuable fuels and chemicals. Thus, future energy production (as conventional crude oil reserves continue to decline) is likely to involve co-processing alternative energy sources in which any of the highly viscous feedstocks (such as crude oil residua, heavy crude oil/extra heavy crude oil, and/or tar sand bitumen) is processed with other energy sources and requires a new degree of refinery flexibility as the key target, especially when related to the increased use of renewable energy sources such as biomass.

7.4.1 THE BIOREFINERY

Whatever the rationale and however the numbers are manipulated, the supply of crude oil, the basic feedstock for refineries and for the petrochemicals industry, is finite and supply/demand issues will continue to deplete crude oil reserves. Although the supplied of heavy oil, extra heavy oil, and tar sand bitumen can be moved into the breach, the situation can be further mitigated to some extent by the exploitation of more technically challenging fossil fuel resources and the introduction of new technologies for fuels and chemicals production from coal and oil shale (Speight, 2011a, 2011b, 2013). In addition, there is a substantial interest in the utilization of plant based matter (biomass) as a raw material feedstock for the chemicals industry. Plants accumulate carbon from the atmosphere via photosynthesis and the widespread utilization of these materials as feedstocks for the generation of power, fuels, and chemicals.

In terms of the use of plant matter as energy-producing feedstocks, there are two terms that need to be defined and these are (1) biomass and (2) biofuels. Biomass is a renewable energy source – unlike the fossil fuel resources: crude oil, coal, and natural gas – and is derived from recently living organisms and/or their metabolic byproducts. An advantage of fuel from biomass (often referred to as biofuel), in comparison to most other fuel types, is it is biodegradable, and thus relatively harmless to the environment if spilled. Typically, a biofuel is any fuel that is derived from biomass but has been further defined as any fuel with a minimum content ($\geq$80% v/v) of materials derived from living organisms harvested within the 10 years preceding its manufacture.

Plants offer a unique and diverse feedstock for energy production and for the production of chemicals. Plant biomass can be gasified to produce synthesis gas – a basic chemical feedstock for the production of hydrocarbons and also a source of hydrogen for a future hydrogen economy. More generally, biomass feedstocks are recognized (and/or defined) by the specific chemical content of the feedstock or the manner in which the feedstock is produced (Speight, 2020a, 2020c). For example, primary biomass feedstocks that are currently being used for bioenergy include grains and oilseed crops used for transportation fuel production, plus some crop residues (such as orchard trimmings and nut hulls) and some residues from logging and forest operations that are currently used for heat and power production (Table 17.1). In the future, it is anticipated that a larger proportion of the residues inherently generated from food crop harvesting, as well as a larger proportion of the residues generated from ongoing logging and forest operations, will be used for bioenergy.

Secondary biomass feedstocks differ from primary biomass feedstocks insofar as the so-named secondary biomass feedstocks are a by-product of processing of the primary biomass feedstocks. Specific examples of secondary biomass include sawdust from sawmills, black liquor (which is a byproduct of paper making), and cheese whey (which is a by-product of cheese making processes). Vegetable oils used for biodiesel that are derived directly from the processing of oilseeds for various uses are also a secondary biomass resource. Tertiary biomass feedstocks include fats, greases, oils, construction and demolition of wood debris, other waste wood from the urban environments, as well as packaging wastes, municipal solid wastes, and landfill gases. A category other wood waste from the urban environment includes trimmings from urban trees.

The simplest, cheapest, and most common method of obtaining energy from biomass is direct combustion. Any organic material, with a water content low enough to allow for sustained combustion, can be burned to produce energy which can be used to provide space or process heat, water heating, or (through the use of a steam turbine) electricity. In the developing world, many types of biomass such as animal dung and agricultural waste materials are burned to produce heat for cooking and warmth. In fact, such organic residues can also be used for energy production through conversion by natural biochemical processes as well as through the auspices of a biorefinery.

The biorefinery concept is analogous to the crude oil refinery and integrates biomass conversion processes and equipment to produce fuels, power, and chemicals from biomass. In addition to applying biological methods to crude oil itself such as the (1) desulfurization of fuels, (2) denitrogenation of fuels, (3) removal of heavy metals, (4) transformation of heavy crudes into light crudes, and (5) the biodegradation and bioremediation of crude oil spills as well as spills of crude oil products, biorefining offers a key method to accessing the integrated production of chemicals, materials, and fuels. While the biorefinery concept is analogous to that of the crude oil refinery in terms of feedstock pretreatment, conversion to products, and product finishing, there are significant differences – particularly in the character and properties of the respective feedstocks.

While the primary function of a biorefinery is to produce biofuels, there are also options to use biomass in various refinery scenarios. The biomass could be supplied by anything from corn, sugarcane, grasses, wood, and soybeans to algae. In place of fossil fuel-based hydrocarbon or hydrocarbonaceous feedstocks, biomass offers sugars, starches, fats, and proteins. Some chemicals will be synthesized using enzymes or genetically engineered microorganisms, and some will be produced using the inorganic catalysts used in traditional chemical processes. Throughout the decision process, consideration must be given to the use of biomass-derived chemicals without perturbing food supplies.

A biorefinery would (in a manner similar to the crude oil refinery) integrate a variety of conversion processes to produce multiple product streams and would combine the essential technologies to transform biological raw materials into a range of industrially useful intermediates. However, there may be the need to differentiate the type of biorefinery on the basis of the feedstock. For example, a *crop biorefinery* would use raw materials such as cereals or maize and a *lignocellulose biorefinery* would use raw material with high cellulose content, such as straw, wood, and paper waste. In addition, a variety of methods techniques can be employed to obtain different product portfolios of bulk chemicals, fuels, and materials (Speight, 2011c, 2020a, 2020b, 2020c). Biotechnology-based conversion processes can be used to ferment the biomass carbohydrate content into sugars that can then be further processed. An alternative is to employ thermochemical conversion processes which use pyrolysis or gasification of biomass to produce a hydrogen-rich synthesis gas which can be used in a wide range of chemical processes. On the other hand, the use of bio-feedstocks in a conventional crude oil refinery cannot be ignored, whether or not they are used as gasifier feedstocks (Speight, 2011c, 2014a, 2014b, 2020a, 2020b, 2020c).

These inherent characteristics and limitations of biomass feedstocks have focused the development of efficient methods of chemically transforming and upgrading biomass feedstocks in a refinery (Figure 17.3). The refinery would be based on two platforms to promote different product slates: (1) the biochemical platform and (2) the thermochemical platform. Using this two-train approach and by analogy with crude oil, every element of the plant feedstock will be utilized including the low-value lignin components. However, the different compositional nature of the biomass feedstock, compared to crude oil, will require the application of a wider variety of processing tools in the biorefinery. Processing of the individual components will utilize conventional thermochemical operations and state-of-the-art bioprocessing techniques.

The biorefinery concept provides a means to significantly reduce production costs such that a substantial substitution of petrochemicals by renewable chemicals becomes possible. However, significant technical challenges remain before the biorefinery concept can be realized.

7.5   EPILOG

There is no one single upgrading solution that fits all refineries and all of the varied refinery feedstocks. Thus, a careful evaluation of the slate of feedstocks into the refinery is necessary and is not always a simple undertaking for an existing refinery.

The evaluation typically starts at the time of selection of the feedstocks that adequately fit the configuration of the refinery, which varies from refinery to refinery. Some refineries may be more oriented toward the production of gasoline (large reforming and/or catalytic cracking), whereas the configuration of other refineries may be more oriented toward the production of middle distillates such as jet fuel and gas oil. Over the past four decades, the refining industry has been challenged because of the changing properties of the feedstocks and product slate which has introduced a high degree of flexibility with improved technologies and improved catalysts.

However, the evolution of the refinery of the future will not be strictly confined to crude oil processes. The major consequence will be a much more environmentally friendly product quality. These will be solved in refinery of the future, the refinery beyond 2020 with the development of deep conversion processing, such as residue hydrocracking and the inclusion of processes to accommodate other feedstocks. The *panacea* (rather than a *Pandora's box*) for a variety of feedstock could well be the *gasification refinery* (Speight, 2011b). This type of refinery approach that of a petrochemical complex, capable of supplying the traditional refined products, but also meeting much more severe specifications, and petrochemical intermediates such as olefins, aromatics, hydrogen and methanol. Furthermore, as already noted above, integrated gasification combined cycle (IGCC) can be used to raise power from feedstocks such as vacuum residua and cracked residua, and in addition to the production of synthesis gas, a major benefit of the IGCC concept is that power can be produced with the lowest sulfur oxide (SOx) and nitrogen oxide (NOx) emissions of any liquid/solid feed power generation technology.

The success of current operations notwithstanding the challenges facing the refining industry will focus on the diversity of the feedstocks. Even within the crude oil family of feedstocks where elemental analysis varies over a relatively narrow range, changes to refining technology are required to produce optimum yield of desired products. Another unique foreseeable disruption coming to the industry is the anticipated inclusion of biomass feedstocks and the changes that this will bring to refining. Indeed, much of the intellectual property embodied in the current refinery operations will have to change as wide variations in feedstock composition occur, and attempts are made to produce the necessary hydrocarbon fuels from a wide variety of biomass feedstocks. Any yet, the refining industry will survive – being one of the most resilient industries to commence operations during the past 150 years.

The refining industry can be regarded as unique insofar as very few industries have to deal with a feedstock-product chain beginning at a natural resource that has to be recovered from a subterranean formation and proceed through the application of a variety of processes all the way through to the end-use consumer. Furthermore, it is imperative for refiners to raise their operations to new levels of performance. Merely extending current process performance incrementally will most likely fail to meet most future performance goals. To do this, it will be necessary to reshape refining technology to be more adaptive to changing feedstocks and product demand and to explore the means by which the technology and methodology of refinery operations can be translated not only into increased profitability but also into survivability.

Part of the future growth will be at or near heavy crude and bitumen production sites to decrease heavy crude viscosity and improve the quality to ease transportation and open markets for crudes of otherwise marginal value. Visbreaking may be considered to be a conversion process rather than a process to produce fuel oil that meets specifications. Coking can be improved by reducing hydrocarbon gas formation and by inhibiting the formation of polynuclear aromatic products that are produced by the process and which are not inherent to the feedstock. Both of these processes would benefit if a higher valued product could be produced.

To this end, hydroconversion will continue to be a necessary and economically justifiable part of a residue conversion project (Parkash, 2003; Gary et al., 2007; Speight, 2014a; Hsu and Robinson, 2017; Speight, 2017). The use of ebullated-bed reactors for hydroconversion is evolving as is the use of slurry-catalyst systems. In addition, the gasification of heavy feedstocks moving closer to the time when it will be a ready choice for the conversion of heavy feedstocks transportation fuels. Heavy feedstocks conversion projects (including power generation using waste streams) could move into positive economics in regions with high local power demand. Moreover, with the ever-increasing and stringent specifications for refinery products, refiners will continue to make efforts to improve the heavy feedstock processing technologies that convert the heavy feedstock into valuable and environment-friendly products. Currently, a variety of residue hydrocracking processes using fixed-bed, moving-bed, or ebullated-bed reactors are available. Furthermore, the economics of the slurry-bed processing technology indicates an attractive rate of returns with the existing crude oils and product price structure.

Several process innovations have been introduced such as the use of using piggy-back techniques (where one process is operative in close conjunction with another process) that will fit into the future refinery. Another significant improvement in the hydroprocessing technologies will be in catalyst design. This follows from recognition that a novel catalyst is one of reasonable design and integration of the active, supporting, and promoting components which allow an optimal combination of activity, surface area, and pore diameter, giving the highest activity.

High conversion refineries will move to gasification of feedstocks for the development of alternative fuels and to enhance equipment usage. A major trend in the refining industry market demand for refined products will be in synthesizing fuels from simple basic reactants (e.g., synthesis gas) when it becomes uneconomical to produce super clean transportation fuels through conventional refining processes. Fischer-Tropsch plants together with IGCC systems will be integrated with, or even into, crude oil refineries, which will offer the advantage of high-quality products (Table 17.2).

REFERENCES

Bhatia, S., and Sharma, D.K. 2006. Emerging Role of Biorefining of Heavier Crude Oils and Integration of Biorefining with Crude Oil Refineries in the Future. *Petroleum Science and Technology*, 24(10): 1125–1159.

Collins, T., and Barletta, A. 2012. Desalting Heavy Canadian Crudes. Digital Refining. http://www.digitalrefining.com/article/1000566,Desalting_heavy_Canadian_crudes.html#.VJiTC14AA

Gary, J.H., Handwerk, G.E., and Kaiser, M.J. 2007. *Petroleum Refining: Technology and Economics*. 5th Edition. CRC Press, Taylor & Francis Group, Boca Raton, Florida.

Hsu, C.S., and Robinson, P.R. (Editors). 2017. *Handbook of Petroleum Technology*. Springer International Publishing AG, Cham, Switzerland.

Parkash, S. 2003. *Refining Processes Handbook*. Gulf Professional Publishing, Elsevier, Amsterdam, Netherlands.

Pruneda, E.F., Escobedo, E.R.B., and Vazquez, F.J.G. 2005. Optimum Temperature in the Electrostatic Desalting of Maya Crude Oil. *Journal of Mexican Chemical Society*, 49(1): 14–19.

Speight, J.G., and Moschopedis, S.E. 1979. The Production of Low-Sulfur Liquids and Coke from Athabasca Bitumen. *Fuel Processing Technology*, 2: 295–302.

Speight, J.G. 2011a. *An Introduction to Petroleum Technology, Economics, and Politics*. Scrivener Publishing, Beverly, Massachusetts.

Speight, J.G. 2011b. *The Refinery of the Future*. Gulf Professional Publishing, Elsevier, Oxford, United Kingdom.

Speight, J.G. 2012. *Shale Oil Production Processes*. Gulf Professional Publishing, Elsevier, Oxford, United Kingdom.

Speight, J.G. 2013. *The Chemistry and Technology of Coal*. 3rd Edition. CRC-Taylor & Francis Group, Boca Raton, Florida.

Speight, J.G. 2014a. *The Chemistry and Technology of Petroleum*. 5th Edition. CRC-Taylor & Francis Group, Boca Raton, Florida.

Speight, J.G. 2014b. *Oil and Gas Corrosion Prevention*. Gulf Professional Publishing Company, Elsevier, Oxford, United Kingdom.

Speight, J.G. 2014c. *High Acid Crudes*. Gulf Professional Publishing, Elsevier, Oxford, United Kingdom.

Speight, J.G. 2015. *Handbook of Petroleum Product Analysis*. 2nd Edition. John Wiley & Sons Inc., Hoboken, New Jersey.

Speight, J.G. 2016. Hydrogen in Refineries. In: *Hydrogen Science and Engineering: Materials, Processes, Systems, and Technology*. D. Stolten and B. Emonts (Editors). Wiley-VCH Verlag GmbH & Co., Weinheim, Germany. Chapter 1. Page 3–18.

Speight, J.G., and Islam, M.R. 2016. *Peak Energy – Myth or Reality*. Scrivener Publishing, Beverly, Massachusetts.

Speight, J.G. 2017. *Handbook of Petroleum Refining*. CRC Press, Taylor & Francis Group, Boca Raton, Florida.

Speight, J.G. 2019. *Handbook of Petrochemical Processes*. CRC Press, Taylor & Francis Group, Boca Raton, Florida.

Speight, J.G. 2020a. *Synthetic Fuels Handbook: Properties, Processes, and Performance*. 2nd Edition. McGraw-Hill, New York.

Speight, J.G. 2020b. *Global Climate Change Demystified*. Scrivener Publishing, Beverly, Massachusetts.

Speight, J.G. 2020c. *Refinery of the Future*. 2nd Edition. Gulf Professional Publishing, Elsevier. Cambridge, Massachusetts.

Glossary

ABN separation: a method of fractionation by which petroleum is separated into acidic, basic, and neutral constituents.

Absorber: see Absorption tower.

Absorption gasoline: gasoline extracted from natural gas or refinery gas by contacting the absorbed gas with an oil and subsequently distilling the gasoline from the higher-boiling components.

Absorption oil: oil used to separate the heavier components from a vapor mixture by absorption of the heavier components during intimate contacting of the oil and vapor; used to recover natural gasoline from wet gas.

Absorption plant: a plant for recovering the condensable portion of natural or refinery gas, by absorbing the higher-boiling hydrocarbons in an absorption oil, followed by separation and fractionation of the absorbed material.

Absorption tower: a tower or column which promotes contact between a rising gas and a falling liquid so that part of the gas may be dissolved in the liquid.

Acetone-benzol process: a dewaxing process in which acetone and benzol (benzene or aromatic naphtha) are used as solvents.

Acid catalyst: a catalyst having acidic character; the alumina minerals are examples of such catalysts.

Acid deposition: acid rain; a form of pollution depletion in which pollutants, such as nitrogen oxides and sulfur oxides, are transferred from the atmosphere to soil or water; often referred to as atmospheric self-cleaning. The pollutants usually arise from the use of fossil fuels.

Acid gas removal: a process for the removal of hydrogen sulfide, other sulfur species, and some carbon dioxide from syngas by absorption in a solvent with subsequent solvent regeneration and production of a hydrogen sulfide (H_2S)-rich stream for sulfur recovery.

Acid number: a measure of the reactivity of petroleum with a caustic solution and given in terms of milligrams of potassium hydroxide that are neutralized by 1 g of petroleum.

Acid rain: the precipitation phenomenon that incorporates anthropogenic acids and other acidic chemicals from the atmosphere to the land and water (see Acid deposition).

Acid sludge: the residue left after treating petroleum oil with sulfuric acid for the removal of impurities; a black, viscous substance containing the spent acid and impurities.

Acid treating: a process in which unfinished petroleum products, such as gasoline, kerosene, and lubricating-oil stocks, are contacted with sulfuric acid to improve their color, odor, and other properties.

Acidity: the capacity of an acid to neutralize a base such as a hydroxyl ion (OH^-).

Acidizing: a technique for improving the permeability of a reservoir by injecting acid.

Acoustic log: see Sonic log.

Acre-foot: a measure of bulk rock volume where the area is 1 acre and the thickness is 1 ft.

Add-on control methods: the use of devices that remove refinery process emissions after they are generated but before they are discharged to the atmosphere.

Additive: a material added to a reaction mixture (usually in small amounts) in order to (1) enhance the reaction, (2) enhance the desirable properties of the products, or (3) to suppress undesirable properties of the products. Different additives, even when added for identical purposes, may be incompatible with each other, for example react and form new compounds. See also Incompatibility, Instability.

Adsorption: transfer of a substance from a solution to the surface of a solid resulting in relatively high concentration of the substance at the place of contact; see also Chromatographic adsorption.

Adsorption gasoline: natural gasoline obtained by the adsorption process from wet gas.

After flow: flow from the reservoir into the wellbore that continues for a period after the well has been shut in; after-flow can complicate the analysis of a pressure transient test.

Afterburn: the combustion of carbon monoxide (CO) to carbon dioxide (CO_2); usually in the cyclones of a catalyst regenerator.

Air-blown asphalt: asphalt produced by blowing air through residua at elevated temperatures.

Air injection: an oil recovery technique using air to force oil from the reservoir into the wellbore.

Air pollution: the discharge of toxic gases and particulate matter introduced into the atmosphere, principally as a result of human activity.

Air separation unit (ASU): a plant that separates oxygen and nitrogen from air, usually by cryogenic distillation.

Air sweetening: a process in which air or oxygen is used to oxidize lead mercaptan derivatives (RSH) to disulfide derivatives (RSSR) instead of using elemental sulfur.

Air toxics: hazardous air pollutants.

Airlift thermofor catalytic cracking: a moving-bed continuous catalytic process for conversion of heavy gas oils into lighter products; the catalyst is moved by a stream of air.

Albertite: a black, brittle, natural hydrocarbon possessing a conchoidal fracture and a specific gravity of approximately 1.1.

Alcohol: the family name of a group of organic chemical compounds composed of carbon, hydrogen, and oxygen. The molecules in the series vary in chain length and are composed of a hydrocarbon plus a hydroxyl group. Alcohol includes methanol and ethanol.

Alicyclic hydrocarbon: a compound containing carbon and hydrogen only which has a cyclic structure (e.g., cyclohexane); also collectively called naphthenes.

Aliphatic hydrocarbon: a compound containing carbon and hydrogen only which has an open-chain structure (e.g., as ethane, butane, octane, butene) or a cyclic structure (e.g., cyclohexane).

Aliquot: that quantity of material of proper size for measurement of the property of interest; test portions may be taken from the gross sample directly, but often preliminary operations such as mixing or further reduction in particle size are necessary.

Alkali treatment: see Caustic wash.

Alkali wash: see Caustic wash.

Alkaline: a high pH usually of an aqueous solution; aqueous solutions of sodium hydroxide, sodium orthosilicate, and sodium carbonate are typical alkaline materials used in enhanced oil recovery.

Alkaline flooding: see EOR process.

Alkalinity: the capacity of a base to neutralize the hydrogen ion (H^+).

Alkanes: hydrocarbons that contain only single carbon-hydrogen bonds. The chemical name indicates the number of carbon atoms and ends with the suffix "ane."

Alkenes: hydrocarbons that contain carbon-carbon double bonds. The chemical name indicates the number of carbon atoms and ends with the suffix "ene."

Alkyl groups: a group of carbon and hydrogen atoms that branch from the main carbon chain or ring in a hydrocarbon molecule. The simplest alkyl group, a methyl group, is a carbon atom attached to three hydrogen atoms.

Alkylate: the product of an alkylation process.

Alkylate bottoms: residua from fractionation of alkylate; the alkylate product which boils higher than the aviation gasoline range; sometimes called heavy alkylate or alkylate polymer.

Alkylation: in the petroleum industry, a process by which an olefin (e.g., ethylene) is combined with a branched-chain hydrocarbon (e.g., *iso*-butane); alkylation may be accomplished as a thermal or as a catalytic reaction.

Alpha-scission: the rupture of the aromatic carbon-aliphatic carbon bond that joins an alkyl group to an aromatic ring.

Alumina (Al_2O_3): used in separation methods as an adsorbent and in refining as a catalyst.

American Society for Testing and Materials (ASTM): the official organization in the United States for designing standard tests for petroleum and other industrial products.

Amine washing: a method of gas cleaning whereby acidic impurities such as hydrogen sulfide and carbon dioxide are removed from the gas stream by washing with an amine (usually an alkanolamine).

Anaerobic digestion: decomposition of biological wastes by micro-organisms, usually under wet conditions, in the absence of air (oxygen), to produce a gas comprising mostly methane and carbon dioxide.

Analyte: the chemical for which a sample is tested or analyzed.

Analytical equivalence: the acceptability of the results obtained from the different laboratories; a range of acceptable results.

Antibody: a molecule having chemically reactive sites specific for certain other molecules.

Aniline point: the temperature, usually expressed in °F, above which equal volumes of a petroleum product are completely miscible; a qualitative indication of the relative proportions of paraffins in a petroleum product which are miscible with aniline only at higher temperatures; a high aniline point indicates low aromatics.

Annual removals: the net volume of growing stock trees removed from the inventory during a specified year by harvesting, cultural operations such as timber stand improvement, or land clearing.

Antiknock: resistance to detonation or pinging in spark-ignition engines.

Antiknock agent: a chemical compound such as tetraethyl lead which, when added in small amount to the fuel charge of an internal-combustion engine, tends to lessen knocking.

Antistripping agent: an additive used in an asphaltic binder to overcome the natural affinity of an aggregate for water instead of asphalt.

API gravity: a measure of the *lightness* or *heaviness* of petroleum which is related to density and specific gravity.

$$°API = (141.5 \ / \ sp \ gr @ 60°F) - 131.5$$

Apparent bulk density: the density of a catalyst as measured; usually loosely compacted in a container.

Apparent viscosity: the viscosity of a fluid, or several fluids flowing simultaneously, measured in a porous medium (rock) and subject to both viscosity and permeability effects; also called effective viscosity.

Aquifer: a subsurface rock interval that will produce water; often the underlay of a petroleum reservoir.

Areal sweep efficiency: the fraction of the flood pattern area that is effectively swept by the injected fluids.

Aromatic hydrocarbon: a hydrocarbon characterized by the presence of an aromatic ring or condensed aromatic rings; benzene and substituted benzene, naphthalene and substituted naphthalene, phenanthrene and substituted phenanthrene, as well as the higher condensed ring systems; compounds that are distinct from those of aliphatic compounds or alicyclic compounds.

Aromatics: see Aromatic hydrocarbon.

Aromatization: the conversion of non-aromatic hydrocarbons to aromatic hydrocarbons by: (1) rearrangement of aliphatic (noncyclic) hydrocarbons into aromatic ring structures; and (2) dehydrogenation of alicyclic hydrocarbons (naphthenes).

Arosorb process: a process for the separation of aromatic derivatives from non-aromatic derivatives by adsorption on a gel from which they are recovered by desorption.

Asphalt: the nonvolatile product obtained by distillation and treatment of an asphaltic crude oil; a manufactured product.

Asphalt cement: asphalt especially prepared as to quality and consistency for direct use in the manufacture of bituminous pavements.

Asphalt emulsion: an emulsion of asphalt cement in water containing a small amount of emulsifying agent.

Asphalt flux: an oil used to reduce the consistency or viscosity of hard asphalt to the point required for use.

Asphalt primer: a liquid asphaltic material of low viscosity which upon application to a non-bituminous surface to waterproof the surface and prepare it for further construction.

Asphaltene (asphaltenes): the brown to black powdery material produced by treatment of petroleum, petroleum residua, or bituminous materials with a low-boiling liquid hydrocarbon, e.g., pentane or heptane; soluble in benzene (and other aromatic solvents), carbon disulfide, and chloroform (or other chlorinated hydrocarbon solvents).

Asphaltene association factor: the number of individual asphaltene species which associate with non-polar solvents as measured by molecular weight methods; the molecular weight of asphaltenes in toluene divided by the molecular weight in a polar non-associating solvent, such as dichlorobenzene, pyridine, or nitrobenzene.

Asphaltic pyrobitumen: see Asphaltoid.

Asphaltic road oil: a thick, fluid solution of asphalt; usually a residual oil; see also Nonasphaltic road oil.

Asphaltite: a variety of naturally occurring, dark brown to black, solid, nonvolatile bituminous material that is differentiated from bitumen primarily by a high content of material insoluble in n-pentane (asphaltene) or other liquid hydrocarbons.

Asphaltoid: a group of brown to black, solid bituminous materials of which the members are differentiated from asphaltites by their infusibility and low solubility in carbon disulfide.

Asphaltum: see Asphalt.

Associated molecular weight: the molecular weight of asphaltenes in an associating (non-polar) solvent, such as toluene.

Atmospheric residuum: a residuum obtained by distillation of a crude oil under atmospheric pressure and which boils above 350°C (660°F).

Atmospheric equivalent boiling point (AEBP): a mathematical method of estimating the boiling point at atmospheric pressure of non-volatile fractions of petroleum.

Attainment area: a geographical area that meets NAAQS for criteria air pollutants (See also Non-attainment area).

Attapulgus clay: see Fuller's earth.

Autofining: a catalytic process for desulfurizing distillates.

Average particle size: the weighted average particle diameter of a catalyst.

Aviation gasoline: any of the special grades of gasoline suitable for use in certain airplane engines.

Aviation turbine fuel: see Jet fuel.

Back mixing: the phenomenon observed when a catalyst travels at a slower rate in the riser pipe than the vapors.

BACT: best available control technology.

Baghouse: a filter system for the removal of particulate matter from gas streams; so-called because of the similarity of the filters to coal bags.

Bank: the concentration of oil (oil bank) in a reservoir that moves cohesively through the reservoir.

Bari-Sol process: a dewaxing process which employs a mixture of ethylene dichloride and benzol as the solvent.

Barrel (bbl): the unit of measure used by the petroleum industry; equivalent to approximately 42 US gallons or approximately 34 (33.6) Imperial gallons or 159 L; 7.2 barrels are equivalent to 1 tonne of oil (metric).

Barrel of oil equivalent (boe): the amount of energy contained in a barrel of crude oil, i.e., approximately 6.1 GJ (5.8 million Btu), equivalent to 1,700 kWh.

Base number: the quantity of acid, expressed in milligrams of potassium hydroxide per gram of sample that is required to titrate a sample to a specified end-point.

Base stock: a primary refined petroleum fraction into which other oils and additives are added (blended) to produce the finished product.

Basic nitrogen: nitrogen (in petroleum) which occurs in pyridine form

Basic sediment and water (BS&W, BSW): the material which collects in the bottom of storage tanks, usually composed of oil, water, and foreign matter; also called bottoms, bottom settlings.

Batch blending: a blending process (also called the in-tank process) in which specific (calculated and measured) volumes of the different crude oils that have been stored in separate tanks are loaded into a blending tank where they are mixed (by a mechanical stirring operation).

Battery: a series of stills or other refinery equipment operated as a unit.

Baumé gravity: the specific gravity of liquids expressed as degrees on the Baumé (°Bé) scale; for liquids lighter than water:

$$\text{Sp gr } 60°\text{F} = 140 / (130 + °\text{Bé})$$

For liquids heavier than water:

$$\text{Sp gr } 60°\text{F} = 145 / (145 - °\text{Bé})$$

Bauxite: mineral matter used as a treating agent; hydrated aluminum oxide formed by the chemical weathering of igneous rock.

Bbl: see Barrel.

Bell cap: a hemispherical or triangular cover placed over the riser in a (distillation) tower to direct the vapors through the liquid layer on the tray; see Bubble cap.

Bender process: a chemical treating process using lead sulfide catalyst for sweetening light distillates by which mercaptan derivatives (RSH) are converted to disulfide derivatives (RSSR) by oxidation.

Bentonite: montmorillonite (a magnesium-aluminum silicate); used as a treating agent.

Benzene: a colorless aromatic liquid hydrocarbon (C_6H_6).

Benzin: a refined light naphtha used for extraction purposes.

Benzine: an obsolete term for light petroleum distillates covering the gasoline and naphtha range; see Ligroine.

Benzol: the general term which refers to commercial or technical (not necessarily pure) benzene; also the term used for aromatic naphtha.

Beta-scission: the rupture of a carbon-carbon bond two bonds removed from an aromatic ring.

Billion: 1×10^9

Biochemical conversion: The use of fermentation or anaerobic digestion to produce fuels and chemicals from organic sources.

Biocide: any chemical capable of killing bacteria and biorganisms.

Biodiesel: a fuel derived from biological sources that can be used in diesel engines instead of petroleum-derived diesel; through the process of transesterification, the triglycerides in the biologically derived oils are separated from the glycerin, creating a clean-burning, renewable fuel.

Bioenergy: useful, renewable energy produced from organic matter – the conversion of the complex carbohydrates in organic matter to energy; organic matter may either be used directly as a fuel, processed into liquids and gasses, or be a residual of processing and conversion.

Bioethanol: ethanol produced from biomass feedstocks; includes ethanol produced from the fermentation of crops, such as corn, as well as cellulosic ethanol produced from woody plants or grasses.

Biofuels: a generic name for liquid or gaseous fuels that are not derived from petroleum-based fossils fuels or contain a proportion of non-fossil fuel; fuels produced from plants and crops such as sugar beet and rape seed oil or re-processed vegetable oils or fuels made from gasified biomass; fuels made from renewable biological sources and include ethanol, methanol, and biodiesel; sources include, but are not limited to: corn, soybeans, flaxseed, rapeseed, sugarcane, palm oil, raw sewage, food scraps, animal parts, and rice.

Biogas: a combustible gas derived from decomposing biological waste under anaerobic conditions. Biogas normally consists of 50%–60% methane. See also Landfill gas.

Biogenic: material derived from bacterial or vegetation sources.

Biological lipid: any biological fluid that is miscible with a nonpolar solvent. These materials include waxes, essential oils, chlorophyll, etc.

Biological oxidation: the oxidative consumption of organic matter by bacteria by which the organic matter is converted into gases.

Biomass: any organic matter that is available on a renewable or recurring basis, including agricultural crops and trees, wood and wood residues, plants (including aquatic plants), grasses, animal manure, municipal residues, and other residue materials. Biomass is generally produced in a sustainable manner from water and carbon dioxide by photosynthesis. There are three main categories of biomass – primary, secondary, and tertiary.

Biopower: the use of biomass feedstock to produce electric power or heat through direct combustion of the feedstock, through gasification and then combustion of the resultant gas, or through other thermal conversion processes. Power is generated with engines, turbines, fuel cells, or other equipment.

Biomass to liquid (BTL): the process of converting biomass to liquid fuels. Hmm, that seems painfully obvious when you write it out.

Biopolymer: a high-molecular weight carbohydrate produced by bacteria.

Biorefinery: a facility that processes and converts biomass into value-added products. These products can range from biomaterials to fuels such as ethanol or important feedstocks for the production of chemicals and other materials.

Bitumen: also, on occasion, referred to as native asphalt, and extra heavy oil; a naturally occurring material that has little or no mobility under reservoir conditions and which cannot be recovered through a well by conventional oil well production methods including currently used enhanced recovery techniques; current methods involve mining for bitumen recovery.

Bituminous: containing bitumen or constituting the source of bitumen.

Bituminous rock: see Bituminous sand.

Bituminous sand: a formation in which the bituminous material (see Bitumen) is found as a filling in veins and fissures in fractured rock or impregnating relatively shallow sand, sandstone, and limestone strata; a sandstone reservoir that is impregnated with a heavy, viscous black petroleum-like material that cannot be retrieved through a well by conventional production techniques.

Black acid(s): a mixture of the sulfonates found in acid sludge which are insoluble in naphtha, benzene, and carbon tetrachloride; very soluble in water but insoluble in 30% sulfuric acid; in the dry, oil-free state, the sodium soaps are black powders.

Black liquor: solution of lignin-residue and the pulping chemicals used to extract lignin during the manufacture of paper.

Black oil: any of the dark-colored oils; a term now often applied to heavy oil.

Black soap: see Black acid.

Black strap: the black material (mainly lead sulfide) formed in the treatment of sour light oils with doctor solution and found at the interface between the oil and the solution.

Blended crude: a mixture of crude oils blended in the pipeline to create a crude with specific physical properties. For example, heavy crude oil, extra heavy crude oil, and tar sand bitumen cannot flow from the field to the refinery in their original state, and at normal surface temperatures, they are blended with lighter crude oils primarily to reduce the viscosity, thereby enabling transportation to a refinery. In addition, the blend can provide a blended refinery feedstock crude oil that has significantly higher value than the raw heavy feedstocks in which case the blend is usually constructed so that the value of the overall blended volume is greater than the summed value of the initial volumes of the individual heavy and light components of the blend.

Blending: in the current context, the process of mixing two or more crude oils without producing a chemical reaction or a physical reaction (i.e., phase separation of asphaltene constituents) to create a composite feedstock that is still suitable for refining.

Blown asphalt: the asphalt prepared by air blowing a residuum or an asphalt.

Bogging: a condition that occurs in a coking reactor when the conversion to coke and light ends is too slow causing the coke particles to agglomerate.

Boiling point: a characteristic physical property of a liquid at which the vapor pressure is equal to that of the atmosphere and the liquid is converted to a gas.

Boiling range: the range of temperature, usually determined at atmospheric pressure in standard laboratory apparatus, over which the distillation of an oil commences, proceeds, and finishes.

Bone dry: having 0% moisture content. Wood heated in an oven at a constant temperature of 100°C (212°F) or above until its weight stabilizes is considered bone dry or oven dry.

Bottled gas: usually butane or propane, or butane-propane mixtures, liquefied and stored under pressure for domestic use; see also Liquefied petroleum gas.

Bottoming cycle: a cogeneration system in which steam is used first for process heat and then for electric power production.

Bottoms: the liquid which collects in the bottom of a vessel (tower bottoms, tank bottoms) either during distillation; also the deposit or sediment formed during storage of petroleum or a petroleum product; see also Residuum and Basic sediment and water.

Bright stock: refined, high-viscosity lubricating oils usually made from residual stocks by processes such as a combination of acid treatment or solvent extraction with dewaxing or clay finishing.

British thermal unit: see Btu.

Bromine number: the number of grams of bromine absorbed by 100 g of oil which indicates the percentage of double bonds in the material.

Brown acid: oil-soluble petroleum sulfonates found in acid sludge which can be recovered by extraction with naphtha solvent. Brown-acid sulfonates are somewhat similar to mahogany sulfonates but are more water-soluble. In the dry, oil-free state, the sodium soaps are light-colored powders.

Brønsted acid: a chemical species which can act as a source of protons.

Brønsted base: a chemical species which can accept protons.

Brown soap: see Brown acid.

BS&W: see Basic sediment and water.

BTEX: benzene, toluene, ethylbenzene, and the xylene isomers.

Btu (British thermal unit): the energy required to raise the temperature of 1 lb of water 1°F.

Bubble cap: an inverted cup with a notched or slotted periphery to disperse the vapor in small bubbles beneath the surface of the liquid on the bubble plate in a distillation tower.

Bubble plate: a tray in a distillation tower.

Bubble point: the temperature at which incipient vaporization of a liquid in a liquid mixture occurs, corresponding with the equilibrium point of 0% vaporization or 100% condensation.

Bubble tower: a fractionating tower so constructed that the vapors rising pass up through layers of condensate on a series of plates or trays (see Bubble plate); the vapor passes from one plate to the next above by bubbling under one or more caps (see Bubble cap) and out through the liquid on the plate where the less volatile portions of vapor condense in bubbling through the liquid on the plate, overflow to the next lower plate, and ultimately back into the reboiler thereby effecting fractionation.

Bubble tray: a circular, perforated plates having the internal diameter of a bubble tower, set at specified distances in a tower to collect the various fractions produced during distillation.

Buckley-Leverett method: a theoretical method of determining frontal advance rates and saturations from a fractional flow curve.

Bumping: the knocking against the walls of a still occurring during distillation of petroleum or a petroleum product which usually contains water.

Bunker: a storage tank.

Bunker C oil: see No. 6 Fuel oil.

Burner fuel oil: any petroleum liquid suitable for combustion.

Burning oil: an illuminating oil, such as kerosene (kerosine) suitable for burning in a wick lamp.

Burning point: see Fire point.

Burning-quality index: an empirical numerical indication of the likely burning performance of a furnace or heater oil; derived from the distillation profile and the API gravity, and generally recognizing the factors of paraffin character and volatility.

Burton process: an older thermal cracking process in which oil was cracked in a pressure still and any condensation of the products of cracking also took place under pressure.

Butane dehydrogenation: a process for removing hydrogen from butane to produce butenes and, on occasion, butadiene.

Butane vapor-phase isomerization: a process for isomerizing n-butane to *iso*-butane using aluminum chloride catalyst on a granular alumina support and with hydrogen chloride as a promoter.

Butanol: though generally produced from fossil fuels, this four-carbon alcohol can also be produced through bacterial fermentation of alcohol.

C_1, C_2, C_3, C_4, C_5 fractions: a common way of representing fractions containing a preponderance of hydrocarbons having 1, 2, 3, 4, or 5 carbon atoms, respectively, and without reference to hydrocarbon type.

CAA: Clean Air Act; this act is the foundation of air regulations in the United States.

Calcining: heating a metal oxide or an ore to decompose carbonates, hydrates, or other compounds often in a controlled atmosphere.

Capillary forces: interfacial forces between immiscible fluid phases, resulting in pressure differences between the two phases.

Capillary number: N_c, the ratio of viscous forces to capillary forces, and equal to viscosity times velocity divided by interfacial tension.

Carbene: the pentane- or heptane-insoluble material that is insoluble in benzene or toluene but which is soluble in carbon disulfide (or pyridine); a type of rifle used for hunting bison.

Carboid: the pentane- or heptane-insoluble material that is insoluble in benzene or toluene and which is also insoluble in carbon disulfide (or pyridine).

Carbon dioxide augmented waterflooding: injection of carbonated water, or water and carbon dioxide, to increase water flood efficiency; see immiscible carbon dioxide displacement.

Carbon dioxide miscible flooding: see EOR process.

Carbon-forming propensity: see Carbon residue.

Carbon monoxide (CO): a lethal gas produced by incomplete combustion of carbon-containing fuels in internal combustion engines. It is colorless, odorless, and tasteless. (As in flavorless, we mean, though it's also been known to tell a bad joke or two.)

Carbon-oxygen log: information about the relative abundance of elements such as carbon, oxygen, silicon, and calcium in a formation; usually derived from pulsed neutron equipment.

Carbon rejection: upgrading processes in which coke is produced, e.g., coking.

Carbon residue: the amount of carbonaceous residue remaining after thermal decomposition of petroleum, a petroleum fraction, or a petroleum product in a limited amount of air; also called the *coke-* or *carbon-forming propensity*; often prefixed by the terms Conradson or Ramsbottom in reference to the inventor of the respective tests.

Carbon sink: a geographical area whose vegetation and/or soil soaks up significant carbon dioxide from the atmosphere. Such areas, typically in tropical regions, are increasingly being sacrificed for energy crop production.

Carbonate washing: processing using a mild alkali (e.g., potassium carbonate) process for emission control by the removal of acid gases from gas streams.

Carbonization: the conversion of an organic compound into char or coke by heat in the substantial absence of air; often used in reference to the destructive distillation (with simultaneous removal of distillate) of coal.

CAS: Chemical Abstract Service.

Cascade tray: a fractionating device consisting of a series of parallel troughs arranged on stair-step fashion in which liquid frown the tray above enters the uppermost trough and liquid thrown from this trough by vapor rising from the tray below impinges against a plate and a perforated baffle and liquid passing through the baffle enters the next longer of the troughs.

Casinghead gas: natural gas which issues from the casinghead (the mouth or opening) of an oil well.

Casinghead gasoline: the liquid hydrocarbon product extracted from casinghead gas by one of three methods: compression, absorption, or refrigeration; see also Natural gasoline.

Cat cracking: see Catalytic cracking.

Catagenesis: the alteration of organic matter during the formation of petroleum that may involve temperatures in the range 50°C (120°F) to 200°C (390°F); see also Diagenesis and Metagenesis.

Catalyst: a chemical agent which when added to a reaction (process) will enhance the conversion of a feedstock without being consumed in the process.

Catalyst selectivity: the relative activity of a catalyst with respect to a particular compound in a mixture, or the relative rate in competing reactions of a single reactant.

Catalyst stripping: the introduction of steam, at a point where spent catalyst leaves the reactor, in order to strip, i.e., remove, deposits retained on the catalyst.

Catalytic activity: the ratio of the space velocity of the catalyst under test to the space velocity required for the standard catalyst to give the same conversion as the catalyst being tested; usually multiplied by 100 before being reported.

Catalytic cracking: the conversion of high-boiling feedstocks into lower boiling products by means of a catalyst which may be used in a fixed bed or fluid bed.

Catalytic reforming: rearranging hydrocarbon molecules in a gasoline-boiling-range feedstock to produce other hydrocarbons having a higher antiknock quality; isomerization of paraffins, cyclization of paraffins to naphthenes, dehydrocyclization of paraffins to aromatics.

Catforming: a process for reforming naphtha using a platinum-silica-alumina catalyst which permits relatively high space velocities and results in the production of high-purity hydrogen.

Caustic consumption: the amount of caustic lost from reacting chemically with the minerals in the rock, the oil, and the brine.

Caustic wash: the process of treating a product with a solution of caustic soda to remove minor impurities; often used in reference to the solution itself.

Ceresin: a hard, brittle wax obtained by purifying ozokerite; see Microcrystralline wax and Ozokerite.

Cetane index: an approximation of the cetane number calculated from the density and mid-boiling point temperature; see also Diesel index.

Cetane number: a number indicating the ignition quality of diesel fuel; a high cetane number represents a short ignition delay time; the ignition quality of diesel fuel can also be estimated from the following formula:

CFR: Code of Federal Regulations; Title 40 (40 CFR) contains the regulations for protection of the environment.

Characterization factor: the UOP characterization factor K, defined as the ratio of the cube root of the molal average boiling point, T_B, in degrees Rankine ($^\circ R = ^\circ F + 460$), to the specific gravity at 60°F /60°F:

$$K = (T_B)^{1/3} / \text{sp gr}$$

The value ranges from 12.5 for paraffin stocks to 10.0 for the highly aromatic stocks; also called the Watson characterization factor.

Cheesebox still: an early type of vertical cylindrical still designed with a vapor dome.

Chelating agents: complex-forming agents having the ability to solubilize heavy metals.

Chemical flooding: see EOR process.

Chemical octane number: the octane number added to gasoline by refinery processes or by the use of octane number improvers such as tetraethyl lead.

Chemical waste: any solid, liquid, or gaseous material discharged from a process and that may pose substantial hazards to human health and environment.

Chlorex process: a process for extracting lubricating-oil stocks in which the solvent used is Chlorex (ß- ß -dichlorodiethyl ether).

Chromatographic adsorption: selective adsorption on materials such as activated carbon, alumina, or silica gel; liquid or gaseous mixtures of hydrocarbons are passed through the adsorbent in a stream of diluent, and certain components are preferentially adsorbed.

Chromatographic separation: the separation of different species of compounds according to their size and interaction with the rock as they flow through a porous medium.

Chromatography: a method of separation based on selective adsorption; see also Chromatographic adsorption.

Clarified oil: the heavy oil which has been taken from the bottom of a fractionator in a catalytic cracking process and from which residual catalyst has been removed.

Clarifier: equipment for removing the color or cloudiness of an oil or water by separating the foreign material through mechanical or chemical means; may involve centrifugal action, filtration, heating, or treatment with acid or alkali.

Clay: silicate minerals that also usually contain aluminum and have particle sizes are less than 0.002 μ; used in separation methods as an adsorbent and in refining as a catalyst.

Clay contact process: see Contact filtration.

Clay refining: a treating process in which vaporized gasoline or other light petroleum product is passed through a bed of granular clay such as fuller's earth.

Clay regeneration: a process in which spent coarse-grained adsorbent clay minerals from percolation processes are cleaned for reuse by de-oiling the clay minerals with naphtha, steaming out the excess naphtha, and then roasting in a stream of air to remove carbonaceous matter.

Clay treating: see Gray clay treating.

Clay wash: light oil, such as kerosene (kerosine) or naphtha, used to clean fuller's earth after it has been used in a filter.

Clastic: composed of pieces of pre-existing rock.

Cleanup: a preparatory step following extraction of a sample media designed to remove components that may interfere with subsequent analytical measurements.

Closed-loop biomass: crops grown, in a sustainable manner, for the purpose of optimizing their value for bioenergy and bioproduct uses. This includes annual crops such as maize and wheat, and perennial crops such as trees, shrubs, and grasses such as switch grass.

Cloud point: the temperature at which paraffin wax or other solid substances begin to crystallize or separate from the solution, imparting a cloudy appearance to the oil when the oil is chilled under prescribed conditions.

Coal: an organic rock.

Coal tar: the specific name for the tar produced from coal.

Coal tar pitch: the specific name for the pitch produced from coal.

Coalescence: the union of two or more droplets to form a larger droplet and, ultimately, a continuous phase.

Coarse materials: wood residues suitable for chipping, such as slabs, edgings, and trimmings.

COFCAW: an EOR process that combines forward combustion and water flooding.

Cogeneration: an energy conversion method by which electrical energy is produced along with steam generated for EOR use.

Coke: a gray to black solid carbonaceous material produced from petroleum during thermal processing; characterized by having a high carbon content (95%+ by weight) and a honeycomb type of appearance and is insoluble in organic solvents.

Coke drum: a vessel in which coke is formed and which can be cut oil from the process for cleaning.

Coke number: used, particularly in Great Britain, to report the results of the Ramsbottom carbon residue test, which is also referred to as a coke test.

Coker: the processing unit in which coking takes place.

Coking: a process for the thermal conversion of petroleum in which gaseous, liquid, and solid (coke) products are formed.

Cold pressing: the process of separating wax from oil by first chilling (to help form wax crystals) and then filtering under pressure in a plate and frame press.

Cold settling: processing for the removal of wax from high-viscosity stocks, wherein a naphtha solution of the waxy oil is chilled and the wax crystallizes out of the solution.

Color stability: the resistance of a petroleum product to color change due to light, aging, etc.

Combined cycle: a combustion (gas) turbine equipped with a heat recovery steam generator that produces steam for the steam turbine; power is produced from both the gas and steam turbines – hence the term combined cycle.

Combustible liquid: a liquid with a flash point in excess of 37.8°C (100°F) but below 93.3°C (200°F).

Combustion zone: the volume of reservoir rock wherein petroleum is undergoing combustion during enhanced oil recovery.

Composition: the general chemical make-up of petroleum.

Completion interval: the portion of the reservoir formation placed in fluid communication with the well by selectively perforating the wellbore casing.

Composition map: a means of illustrating the chemical make-up of petroleum using chemical and/or physical property data.

Con Carbon: see Carbon residue.

Condensate: a mixture of light hydrocarbon liquids obtained by condensation of hydrocarbon vapors: predominately butane, propane, and pentane with some heavier hydrocarbons and relatively little methane or ethane; see also Natural gas liquids.

Conductivity: a measure of the ease of flow through a fracture, perforation, or pipe.

Conformance: the uniformity with which a volume of the reservoir is swept by injection fluids in area and vertical directions.

Connate water: water trapped in the pores of a rock during formation of the rock – also described as fossil water. The chemistry of connate water can change in composition throughout the history of the rock; connate water can be dense and saline compared with seawater. On the other hand, formation water or interstitial water is water found in the pore spaces of a rock and might not have been present when the rock was formed.

Conradson carbon residue: see Carbon residue.

Contact filtration: a process in which finely divided adsorbent clay is used to remove color bodies from petroleum products.

Contaminant: a substance that causes deviation from the normal composition of an environment.

Continuous contact coking: a thermal conversion process in which petroleum-wetted coke particles move downward into the reactor in which cracking, coking, and drying take place to produce coke, gas, gasoline, and gas oil.

Continuous contact filtration: a process to finish lubricants, waxes, or special oils after acid treating, solvent extraction, or distillation.

Conventional crude oil (conventional petroleum): crude oil that is pumped from the ground and recovered using the energy inherent in the reservoir; also recoverable by application of secondary recovery techniques.

Conventional recovery: primary and/or secondary recovery.

Conversion: the thermal treatment of petroleum which results in the formation of new products by the alteration of the original constituents.

Conversion cost: the cost of changing a production well to an injection well, or some other change in the function of an oilfield installation.

Conversion factor: the percentage of feedstock converted to light ends, gasoline, other liquid fuels, and coke.

Copper sweetening: processes involving the oxidation of mercaptan derivatives (RSH) to disulfide derivatives (RSSR) by oxygen in the presence of cupric chloride ($CuCl_2$).

Cord: a stack of wood comprising 128 ft^3 (3.62 m^3); standard dimensions are $4\times4\times8$ ft, including air space and bark. One cord contains approx. 1.2 U.S. tons (oven-dry) = 2,400 lb = 1,089 kg.

Core floods: laboratory flow tests through samples (cores) of porous rock.

Co-surfactant: a chemical compound, typically alcohol that enhances the effectiveness of a surfactant.

Cp (centipoise): a unit of viscosity.

Craig-Geffen-Morse method: a method for predicting oil recovery by water flood.

Cracked residua: residua that have been subjected to temperatures above 350°C (660°F) during the distillation process.

Cracking: a secondary refining process that uses heat and/or a catalyst to break down high molecular weight chemical components into lower molecular weight products which can be used as blending components for fuels.

Cracking activity: see Catalytic activity.

Cracking coil: equipment used for cracking heavy petroleum products consisting of a coil of heavy pipe running through a furnace so that the oil passing through it is subject to high temperature.

Cracking still: the combined equipment-furnace, reaction chamber, fractionator for the thermal conversion of heavier feedstocks to lighter products.

Cracking temperature: the temperature (350°C; 660°F) at which the rate of thermal decomposition of petroleum constituents becomes significant.

Criteria air pollutants: air pollutants or classes of pollutants regulated by the Environmental Protection Agency; the air pollutants are (including VOCs) ozone, carbon monoxide, particulate matter, nitrogen oxides, sulfur dioxide, and lead.

Cropland: total cropland includes five components: cropland harvested, crop failure, cultivated summer fallow, cropland used only for pasture, and idle cropland.

Cropland pasture: land used for long-term crop rotation. However, some cropland pasture is marginal for crop uses and may remain in pasture indefinitely. This category also includes land that was used for pasture before crops reached maturity and some land used for pasture that could have been cropped without additional improvement.

Cross-linking: combining of two or polymer molecules by use of a chemical that mutually bonds with a part of the chemical structure of the polymer molecules.

Crude assay: a procedure for determining the general distillation characteristics (e.g., distillation profile, *q.v.*) and other quality information of crude oil.

Crude oil: see Petroleum.

Crude scale wax: the wax product from the first sweating of the slack wax.

Crude still: distillation equipment in which crude oil is separated into various products.

Cull tree: a live tree, 5.0 in. in diameter at breast height (d.b.h.) or larger that is non-merchantable for saw logs now or prospectively because of rot, roughness, or species. (See definitions for rotten and rough trees.)

Cultivated summer fallow: cropland cultivated for one or more seasons to control weeds and accumulate moisture before small grains are planted.

Cumene: a colorless liquid [$C_6H_5CH(CH_3)_2$] used as an aviation gasoline blending component and as an intermediate in the manufacture of chemicals.

Cut point: the boiling-temperature division between distillation fractions of petroleum.

Cutback: the term applied to the products from blending heavier feedstocks or products with lighter oils to bring the heavier materials to the desired specifications.

Cutback asphalt: asphalt liquefied by the addition of a volatile liquid such as naphtha or kerosene which, after application and on exposure to the atmosphere, evaporates leaving the asphalt.

Cutting oil: an oil to lubricate and cool metal-cutting tools; also called cutting fluid, cutting lubricant.

Cycle stock: the product taken from some later stage of a process and recharged (recycled) to the process at some earlier stage.

Cyclic steams injection: the alternating injection of steam and production of oil with condensed steam from the same well or wells.

Cyclization: the process by which an open-chain hydrocarbon structure is converted to a ring structure, e.g., hexane to benzene.

Cyclone: a device for extracting dust from industrial waste gases. It is in the form of an inverted cone into which the contaminated gas enters tangential from the top; the gas is propelled down a helical pathway, and the dust particles are deposited by means of centrifugal force onto the wall of the scrubber.

Deactivation: reduction in catalyst activity by the deposition of contaminants (e.g., coke, metals) during a process.

Dealkylation: the removal of an alkyl group from aromatic compounds.

Deasphaltened oil: the fraction of petroleum after the asphaltene constituents have been removed.

Deasphaltening: removal of a solid powdery asphaltene fraction from petroleum by the addition of the low-boiling liquid hydrocarbons such as n-pentane or n-heptane under ambient conditions.

Deasphalting: the removal of the asphaltene fraction from petroleum by the addition of a low-boiling hydrocarbon liquid such as n-pentane or n-heptane; more correctly the removal asphalt (tacky, semi-solid) from petroleum (as occurs in a refinery asphalt plant) by the addition of liquid propane or liquid butane under pressure.

Debutanization: distillation to separate butane and lighter components from higher boiling components.

Decant oil: the highest boiling product from a catalytic cracker; also referred to as slurry oil, clarified oil, or bottoms.

Decarbonizing: a thermal conversion process designed to maximize coker gas-oil production and minimize coke and gasoline yields; operated at essentially lower temperatures and pressures than delayed coking.

Decoking: removal of petroleum coke from equipment such as coking drums; hydraulic decoking uses high-velocity water streams.

Decolorizing: removal of suspended, colloidal, and dissolved impurities from liquid petroleum products by filtering, adsorption, chemical treatment, distillation, bleaching, etc.

De-ethanization: distillation to separate ethane and lighter components from propane and higher-boiling components; also called de-ethanation.

Degradation: the loss of desirable physical properties of EOR fluids, e.g., the loss of viscosity of polymer solutions.

Dehydrating agents: substances capable of removing water (drying, q.v.) or the elements of water from another substance.

Dehydrocyclization: any process by which both dehydrogenation and cyclization reactions occur.

Dehydrogenation: the removal of hydrogen from a chemical compound; for example, the removal of two hydrogen atoms from butane to make butene(s) as well as the removal of additional hydrogen to produce butadiene.

Delayed coking: a coking process in which the thermal reaction are allowed to proceed to completion to produce gaseous, liquid, and solid (coke) products.

Demethanization: the process of distillation in which methane is separated from the higher boiling components; also called demethanation.

Demulsifiers: a group of chemicals or surfactants used to separate water content in the water-in-oil and oil-in-water emulsions usually at low concentrations. See: Emulsion.

Density: the mass (or weight) of a unit volume of any substance at a specified temperature; see also Specific gravity.

Deoiling: reduction in quantity of liquid oil entrained in solid wax by draining (sweating) or by a selective solvent; see MEK deoiling.

Depentanizer: a fractionating column for the removal of pentane and lighter fractions from a mixture of hydrocarbons.

Depropanization: distillation in which lighter components are separated from butanes and higher boiling material; also called depropanation.

Desalting: the removal of mineral salts (mostly chlorides) from crude oils; the first refining process applied to crude oil which removes salt, water, and solid particles that would otherwise lead to operational problems during refining such as corrosion, fouling of equipment, or poisoning of catalysts.

Desorption: the reverse process of adsorption whereby adsorbed matter is removed from the adsorbent; also used as the reverse of absorption.

Desulfurization: the removal of sulfur or sulfur compounds from a feedstock.

Detergent oil: lubricating oil possessing special sludge-dispersing properties for use in internal-combustion engines.

Dewaxing: see Solvent dewaxing.

Devolatilized fuel: smokeless fuel; coke that has been reheated to remove all of the volatile material.

Diagenesis: the concurrent and consecutive chemical reactions which commence the alteration of organic matter (at temperatures up to 50°C (120°F) and ultimately result in the formation of petroleum from the marine sediment; see also Catagenesis and Metagenesis.

Diagenetic rock: rock formed by conversion through pressure or chemical reaction) from a rock, e.g., sandstone is a diagenetic.

Diesel cycle: a repeated succession of operations representing the idealized working behavior of the fluids in a diesel engine.

Diesel engine: named for the German engineer Rudolph Diesel, this internal-combustion, compression-ignition engine works by heating fuels and causing them to ignite; can use either petroleum or bio-derived fuel.

Diesel fuel: fuel used for internal combustion in diesel engines; usually that fraction which distills after kerosene.

Diesel index: an approximation of the cetane number of diesel fuel calculated from the density and aniline point.

$$DI = (\text{aniline point } (°F) \times \text{API gravity})100$$

Diesel knock: the -result of a delayed period of ignition is long and the accumulated of diesel fuel in the engine.

Differential-strain analysis: measurement of thermal stress relaxation in a recently cut well.

Digester: an airtight vessel or enclosure in which bacteria decomposes biomass in water to produce biogas.

Direct-injection engine: a diesel engine in which fuel is injected directly into the cylinder.

Dispersion: a measure of the convective mi fluids due to flow in a reservoir.

Displacement efficiency: the ratio of the amount of oil moved from the zone swept by the reprocess to the amount of oil present in the zone prior to start of the process.

Distribution coefficient: a coefficient that describes the distribution of a chemical in reservoir fluids, usually defined as the equilibrium concentrations in the aqueous phases.

Distillate: any petroleum product produced by boiling crude oil and collecting the vapors produced as a condensate in a separate vessel, for example gasoline (light distillate), gas oil (middle distillate), or fuel oil (heavy distillate).

Distillation: the primary refining process which uses high temperature to separate crude oil into vapor and fluids (bulk products) which can then be fed into a variety of processes to produce saleable products.

Distillation curve: see Distillation profile.

Distillation loss: the difference, in a laboratory distillation, between the volume of liquid originally introduced into the distilling flask and the sum of the residue and the condensate recovered.

Distillation range: the difference between the temperature at the initial boiling point and at the end point, as obtained by the distillation test.

Distillation profile: the distillation characteristics of petroleum or petroleum products showing the temperature and the per cent distilled.

Doctor solution: a solution of sodium plumbite used to treat gasoline or other light petroleum distillates to remove mercaptan sulfur; see also Doctor test.

Doctor sweetening: a process for sweetening gasoline, solvents, and kerosene by converting mercaptan derivatives (RSH) to disulfide derivatives (RSSR) using sodium plumbite (Na_2PbO_2) and sulfur.

Doctor test: a test used for the detection of compounds in light petroleum distillates which react with sodium plumbite; see also Doctor solution.

Domestic heating oil: see No. 2 Fuel Oil.

Donor solvent process: a conversion process in which hydrogen donor solvent is used in place of or to augment hydrogen.

Downcomer: a means of conveying liquid from one tray to the next below in a bubble tray column.

Downdraft gasifier: a gasifier in which the product gases pass through a combustion zone at the bottom of the gasifier.

Downhole steam generator: a generator installed downhole in an oil well to which oxygen-rich air, fuel, and water are supplied for the purposes of generating steam for it into the reservoir. Its major advantage over a surface steam generating facility is the losses to the wellbore and surrounding formation are eliminated.

Dropping point: the temperature at which grease passes from a semisolid to a liquid state under prescribed conditions.

Dry gas: a gas which does not contain fractions that may easily condense under normal atmospheric conditions.

Dry point: the temperature at which the last drop of petroleum fluid evaporates in a distillation test.

Drying: removal of a solvent or water from a chemical substance; also referred to as the removal of solvent from a liquid or suspension.

Dualayer distillate process: a process for removing mercaptan derivatives (RSH) and oxygenated compounds from distillate fuel oils and similar products, using a combination of treatment with concentrated caustic solution and electrical precipitation of the impurities.

Dualayer gasoline process: a process for extracting mercaptan derivatives (RSH) and other objectionable acidic compounds from petroleum distillates; see also Dualayer solution.

Dualayer solution: a solution which consists of concentrated potassium or sodium hydroxide containing a solubilizer; see also Dualayer gasoline process.

Dubbs cracking: an older continuous, liquid-phase thermal cracking process formerly used.

Dutch oven furnace: One of the earliest types of furnaces, having a large, rectangular box lined with firebrick (refractory) on the sides and top; commonly used for burning wood.

Dykstra-Parsons coefficient: an index of reservoir heterogeneity arising from permeability variation and stratification.

E85: an alcohol fuel mixture containing 85% ethanol and 15% gasoline by volume, and the current alternative fuel of choice of the U.S. government.

Ebullated bed: a process in which the catalyst bed is in a suspended state in the reactor by means of a feedstock recirculation pump which pumps the feedstock upward at sufficient speed to expand the catalyst bed at approximately 35% above the settled level.

Edeleanu process: a process for refining oils at low temperature with liquid sulfur dioxide (SO_2), or with liquid sulfur dioxide and benzene; applicable to the recovery of aromatic concentrates from naphtha and heavier petroleum distillates.

Effective viscosity: see Apparent viscosity.

Effluent: any contaminating substance, usually a liquid, which enters the environment via a domestic industrial, agricultural, or sewage plant outlet.

Electric desalting: a continuous process to remove inorganic salts and other impurities from crude oil by settling out in an electrostatic field.

Electrical precipitation: a process using an electrical field to improve the separation of hydrocarbon reagent dispersions. May be used in chemical treating processes on a wide variety of refinery stocks.

Electrofining: a process for contacting a light hydrocarbon stream with a treating agent (acid, caustic, doctor, etc.), then assisting the action of separation of the chemical phase from the hydrocarbon phase by an electrostatic field.

Electrolytic mercaptan process: a process in which aqueous caustic solution is used to extract mercaptan derivatives (RSH) from refinery streams.

Electrostatic precipitators: devices used to trap fine dust particles (usually in the size range 30–60 µ) that operate on the principle of imparting an electric charge to particles in an incoming air stream and which are then collected on an oppositely charged plate across a high voltage field.

Eluate: the solutes, or analytes, moved through a chromatographic column (see *elution*).

Eluent: solvent used to elute sample.

Elution: a process whereby a solute is moved through a chromatographic column by a solvent (liquid or gas) or eluent.

Emission control: the use of gas cleaning processes to reduce emissions.

Emission standard: the maximum amount of a specific pollutant permitted to be discharged from a particular source in a given environment.

Emissions: substances discharged into the air during combustion.

Emulsion: a two-phase system comprising of two liquids which are not homogenous when mixed. In the emulsion, one of the liquid is the constantly dispersed (the dispersed phase) as globules in the second phase (the continuous phase), such as oil in water. See: Demulsifiers.

Emulsion breaking: the settling or aggregation of colloidal-sized emulsions from suspension in a liquid medium.

End-of-pipe emission control: the use of specific emission control processes to clean gases after production of the gases.

Energy: the capacity of a body or system to do work, measured in joules (SI units); also the output of fuel sources.

Energy balance: the difference between the energy produced by a fuel and the energy required to obtain it through agricultural processes, drilling, refining, and transportation.

Energy crops: crops grown specifically for their fuel value; include food crops such as corn and sugarcane, and nonfood crops such as poplar trees and switch grass.

Energy-efficiency ratio: a number representing the energy stored in a fuel as compared to the energy required to produce, process, transport, and distribute that fuel.

Energy from biomass: the production of energy from biomass.

Engler distillation: a standard test for determining the volatility characteristics of a gasoline by measuring the percent distilled at various specified temperatures.

Enhanced oil recovery (EOR): petroleum recovery following recovery by conventional (i.e., primary and/or secondary) methods.

Enhanced oil recovery (EOR) process: a method for recovering additional oil from a petroleum reservoir beyond that economically recoverable by conventional primary and secondary recovery methods. EOR methods are usually divided into three main categories: (1) *chemical flooding:* injection of water with added chemicals into a petroleum reservoir. The chemical processes include: surfactant flooding, polymer flooding, and alkaline flooding; (2) *miscible flooding:* injection into a petroleum reservoir of a material that is miscible, or can become miscible, with the oil in the reservoir. Carbon dioxide, hydrocarbons, and nitrogen are used; (3) *thermal recovery:* injection of steam into a petroleum reservoir, or propagation of a combustion zone through a reservoir by air or oxygen-enriched air injection. The thermal processes include: steam drive, cyclic steam injection, and in situ combustion.

Entrained bed: a bed of solid particles suspended in a fluid (liquid or gas) at such a rate that some of the solid is carried over (entrained) by the fluid.

Entrainer: the separating agent used to enhance the separation of closely boiling compounds in azeotropic distillation or in extractive distillation.

EPA: Environmental Protection Agency.

Ester: a compound formed by the reaction between an organic acid and an alcohol. (i.e., alcohols having ethylene oxide functional groups attached to the alcohol molecule).

Ethanol (ethyl alcohol, alcohol, or grain-spirit): a clear, colorless, flammable oxygenated hydrocarbon; used as a vehicle fuel by itself (E100 is 100% ethanol by volume), blended with gasoline (E85 is 85% ethanol by volume), or as a gasoline octane enhancer and oxygenate (10% by volume); formed during fermentation of sugars; used as an intoxicant and as a fuel.

Evaporation: a process for concentrating nonvolatile solids in a solution by boiling off the liquid portion of the waste stream.

Expanding clays: clays that expand or swell on contact with water, e.g., montmorillonite.

Explosive limits: the limits of percentage composition of mixtures of gases and air within which an explosion takes place when the mixture is ignited.

Extract: the portion of a sample preferentially dissolved by the solvent and recovered by physically separating the solvent.

Extractive distillation: the separation of different components of mixtures which have similar vapor pressures by flowing a relatively high-boiling solvent, which is selective for one of the components in the feed, down a distillation column as the distillation proceeds; the selective solvent scrubs the soluble component from the vapor.

Fabric filters: filters made from fabric materials and used for removing particulate matter from gas streams (see Baghouse).

Facies: one or more layers of rock that differs from other layers in composition, age, or content.

FAST: fracture-assisted steamflood technology.

Fast pyrolysis: a pyrolysis process that involves heating a feedstock, such as biomass rapidly (2 seconds) at temperatures ranging on the order of from 350°C to 650°C (650°F–1,200°F).

Fat oil: the bottom or enriched oil drawn from the absorber as opposed to lean oil.

Faujasite: a naturally occurring silica-alumina (SiO_2–Al_2O_3) mineral.

FCC: fluid catalytic cracking.

FCCU: fluid catalytic cracking unit.

Feedstock: petroleum as it is fed to the refinery; a refinery product that is used as the raw material for another process; biomass used in the creation of a particular biofuel (e.g., corn or sugarcane for ethanol, soybeans or rapeseed for biodiesel); the term is also generally applied to raw materials used in other industrial processes.

Fermentation: conversion of carbon-containing compounds by micro-organisms for the production of fuels and chemicals such as alcohols, acids or energy-rich gases.

Ferrocyanide process: a regenerative chemical treatment for mercaptan removal using caustic-sodium ferrocyanide reagent.

Field-scale: the application of EOR processes to a significant portion of a field.

Filtration: the use of an impassable barrier to collect solids but which allows liquids to pass.

Fiber products: products derived from fibers of herbaceous and woody plant materials; examples include pulp, composition board products, and wood chips for export.

Fine materials: wood residues not suitable for chipping, such as planer shavings and sawdust.

Fingering: the formation of finger-shaped irregularities at the leading edge of a displacing fluid in a porous medium which move out ahead of the main body of fluid.

Fire point: the lowest temperature at which, under specified conditions in standardized apparatus, a petroleum product vaporizes sufficiently rapidly to form above its surface an air-vapor mixture which burns continuously when ignited by a small flame.

First contact miscibility: see miscibility.

Fischer-Tropsch process: a process for synthesizing hydrocarbons and oxygenated chemicals from a mixture of hydrogen and carbon monoxide.

Five-spot: an arrangement or pattern of wells with four injection wells at the comers of a square and a producing well in the center of the square.

Fixed bed: a stationary bed (of catalyst) to accomplish a process (see Fluid bed).

Flammability range: the range of temperature over which a chemical is flammable.

Flammable: a substance that will burn readily.

Flammable liquid: a liquid having a flash point below 37.8°C (100°F).

Flammable solid: a solid that can ignite from friction or from heat remaining from its manufacture, or which may cause a serious hazard if ignited.

Flash point: the lowest temperature to which the product must be heated under specified conditions to give off sufficient vapor to form a mixture with air that can be ignited momentarily by a flame.

Flash vaporization: a simple unit operation in which a heated liquid mixture is throttled through a valve in order to vaporize the liquid mixture. The vaporization is done in order to separate the constituents of the liquid mixture. Flash vaporization can be seen as a single stage.

Flexible-fuel vehicle (flex-fuel vehicle): a vehicle that can run alternately on two or more sources of fuel includes cars capable of running on gasoline and gasoline/ethanol mixtures, as well as cars that can run on both gasoline and natural gas.

Flexicoking: a modification of the fluid coking process insofar as the process also includes a gasifier adjoining the burner/regenerator to convert excess coke to a clean fuel gas.

Floc point: the temperature at which wax or solids separate as a definite floc.

Flocculants: chemicals that flocculate the water droplets and facilitate coalescence.

Flocculation threshold: the point at which constituents of a solution (e.g., asphaltene constituents or coke precursors) will separate from the solution as a separate (solid) phase.

Flood, flooding: the process of displacing petroleum from a reservoir by the injection of fluids.

Flue gas: gas from the combustion of fuel, the heating value of which has been substantially spent and which is, therefore, discarded to the flue or stack.

Flue gases: the gaseous products of the combustion process mostly comprised of carbon dioxide, nitrogen, and water vapor.

Fluid: a reservoir gas or liquid.

Fluid-bed: a bed (of catalyst) that is agitated by an upward passing gas in such a manner that the particles of the bed simulate the movement of a fluid and has the characteristics associated with a true liquid; c.f. Fixed bed.

Fluid catalytic cracking: cracking in the presence of a fluidized bed of catalyst.

Fluid coking: a continuous fluidized solids process that cracks feed thermally over heated coke particles in a reactor vessel to gas, liquid products, and coke.

Fluidized bed combustion: a process used to burn low-quality solid fuels in a bed of small particles suspended by a gas stream (usually air that will lift the particles but not blow them out of the vessel). Rapid burning removes some of the offensive by-products of combustion from the gases and vapors that result from the combustion process.

Fluidized-bed boiler: a large, refractory-lined vessel with an air distribution member or plate in the bottom, a hot gas outlet in or near the top, and some provisions for introducing fuel; the fluidized bed is formed by blowing air up through a layer of inert particles (such as sand or limestone) at a rate that causes the particles to go into suspension and continuous motion.

Fly ash: particulate matter produced from mineral matter in coal that is converted during combustion to finely divided inorganic material and which emerges from the combustor in the gases.

Foots oil: the oil sweated out of slack wax; named from the fact that the oil goes to the foot, or bottom, of the pan during the sweating operation.

Forest health: a condition of ecosystem sustainability and attainment of management objectives for a given forest area; usually considered to include green trees, snags, resilient stands growing at a moderate rate, and endemic levels of insects and disease.

Forest land: land at least 10% stocked by forest trees of any size, including land that formerly had such tree cover and that will be naturally or artificially regenerated; includes transition zones, such as areas between heavily forested and non-forested lands that are at least 10% stocked with forest trees and forest areas adjacent to urban and built-up lands; also included are pinyon-juniper and chaparral areas; minimum area for classification of forest land is 1 acre.

Forest residues: material not harvested or removed from logging sites in commercial hardwood and softwood stands as well as material resulting from forest management operations such as precommercial thinnings and removal of dead and dying trees.

Formation: an interval of rock with distinguishable geologic characteristics.

Formation volume factor: the volume that one stock tank barrel occupies in the formation at reservoir temperature and with the solution gas that is held in the oil at reservoir pressure.

Fossil fuel resources: a gaseous, liquid, or solid fuel material formed in the ground by chemical and physical changes (diagenesis, q.v.) in plant and animal residues over geological time; natural gas, petroleum, coal, and oil shale.

Fouling: the name applied to the disposition of solids or phase separation which typically commences with the introduction into the refinery of a feedstock (crude oil or the blends of crude oils) that is incompatible with (in the current context) the other components of a blend of feedstocks. Fouling includes any of the following events that can occur during or after processing which include (1) deposit formation, (2) encrustation, (3) deposition, (4) scaling, (5) scale formation, (6) slagging, and (7) sludge formation, all of which can have an adverse effect on recovery operations, refinery operations, as well as transportation of crude oil and crude oil products. See also Macrofouling, Microfouling.

Fractional composition: the composition of petroleum as determined by fractionation (separation) methods.

Fractional distillation: the separation of the components of a liquid mixture by vaporizing and collecting the fractions, or cuts, which condense in different temperature ranges.

Fractional flow: the ratio of the volumetric flow rate of one fluid phase to the total fluid volumetric flow rate within a volume of rock.

Fractional flow curve: the relationship between the fractional flow of one fluid and its saturator during simultaneous flow of fluids through rock.

Fractionating column: a column arranged to separate various fractions of petroleum by a single distillation and which may be tapped at different points along its length to separate various fractions in the order of their boiling points.

Fractionation: the separation of petroleum into the constituent fractions using solvent or adsorbent methods; chemical agents such as sulfuric acid may also be used.

Fracture: a natural or man-made crack in a reservoir rock.

Fracturing: the breaking apart of reservoir rock by applying very high fluid pressure at the rock face.

Frasch process: a process formerly used for removing sulfur by distilling oil in the presence of copper oxide.

Fuel cell: a device that converts the energy of a fuel directly to electricity and heat, without combustion.

Fuel cycle: the series of steps required to produce electricity. The fuel cycle includes mining or otherwise acquiring the raw fuel source, processing and cleaning the fuel, transport, electricity generation, waste management, and plant decommissioning.

Fuel oil: also called heating oil is a distillate product that covers a wide range of properties; see also No. 1 to No. 4 Fuel oils.

Fuel treatment evaluator (FTE): a strategic assessment tool capable of aiding the identification, evaluation, and prioritization of fuel treatment opportunities.

Fuel wood: wood used for conversion to some form of energy, primarily for residential use.

Fuller's earth: a clay which has high adsorptive capacity for removing color from oils; Attapulgus clay is a widely used fuller's earth.

Functional group: the portion of a molecule that is characteristic of a family of compounds and determines the properties of these compounds.

Furfural extraction: a single-solvent process in which furfural is used to remove aromatic, naphthene, olefin, and unstable hydrocarbons from a lubricating-oil charge stock.

Furnace: an enclosed chamber or container used to burn biomass in a controlled manner to produce heat for space or process heating.

Furnace oil: a distillate fuel primarily intended for use in domestic heating equipment.

Gas cap: a part of a hydrocarbon reservoir at the top that will produce only gas.

Gas oil: a petroleum distillate with a viscosity and boiling range between those of kerosene and lubricating oil.

Gas-oil ratio: ratio of the number of cubic feet of gas measured at atmospheric (standard) conditions to barrels of produced oil measured at stock tank conditions.

Gas-oil sulfonate: sulfonate made from a specific refinery stream, in this case the gas-oil stream.

Gasoline: fuel for the internal combustion engine that is commonly, but improperly, referred to simply as gas.

Gas reversion: a combination of thermal cracking or reforming of naphtha with thermal polymerization or alkylation of hydrocarbon gases carried out in the same reaction zone.

Gas to liquids (GTL): the process of refining natural gas and other hydrocarbons into longer-chain hydrocarbons, which can be used to convert gaseous waste products into fuels.

Gas turbine: a device in which fuel is combusted at pressure and the products of combustion expanded through a turbine to generate power (the Brayton Cycle); it is based on the same principle as the jet engine.

Gaseous pollutants: gases released into the atmosphere that act as primary or secondary pollutants.

Gasification: a chemical or thermal process used to convert carbonaceous material (such as coal, petroleum, and biomass) into gaseous components such as carbon monoxide and hydrogen; a process for converting a solid or liquid fuel into a gaseous fuel useful for power generation or chemical feedstock with an oxidant and steam.

Gasifier: a device for converting solid fuel into gaseous fuel; in biomass systems, the process is referred to as pyrolitic distillation.

Gasifier cold gas efficiency (CGE): the percentage of the coal heating value that appears as chemical heating value in the gasifier product gas.

Gasohol: a mixture of 10% v/v anhydrous ethanol and 90% v/v gasoline; 7.5% v/v anhydrous ethanol and 92.5% v/v gasoline; or 5.5% v/v anhydrous ethanol and 94.5% v/v gasoline; a term for motor vehicle fuel comprising between 80%–90% v/v unleaded gasoline and 10%–20% v/v ethanol (see also Ethyl alcohol).

Gel point: the point at which a liquid fuel cools to the consistency of petroleum jelly.

Genetically modified organism (GMO): an organism whose genetic material has been modified through recombinant DNA technology, altering the phenotype of the organism to meet desired specifications.

Gilsonite: an asphaltite that is >90% bitumen.

Girbotol process: a continuous, regenerative process to separate hydrogen sulfide, carbon dioxide, and other acid impurities from natural gas, refinery gas, etc., using mono-, di-, or triethanolamine as the reagent.

Glance pitch: an asphaltite.

Glycol-amine gas treating: a continuous, regenerative process to simultaneously dehydrate and remove acid gases from natural gas or refinery gas.

Grahamite: an asphaltite.

Grassland pasture and range: all open land used primarily for pasture and grazing, including shrub and brush land types of pasture; grazing land with sagebrush and scattered mesquite; and all tame and native grasses, legumes, and other forage used for pasture or grazing; because of the diversity in vegetative composition, grassland pasture and range are not always clearly distinguishable from other types of pasture and range; at one extreme, permanent grassland may merge with cropland pasture, or grassland may often be found in transitional areas with forested grazing land.

Grain alcohol: see Ethyl alcohol.

Gravimetric: gravimetric methods weigh a residue.

Gravity: see API gravity.

Gravity drainage: the movement of oil in a reservoir that results from the force of gravity.

Gravity segregation: partial separation of fluids in a reservoir caused by the gravity force acting on differences in density.

Gravity-stable displacement: the displacement of oil from a reservoir by a fluid of a different density, where the density difference is utilized to prevent gravity segregation of the injected fluid.

Gray clay treating: a fixed-bed, usually fuller's earth, vapor-phase treating process to selectively polymerize unsaturated gum-forming constituents (diolefins) in thermally cracked gasoline.

Grease car: a diesel-powered automobile rigged post-production to run on used vegetable oil.

Greenhouse effect: the effect of certain gases in the Earth's atmosphere in trapping heat from the sun.

Greenhouse gases: gases that trap the heat of the sun in the Earth's atmosphere, producing the greenhouse effect. The two major greenhouse gases are water vapor and carbon dioxide. Other greenhouse gases include methane, ozone, chlorofluorocarbons, and nitrous oxide.

Grid: an electric utility company's system for distributing power.

Growing stock: a classification of timber inventory that includes live trees of commercial species meeting specified standards of quality or vigor; cull trees are excluded.

Guard bed: a bed of an adsorbent (such as, for example, bauxite) that protects a catalyst bed by adsorbing species detrimental to the catalyst.

Gulf HDS process: a fixed-bed process for the catalytic hydrocracking of heavy stocks to lower-boiling distillates with accompanying desulfurization.

Gulfining: a catalytic hydrogen treating process for cracked and straight-run distillates and fuel oils, to reduce sulfur content; improve carbon residue, color, and general stability; and effect a slight increase in gravity.

Gum: an insoluble tacky semi-solid material formed as a result of the storage instability and/or the thermal instability of petroleum and petroleum products.

Habitat: the area where a plant or animal lives and grows under natural conditions. Habitat includes living and non-living attributes and provides all requirements for food and shelter.

HAP(s): hazardous air pollutant(s).

Hardness: the concentration of calcium and magnesium in brine.

Hardwoods: usually broad-leaved and deciduous trees.

HCPV: hydrocarbon pore volume.

Headspace: the vapor space above a sample into which volatile molecules evaporate. Certain methods sample this vapor.

Hearn method: a method used in reservoir simulation for calculating a pseudo relative permeability curve that reflects reservoir stratification.

Heat recovery steam generator: a heat exchanger that generates steam from the hot exhaust gases from a combustion turbine.

Heating oil: see Fuel oil.

Heating value: the maximum amount of energy that is available from burning a substance.

Heavy ends: the highest boiling portion of a petroleum fraction; see also Light ends.

Heavy (crude) oil: oil that is more viscous that conventional crude oil has a lower mobility in the reservoir but can be recovered through a well from the reservoir by the application of a secondary or enhanced recovery methods; sometimes petroleum having an API gravity of less than 20°.

Heavy fuel oil: fuel oil having a high density and viscosity; generally residual fuel oil such as No. 5 and No 6. fuel oil.

Heavy petroleum: see Heavy oil.

Hectare: common metric unit of area, equal to 2.47 acres. 100 ha = 1 km^2.

Herbaceous: non-woody type of vegetation, usually lacking permanent strong stems, such as grasses, cereals, and canola (rape).

Heteroatom compounds: chemical compounds which contain nitrogen and/or oxygen and/or sulfur and /or metals bound within their molecular structure(s).

Heterogeneity: lack of uniformity in reservoir properties such as permeability.

HF alkylation: an alkylation process whereby olefins (C_3, C_4, C_5) are combined with *iso*-butane in the presence of hydrofluoric acid catalyst.

Higgins-Leighton model: stream tube computer model used to simulate waterflood.

Hortonsphere (Horton sphere): a spherical pressure-type tank used to store a volatile liquid which prevents the excessive evaporation loss that occurs when such products are placed in conventional storage tanks.

Hot filtration test: a test for the stability of a petroleum product.

Hot spot: an area of a vessel or line wall appreciably above normal operating temperature, usually as a result of the deterioration of an internal insulating liner which exposes the line or vessel shell to the temperature of its contents.

Houdresid catalytic cracking: a continuous moving-bed process for catalytically cracking reduced crude oil to produce high-octane gasoline and light distillate fuels.

Houdriflow catalytic cracking: a continuous moving-bed catalytic cracking process employing an integrated single vessel for the reactor and regenerator kiln.

Houdriforming: a continuous catalytic reforming process for producing aromatic concentrates and high-octane gasoline from low-octane straight naphtha.

Houdry butane dehydrogenation: a catalytic process for dehydrogenating light hydrocarbons to their corresponding mono- or diolefins.

Houdry fixed-bed catalytic cracking: a cyclic regenerable process for cracking of distillates.

Houdry hydrocracking: a catalytic process combining cracking and desulfurization in the presence of hydrogen.

Huff-and-puff: a cyclic EOR method in which steam or gas is injected into a production well; after a short shut-in period, oil and the injected fluid are produced through the same well.

Hydration: the association of molecules of water with a substance.

Hydraulic fracturing: the opening of fractures in a reservoir by high-pressure, high-volume injection of liquids through an injection well.

Hydrocarbon compounds: chemical compounds containing only carbon and hydrogen.

Hydrocarbon-producing resource: a resource such as coal and oil shale (kerogen) which produces derived hydrocarbons by the application of conversion processes; the hydrocarbons so-produced are not naturally-occurring materials.

Hydrocarbon resource: resources such as petroleum and natural gas which can produce naturally-occurring hydrocarbons without the application of conversion processes.

Hydrocarbonaceous material: a material such as bitumen that is composed of carbon and hydrogen with other elements (heteroelements) such as nitrogen, oxygen, sulfur, and metals chemically combined within the structures of the constituents; even though carbon and hydrogen may be the predominant elements, there may be very few true hydrocarbons.

Hydrocarbons: organic compounds containing only hydrogen and carbon.

Hydrolysis: a chemical reaction in which water reacts with another substance to form one or more new substances.

Hydroconversion: a term often applied to hydrocracking

Hydrocracking: a catalytic high-pressure high-temperature process for the conversion of petroleum feedstocks in the presence of fresh and recycled hydrogen; carbon-carbon bonds are cleaved in addition to the removal of heteroatomic species.

Hydrocracking catalyst: a catalyst used for hydrocracking which typically contains separate hydrogenation and cracking functions.

Hydrodenitrogenation: the removal of nitrogen by hydrotreating.

Hydrodesulfurization: the removal of sulfur by hydrotreating.

Hydrofining: a fixed-bed catalytic process to desulfurize and hydrogenate a wide range of charge stocks from gases through waxes.

Hydroforming: a process in which naphtha is passed over a catalyst at elevated temperatures and moderate pressures, in the presence of added hydrogen or hydrogen-containing gases, to form high-octane motor fuel or aromatics.

Hydrogen addition: an upgrading process in the presence of hydrogen, e.g., hydrocracking; see Hydrogenation.

Hydrogen blistering: blistering of steel caused by trapped molecular hydrogen formed as atomic hydrogen during corrosion of steel by hydrogen sulfide.

Hydrogen transfer: the transfer of inherent hydrogen within the feedstock constituents and products during processing.

Hydrogenation: the chemical addition of hydrogen to a material. In nondestructive hydrogenation, hydrogen is added to a molecule only if, and where, unsaturation with respect to hydrogen exists.

Hydroprocesses: refinery processes designed to add hydrogen to various products of refining.

Hydroprocessing: a term often equally applied to hydrotreating and to hydrocracking; also often collectively applied to both.

Hydrotreating: the removal of heteroatomic (nitrogen, oxygen, and sulfur) species by treatment of a feedstock or product at relatively low temperatures in the presence of hydrogen.

Hydrovisbreaking: a non-catalytic process, conducted under similar conditions to visbreaking, which involves treatment with hydrogen to reduce the viscosity of the feedstock and produce more stable products than is possible with visbreaking.

Hydropyrolysis: a short residence time high temperature process using hydrogen.

Hyperforming: a catalytic hydrogenation process for improving the octane number of naphtha through removal of sulfur and nitrogen compounds.

Hypochlorite sweetening: the oxidation of mercaptan derivatives (RSH) in a sour feedstock by agitation with aqueous, alkaline hypochlorite solution; used where avoidance of free-sulfur addition is desired, because of a stringent copper strip requirements and minimum expense is not the primary object.

Idle cropland: land in which no crops were planted; acreage diverted from crops to soil-conserving uses (if not eligible for and used as cropland pasture) under federal farm programs is included in this component.

Ignitability: characteristic of liquids whose vapors are likely to ignite in the presence of ignition source; also characteristic of non-liquids that may catch fire from friction or contact with water and that burn vigorously.

Illuminating oil: oil used for lighting purposes.

Immiscible: two or more fluids that do not have complete mutual solubility and co-exist as separate phases.

Immiscible carbon dioxide displacement: injection of carbon dioxide into an oil reservoir to effect oil displacement under conditions in which miscibility with reservoir oil is not obtained; see Carbon dioxide augmented waterflooding.

Immiscible displacement: a displacement of oil by a fluid (gas or water) that is conducted under conditions so that interfaces exist between the driving fluid and the oil.

Immunoassay: portable tests that take advantage of an interaction between an antibody and a specific analyte. Immunoassay tests are semi-quantitative and usually rely on color changes of varying intensities to indicate relative concentrations.

In-line blending: the controlled proportioning of two or more component streams to produce a final blended product of closely defined quality from the beginning to the end of the batch which permits the blended product to be used immediately for the prescribed purpose.

In situ: in its original place; in the reservoir.

In situ combustion: an EOR process consisting of injecting air or oxygen-enriched air into a reservoir under conditions that favor burning part of the in situ petroleum, advancing this burning zone, and recovering oil heated from a nearby producing well.

Incinerator: any device used to burn solid or liquid residues or wastes as a method of disposal.

Incompatibility: the *immiscibility* of petroleum products and also of different crude oils which is often reflected in the formation of a separate phase after mixing and/or storage; the phenomenon may involve any one of two of several possible events which are (1) phase separation, (2) precipitation of asphaltene constituents when a paraffinic crude oil is blended with a viscous crude oil, (3) precipitation of asphaltene constituents when a paraffinic crude oil product is blended with a viscous crude oil, (4) when the blend is heated in pipes leading to a reactor, and also through (5) the formation of degradation products and other undesirable changes in the original properties of crude oil products. When such phenomena occur, which may not be immediately at the time of the blend but can occur after an induction period, it is often referred to as instability of the blends. See also: Instability.

Inclined grate: a type of furnace in which fuel enters at the top part of a grate in a continuous ribbon, passes over the upper drying section where moisture is removed, and descends into the lower burning section. Ash is removed at the lower part of the grate.

Incremental ultimate recovery: the difference between the quantity of oil that can be recovered by EOR methods and the quantity of oil that can be recovered by conventional recovery methods.

Indirect-injection engine: an older model of diesel engine in which fuel is injected into a pre-chamber, partly combusted, and then sent to the fuel-injection chamber.

Indirect liquefaction: conversion of biomass to a liquid fuel through a synthesis gas intermediate step.

Industrial wood: all commercial round wood products except fuel wood.

Infill drilling: drilling additional wells within an established pattern.

Infrared spectroscopy: an analytical technique that quantifies the vibration (stretching and bending) that occurs when a molecule absorbs (heat) energy in the infrared region of the electromagnetic spectrum.

Inhibitor: a substance, the presence of which, in small amounts, in a petroleum product prevents or retards undesirable chemical changes from taking place in the product, or in the condition of the equipment in which the product is used.

Inhibitor sweetening: a treating process to sweeten gasoline of low mercaptan content, using a phenylenediamine type of inhibitor, air, and caustic.

Initial boiling point: the recorded temperature when the first drop of liquid falls from the end of the condenser.

Initial vapor pressure: the vapor pressure of a liquid of a specified temperature and 0% evaporated.

Injection profile: the vertical flow rate distribution of fluid flowing from the wellbore into a reservoir.

Injection well: a well in an oil field used for injecting fluids into a reservoir.

Injectivity: the relative ease with which a fluid is injected into a porous rock.

Instability: the inability of a petroleum product to exist for periods of time without change to the product. See also Incompatibility.

Integrated gasification combine cycle (IGCC): a power plant in which a gasification process provides syngas to a combined cycle under an integrated control system.

Integrity: maintenance of a slug or bank at its preferred composition without too much dispersion or mixing.

Interface: the thin surface area separating two immiscible fluids that are in contact with each other.

Interfacial film: a thin layer of material at the interface between two fluids which differs in composition from the bulk fluids.

Interfacial tension: the strength of the film separating two immiscible fluids, e.g., oil and water or microemulsion and oil; measured in dynes (force) per centimeter or milli-dynes per centimeter.

Interfacial viscosity: the viscosity of the interfacial film between two immiscible liquids.

Interference testing: a type of pressure transient test in which pressure is measured over time in a closed-in well while nearby wells are produced; flow and communication between wells can sometimes be deduced from an interference test.

Interphase mass transfer: the net transfer of chemical compounds between two or more phases.

Iodine number: a measure of the iodine absorption by oil under standard conditions; used to indicate the quantity of unsaturated compounds present; also called iodine value.

Ion exchange: a means of removing cations or anions from solution onto a solid resin.

Ion exchange capacity: a measure of the capacity of a mineral to exchange ions in amount of material per unit weight of solid.

Ions: chemical substances possessing positive or negative charges in solution.

Iso-Kel process: a fixed-bed, vapor-phase isomerization process using a precious metal catalyst and external hydrogen.

Iso-plus Houdriforming: a combination process using a conventional Houdriformer operated at moderate severity, in conjunction with one of three possible alternatives – including the use of an aromatic recovery unit or a thermal reformer; see Houdriforming.

Isocracking: a hydrocracking process for conversion of hydrocarbons which operates at relatively low temperatures and pressures in the presence of hydrogen and a catalyst to produce more valuable, lower-boiling products.

Isoforming: a process in which olefinic naphtha is contacted with an alumina catalyst at high temperature and low pressure to produce isomers of higher octane number.

Isomate process: a continuous, non-regenerative process for isomerizing C_5-C_8 normal paraffin hydrocarbons, using aluminum chloride-hydrocarbon catalyst with anhydrous hydrochloric acid as a promoter.

Isomerate process: a fixed-bed isomerization process to convert pentane, heptane, and heptane to high-octane blending stocks.

Isomerization: the conversion of a *normal* (straight-chain) paraffin hydrocarbon into an *iso* (branched-chain) paraffin hydrocarbon having the same atomic composition.

Isopach: a line on a map designating points of equal formation thickness.

Jet fuel: fuel meeting the required properties for use in jet engines and aircraft turbine engines.

Joule: metric unit of energy, equivalent to the work done by a force of 1 N applied over distance of 1 m (= 1 kg m^2/s^2). 1 Joule (J) = 0.239 calories (1 calorie = 4.187 J).

K-factor: see Characterization factor.

Kaolinite: a clay mineral formed by hydrothermal activity at the time of rock formation or by chemical weathering of rock with high feldspar content; usually associated with intrusive granite rock with high feldspar content.

Kata-condensed aromatic compounds: compounds based on linear condensed aromatic hydrocarbon systems, e.g., anthracene and naphthacene (tetracene).

Kauri butanol number: a measurement of solvent strength for hydrocarbon solvents; the higher the kauri-butanol (KB) value, the stronger the solvency; the test method (ASTM D1133) is based on the principle that kauri resin is readily soluble in butyl alcohol but not in hydrocarbon solvents, and the resin solution will tolerate only a certain amount of dilution and is reflected as a cloudiness when the resin starts to come out of solution; solvents such as toluene can be added in a greater amount (and thus have a higher KB value) than weaker solvents like hexane.

Kerogen: a complex carbonaceous (organic) material that occurs in sedimentary rock and shale; generally insoluble in common organic solvents.

Kerosene (kerosine): a fraction of petroleum that was initially sought as an illuminant in lamps; a precursor to diesel fuel.

Kilowatt(kW): a measure of electrical power equal to 1,000 W. 1 kW = 3,412 Btu/h = 1.341 horsepower.

Kilowatt hour - (kWh): a measure of energy equivalent to the expenditure of 1 kWh for 1 hour. For example, 1 kWh will light a 100-W light bulb for 10 hours. 1 kWh = 3,412 Btu.

Kinematic viscosity: the ratio of viscosity to density, both measured at the same temperature.

Knock: the noise associated with self-ignition of a portion of the fuel-air mixture ahead of the advancing flame front.

Kriging: a technique used in reservoir description for interpolation of reservoir parameters between wells based on random field theory.

LAER: lowest achievable emission rate; the required emission rate in non-attainment permits.

Lamp burning: a test of burning oils in which the oil is burned in a standard lamp under specified conditions in order to observe the steadiness of the flame, the degree of encrustation of the wick, and the rate of consumption of the kerosene.

Lamp oil: see Kerosene.

Landfill gas: a type of biogas that is generated by decomposition of organic material at landfill disposal sites. Landfill gas is approximately 50% methane. See also Biogas.

Laterite: the name applied to both a soil and a rock type rich in iron and aluminum and is commonly considered to have formed in hot and wet tropical areas. Nearly all laterites are of rusty-red coloration, because of high iron oxide content.

Leaded gasoline: gasoline containing tetraethyl lead or other organometallic lead antiknock compounds.

Lean gas: the residual gas from the absorber after the condensable gasoline has been removed from the wet gas.

Lean oil: absorption oil from which gasoline fractions have been removed; oil leaving the stripper in a natural-gasoline plant.

Lewis acid: a chemical species which can accept an electron pair from a base.

Lewis base: a chemical species which can donate an electron pair.

Light ends: the lower-boiling components of a mixture of hydrocarbons; see also Heavy ends, Light hydrocarbons.

Light hydrocarbons: hydrocarbons with molecular weights less than that of heptane (C_7H_{16}).

Light oil: the products distilled or processed from crude oil up to, but not including, the first lubricating-oil distillate.

Light petroleum: petroleum having an API gravity greater than 20°.

Lignin: structural constituent of wood and (to a lesser extent) other plant tissues, which encrusts the walls and cements the cells together.

Ligroine (Ligroin): a saturated petroleum naphtha boiling in the range of 20°C–135°C (68°F–275°F) and suitable for general use as a solvent; also called benzine or petroleum ether.

Linde copper sweetening: a process for treating gasoline and distillates with a slurry of clay and cupric chloride.

Liquid petrolatum: see White oil.

Liquefied petroleum gas: propane, butane, or mixtures thereof, gaseous at atmospheric temperature and pressure, held in the liquid state by pressure to facilitate storage, transport, and handling.

Liquid chromatography: a chromatographic technique that employs a liquid mobile phase.

Liquid/liquid extraction: an extraction technique in which one liquid is shaken with or contacted by an extraction solvent to transfer molecules of interest into the solvent phase.

Liquid sulfur dioxide-benzene process: a mixed-solvent process for treating lubricating-oil stocks to improve viscosity index; also used for dewaxing.

Lithology: the geological characteristics of the reservoir rock.

Live cull: a classification that includes live cull trees; when associated with volume, it is the net volume in live cull trees that are 5.0 in. in diameter and larger.

Live steam: steam coming directly from a boiler before being utilized for power or heat.

Liver: the intermediate layer of dark-colored, oily material, insoluble in weak acid and in oil, which is formed when acid sludge is hydrolyzed.

Logging residues: the unused portions of growing-stock and non-growing-stock trees cut or killed logging and left in the woods.

Lorenz coefficient: a permeability heterogeneity factor.

Lower-phase micro emulsion: a microemulsion phase containing a high concentration of water that, when viewed in a test tube, resides near the bottom with oil phase on top.

Lube: see Lubricating oil.

Lube cut: a fraction of crude oil of suitable boiling range and viscosity to yield lubricating oil when completely refined; also referred to as lube oil distillates or lube stock.

Lubricating oil: a fluid lubricant used to reduce friction between bearing surfaces.

M85: an alcohol fuel mixture containing 85% methanol and 15% gasoline by volume. Methanol is typically made from natural gas but can also be derived from the fermentation of biomass.

Macrofouling: fouling of refinery equipment and pipes by the deposition of coarse matter from either organic or biological or inorganic origin; these deposits foul the surfaces of heat exchangers and may cause deterioration of the relevant heat transfer coefficient as well as flow blockages. See also: Fouling, Microfouling.

MACT: maximum achievable control technology. Applies to major sources of hazardous air pollutants.

Mahogany acids: oil-soluble sulfonic acids formed by the action of sulfuric acid on petroleum distillates. They may be converted to their sodium soaps (mahogany soaps) and extracted from the oil with alcohol for use in the manufacture of soluble oils, rust preventives, and special greases. The calcium and barium soaps of these acids are used as detergent additives in motor oils; see also Brown acids and Sulfonic acids.

Major source: a source that has a potential to emit for a regulated pollutant that is at or greater than an emission threshold set by regulations.

Maltene fraction (maltenes): that fraction of petroleum that is soluble in, for example, pentane or heptane; deasphaltened oil; also the term arbitrarily assigned to the pentane-soluble portion of petroleum that is relatively high boiling (>300°C, 760 mm) (see also Petrolenes).

Marine engine oil: oil used as a crankcase oil in marine engines.

Marine gasoline: fuel for motors in marine service.

Marine sediment: the organic biomass from which petroleum is derived.

Marsh: an area of spongy waterlogged ground with large numbers of surface water pools. Marshes usually result from: (1) an impermeable underlying bedrock; (2) surface deposits of glacial boulder clay; (3) a basin-like topography from which natural drainage is poor; (4) very heavy rainfall in conjunction with a correspondingly low evaporation rate; (5) low-lying land, particularly at estuarine sites at or below sea level.

Marx-Langenheim model: mathematical equations for calculating heat transfer in a hot water or steam flood.

Mass spectrometer: an analytical technique that *fractures* organic compounds into characteristic "fragments" based on functional groups that have a specific mass-to-charge ratio.

Mayonnaise: low-temperature sludge; a black, brown, or gray deposit having a soft, mayonnaise-like consistency; not recommended as a food additive!

MCL: maximum contaminant level as dictated by regulations.

Medicinal oil: highly refined, colorless, tasteless, and odorless petroleum oil used as a medicine in the nature of an internal lubricant; sometimes called liquid paraffin.

Megawatt (MW): a measure of electrical power equal to 1 million watts (1,000 kW).

Membrane technology: gas separation processes utilizing membranes that permit different components of a gas to diffuse through the membrane at significantly different rates.

MDL: see Method detection limit.

MEK-(methyl ethyl ketone): a colorless liquid ($CH_3COCH_2CH_3$) used as a solvent; as a chemical intermediate; and in the manufacture of lacquers, celluloid, and varnish removers.

MEK deoiling: a wax-deoiling process in which the solvent is generally a mixture of methyl ethyl ketone and toluene.

MEK dewaxing: a continuous solvent dewaxing process in which the solvent is generally a mixture of methyl ethyl ketone and toluene.

MEOR: microbial-enhanced oil recovery.

Mercapsol process: a regenerative process for extracting mercaptan derivatives (RSH), utilizing aqueous sodium (or potassium) hydroxide containing mixed cresols as solubility promoters.

Mercaptans: organic compounds having the general formula R-SH.

Metagenesis: the alteration of organic matter during the formation of petroleum that may involve temperatures above 200°C (390°F); see also Catagenesis and Diagenesis.

Methanol: see Methyl alcohol.

Method Detection Limit: the smallest quantity or concentration of a substance that the instrument can measure.

Methyl t-butyl ether: an ether added to gasoline to improve its octane rating and to decrease gaseous emissions; see Oxygenate.

Methyl alcohol (methanol; wood alcohol): a colorless, volatile, inflammable, and poisonous alcohol (CH_3OH) traditionally formed by destructive distillation of wood or, more recently, as a result of synthetic distillation in chemical plants; a fuel typically derived from natural gas, but which can be produced from the fermentation of sugars in biomass.

Methyl ethyl ketone: see MEK.

Mica: a complex aluminum silicate mineral that is transparent, tough, flexible, and elastic.

Micellar fluid (surfactant slug): an aqueous mixture of surfactants, co-surfactants, salts, and hydrocarbons. The term micellar is derived from the word micelle, which is a submicroscopic aggregate of surfactant molecules and associated fluid.

Micelle: the structural entity by which asphaltene constituents are dispersed in petroleum.

Microcarbon residue: the carbon residue determined using a themogravimetric method. See also Carbon residue.

Microcrystalline wax: wax extracted from certain petroleum residua and having a finer and less apparent crystalline structure than paraffin wax.

Microemulsion: a stable, finely dispersed mixture of oil, water, and chemicals (surfactants and alcohols).

Microemulsion or micellar/emulsion flooding: an augmented waterflooding technique in which a surfactant system is injected in order to enhance oil displacement toward producing wells.

Microfouling: deposition of solids in refinery equipment or pipes in which the solid is (1) particulate fouling, which is the accumulation of particles on a surface; (2) chemical reaction fouling, such as decomposition of organic matter on heating surfaces; (3) solidification fouling, which occurs when components of a flowing fluid with a high-melting point freeze onto a subcooled surface; (4) corrosion fouling, which is caused by corrosion; (5) biofouling, which can often ensure after biocorrosion which is due to the action of bacteria or algae; and (6) composite fouling, whereby fouling involves more than one foulant or fouling mechanism. See also: Fouling, Macrofouling.

Microorganisms: animals or plants of microscopic size, such as bacteria.

Microscopic displacement efficiency: the efficiency with which an oil displacement process removes the oil from individual pores in the rock.

Mid-boiling point: the temperature at which approximately 50% of a material has distilled under specific conditions.

Middle distillate: distillate boiling between the kerosene and lubricating oil fractions.

Middle-phase micro emulsion: a micro emulsion phase containing a high concentration of both oil and water that, when viewed in a test tube, resides in the middle with the oil phase above it and the water phase below it.

Migration (primary): the movement of hydrocarbons (oil and natural gas) from mature, organic-rich source rocks to a point where the oil and gas can collect as droplets or as a continuous phase of liquid hydrocarbon.

Migration (secondary): the movement of the hydrocarbons as a single, continuous fluid phase through water-saturated rocks, fractures, or faults followed by accumulation of the oil and gas in sediments (traps, *q.v.*) from which further migration is prevented.

Mill residue: wood and bark residues produced in processing logs into lumber, plywood, and paper.

Mineral hydrocarbons: petroleum hydrocarbons, considered *mineral* because they come from the earth rather than from plants or animals.

Mineral oil: the older term for petroleum; the term was introduced in the 19th century as a means of differentiating petroleum (rock oil) from whale oil which, at the time, was the predominant illuminant for oil lamps.

Mineral seal oil: a distillate fraction boiling between kerosene and gas oil.

Mineral wax: yellow to dark brown, solid substances that occur naturally and are composed largely of paraffins; usually found associated with considerable mineral matter, as a filling in veins and fissures or as an interstitial material in porous rocks.

Minerals: naturally occurring inorganic solids with well-defined crystalline structures.

Minimum miscibility pressure (MMP): see Miscibility.

Miscibility: an equilibrium condition, achieved after mixing two or more fluids, which is characterized by the absence of interfaces between the fluids: (1) *first-contact miscibility:* miscibility in the usual sense, whereby two fluids can be mixed in all proportions without any interfaces forming. Example: At room temperature and pressure, ethyl alcohol and water are first-contact miscible; (2) *multiple-contact miscibility (dynamic miscibility):* miscibility that is developed by repeated enrichment of one fluid phase with components from a second fluid phase with which it comes into contact; (3) *minimum miscibility* pressure: the minimum pressure above which two fluids become miscible at a given temperature, or can become miscible, by dynamic processes.

Miscible flooding: see EOR process.

Miscible fluid displacement (miscible displacement): is an oil displacement process in which an alcohol, a refined hydrocarbon, a condensed petroleum gas, carbon dioxide, liquefied natural gas, or even exhaust gas is injected into an oil reservoir, at pressure levels such that the injected gas or fluid and reservoir oil are miscible; the process may include the concurrent, alternating, or subsequent injection of water.

Mitigation: identification, evaluation, and cessation of potential impacts of a process product or by-product.

Mixed-phase cracking: the thermal decomposition of higher-boiling hydrocarbons to gasoline components.

Mobility: a measure of the ease with which a fluid moves through reservoir rock; the ratio of rock permeability to apparent fluid viscosity.

Mobility buffer: the bank that protects a chemical slug from water invasion and dilution and assures mobility control.

Mobility control: ensuring that the mobility of the displacing fluid or bank is equal to or less than that of the displaced fluid or bank.

Mobility ratio: ratio of mobility of an injection fluid to mobility of fluid being displaced.

Modified alkaline flooding: the addition of a co-surfactant and/or polymer to the alkaline flooding process.

Modified naphtha insolubles (MNI): an insoluble fraction obtained by adding naphtha to petroleum; usually the naphtha is modified by adding paraffin constituents; the fraction might be equated to asphaltenes *if* the naphtha is equivalent to n-heptane, but usually it is not

Modified/unmodified diesel engine: traditional diesel engines must be *modified* to heat the oil before it reaches the fuel injectors in order to handle straight vegetable oil. Modified, any diesel engine can run on veggie oil; without modification, the oil must first be converted to biodiesel.

Moisture content (MC): the weight of the water contained in wood, usually expressed as a percentage of weight, either oven-dry or as received.

Moisture content, dry basis: moisture content expressed as a percentage of the weight of oven-wood, i.e.: [(weight of wet sample – weight of dry sample) / weight of dry sample] × 100.

Moisture content, wet basis: moisture content expressed as a percentage of the weight of wood as-received, i.e.: [(weight of wet sample – weight of dry sample) / weight of wet sample] × 100.

Molecular sieve: a synthetic zeolite mineral having pores of uniform size; it is capable of separating molecules, on the basis of their size, structure, or both, by absorption or sieving.

Motor Octane Method: a test for determining the knock rating of fuels for use in spark-ignition engines; see also Research Octane Method.

Moving-bed catalytic cracking: a cracking process in which the catalyst is continuously cycled between the reactor and the regenerator.

MSDS: material safety data sheet.

MTBE: methyl tertiary butyl ether is highly refined high octane light distillate used in the blending of gasoline.

NAAQS: National Ambient Air Quality Standards; standards exist for the pollutants known as the criteria air pollutants: nitrogen oxides (NO_x), sulfur oxides (SO_x), lead, ozone, particulate matter, less than 10 μ in diameter, and carbon monoxide (CO).

Naft: pre-Christian era (Greek) term for naphtha.

Napalm: a thickened gasoline used as an incendiary medium that adheres to the surface it strikes.

Naphtha: a generic term applied to refined, partly refined, or unrefined petroleum products and liquid products of natural gas, the majority of which distills below 240°C (464°F); the volatile fraction of petroleum which is used as a solvent or as a precursor to gasoline.

Naphthenes: cycloparaffins.

Natural asphalt: see Bitumen.

Natural gas: the naturally occurring gaseous constituents that are found in many petroleum reservoirs; also there are also those reservoirs in which natural gas may be the sole occupant.

Natural gas liquids (NGL): the hydrocarbon liquids that condense during the processing of hydrocarbon gases that are produced from oil or gas reservoir; see also Natural gasoline.

Natural gasoline: a mixture of liquid hydrocarbons extracted from natural gas suitable for blending with refinery gasoline.

Natural gasoline plant: a plant for the extraction of fluid hydrocarbon, such as gasoline and liquefied petroleum gas, from natural gas.

NESHAP: National Emissions Standards for Hazardous Air Pollutants; emission standards for specific source categories that emit or have the potential to emit one or more hazardous air pollutants; the standards are modeled on the best practices and most effective emission reduction methodologies in use at the affected facilities.

Neutral oil: a distillate lubricating oil with viscosity usually not above 200 seconds at 100°F.

Neutralization: a process for reducing the acidity or alkalinity of a waste stream by mixing acids and bases to produce a neutral solution; also known as pH adjustment.

Neutralization number: the weight, in milligrams, of potassium hydroxide needed to neutralize the acid in 1 g of oil; an indication of the acidity of an oil.

Nitrogen fixation: the transformation of atmospheric nitrogen into nitrogen compounds that can be used by growing plants.

Nitrogen oxides (NOx): products of combustion that contribute to the formation of smog and ozone.

No. 1 Fuel oil: very similar to kerosene and is used in burners where vaporization before burning is usually required and a clean flame is specified.

No. 2 Fuel oil: also called domestic heating oil; has properties similar to diesel fuel and heavy jet fuel; used in burners where complete vaporization is not required before burning.

No. 4 Fuel oil: a light industrial heating oil and is used where preheating is not required for handling or burning; there are two grades of No. 4 fuel oil, differing in safety (flash point) and flow (viscosity) properties.

No. 5 Fuel oil: a heavy industrial fuel oil which requires preheating before burning.

No. 6 Fuel oil: a heavy fuel oil and is more commonly known as Bunker C oil when it is used to fuel ocean-going vessels; preheating is always required for burning this oil.

Non-forest land: land that has never supported forests and lands formerly forested where use of timber management is precluded by development for other uses; if intermingled in forest areas, unimproved roads and non-forest strips must be more than 120 ft wide, and clearings, etc., must be more than 1 acre in area to qualify as non-forest land.

Non-asphaltic road oil: any of the nonhardening petroleum distillates or residual oils used as dust layers. They have sufficiently low viscosity to be applied without heating and, together with asphaltic road oils, are sometimes referred to as dust palliatives.

Non-attainment area: any area that does not meet the national primary or secondary ambient air quality standard established (by the Environmental Protection Agency) for designated pollutants, such as carbon monoxide and ozone.

Non-industrial private: an ownership class of private lands where the owner does not operate wood processing plants.

Non-ionic surfactant: a surfactant molecule containing no ionic charge.

Non-Newtonian: a fluid that exhibits a change of viscosity with flow rate.

NOx: the oxides of nitrogen.

Nuclear magnetic resonance spectroscopy: an analytical procedure that permits the identification of complex molecules based on the magnetic properties of the atoms they contain.

Observation wells: wells that are completed and equipped to measure reservoir conditions and/or sample reservoir fluids, rather than to inject produced reservoir fluids.

Octane barrel yield: a measure used to evaluate fluid catalytic cracking processes; defined as (RON+MON)/2 times the gasoline yield, where RON is the research octane number and MON is the motor octane number.

Octane number: a number indicating the anti-knock characteristics of gasoline.

Oil bank: see Bank.

Oil breakthrough (time): the time at which the oil-water bank arrives at the producing well.

Oil from tar sand: synthetic crude oil.

Oil mining: application of a mining method to the recovery of bitumen.

Oil originally in place (OOIP): the quantity of petroleum existing in a reservoir before oil recovery operations begin.

Oils: that portion of the maltenes that is not adsorbed by a surface-active material such as clay or alumina.

Oil sand: see Tar sand.

Oil shale: a fine-grained impervious sedimentary rock which contains an organic material called kerogen.

Olefin: synonymous with *alkene*.

OOIP: see Oil originally in place.

Open-loop biomass: biomass that can be used to produce energy and bioproducts even though it was not grown specifically for this purpose; include agricultural livestock waste, residues from forest harvesting operations and crop harvesting.

Optimum salinity: the salinity at which a middle-phase microemulsion containing equal concentrations of oil and water results from the mixture of a micellar fluid (surfactant slug) with oil.

Organic sedimentary rocks: rocks containing organic material such as residues of plant and animal remains/decay.

Orifice-plate mixer (orifice mixer): a mixer in which two or more liquids are pumped through an orifice constriction to cause turbulence and consequent mixing action.

Overhead: that portion of the feedstock which is vaporized and removed during distillation.

Override: the gravity-induced flow of a lighter fluid in a reservoir above another heavier fluid.

Oxidation: a process which can be used for the treatment of a variety of inorganic and organic substances.

Oxidized asphalt: see Air-blown asphalt.

Oxygen scavenger: a chemical which reacts with oxygen in injection water, used to prevent degradation of polymer.

Oxygenate: an oxygen-containing compound that is blended into gasoline to improve its octane number and to decrease gaseous emissions; a substance which, when added to gasoline, increases the amount of oxygen in that gasoline blend; includes fuel ethanol, methanol, and methyl tertiary butyl ether (MTBE).

Oxygenated gasoline: gasoline with added ethers or alcohols, formulated according to the Federal Clean Air Act to reduce carbon monoxide emissions during winter months.

Ozokerite (Ozocerite): a naturally occurring wax; when refined also known as ceresin.

Pale oil: a lubricating oil or a process oil refined until its color, by transmitted light, is straw to pale yellow.

Paraffin wax: the colorless, translucent, highly crystalline material obtained from the light lubricating fractions of paraffin crude oils (wax distillates).

Paraffinum liquidum: see Liquid petrolatum.

Particle density: the density of solid particles.

Particulate: a small, discrete mass of solid or liquid matter that remains individually dispersed in gas or liquid emissions.

Particulate emissions: particles of a solid or liquid suspended in a gas, or the fine particles of carbonaceous soot and other organic molecules discharged into the air during combustion.

Particulate matter (particulates): particles in the atmosphere or on a gas stream that may be organic or inorganic and originate from a wide variety of sources and processes.

Particle size distribution: the particle size distribution (of a catalyst sample) expressed as a percent of the whole.

Partition ratios, K: the ratio of total analytical concentration of a solute in the stationary phase, CS, to its concentration in the mobile phase, CM.

Partitioning: in chromatography, the physical act of a solute having different affinities for the stationary and mobile phases.

Pattern: the areal pattern of injection and producing wells selected for a secondary or enhanced recovery project.

Pattern life: the length of time a flood pattern participates in oil recovery.

Pay zone thickness: the depth of a tar sand deposit from which bitumen (or a product) can be recovered.

Penex process: a continuous, non-regenerative process for isomerization of C_5 and/or C_6 fractions in the presence of hydrogen (from reforming) and a platinum catalyst.

Pentafining: a pentane isomerization process using a regenerable platinum catalyst on a silica-alumina support and requiring outside hydrogen.

Pepper sludge: the fine particles of sludge produced in acid treating which may remain in suspension.

Peri-condensed aromatic compounds: compounds based on angular condensed aromatic hydrocarbon systems, e.g., phenanthrene, chrysene, picene, etc.

Permeability: the ease of flow of the water through the rock.

Petrol: a term commonly used in some countries for gasoline.

Petrolatum: a semisolid product, ranging from white to yellow in color, produced during refining of residual stocks; see Petroleum jelly.

Petrolenes: the term applied to that part of the pentane-soluble or heptane-soluble material that is low boiling (<300°C , <570°F, 760 mm) and can be distilled without thermal decomposition (see also Maltenes).

Petroleum (crude oil): a naturally occurring mixture of gaseous, liquid, and solid hydrocarbon compounds usually found trapped deep underground beneath impermeable cap rock and above a lower dome of sedimentary rock such as shale; most petroleum reservoirs occur in sedimentary rocks of marine, deltaic, or estuarine origin.

Petroleum asphalt: see Asphalt.

Petroleum ether: see Ligroine.

Petroleum jelly: a translucent, yellowish to amber or white, hydrocarbon substance (melting point: 38°C–54°C) having almost no odor or taste, derived from petroleum and used principally in medicine and pharmacy as a protective dressing and as a substitute for fats in ointments and cosmetics; also used in many types of polishes and in lubricating greases, rust preventives, and modeling clay; obtained by dewaxing heavy lubricating-oil stocks.

Petroleum refinery: see Refinery.

Petroleum refining: a complex sequence of events that result in the production of a variety of products.

Petroleum sulfonate: a surfactant used in chemical flooding prepared by sulfonating selected crude oil fractions.

Petroporphyrins: see Porphyrins.

Phase: a separate fluid that co-exists with other fluids; gas, oil, water, and other stable fluids such as micro emulsions are all called phases in EOR research.

Phase behavior: the tendency of a fluid system to form phases as a result of changing temperature, pressure, or the bulk composition of the fluids or of individual fluid phases.

Phase diagram: a graph of phase behavior. In chemical flooding a graph showing the relative volume of oil, brine, and sometimes one or more micro emulsion phases. In carbon dioxide flooding, conditions for formation of various liquid, vapor, and solid phases.

Phase properties: types of fluids, compositions, densities, viscosities, and relative amounts of oil, microemulsion, or solvent, and water formed when a micellar fluid (surfactant slug) or miscible solvent (e.g., carbon dioxide) is mixed with oil.

Phase separation: the formation of a separate phase that is usually the prelude to coke formation during a thermal process; the formation of a separate phase as a result of the instability/incompatibility of petroleum and petroleum products; the precipitation of asphaltene constituents when a paraffinic crude oil is blended with a viscous crude oil or when the blend is heated in pipes leading to a reactor leading to the formation of degradation products and other undesirable changes in the original properties of crude oil products. When such phenomena occur, which may not be immediately at the time of the blend but can occur after an induction period, it is often referred to as instability of the blends. See also: Incompatibility.

pH adjustment: neutralization.

Phosphoric acid polymerization: a process using a phosphoric acid catalyst to convert propene, butene, or both, to gasoline or petrochemical polymers.

Photoionization: a gas chromatographic detection system that utilizes an *detector (PID)* ultraviolet lamp as an ionization source for analyte detection. It is usually used as a selective detector by changing the photon energy of the ionization source.

Photosynthesis: process by which chlorophyll-containing cells in green plants concert incident light to chemical energy, capturing carbon dioxide in the form of carbohydrates.

PINA analysis: a method of analysis for paraffins, *iso*-paraffins, naphthenes, and aromatics.

PIONA analysis: a method of analysis for paraffins, *iso*-paraffins, olefins, naphthenes, and aromatics.

Pipe still: a still in which heat is applied to the oil while being pumped through a coil or pipe arranged in a suitable firebox.

Pipestill gas: the most volatile fraction that contains most of the gases that are generally dissolved in the crude. Also known as pipestill light ends.

Pipestill light ends: see *Pipestill gas.*

Pitch: the nonvolatile, brown to black, semi-solid to solid viscous product from the destructive distillation of many bituminous or other organic materials, especially coal.

Platforming: a reforming process using a platinum-containing catalyst on an alumina base.

PNA: a polynuclear aromatic compound.

PNA analysis: a method of analysis for paraffins, naphthenes, and aromatics.

Polar aromatics: resins; the constituents of petroleum that are predominantly aromatic in character and contain polar (nitrogen, oxygen, and sulfur) functions in their molecular structure(s).

Pollutant: a chemical (or chemicals) introduced into the land water and air systems of that is (are) not indigenous to these systems; also an indigenous chemical (or chemicals) introduced into the land water and air systems in amounts greater than the natural abundance.

Pollution: the introduction into the land water and air systems of a chemical or chemicals that are not indigenous to these systems or the introduction into the land water and air systems of indigenous chemicals in greater-than-natural amounts.

Polyacrylamide: very high molecular weight material used in polymer flooding.

Polycyclic aromatic hydrocarbons (PAHs): polycyclic aromatic hydrocarbons are a suite of compounds comprised of two or more condensed aromatic rings. They are found in many petroleum mixtures, and they are predominantly introduced to the environment through natural and anthropogenic combustion processes.

Polyforming: a process charging both C_3 and C_4 gases with naphtha or gas oil under thermal conditions to produce gasoline.

Polymer: in EOR, any very high molecular weight material that is added to water to increase viscosity for polymer flooding.

Polymer augmented waterflooding: waterflooding in which organic polymers are injected with the water to improve areal and vertical sweep efficiency.

Polymer gasoline: the product of polymerization of gaseous hydrocarbons to hydrocarbons boiling in the gasoline range.

Polymer stability: the ability of a polymer to resist degradation and maintain its original properties.

Polymerization: the combination of two olefin molecules to form a higher molecular weight paraffin.

Polynuclear aromatic compound: an aromatic compound having two or more fused benzene rings, e.g., naphthalene, phenanthrene.

Polysulfide treating: a chemical treatment used to remove elemental sulfur from refinery liquids by contacting them with a non-regenerable solution of sodium polysulfide.

PONA analysis: a method of analysis for paraffins (P), olefins (O), naphthenes (N), and aromatics (A).

Pore diameter: the average pore size of a solid material, e.g., catalyst.

Pore space: a small hole in reservoir rock that contains fluid or fluids; a 4 in. cube of reservoir rock may contain millions of inter-connected pore spaces.

Pore volume: total volume of all pores and fractures in a reservoir or part of a reservoir; also applied to catalyst samples.

Porosity: the percentage of rock volume available to contain water or other fluid.

Porphyrins: organometallic constituents of petroleum that contain vanadium or nickel; the degradation products of chlorophyll that became included in the protopetroleum.

Positive bias: a result that is incorrect and too high.

Possible reserves: reserves where there is an even greater degree of uncertainty but about which there is some information.

Potential reserves: reserves based upon geological information about the types of sediments where such resources are likely to occur and they are considered to represent an educated guess.

Pour point: the lowest temperature at which oil will pour or flow when it is chilled without disturbance under definite conditions.

Power-law exponent: an exponent used to model the degree of viscosity change of some non-Newtonian liquids.

Powerforming: a fixed-bed naphtha-reforming process using a regenerable platinum catalyst.

Precipitation number: the number of milliliters of precipitate formed when 10 mL of lubricating oil is mixed with 90 mL of petroleum naphtha of a definite quality and centrifuged under definitely prescribed conditions.

Preflush: a conditioning slug injected into a reservoir as the first step of an EOR process.

Pressure cores: cores cut into a special coring barrel that maintains reservoir pressure when brought to the surface; this prevents the loss of reservoir fluids that usually accompanies a drop in pressure from reservoir to atmospheric conditions.

Pressure gradient: rate of change of pressure with distance.

Pressure maintenance: augmenting the pressure (and energy) in a reservoir by injecting gas and/or water through one or more wells.

Pressure pulse test: a technique for determining reservoir characteristics by injecting a sharp pulse of pressure in one well and detecting it surrounding wells.

Pressure transient testing: measuring the effect of changes in pressure at a well in a field.

Primary oil recovery: oil recovery utilizing only naturally occurring forces.

Primary structure: the chemical sequence of atoms in a molecule.

Primary tracer: a chemical that, when inject into a test well, reacts with reservoir fluids form a detectable chemical compound.

Primary wood-using mill: a mill that converts round wood products into other wood products; common examples are sawmills that convert saw logs into lumber and pulp mills that convert pulpwood round wood into wood pulp.

Probable reserves: mineral reserves that are nearly certain but about which a slight doubt exists.

Process heat: heat used in an industrial process rather than for space heating or other housekeeping purposes.

Producer gas: fuel gas high in carbon monoxide (CO) and hydrogen (H_2), produced by burning a solid fuel with insufficient air or by passing a mixture of air and steam through a burning bed of solid fuel.

Producibility: the rate at which oil or gas can produced from a reservoir through a wellbore.

Producing well: a well in an oil field used for removing fluids from a reservoir.

Propane asphalt: see Solvent asphalt.

Propane deasphalting: solvent deasphalting using propane as the solvent.

Propane decarbonizing: a solvent extraction process used to recover catalytic cracking feed from heavy fuel residues

Propane dewaxing: a process for dewaxing lubricating oils in which propane serves as solvent.

Propane fractionation: a continuous extraction process employing liquid propane as the solvent; a variant of propane deasphalting.

Protopetroleum: a generic term used to indicate the initial product formed changes have occurred to the precursors of petroleum.

Proved reserves: mineral reserves that have been positively identified as recoverable with current technology.

PSD: prevention of significant deterioration.

PTE: potential to emit; the maximum capacity of a source to emit a pollutant, given its physical or operation design, and considering certain controls and limitations.

Pulpwood: round wood, whole-tree chips, or wood residues that are used for the production of wood pulp.

Pulse-echo ultrasonic borehole televiewer: well-logging system wherein a pulsed, narrow acoustic beam scans the well as the tool is pulled up the borehole; the amplitude of the reflecting beam is displayed on a cathode-ray tube resulting in a pictorial representation of wellbore.

Purge and trap: a chromatographic sample introduction technique in volatile components that are purged from a liquid medium by bubbling gas through it. The components are then concentrated by "trapping" them on a short intermediate column, which is subsequently heated to drive the components on to the analytical column for separation.

Purge gas: typically helium or nitrogen, used to remove analytes from the sample matrix in purge/trap extractions.

Pyrobitumen: see Asphaltoid.

Pyrolysis: the thermal decomposition of biomass at high temperatures (greater than 400°F, or 200°C) in the absence of air; the end product of pyrolysis is a mixture of solids (char), liquids (oxygenated oils), and gases (methane, carbon monoxide, and carbon dioxide) with proportions determined by operating temperature, pressure, oxygen content, and other conditions; exposure of a feedstock to high temperatures in an oxygen-poor environment.

Pyrophoric: substances that catch fire spontaneously in air without an ignition source.

Quad: 1 quadrillion Btu (10^{15} Btu)$= 1.055$ exajoules (EJ), or approximately 172 million barrels of oil equivalent.

Quadrillion: 1×10^{15}

Quench: the sudden cooling of hot material discharging from a thermal reactor.

RACT: Reasonably Available Control Technology standards; implemented in areas of non-attainment to reduce emissions of volatile organic compounds and nitrogen oxides.

Raffinate: that portion of the oil which remains undissolved in a solvent refining process.

Ramsbottom carbon residue: see Carbon residue.

Raw crude oil: crude oil direct from the wellbore, before it is treated in a gas separation plant; typically contains nonhydrocarbon contaminants.

Raw materials: minerals extracted from the earth prior to any refining or treating.

Recovery boiler: a pulp mill boiler in which lignin and spent cooking liquor (black liquor) is burned to generate steam.

Rectification zone: the zone in a distillation tower in which the more volatile component(s) is (are) removed through contacting the rising vapor with the down-flowing liquid; at the stripping section, the down-flowing liquid is stripped of the more volatile component by the rising vapor; the distillation tower can be equipped with trays or packings or both; also known as the rectification section or the rectifying section.

Recycle ratio: the volume of recycle stock per volume of fresh feed; often expressed as the volume of recycle divided by the total charge.

Recycle stock: the portion of a feedstock which has passed through a refining process and is recirculated through the process.

Recycling: the use or reuse of chemical waste as an effective substitute for commercial products or as an ingredient or feedstock in an industrial process.

Reduced crude: a residual product remaining after the removal, by distillation or other means, of an appreciable quantity of the more volatile components of crude oil.

Refinery: a series of integrated unit processes by which petroleum can be converted to a slate of useful (salable) products.

Refinery gas: a gas (or a gaseous mixture) produced as a result of refining operations.

Refining: the processes by which petroleum is distilled and/or converted by application of physical and chemical processes to form a variety of products are generated.

Reformate: the liquid product of a reforming process.

Reformed gasoline: gasoline made by a reforming process.

Reforming: the conversion of hydrocarbons with low octane numbers into hydrocarbons having higher octane numbers; e.g., the conversion of an n-paraffin into an iso-paraffin.

Reformulated gasoline (RFG): gasoline designed to mitigate smog production and to improve air quality by limiting the emission levels of certain chemical compounds such as benzene and other aromatic derivatives; often contains oxygenates.

Refractory lining: a lining, usually of ceramic, capable of resisting and maintaining high temperatures.

Refuse-derived fuel (RDF): fuel prepared from municipal solid waste; non-combustible materials such as rocks, glass, and metals are removed, and the remaining combustible portion of the solid waste is chopped or shredded; the combustible portion of municipal solid waste after removal of glass and metals.

Regeneration: the reactivation of a catalyst by burning off the coke deposits.

Regenerator: a reactor for catalyst reactivation.

Reid vapor pressure: a measure of the volatility of liquid fuels, especially gasoline.

Relative permeability: the permeability of rock to gas, oil, or water, when any two or more are present, expressed as a fraction of the sir phase permeability of the rock.

Renewable energy sources: solar, wind, and other non-fossil fuel energy sources.

Rerunning: the distillation of an oil which has already been distilled.

Research Octane Method: a test for determining the knock rating, in terms octane numbers, of fuels for use in spark-ignition engines; see also Motor Octane Method.

Reserves: well-identified resources that can be profitably extracted and utilized with existing technology.

Reservoir: a rock formation below the earth's surface containing petroleum or natural gas; a domain where a pollutant may reside for an indeterminate time.

Reservoir simulation: analysis and prediction of reservoir performance with a computer model.

Residual asphalt: see Straight-run asphalt.

Residual fuel oil: obtained by blending the residual product(s) from various refining processes with suitable diluent(s) (usually middle distillates) to obtain the required fuel oil grades.

Residual oil: see Residuum; petroleum remaining in situ after oil recovery.

Residual resistance factor: the reduction in permeability of rock to water caused by the adsorption of polymer.

Residues: bark and woody materials that are generated in primary wood-using mills when round wood products are converted to other products.

Residuum (resid; *pl.*: **residua):** the residue obtained from petroleum after nondestructive distillation has removed all the volatile materials from crude oil, e.g., an atmospheric (345°C, 650°F+) residuum.

Resins: that portion of the maltenes that is adsorbed by a surface-active material such as clay or alumina; the fraction of deasphaltened oil that is insoluble in liquid propane but soluble in n-heptane.

Resistance factor: a measure of resistance to flow of a polymer solution relative to the resistance to flow of water.

Resource: the total amount of a commodity (usually a mineral but can include non-minerals such as water and petroleum) that has been estimated to be ultimately available.

Retention: the loss of chemical components due to adsorption onto the rock's surface, precipitation, or to trapping within the reservoir.

Retention time: the time it takes for an eluate to move through a chromatographic system and reach the detector. Retention times are reproducible and can therefore be compared to a standard for analyte identification.

Rexforming: a process combining platforming with aromatics extraction, wherein low-octane raffinate is recycled to the platformer.

Rich oil: absorption oil containing dissolved natural gasoline fractions.

Riser: the part of the bubble-plate assembly which channels the vapor and causes it to flow downward to escape through the liquid; also the vertical pipe where fluid catalytic cracking reactions occur.

Rock asphalt: bitumen which occurs in formations that have a limiting ratio of bitumen-to-rock matrix.

Rock matrix: the granular structure of a rock or porous medium.

Rotation: period of years between establishment of a stand of timber and the time when it is considered ready for final harvest and regeneration.

Round wood products: logs and other round timber generated from harvesting trees for industrial or consumer use.

Run-of-the-river reservoirs: reservoirs with a large rate of flow-through compared to their volume.

Salinity: the concentration of salt in water.

Sand: a course granular mineral mainly comprising quartz grains that is derived from the chemical and physical weathering of rocks rich in quartz, notably sandstone and granite.

Sand face: the cylindrical wall of the wellbore through which the fluids must flow to or from the reservoir.

Sandstone: a sedimentary rock formed by compaction and cementation of sand grains; can be classified according to the mineral composition of the sand and cement.

SARA analysis: a method of fractionation by which petroleum is separated into saturates, aromatics, resins, and asphaltene fractions.

SARA separation: see SARA analysis.

Saturated steam: steam at boiling temperature for a given pressure.

Saturates: paraffins and cycloparaffins (naphthenes).

Saturation: the ratio of the volume of a single fluid in the pores to pore volume, expressed as a percent and applied to water, oil, or gas separately; the sum of the saturations of each fluid in a pore volume is 100%.

Saybolt Furol viscosity: the time, in seconds (Saybolt Furol Seconds, SFS), for 60 mL of fluid to flow through a capillary tube in a Saybolt Furol viscometer at specified temperatures between 70°F and 210°F; the method is appropriate for high-viscosity oils such as transmission, gear, and heavy fuel oils.

Saybolt Universal viscosity: the time, in seconds (Saybolt Universal Seconds, SUS), for 60 mL of fluid to flow through a capillary tube in a Saybolt Universal viscometer at a given temperature.

Scale wax: the paraffin derived by removing the greater part of the oil from slack wax by sweating or solvent deoiling.

Screen factor: a simple measure of the viscoelastic properties of polymer solutions.

Screening guide: a list of reservoir rock and fluid properties critical to an EOR process.

Scrubber: a device that uses water and chemicals to clean air pollutants from combustion exhaust.

Scrubbing: purifying a gas by washing with water or chemical; less frequently, the removal of entrained materials.

Secondary oil recovery: application of energy (e.g., water flooding) to recovery of crude oil from a reservoir after the yield of crude oil from primary recovery diminishes.

Secondary pollutants: a pollutant (chemical species) produced by interaction of a primary pollutant with another chemical or by dissociation of a primary pollutant or by other effects within a particular ecosystem.

Secondary recovery: oil recovery resulting from injection of water, or an immiscible gas at moderate pressure, into a petroleum reservoir after primary depletion.

Secondary structure: the ordering of the atoms of a molecule in space relative to each other.

Secondary tracer: the product of the chemical reaction between reservoir fluids and an injected primary tracer.

Secondary wood processing mills: a mill that uses primary wood products in the manufacture of finished wood products, such as cabinets, moldings, and furniture.

Sediment: an insoluble solid formed as a result of the storage instability and/or the thermal instability of petroleum and petroleum products.

Sedimentary: formed by or from deposits of sediments, especially from sand grains or silts transported from their source and deposited in water, as sandstone and shale; or from calcareous remains of organisms, as limestone.

Sedimentary strata: typically consist of mixtures of clay, silt, sand, organic matter, and various minerals; formed by or from deposits of sediments, especially from sand grains or silts transported from their source and deposited in water, such as sandstone and shale; or from calcareous remains of organisms, such as limestone.

Selective solvent: a solvent which, at certain temperatures and ratios, will preferentially dissolve more of one component of a mixture than of another and thereby permit partial separation.

Separation process: an upgrading process in which the constituents of petroleum are separated, usually without thermal decomposition, e.g., distillation and deasphalting.

Separator-Nobel dewaxing: a solvent (tricholoethylene) dewaxing process.

Separatory funnel: glassware shaped like a funnel with a stoppered rounded top and a valve at the tapered bottom, used for liquid/liquid separations.

Sepiolite: sepiolite (also known by the German name meerschaum) is a soft white clay mineral, often used to make tobacco pipes. Chemically, the mineral is a complex magnesium silicate, a typical chemical formula for which is $Mg_4Si_6O_{15}(OH)_2 \cdot 6H_2O$; it can be present in fibrous, fine-particulate, and solid forms.

Shear: mechanical deformation or distortion, or partial destruction of a polymer molecule as it flows at a high rate.

Shear rate: a measure of the rate of deformation of a liquid under mechanical stress.

Shear-thinning: the characteristic of a fluid whose viscosity decreases as the shear rate Increases.

Shell fluid catalytic cracking: a two-stage fluid catalytic cracking process in which the catalyst is regenerated.

Shell still: a still formerly used in which the oil was charged into a closed, cylindrical shell and the heat required for distillation was applied to the outside of the bottom from a firebox.

Sidestream: a liquid stream taken from any one of the intermediate plates of a bubble tower.

Sidestream stripper: a device used to perform further distillation on a liquid stream from any one of the plates of a bubble tower, usually by the use of steam.

Single well tracer: a technique for determining residual oil saturation by injecting an ester, allowing it to hydrolyze; following dissolution of some of the reaction product in residual oil the injected solutions produced back and analyzed.

Slack wax: the soft, oily crude wax obtained from the pressing of paraffin distillate or wax distillate.

Slim tube testing: laboratory procedure for the determination of minimum miscibility pressure using long, small-diameter, sand-packed, oil- saturated, stainless steel tube.

Slime: a name used for petroleum in ancient texts.

Sludge: a semi-solid to solid product which results from the storage instability and/or the thermal instability of petroleum and petroleum products.

Slug: a quantity of fluid injected into a reservoir during enhanced oil recovery.

Slurry hydroconversion process: a process in which the feedstock is contacted with hydrogen under pressure in the presence of a catalytic coke-inhibiting additive.

Slurry phase reactors: tanks into which wastes, nutrients, and microorganisms are placed.

Smoke point: a measure of the burning cleanliness of jet fuel and kerosine.

Sodium hydroxide treatment: see Caustic wash.

Sodium plumbite: a solution prepared front a mixture of sodium hydroxide, lead oxide, and distilled water; used in making the doctor test for light oils such as gasoline and kerosine.

Solubility parameter: a measure of the solvent power and polarity of a solvent.

Solutizer-steam regenerative process: a chemical treating process for extracting mercaptan derivatives (RSH) from gasoline or naphtha, using solutizers (potassium isobutyrate, potassium alkyl phenolate) in strong potassium hydroxide solution.

Solvent: a liquid in which certain kinds of molecules dissolve. While they typically are liquids with low boiling points, they may include high-boiling liquids, supercritical fluids, or gases.

Solvent asphalt: the asphalt produced by solvent extraction of residua or by light hydrocarbon (propane) treatment of a residuum or an asphaltic crude oil.

Solvent deasphalting: a process for removing asphaltic and resinous materials from reduced crude oils, lubricating-oil stocks, gas oils, or middle distillates through the extraction or precipitant action of low-molecular-weight hydrocarbon solvents; see also Propane deasphalting.

Solvent decarbonizing: see Propane decarbonizing.

Solvent deresining: see Solvent deasphalting.

Solvent dewaxing: a process for removing wax from oils by means of solvents usually by chilling a mixture of solvent and waxy oil, filtration or by centrifuging the wax which precipitates, and solvent recovery.

Solvent extraction: a process for separating liquids by mixing the stream with a solvent that is immiscible with part of the waste but that will extract certain components of the waste stream.

Solvent gas: an injected gaseous fluid that becomes miscible with oil under reservoir conditions and improves oil displacement.

Solvent naphtha: a refined naphtha of restricted boiling range used as a solvent; also called petroleum naphtha; petroleum spirits.

Solvent refining: see Solvent extraction.

Sonication: a physical technique employing ultrasound to intensely vibrate a sample media in extracting solvent and to maximize solvent/analyte interactions.

Sonic log: a well log based on the time required for sound to travel through rock, useful in determining porosity.

Sour crude oil: crude oil containing an abnormally large amount of sulfur compounds; see also Sweet crude oil.

SOx: the oxides of sulfur.

Soxhlet extraction: an extraction technique for solids in which the sample is repeatedly contacted with solvent over several hours, increasing extraction efficiency.

Spalling: the term used to describe the breaking of a solid (such as coke) into smaller pieces.

Specific gravity: the mass (or weight) of a unit volume of any substance at a specified temperature compared to the mass of an equal volume of pure water at a standard temperature; see also Density.

Spent catalyst: catalyst that has lost much of its activity due to the deposition of coke and metals.

Spontaneous ignition: ignition of a fuel, such as coal, under normal atmospheric conditions; usually induced by climatic conditions.

Stabilization: the removal of volatile constituents from a higher boiling fraction or product (q.v. stripping); the production of a product which, to all intents and purposes, does not undergo any further reaction when exposed to the air.

Stabilizer: a fractionating tower for removing light hydrocarbons from an oil to reduce vapor pressure particularly applied to gasoline.

Stand (of trees): a tree community that possesses sufficient uniformity in composition, constitution, age, spatial arrangement, or condition to be distinguishable from adjacent communities.

Standpipe: the pipe by which catalyst is conveyed between the reactor and the regenerator.

Stationary phase: in chromatography, the porous solid or liquid phase through which an introduced sample passes. The different affinities the stationary phase has for a sample allow the components in the sample to be separated, or resolved.

Steam cracking: a conversion process in which the feedstock is treated with superheated steam.

Steam distillation: distillation in which vaporization of the volatile constituents is effected at a lower temperature by introduction of steam (open steam) directly into the charge.

Steam drive injection (steam injection): EOR process in which steam is continuously injected into one set of wells (injection wells) or other injection source to effect oil displacement toward and production from a second set of wells (production wells); steam stimulation of production wells is *direct steam stimulation*, whereas steam drive by steam injection to increase production from other wells is *indirect steam stimulation*.

Steam stimulation: injection of steam into a well and the subsequent production of oil from the same well.

Steam turbine: a device for converting energy of high-pressure steam (produced in a boiler) into mechanical power which can then be used to generate electricity.

Stiles method: a simple approximate method for calculating oil recovery by waterflood that assumes separate layers (stratified reservoirs) for the permeability distribution.

Storage stability (or storage instability): the ability (inability) of a liquid to remain in storage over extended periods of time without appreciable deterioration as measured by gum formation and the depositions of insoluble material (sediment).

Straight-run asphalt: the asphalt produced by the distillation of asphaltic crude oil.

Straight-run products: obtained from a distillation unit and used without further treatment.

Straight vegetable oil (SVO): any vegetable oil that has not been optimized through the process of transesterification.

Strata: layers including the solid iron-rich inner core, molten outer core, mantle, and crust of the earth.

Straw oil: pale paraffin oil of straw color used for many process applications.

Stripper well: a well that produces (strips from the reservoir) oil or gas.

Stripping: a means of separating volatile components from less volatile ones in a liquid mixture by the partitioning of the more volatile materials to a gas phase of air or steam (q.v. stabilization).

Sulfonic acids: acids obtained by of petroleum or a petroleum product with strong sulfuric acid.

Sulfuric acid alkylation: an alkylation process in which olefins (C_3, C_4, and C_5) combine with *iso*-butane in the presence of a catalyst (sulfuric acid) to form branched chain hydrocarbons used especially in gasoline blending stock.

Supercritical fluid: an extraction method where the extraction fluid is present at a pressure and temperature above its critical point.

Superheated steam: steam which is hotter than boiling temperature for a given pressure.

Surface active material: a chemical compound, molecule, or aggregate of molecules with physical properties that cause it to adsorb at the interface between *two* immiscible liquids, resulting in a reduction of interfacial tension or the formation of a microemulsion.

Surfactant: a type of chemical, characterized as one that reduces interfacial resistance to mixing between oil and water or changes the degree to which water wets reservoir rock.

Suspensoid catalytic cracking: a non-regenerative cracking process in which cracking stock is mixed with slurry of catalyst (usually clay) and cycle oil and passed through the coils of a heater.

Sustainable: an ecosystem condition in which biodiversity, renewability, and resource productivity are maintained over time.

SW-846: an EPA multi-volume publication entitled *Test Methods for Evaluating Solid Waste, Physical/Chemical Methods*; the official compendium of analytical and sampling methods that have been evaluated and approved for use in complying with the RCRA regulations and that functions primarily as a guidance document setting forth acceptable, although not required, methods for the regulated and regulatory communities to use in responding to RCRA-related sampling and analysis requirements. SW-846 changes over time as new information and data are developed.

Sweated wax: a crude petroleum-based wax that has been freed from oil by having been passed through a sweater.

Sweating: the separation of paraffin oil and low-melting wax from paraffin wax.

Sweep efficiency: the ratio of the pore volume of reservoir rock contacted by injected fluids to the total pore volume of reservoir rock in the project area. (*See also* areal sweep efficiency *and* vertical sweep efficiency.)

Sweet crude oil: crude oil containing little sulfur; see also Sour crude oil.

Sweetening: the process by which petroleum products are improved in odor and color by oxidizing or removing the sulfur-containing and unsaturated compounds.

Swelling: increase in the volume of crude oil caused by absorption of EOR fluids, especially carbon dioxide. Also increase in volume of clays when exposed to brine.

Swept zone: the volume of rock that is effectively swept by injected fluids.

Synthesis gas (syngas): a gas produced by the gasification of a solid or liquid fuel that consists primarily of carbon monoxide and hydrogen.

Synthetic crude oil (syncrude): a hydrocarbon product produced by the conversion of coal, oil shale, or tar sand bitumen that resembles conventional crude oil; can be refined in a petroleum refinery.

Synthetic ethanol: ethanol produced from ethylene, a petroleum by-product.

Tar: the volatile, brown to black, oily, viscous product from the destructive distillation of many bituminous or other organic materials, especially coal; a name used arbitrarily for petroleum in ancient texts.

Tar sand (bituminous sand): a formation in which the bituminous material (bitumen) is found as a filling in veins and fissures in fractured rocks or impregnating relatively shallow sand, sandstone, and limestone strata; a sandstone reservoir that is impregnated with a heavy, extremely viscous, black hydrocarbonaceous, petroleum-like material that cannot be retrieved through a well by conventional or enhanced oil recovery techniques; (FE 76-4): The several rock types that contain an extremely viscous hydrocarbon which is not recoverable in its natural state by conventional oil well production methods including currently used enhanced recovery techniques; see Bituminous sand.

Target analyte: target analytes are compounds that are required analytes in U.S. EPA analytical methods. BTEX and PAHs are examples of petroleum-related compounds that are target analytes in U.S. EPA methods.

Tertiary structure: the three-dimensional structure of a molecule.

Tetraethyl lead (TEL): an organic compound of lead, $Pb(CH_3)_4$, which, when added in small amounts, increases the antiknock quality of gasoline.

Thermal coke: the carbonaceous residue formed as a result of a non-catalytic thermal process; the Conradson carbon residue; the Ramsbottom carbon residue.

Thermal cracking: a process which decomposes, rearranges, or combines hydrocarbon molecules by the application of heat, without the aid of catalysts.

Thermal polymerization: a thermal process to convert light hydrocarbon gases into liquid fuels.

Thermal process: any refining process which utilizes heat, without the aid of a catalyst.

Thermal recovery: see EOR process.

Thermal reforming: a process using heat (but no catalyst) to effect molecular rearrangement of low-octane naphtha into gasoline of higher antiknock quality.

Thermal stability (thermal instability): the ability (inability) of a liquid to withstand relatively high temperatures for short periods of time without the formation of carbonaceous deposits (sediment or coke).

Thermochemical conversion: use of heat to chemically change substances from one state to another, e.g., to make useful energy products.

Thermofor catalytic cracking: a continuous, moving-bed catalytic cracking process.

Thermofor catalytic reforming: a reforming process in which the synthetic, bead-type catalyst of coprecipitated chromia (Cr_2O_3) and alumina (Al_2O_3) flows down through the reactor concurrent with the feedstock.

Thermofor continuous percolation: a continuous clay treating process to stabilize and decolorize lubricants or waxes.

Thief zone: any geologic stratum not intended to receive injected fluids in which significant amounts of injected fluids are lost; fluids may reach the thief zone due to an improper completion or a faulty cement job.

Thin layer chromatography (TLC): a chromatographic technique employing a porous medium of glass coated with a stationary phase. An extract is spotted near the bottom of the medium and placed in a chamber with solvent (mobile phase). The solvent moves up the medium and separates the components of the extract, based on affinities for the medium and solvent.

Throttling device: the generic name of any device or process that dissipates pressure energy by irreversibly converting it into thermal energy. See also: Throttling valve.

Throttling valve: a type of valve that can be used to start, stop, and regulate the flow of fluid through a rotodynamic pump. When the flow of a pump is regulated using a throttling valve, the system curve is changed. The operating point moves to the left on the pump curve when the flow is decreased. See also: Throttling devicevalve.

Timberland: forest land that is producing or is capable of producing crops of industrial wood, and that is not withdrawn from timber utilization by statute or administrative regulation.

Time-lapse logging: the repeated use of calibrated well logs to quantitatively observe changes in measurable reservoir properties over time.

Tipping fee: a fee for disposal of waste.

Ton (short ton): 2,000 lbs.

Tonne (Imperial ton, long ton, shipping ton): 2,240 lbs; equivalent to 1,000 kg or in crude oil terms approximately 7.5 barrels of oil.

Topped crude: petroleum that has had volatile constituents removed up to a certain temperature, e.g., 250°C+ (480°F+) topped crude; not always the same as a residuum.

Topping: the distillation of crude oil to remove light fractions only

Topping and back pressure turbines: turbines which operate at exhaust pressure considerably higher than atmospheric (non-condensing turbines); often multistage with relatively high efficiency.

Topping cycle: a cogeneration system in which electric power is produced first. The reject heat from power production is then used to produce useful process heat.

Total petroleum hydrocarbons (TPH): the family of several hundred chemical compounds that originally come from petroleum.

Tower: equipment for increasing the degree of separation obtained during the distillation of oil in a still.

TPH E: gas chromatographic test for TPH extractable organic compounds.

TPH V: gas chromatographic test for TPH volatile organic compounds.

TPH-D(DRO): gas chromatographic test for TPH diesel-range organics.

TPH-G(GRO): gas chromatographic test for TPH gasoline-range organics.

Trace element: those elements that occur at very low levels in a given system.

Tracer test: a technique for determining fluid flow paths in a reservoir by adding small quantities of easily detected material (often radioactive) to the flowing fluid, and monitoring their appearance at production wells. Also used in cyclic injection to appraise oil saturation.

Transesterification: the chemical process in which an alcohol reacts with the triglycerides in vegetable oil or animal fats, separating the glycerin and producing biodiesel.

Transmissibility (transmissivity): an index of producibility of a reservoir or zone, the product of permeability and layer thickness.

Traps: sediments in which oil and gas accumulate from which further migration is prevented.

Traveling grate: a type of furnace in which assembled links of grates are joined together in a perpetual belt arrangement. Fuel is fed in at one end and ash is discharged at the other.

Treatment: any method, technique, or process that changes the physical and/or chemical character of petroleum.

Triaxial borehole seismic survey: a technique for detecting the orientation of hydraulically induced fractures, wherein a tool holding three mutually seismic detectors is clamped in the borehole during fracturing; fracture orientation is deduced through analysis of the detected microseismic perpendicular events that are generated by the fracturing process.

Trickle hydrodesulfurization: a fixed-bed process for desulfurizing middle distillates.

Trillion: 1×10^{12}

True boiling point (True boiling range): the boiling point (boiling range) of a crude oil fraction or a crude oil product under standard conditions of temperature and pressure.

True boiling point curve (TBP curve): the composition of any crude oil sample is approximated by a true boiling point curve. The method used is a batch distillation operation, using a large number of stages – typically in excess of 60, and high reflux to distillate ratio (in excess of 5). The temperature at any point on the temperature-volumetric yield curve represents the true boiling point of the hydrocarbon material present at the given volume percent point distilled. True boiling point distillation curves are generally determined for the whole crude oil and not for any of the crude oil products.

Tube-and-tank cracking: an older liquid-phase thermal cracking process.

Turbine: a machine for converting the heat energy in steam or high temperature gas into mechanical energy. In a turbine, a high velocity flow of steam or gas passes through successive rows of radial blades fastened to a central shaft.

Turn down ratio: the lowest load at which a boiler will operate efficiently as compared to the boiler's maximum design load.

Ultimate analysis: elemental composition.

Ultimate recovery: the cumulative quantity of oil that will be recovered when revenues from further production no longer justify the costs of the additional production.

Ultrafining: a fixed-bed catalytic hydrogenation process to desulfurize naphtha and upgrade distillates by essentially removing sulfur, nitrogen, and other materials.

Ultraforming: a low-pressure naphtha-reforming process employing onstream regeneration of a platinum-on-alumina catalyst and producing high yields of hydrogen and high-octane-number reformate.

Unassociated molecular weight: the molecular weight of asphaltenes in a non-associating (polar) solvent, such as dichlorobenzene, pyridine, or nitrobenzene.

Unconformity: a surface of erosion that separates younger strata from older rocks.

Unifining: a fixed-bed catalytic process to desulfurize and hydrogenate refinery distillates.

Unisol process: a chemical process for extracting mercaptan sulfur and certain nitrogen compounds from sour gasoline or distillates using regenerable aqueous solutions of sodium or potassium hydroxide containing methanol.

Unit process: one of a grouped operation in a refinery system that can be defined and separated from others.

Universal viscosity: see Saybolt Universal viscosity.

Unresolved complex: the thousands of compounds that a gas chromatograph *mixture (UCM)* is unable to fully separate.

Unstable: usually refers to a petroleum product that has more volatile constituents present or refers to the presence of olefin and other unsaturated constituents.

UOP alkylation: a process using hydrofluoric acid (which can be regenerated) as a catalyst to unite olefins with *iso*-butane.

UOP copper sweetening: a fixed-bed process for sweetening gasoline by converting mercaptan derivatives (RSH) to disulfide derivatives (RSSR) by contact with ammonium chloride and copper sulfate in a bed.

UOP fluid catalytic cracking: a fluid process of using a reactor-over-regenerator design.

Upgrading: the conversion of petroleum to value-added salable products.

Upper-phase microemulsion: a microemulsion phase containing a high concentration of oil that, when viewed in a test tube, resides on top of a water phase.

Urea dewaxing: a continuous dewaxing process for producing low-pour-point oils, and using urea which forms a solid complex (adduct) with the straight-chain wax paraffins in the stock; the complex is readily separated by filtration.

Vacuum distillation: a secondary distillation process which uses a partial vacuum to lower the boiling point of residues from primary distillation and extract further blending components; distillation under reduced pressure.

Vacuum residuum: a residuum obtained by distillation of a crude oil under vacuum (reduced pressure); that portion of petroleum which boils above a selected temperature such as 510°C (950°F) or 565°C (1,050°F).

Vapor-phase cracking: a high-temperature, low-pressure conversion process.

Vapor-phase hydrodesulfurization: a fixed-bed process for desulfurization and hydrogenation of naphtha.

Vertical sweep efficiency: the fraction of the layers or vertically distributed zones of a reservoir that are effectively contacted by displacing fluids.

Visbreaking: a process for reducing the viscosity of heavy feedstocks by controlled thermal decomposition.

Viscosity: a measure of the ability of a liquid to flow or a measure of its resistance to flow; the force required to move a plane surface of area $1 m^2$ over another parallel plane surface 1 m away at a rate of 1 m/s when both surfaces are immersed in the fluid; the higher the viscosity, the slower the liquid flows.

VGC (viscosity-gravity constant): an index of the chemical composition of crude oil defined by the general relation between specific gravity, sg, at 60°F and Saybolt Universal viscosity, SUV, at 100°F:

$$a = 10sg - 1.0752 \log (SUV - 38) / 10sg - \log(SUV - 38)$$

The constant, a, is low for the paraffin crude oils and high for the naphthenic crude oils.

VI (Viscosity index): an arbitrary scale used to show the magnitude of viscosity changes in lubricating oils with changes in temperature.

Viscosity-gravity constant: see VGC.

Viscosity index: see VI.

VOC (VOCs): volatile organic compound(s); volatile organic compounds are regulated because they are precursors to ozone; carbon-containing gases and vapors from incomplete gasoline combustion and from the evaporation of solvents.

Volatile compounds: a relative term that may mean (1) any compound that will purge, (2) any compound that will elute before the solvent peak (usually those <C6), or (3) any compound that will not evaporate during a solvent removal step.

Volatile Organic Compounds (VOCs): name given to light organic hydrocarbons which escape as vapor from fuel tanks or other sources, and during the filling of tanks. VOCs contribute to smog.

Volumetric sweep: the fraction of the total reservoir volume within a flood pattern that is effectively contacted by injected fluids.

VSP: vertical seismic profiling, a method of conducting seismic surveys in the borehole for detailed subsurface information.

Waste streams: unused solid or liquid by-products of a process.

Waste vegetable oil (WVO): grease from the nearest fryer which is filtered and used in modified diesel engines, or converted to biodiesel through the process of transesterification and used in any diesel-fueled vehicle.

Water-cooled vibrating grate: a boiler grate made up of a tuyere grate surface mounted on a grid of water tubes interconnected with the boiler circulation system for positive *cooling; the structure is supported by flexing plates allowing the grid and grate to move in a vibrating action; ash is automatically discharged.

Waterflood: injection of water to displace oil from a reservoir (usually a secondary recovery process).

Waterflood mobility ratio: mobility ratio of water displacing oil during waterflooding. (See *also* mobility ratio.)

Waterflood residual: the waterflood residual oil saturation; the saturation of oil remaining after waterflooding in those regions of the reservoir that have been thoroughly contacted by water.

Watershed: the drainage basin contributing water, organic matter, dissolved nutrients, and sediments to a stream or lake.

Watson characterization factor: see Characterization factor.

Watt: the common base unit of power in the metric system; 1 W equals 1 J/s, or the power developed in a circuit by a current of 1 A flowing through a potential difference of 1 V. 1 W=3.412 Btu/h.

Wax: see Mineral wax and Paraffin wax.

Wax distillate: a neutral distillate containing a high percentage of crystallizable paraffin wax, obtained on the distillation of paraffin or mixed-base crude, and on reducing neutral lubricating stocks.

Wax fractionation: a continuous process for producing waxes of low oil content from wax concentrates; see also MEK deoiling.

Wax manufacturing: a process for producing oil-free waxes.

Weathered crude oil: crude oil which, due to natural causes during storage and handling, has lost an appreciable quantity of its more volatile components; also indicates uptake of oxygen.

Well completion: the complete outfitting of an oil well for either oil production or fluid injection; also the technique used to control fluid communication with the reservoir.

Wellbore: the hole in the earth comprising a well.

Wellhead: that portion of an oil well above the surface of the ground.

Wet gas: gas containing a relatively high proportion of hydrocarbons which are recoverable as liquids; see also Lean gas.

Wet scrubbers: devices in which a counter-current spray liquid is used to remove impurities and particulate matter from a gas stream.

Wettability: the relative degree to which a fluid will spread on (or coat) a solid surface in the presence of other immiscible fluids.

Wettability number: a measure of the degree to which a reservoir rock is water-wet or oil-wet, based on capillary pressure curves.

Wettability reversal: the reversal of the preferred fluid wettability of a rock, e.g., from water-wet to oil-wet, or vice versa.

Wheeling: the process of transferring electrical energy between buyer and seller by way of an intermediate utility or utilities.

White oil: a generic tame applied to highly refined, colorless hydrocarbon oils of low volatility, and covering a wide range of viscosity.

Whole-tree harvesting: a harvesting method in which the whole tree (above the stump) is removed.

Wobbe Index (or Wobbe Number): the calorific value of a gas divided by the specific gravity.

Wood alcohol: see Methyl alcohol.

Yarding: the initial movement of logs from the point of felling to a central loading area or landing.

Zeolite: a crystalline aluminosilicate used as a catalyst and having a particular chemical and physical structure.

Conversion Factors

1 acre = 43,560 ft^2 = 4,046.9 m^2
1 acre foot = 7,758.0 bbl
1 atmosphere = 760 mm Hg = 14.696 psi = 29.91 in. Hg
1 atmosphere = 1.0133 bars = 33.899 ft. H$_2$O
1 barrel (oil) = 42 US gallons = 34.97 UK gallons = 164 L = 5.6146 ft^3
1 barrel (water) = 350 lb at 15.6°C (60°F)
1 barrel per day = 1.84 cm^3/sec
1 Btu = 778.26 ft-lb.
1 centipoise × 2.42 = lb. mass/(ft) (hour), viscosity
1 centipoise × 0.000672 = lb. mass/(ft) (second), viscosity
1 cubic foot = 28,317 cm^3 = 7.4805 gallons
Density of water at 15.6°C (60°F) = 0.999 g/cm^3 = 62.367 lb./ft^3 = 8.337 lb./gallon
1 gallon = 231 in.^{3} = 3,785.4 cm^3 = 0.13368 ft^3
1 horsepower-hour = 0.7457 kWh = 2544.5 Btu
1 horsepower = 550 ft-lb./sec = 745.7 W
1 inch = 2.54 cm
1 meter = 100 cm = 1,000 mm = 10^6 microns = 10^{10} angstroms (Δ)
1 ounce = 28.35 g
1 pound = 453.59 g = 7,000 grains
1 square mile = 640 acres

SI METRIC CONVERSION FACTORS

E = exponent; i.e., E+03 = 10^3 and E − 03 = 10^{-3}

I have expanded the column width for easier reading (JS)

acre-foot × 1.233482	E+03 = meters cubed
barrels × 1.589873	E − 01 = meters cubed
centipoise × 1.000000	E − 03 = pascal seconds
darcy × 9.869233	E − 01 = micro meters squared
feet × 3.048000	E − 01 = meters
pounds/acre-foot × 3.677332	E − 04 = kilograms/meters cubed
pounds/square inch × 6.894757	E+00 = kilo pascals
dyne/cm × 1.000000	E+00 = mN/m
parts per million × 1.000000	E+00 = milligrams/kilograms

Index

Printed in the United States
by Baker & Taylor Publisher Services